METALLIC SYSTEMS

A Quantum Chemist's Perspective

T0225512

METALLIC SYSTEMS

A Quantum Chemist's Perspective

Edited by

Thomas C. Allison

Orkid Coskuner

Carlos A. González

CRC Press
Taylor & Francis Group
Boca Raton London New York

CRC Press is an imprint of the
Taylor & Francis Group, an **informa** business

CRC Press
Taylor & Francis Group
6000 Broken Sound Parkway NW, Suite 300
Boca Raton, FL 33487-2742

First issued in paperback 2017

© 2011 by Taylor and Francis Group, LLC
CRC Press is an imprint of Taylor & Francis Group, an Informa business

No claim to original U.S. Government works

ISBN-13: 978-1-4200-6077-5 (hbk)
ISBN-13: 978-1-138-11209-4 (pbk)

Visit the Taylor & Francis Web site at
http://www.taylorandfrancis.com

and the CRC Press Web site at
http://www.crcpress.com

Contents

Introduction

The chemistry of metals is rich and diverse. It is important in many traditional disciplines such as inorganic chemistry, metallurgy, geochemistry, solid state physics, medicine, and biochemistry among others. It spans many orders of magnitude in size, from nanoclusters containing a few atoms up to extended macroscopic solids. It encompasses studies of the properties of pure metals, alloys, nanoparticles, small metal-containing molecules and biomolecules, as well as the interactions that occur on and between metal surfaces. Metals and metal-containing compounds may be studied in all three phases. Other properties of metals of particular interest are magnetism and electrical conductivity. Bonding with and between metals frequently involves d orbitals, which adds to the complexity of structures they form. This both enhances the variety of chemistry that these compounds may exhibit and confounds simple prediction of their properties. Many metals are of sufficient mass that relativistic effects must be taken into account.

In this book, we explore topics that touch on most, if not all, of the areas listed above. The topic of metals and metal-containing molecules will be presented from the viewpoint of the quantum chemist, that is, we are interested in the calculation of properties of metals and metal-containing molecules using first principles (or semiempirical) methods. From the perspective of the chemist, we are more frequently interested in molecular properties as opposed to properties of extended solids, or, rather, we are interested in the chemistry that occurs on the surfaces of metals. The exploration of this chemistry will take place via calculations of the static electronic structure and via dynamics calculations.

The book begins with two chapters on biological studies involving metals. As biological applications frequently take place in solution, the next chapters consider metals in solution. A chapter on liquid metals leads to the consideration of systems that are mostly or entirely metallic. There are several chapters on metal surfaces followed by a presentation of the theory and application of tight-binding methods that have been used to study metals for a long time. The chapter on tight-binding methods includes a discussion of metallic nanoparticles, which continues through the remainder of the book. The chapters are largely independent and may be read in any order without loss of continuity.

Throughout the chapters, there is presentation of the underlying theory as well as practical advice for the application of these methods. The reader will gain an appreciation for the subtlety and complexity of many calculations while learning how to approach the simulation of such systems.

In the remainder of this introduction, we will briefly discuss the theory underlying the methods presented in this book and give a brief introduction to each of the chapters. We acknowledge that while the material presented in this book covers a wide range of topics, it is by no means a complete discussion of the chemistry of metals. Likewise, it would take several volumes to adequately describe the theory underlying the calculations presented in this book. We will content ourselves to introduce a few of the salient details and provide references that the unfamiliar reader may consult for a deeper discussion of this important topic.

The fundamental equation of quantum chemistry is, of course, the Schrödinger equation

$$H\Psi = E\Psi \tag{I.1}$$

where
   H is the Hamiltonian operator
   Ψ is the wavefunction
   E is the energy

The main occupation of the quantum chemist, then, is to solve this equation for the molecular system of interest. There is lengthy literature on the Schrödinger equation and methods for its solution. The unfamiliar reader may wish to consult an introductory text [1–3] for a solid introduction to the theory and practice of quantum chemistry calculations. For readers desiring a more in-depth treatment of the subject, several advanced texts are available [4–6]. Several chapters in this book also contain discussions of the theory.

Solution of the Schrödinger equation usually involves the Born–Oppenheimer approximation that separates Equation I.1 into an electronic structure problem and a dynamics problem. This approximation gives rise to the concept of the potential energy surface in which the potential energy of a molecular system is expressed as a function of the positions of the nuclei. The electronic structure problem is solved under the assumption that the nuclei remain in a fixed position. The dynamics problem then considers the motion of the nuclei on the potential energy surface.

One may broadly divide the methods for solution of the electronic structure problem into three categories: semiempirical methods, ab initio methods, and density functional theory (DFT) methods. Semiempirical methods include traditional semiempirical methods in quantum chemistry (e.g., AM1, PM3, PM6) as well as extended Hückel and tight-binding methods. These methods generally neglect some terms in the Hamiltonian and approximate others to gain computational speed at the expense of rigorous solution. The loss in accuracy is usually compensated through parameterization to accurate data (either from experiment or higher-level theory). Ab initio is taken to mean those methods based on the Hartree–Fock equations. Examples include the Hartree–Fock method, Møller–Plesset perturbation theory methods, coupled-cluster methods, and multi-reference methods [4,5]. Finally, DFT, typically realized through the Kohn–Sham method [6], will form the basis for most of the calculations presented in this book.

It is difficult to overstate the importance of the Kohn–Sham DFT method in the treatment of metallic systems. There are many functionals available that permit the efficient and accurate treatment of metallic systems. DFT methods are the driving force behind the advances made in this area. Many chapters in this book emphasize the importance of DFT and discuss the underlying theory. The interested reader is directed to two recent reviews for a comprehensive review of many topics related to the subject at this book [7,8].

Analogous to the importance of DFT in the solution of the electronic structure problem, the Car–Parrinello method [9] for performing molecular dynamics calculations is one of the primary tools available to the quantum chemist. Car–Parrinello molecular dynamics (CPMD) calculations, which are based on DFT calculations, are efficient and accurate for a wide variety of chemical studies of metallic systems. These methods are discussed in several chapters.

At the heart of many biological processes are molecules whose function depends on the metal atoms they contain. Chapter 1 is concerned with understanding such molecules with molecular detail. Topics in this chapter include first principles calculations of biomolecules and pharmaceuticals, the incorporation of solvation effects via explicit solvent models, and studies of chemical reactivity. These topics are presented through several interesting applications.

Metal ions in solution can have significant effects on nearby molecules. In Chapter 2, chemical dynamics calculations are used to investigate interactions between metal ions and carbohydrates. The importance of using dynamics methods that are based on ab initio calculations of the potential energy surface (as opposed to classical force field methods) is emphasized. Results are presented both in the gas phase and in solution, thus highlighting the role of the solvent.

The techniques introduced in Chapter 2 are used again in Chapter 3, this time to explore the interactions of iron ions with α-synuclein proteins and the role that this interaction may play in Parkinson's disease. This chapter highlights the value of chemical computation in obtaining molecular detail about important biological processes that may ultimately lead to the development of more effective treatments. Significant attention is given to classical molecular dynamics calculations, and the value and limitations of these calculations versus their quantum counterparts are discussed. Again, the role of the solvent is emphasized.

The discussion of metal ions in aqueous solutions, a central theme in the preceding two chapters, culminates in Chapter 4. In this chapter, the effect of a metal ion on surrounding water molecules is considered. Although on the surface this seems to be simpler than the systems considered in the preceding chapters, the reader will discover the rich and detailed dynamics of these solutions. In particular, topics such as the migration of a proton through aqueous solution and the formation of Zundel and Eigen complexes are discussed. As before, the primary tool is chemical dynamics calculations, but these are augmented with a rare event sampling technique known as transition path sampling that is essential to the study of these systems.

The focus of the book shifts to liquid metal surfaces in Chapter 5. Such systems have considerable industrial importance. In this chapter, dynamics simulations based on DFT calculations are used to obtain molecular level detail of the layering of various liquid metals. Attention is given to the comparison of simulation results with experimental results. Liquid sodium surfaces are presented in detail with an emphasis on the role played by geometrical confinement in the layering of these systems.

Chapter 6 presents a wealth of practical advice on the calculation of solid metal surfaces. This chapter contains a detailed discussion of the importance of DFT in calculations on metals. The two main themes in this chapter are the choice of the exchange-correlation potential and the choice of pseudopotential in these calculations. The development of these topics in this chapter will be of great value to most researchers wishing to perform calculations on metal surfaces.

The theme of metal surfaces is continued in Chapter 7, but this time with an emphasis on the formation and properties of oxides. The role played by quantum chemical computation is illustrated through a discussion of Al_2O_3 and Fe_2O_3 surfaces. The importance of metal oxides is discussed in the context of applications of current relevance in areas such as catalysis, environmental chemistry, and energy.

In Chapter 8, an important semiempirical technique—tight binding—is discussed both from a historical perspective and from an application-oriented point of view. The importance of the tight-binding method to larger molecular systems is discussed. Several practical applications of the method are described such as the determination of global minima of nanoclusters and the computation of optical properties. Since tight-binding methods can be applied at much lower computational expense than methods such as DFT, several aspects of their implementation that impact performance are discussed, namely, parallelization and the role of sparse matrix techniques.

Nanoclusters of metal atoms, introduced in Chapter 8, are the main subject of the remainder of the book. Chapter 9 continues the discussion of metal clusters from the point of view of the developers of a DFT code, deMon. The development of DFT is discussed from its earliest realizations through the various methods implemented in the deMon code and up to the present version of the code, deMon2k. A wide variety of properties and metals are discussed along with present and future capabilities.

Chapter 10 contains a detailed investigation of cobalt and platinum nanoclusters. Topics such as magnetism and spin-orbit coupling effects are investigated. Particular attention is given to the dramatic changes that occur in properties of the nanocluster upon the addition of single atoms, a theme that is also discussed in Chapter 11, and due to changes in the charge of the nanocluster.

The book concludes in Chapter 11 with another perspective on metal clusters. Special attention is given to methods for the optimization of nanoclusters, to the electron localization function for determining chemical bonding, and to the Fukui function for predicting reactivity. This chapter also explores the onset of magnetism in nanoclusters and thin films of gold.

REFERENCES

1. W. A. Harrison, 1999, *Elementary Electronic Structure*, Singapore, World Scientific Publishing Company.
2. C. J. Cramer, 2002, *Essentials of Computational Chemistry, Theories and Models*, Chichester, England, John Wiley & Sons Ltd.
3. R. M. Martin, 2004, *Electronic Structure: Basic Theory and Practical Methods*, Cambridge, U.K., Cambridge University Press.
4. A. Szabo and N. Ostlund, 1982, *Modern Quantum Chemistry: Introduction to Advanced Electronic Structure Theory*, New York, Macmillan Publishing Co.
5. W. J. Hehre, L. Radom, P. R. Schleyer, and J. A. Pople, 1986, *Ab Initio Molecular Orbital Theory*, New York, Wiley.

6. R. G. Parr and W. Yang, 1989, *Density-Functional Theory of Atoms and Molecules*, New York, Oxford University Press.
7. P. Huang and E. A. Carter, Advances in correlated electronic structure methods for solids, surfaces, and nanostructures, *Annu. Rev. Phys. Chem.*, 2008, 59, 261–290.
8. C. J. Cramer and D. G. Truhlar, Density functional theory for transition metals and transition metal chemistry, *Phys. Chem. Chem. Phys.*, 2009, 11, 10757–10816.
9. R. Car and M. Parrinello, Unified approach for molecular dynamics and density-functional theory, *Phys. Rev. Lett.*, 1985, 55, 2471–2474.

Contributors

Thomas C. Allison
Chemical and Biochemical Reference
    Data Division
National Institute of Standards and
    Technology
Gaithersburg, Maryland

Patrizia Calaminici
Departamento de Química
Centro de Investigación y de Estudios
    Avanzados
Avenida Instituto Politécnico Nacional
    2508
Col. San Pedro Zacatenco, México

Paolo Carloni
German Research School for Simulation
    Sciences GmbH
Jülich, Germany

and

International School for Advanced
    Studies
Statistical and Biological Physics Sector
Italian Institute of Technology
Trieste, Italy

Orkid Coskuner
Department of Chemistry
The University of Texas
San Antonio, Texas

Matteo Dal Peraro
Institute of Bioengineering, School of
    Life Sciences
Ecole Polytechnique Federale de
    Lausanne
Swiss Federal Institute of Technology
    Lausanne
Lausanne, Switzerland

Patricio Fuentealba
Departamento de Física
Facultad de Ciencias
Universidad de Chile
Santiago, Chile

Gabriel U. Gamboa
Departamento de Química
Centro de Investigación y de Estudios
    Avanzados
Avenida Instituto Politécnico Nacional
    2508
Col. San Pedro Zacatenco, México

Carlos A. González
Chemical and Biochemical Reference
    Data Division
National Institute of Standards and
    Technology
Gaithersburg, Maryland

Emily A. A. Jarvis
Department of Chemistry
Loyola Marymount University
Los Angeles, California

Andreas M. Köster
Departamento de Química
Centro de Investigación y de Estudios
    Avanzados
Avenida Instituto Politécnico Nacional
    2508
Col. San Pedro Zacatenco, México

Cynthia S. Lo
Department of Energy, Environmental,
    and Chemical Engineering
Washington University
Saint Louis, Missouri

Alessandra Magistrato
Permanent Research Staff
CNR-INFM-Democritos National
    Simulation Center and International
    School for Advanced Studies
    (SISSA/ISAS)
Trieste, Italy

Rudolph J. Magyar
Multiscale Dynamic Materials
    Modeling Department
Sandia National Laboratories
Albuquerque, New Mexico

Nicola Marzari
Department of Materials Science and
    Engineering
Massachusetts Institute of Technology
Cambridge, Massachusetts

Ann E. Mattsson
Multiscale Dynamic Materials
    Modeling Department
Sandia National Laboratories
Albuquerque, New Mexico

Carla Molteni
Physics Department
King's College London
London, United Kingdom

Luis Rincón
Chemical and Biochemical Reference
    Data Division
National Institute of Standards and
    Technology
Gaithersburg, Maryland

and

Departamento de Quimica
Facultad de Ciencias
Universidad de Los Andes
Merida, Venezuela

Arturo Robertazzi
Department of Biology, Chemistry,
    Pharmacy
Free University of Berlin
Berlin, Germany

and

Istituto Nazionale per la Fisica della
    Materia
Consiglio Nazionale delle Ricerche
Laboratorio sardo for Computational
    Materials Science
Dipartimento di Fisica
Università di Cagliari
Monserrato, Italy

and

International School for Advanced
    Studies
Trieste, Italy

Dennis R. Salahub
Department of Chemistry
Institute for Biocomplexity and
    Informatics
Institute for Sustainable Energy,
    Environment, Economy
University of Calgary
Alberta, Canada

Peter A. Schultz
Multiscale Dynamic Materials
    Modeling Department
Sandia National Laboratories
Albuquerque, New Mexico

Ali Sebetci
Computer Engineering Department
Zirve University
Gaziantep, Turkey

Yamil Simón-Manso
Chemical and Biochemical References
    Data Division
National Institute of Standards and
    Technology
Gaithersburg, Maryland

J. Manuel Vásquez
Departamento de Química
Centro de Investigación y de Estudios
    Avanzados
Avenida Instituto Politécnico Nacional
    2508
Col. San Pedro Zacatenco, México

Brent Walker
London Centre for Nanotechnology
University College London
London, United Kingdom

Olivia M. Wise
Department of Chemistry
The University of Texas
San Antonio, Texas

Liang Xu
Department of Chemistry
The University of Texas
San Antonio, Texas

1 First Principles DFT Studies of Metal-Based Biological and Biomimetic Systems

Arturo Robertazzi, Alessandra Magistrato,
Matteo Dal Peraro, and Paolo Carloni

CONTENTS

1.1 INTRODUCTION

Metal elements play a pivotal role in many aspects of living organisms, from the stabilization of nucleic acids (RNA and DNA) to a variety of proteins.[1–5] One of the most peculiar features of metal elements is their ability to be oxidized and bind to (or otherwise interact with) electron-rich biological molecules, such as proteins and DNA. Natural evolution has led living organisms to employ metal elements extensively in a wide variety of tasks. For instance, approximately one-third of known enzymes contains metals.[1,6] Metal ions promote a number of chemical processes including bond cleavage and formation, electron transfer, and radical reactions.[6] These processes are central to many biological phenomena, such as DNA synthesis and replication, cellular energy production, and O_2 generation, sensing and transport.[6] Iron and copper are essential elements in mammalian metabolism: iron is the metal constituent of heme in hemoproteins such as hemoglobin and myoglobin and copper is relatively abundant in the ferroxidases ceruloplasmin and haphaestin and in other enzymes such as cytochrome c oxidase.[7] Magnesium is, with few

1

exceptions, the metal ion of choice in the enzymes involved in the biochemistry of nucleic acids, while potassium is close to DNA in the cell.[8] Zinc plays a structural and catalytic role in a large number of proteins, enzymes, and transcription factors.[9,10] Manganese-containing enzymes have a wide range of functionalities, especially in redox-active biomolecules such as manganese catalase and superoxide dismutase.[11]

Not only are metal elements vital in many biological phenomena, but they can also be exploited in therapeutics: medicinal applications have been known for almost 5000 years.[1] The development of modern medicinal inorganic chemistry has been fueled by the serendipitous discovery of cisplatin as an anticancer agent at the beginning of the 1960s.[12] Since then, the application of metal-based drugs for therapeutics has experienced tremendous and continuous growth.[13] Nowadays, research is fervent with the aim of providing effective anticancer, antibiotic, antibacterial, antiviral, and radio-sensitizing agents exploiting the special features of metal elements.[1] Besides platinum, whose complexes are probably the best-studied metal-based compounds in therapeutics, many other metal complexes are currently being used, tested, or designed.[13] A great hope lies in ruthenium compounds as a valid alternatives to cisplatin for the treatment of cancer.[14] Examples of the latter include NAMI-A,[15] ICR,[16] and Salder's compounds.[17,18] Furthermore, complexes of V,[19] Au,[20] Rh,[21] Ir,[22] and Cu[23] are considered potentially active against tumors featuring a diversity of mechanisms of action. Another important field of bioscience in which metals can play an important role is the research for artificial compounds that can mimic biosystems. One of the most recent and striking examples of this is the molybdenum-based catalyst of Yandulov and Schrock, which promotes nitrogen fixation under mild conditions.[24–30]

The impressive improvements achieved by theoretical methods have provided an important support to biosciences in many fields.[31–34] In particular, the theoretical investigation of biological systems has now achieved the necessary maturity to complement the experimental approach. For example, quantum mechanical (QM) calculations are applied to drug design and to the prediction of spectroscopic phenomena.[35–41] The advantage of QM calculations with respect to force fields based methods is that they account for quantum electronic effects and can therefore describe bond forming/breaking phenomena, polarization effects, and charge transfer along with an accurate estimation of reaction (free) energies.

Unfortunately, the investigation of systems of biological relevance with first principles methods is not straightforward as many issues render their application rather awkward, among which the timescale problem is perhaps the most difficult to solve.[32,42,43] A complete QM description of biomolecules, with the exception of few remarkable instances,[44,45] is, in fact, beyond the possibility of present theories and computer power. First, biological processes occur in aqueous solution (and/or membranes), i.e., to achieve an accurate description of the system, solvation effects cannot be neglected. However, modeling solvation is one of the most critical challenges of QM calculations,[40,46–48] due to the size of the system, the large number of degrees of freedom and the effects of counter ions. Second, most biological phenomena occur around 300 K, and temperature effects should be taken into account. A further complication is the dynamical effects on both short and long time scales that become crucial for many biochemical phenomena that take place in the living cell.[40] The use of Density Functional Theory (DFT), possibly coupled with molecular mechanics or molecular dynamics methods, may help to overcome most of these issues.[35,39,40,48-52] In the first instance, DFT methods represent an excellent trade-off between accuracy and computer requirements, particularly suitable for the investigation of biological systems.[35,40] The simple principle on which DFT is based is that of expressing the total energy of the system as a functional of the electron density.[53,54] This leads to the treatment of correlation effects at a lower computational cost than that of more rigorous post-Hartree–Fock methods. QM (and in particular DFT) methods can be coupled with molecular mechanics in the well-known QM/MM approaches.[55-58] In a typical protocol, a relatively small region of the entire model, e.g., the active site of an enzyme or the solute in an aqueous solution, is described at the DFT level, with the remainder being treated within the MM level. Based on this approach, a myriad of studies have been successfully completed in recent years.[35,36,39,40,44,51,52,59-65]

In this review, we illustrate some examples of how DFT calculations may be employed to address crucial issues of biological simulations. Our contribution is clearly limited to a few examples taken from our studies and does not cover important studies from several other research groups. For other surveys on this topic the reader is directed to the following references.[31,32,34,40,49,50,52]

Our focus is on DFT and on QM(DFT)/MM methods applied to metal-containing biosystems. First, an intriguing example of solvation effects in the Yandulov and Schrock biomimetic nitrogen fixation is discussed.[24-30] Second, the important contribution played by QM/MM calculations for macromolecules such as DNA and proteins is covered. In particular, examples of platinum-containing anticancer compounds interacting with DNA fragments are proposed,[66-68] as well as the reactivity of specific zinc-containing proteins. Each of the following sections describes the limitations as well as the potential of DFT methods with the overall goal of convincing the reader that these calculations are central to modern biosciences and can offer a solid complement to the experimental approach.

1.2 DENSITY FUNCTIONAL THEORY AND ITS APPLICATION TO BIOLOGICAL SYSTEMS

This section illustrates the basic elements of the theoretical tools employed in the selected examples discussed below. It is clearly beyond our aims to propose a rigorous description of DFT-based methods, which is already available in many text books[53,54,69,70] and in recent reviews.[32,34,35,40,42,48,50-52]

The electron correlation, i.e., the quantum interaction of electrons, is central to achieving a quantitative theoretical description of natural processes, especially those for which, for instance, hydrophobic forces, weak H-bonding, transition metals, covalent bond breaking/forming are important. The most rigorous approach to take into account this physical feature is the use of post-Hartree–Fock (post-HF) methods, such as Møller–Plesset or coupled cluster methods. Successful applications of these two systems of biological relevance are reviewed in references.[32,42,43] However, biological systems often contain many heavy atoms, and high-level correlated methods become rapidly prohibitive in computer time and resources. DFT, in contrast, offers a reasonable (mostly based on empirical corrections) treatment of electron correlation at computational cost similar to that of HF calculations:[35,40,51–54] for this reason, DFT is the most common correlated method employed for investigation of proteins and DNA.[35,40,51–54]

Fermi and Thomas developed a model in which they suggested that the ground state energy of a system may be connected to the overall electron density of the system itself.[71,72] The breakthrough came with the work of Hohenberg and Kohn about 40 years later.[73] They showed that the ground-state energy of a system is *uniquely* defined by its overall electron density. In other words, the energy is a unique *functional* of the electron density. Shortly thereafter, Kohn and Sham, by replacing the kinetic energy of the interacting electrons with that of a noninteracting system, proposed the final expression of the ground-state electronic energy.[74] For a rigorous description of DFT methods the reader is directed to the following references.[75,76]

One of the most delicate points of the DFT theory is the expression of the exchange–correlation functional in terms of electron density, which if correctly determined could lead to the exact solution of the many-body problem. To date, no exact analytical form of the exchange–correlation functional in terms of the electron density has been formulated. However, even very simple approximations to the exchange–correlation functional can give satisfactory results. For instance, the simplest local density approximation (LDA), which replaces the exchange–correlation energy density of the real system with that of an homogeneous electron gas of the same density, is known to yield excellent results for solids,[40] but performs poorly for molecules. Thus, when dealing with biological systems, more sophisticated approximations are required. For instance, within the gradient-corrected approximation, the exchange–correlation functional depends on the electron density as well as on its gradient. A typical example is the exchange functional proposed by Becke, used in combination with Lee, Parr, and Yang's correlation functional, i.e., the BLYP functional.[77,78] Many other functionals have been designed that perform well for a wide range of chemical and biochemical systems including BP,[79,80] PBE,[81,82] and PW91.[83–85] However, a further improvement to the accuracy of the exchange correlation functional has been achieved by combining the exchange functional with HF exact exchange. This approach led to the development of the so-called *hybrid* functionals, such as the very popular B3LYP.[77,78]

Biological processes take place in a complex environment consisting of thousands of atoms, and the straightforward application of DFT (or of other QM methods) to DNA oligonucleotides, proteins, and enzymes is difficult or impossible. The most commonly used approach to elude the size limitation is the combination of DFT (or other QM methods) with classical force-field-based molecular mechanics in the so-called QM/MM approaches.[35,36,49,50,86–88] In these methods the system is partitioned in two or more regions (or layers) where different levels of theory can be applied. The chemically relevant part of the system is often described quantum mechanically, whereas the remainder of the system is treated with a more efficient molecular mechanics method. The choice of QM region is not trivial as solvent molecules, counter ions and part of protein/DNA may be directly involved in the process of interest. In addition, a proper treatment of the long-range electrostatic interactions between the QM and MM atoms is crucial to obtain reliable results.[35,40,50–52]

Nowadays, several QM/MM schemes are available: for a detailed description of these methods, see for example.[50,55–58,86,89,90] Among these, the QM/MM approach in combination with the Car–Parrinello molecular dynamics (CPMD) scheme has been extensively employed for the studies reported below.[35,51,52,91] Within this scheme, simulations are typically performed by using plane wave (PW) basis sets rather than localized functions as the former perform better for first principles molecular dynamics of large systems. PW basis sets have further advantages,[40] e.g., convergence can be easily monitored by increasing the number of PWs that, unlike local basis sets, depends on one parameter, i.e., the cutoff energy; PWs are completely unbiased and do not suffer from basis set superposition error which affects local basis sets. Pseudopotentials (which can be applied to local basis sets as well) are employed for the core electrons, providing a great computational advantage as only valence electrons are explicitly described. Relativistic effects, which are crucial for heavy metals may also be included in pseudopotentials. It has to be stressed that the choice of the basis sets is crucial for the reliability of the theoretical description, and special care has to be taken for charged molecules, and for those systems in which intermolecular forces (such as H-bonding and/or hydrophobic interactions, vide infra) play an important role.[53,54]

Albeit widely applied, DFT methods as implemented suffer severe limitations. Even the popular B3LYP and BLYP exchange–correlation functionals tend to underestimate reaction barriers of chemical processes compared to MP2 calculations.[36,92–94] This may be due to the failure of gradient-corrected functionals in describing three-center, two-electron chemical bonds as previously reported.[40] When predicting reaction barriers, the accuracy of DFT functional becomes a crucial consideration and the functional employed should always be compared with higher-level calculations (usually HF-based calculations with explicit electron correlation) and/or experimental data. Moreover, DFT functionals are not capable of rigorously describing dispersive forces which play a fundamental role in many aspects of chemistry and biochemistry.[95–97] For instance, π-stacking is central to the correct function of proteins and in determining the packing of DNA base pairs.[98,99] Current implementations of gradient-corrected exchange–correlation functionals prevent accurate descriptions of π-stacking since long-range electron correlation is not included. Fortunately, recent advances allow such problems to be partially overcome.[100–110]

A further issue that affects DFT calculations is the so-called self-interaction problem.[32,34,35] Unlike HF theory, the self-interaction term in Coulomb interaction is not fully cancelled by the approximate self-exchange term.[75] This may lead to an unphysical localization of the spin densities in open shell systems (such as heavy metals), although a few cases of reasonable agreement between DFT calculations and experimental spin density exist.[111–113] Several schemes to overcome this error have been proposed, but they are generally too computationally demanding to be extensively employed in QM/MM molecular dynamics (QM/MM MD).[40]

The final scenario is that DFT calculations have to be performed carefully and comparison with experimental data and/or higher-level calculations is essential to provide a correct picture of the system of interest. Yet once the model has been built carefully, the exchange–correlation performance rigorously checked, the suitable basis set chosen, and the system properly partitioned into QM and MM layers, DFT becomes a powerful predictive tool, ideal for the study of large biological systems containing metal elements.

1.3 DISCUSSION

1.3.1 SOLVENT EFFECTS ON NITROGEN FIXATION BIOMIMETIC MOLYBDENUM CATALYST

Nitrogen fixation, i.e., the conversion of molecular nitrogen to ammonia, is a key reaction in nature, performed by a number of different prokaryotes including bacteria, actinobacteria and certain types of anaerobic bacteria.[24,114,115] The nitrogenase enzyme catalyzes nitrogen fixation at room temperature through complex reactions that are not fully understood.[116] Artificial nitrogen fixation, instead, requires high temperature and strong pressure and it is industrially achieved by the well-known Haber–Bosch reaction.[117] For years, coordination chemists have been searching for a *nonbiological* catalyst that will render the nitrogen fixation under mild ambient conditions feasible.[118,119] Among many adopted strategies, those based on biomimetics (i.e., compounds that perform a similar chemical reaction to that of natural enzymes) of Fe- and Mo-containing nitrogenase enzyme seemed to be the most promising.[120,121]

The search for the "*Holy Grail* of coordination chemistry"[122] was over in 2002,[30] when a molybdenum triamidoamine chelate complex ([Mo(hiptN3N)], where hiptN3N = tris(hexaisopropylterphenyl)triamidoamine, Figure 1.1A and B)[25–30] was shown to catalyze the reduction of molecular nitrogen to ammonia with high efficiency at room temperature. The catalytic process proceeds with the slow addition of a proton source (2,6-lutidiniumBAr$_4$ where Ar is 3,5-(CF$_3$)$_2$C$_6$H$_3$, hereafter referred to as LutH$^+$ and Lut in the protonated and deprotonated forms respectively, Figure 1.1C) and a reducing agent (decamethylcromocene) to a solution of the catalyst in liquid heptane at ambient conditions (Figure 1.1D). Clearly, a full description of the single steps of the whole process is extremely important not only to comprehend the Yandulov and Schrock cycle in detail, but also to obtain hints about the intricate functioning of the enzymes that naturally perform nitrogen fixation. While many intermediates of the cycle were experimentally characterized,[26,27,29] an exhaustive understanding of the molecular processes is still lacking.[24] To support the experimental research, DFT calculations, *in vacuo* and corrected with implicit solvation models,[123] were performed and provided preliminary insights into these reactions.[122,124,125] While structural characterization of the compounds was satisfactory, energetics of the catalytic cycle turned out to differ remarkably from the available experimental data.[26,27,125] In particular the exothermic character of the two protonation steps for which experimental data are available (step **I** = MoNNH + H$^+$ → MoNNH$_2^+$ and step **II** = MoN + H$^+$ → MoNH$^+$) was overestimated (Figure 1.1D and Table 1.1), e.g., the free energy was predicted to be too negative by these calculations.[27,125] Such a discrepancy (that is well beyond the standard DFT error, 3–5 kcal mol^{-1})[94] may be caused by many factors including (i) the choice of the exchange and correlation functionals, (ii) the solvation model, and (iii) the lack of a rigorous treatment of entropic effects. In this section we shall describe an extensive study aimed at finding the reasons of the apparent failure of

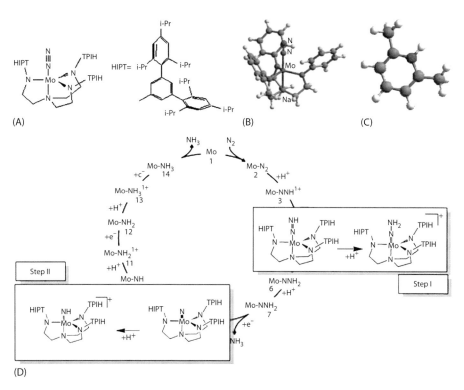

FIGURE 1.1 (A) Experimental catalyst and (B) theoretical model. (C) Proton source, LutH⁺. (D) Reactions **I** and **II** from the putative nitrogen fixation cycle. (Reprinted with permission from Alessandra Magistrato et al., *J. Chem. Theory Comput.*, 3(5), 1708. Copyright 2007 American Chemical Society.)

TABLE 1.1
Reaction Energies in Solution (ΔE_R, kcal/mol) of Step I and Step II: DFT vs. Experimental Results

	DFT						Expt.
	C_6H_6			C_7H_{16}			C_6H_6
	B3LYP (PCM)[a]	BH&H[a,d]	BH&H[c,d]	B3LYP (PCM)[b]	BH&H[a,d]	BH&H[c,d]	
Step **I**	−8.7	4.7	2.9	−10.2	−5.3	−4.7	0
Step **II**	−7.9	6.4	4.6	−13.2	−3.6	−3.0	−1

[a] With 6–31+G(d).
[b] B3LYP/TZVP, Ref. 125.
[c] With BH&H/6–311++G(d,p).
[d] Considering solvated LutH⁺ and Lut with four solvent molecules.

DFT on steps **I** and **II** of the Yandulov and Schrock cycle for which experimental results are known (Figure 1.1).

From the comparison of calculated structural and electronic properties with experimental data we concluded,[126] in agreement with experimental findings, that the large hexaisopropylterphenyl (HIPT) substituents mainly play a steric role,[115] preventing the formation of catalyst dimers. Indeed, further DFT calculations confirmed that the chosen model (Figure 1.1B) without the HIPT substituents is the best trade-off between accuracy and computational costs to explore the energetic profile of dinitrogen reduction at a single Mo center. In order to study the interaction between solvent molecules and the solutes, many DFT functionals were used; among these Becke's "half-and-half" functional (BH&H) was extensively employed to account for possible dispersive interactions.[77] This functional, combined with medium-sized basis sets (such as 6–31+G*, for instance), can reproduce both geometries and energies of correlated *ab initio* methods for several archetypal π-stacked complexes.[100] Many recent studies have further confirmed that the BH&H functional can provide crucial information for molecular systems dominated by dispersive forces.[101,106,109,127,128] The atoms in molecules (AIM) theory[129,130] were then employed to characterise intermolecular interactions between solute and solvent molecules of the studied systems. It has to be stressed that electron density, at the so-called bond critical points (BCPs) of the AIM topology is a valuable tool to quantify chemical bonds and intermolecular interactions.[100,129–132]

In agreement with previous findings, geometries of molybdenum complexes were well reproduced by several DFT functionals,[126] whereas reaction energies (ΔE_R) of steps **I** and **II** calculated in the gas-phase strongly differed from experimental values, with an overestimation larger than ~15 kcal mol^{-1} (data not shown).[122,125] The inclusion of solvent effects with an implicit solvent model (i.e., PCM)[123] as previously suggested[125] hardly improved DFT performance (Table 1.1). Since reaction energies have been demonstrated to be rather independent of the exchange correlation functional used,[122,125] and PCM was unable to significantly improve the *in vacuo* energies (Table 1.1),[125] the lack of an explicit description of solvent effects seemed to be one of the most likely reasons for the poor energetic agreement. Both molybdenum compounds and the proton source have aromatic rings which may interact significantly with apolar solvent molecules *via* C—H...π and π...π interactions. While individually weak, these interactions may play a significant role in the energetics of reactions **I** and **II**. In order to address this issue, the effect of an increasing number of explicit solvent molecules (from one up to four) on the energetics of steps **I** and **II** was explored. Both benzene and heptane molecules were chosen: the former is the solvent in which available experiments were carried out and the latter is the most suitable solvent to perform the entire catalytic cycle.[25–30] For this preliminary study, no more than four solvent molecules were considered as the size of the system rapidly becomes prohibitive for DFT calculations combined with medium/large basis sets such as 6–31+G*. Nevertheless, we were confident that four solvent molecules were a good trade-off between computational costs and a reasonable description of solvation effects. Indeed, four solvent molecules occupied the entire space around the solute, i.e., further solvent molecules would probably interact with one of the solvent molecules rather than with the solute.[126] In addition, solvation energies estimated with three and four solvent molecules hardly differed.[126]

This approach clearly showed that the strongest solvation effects (with either benzene or heptane molecules) were occurring for the proton source, rather than for the catalyst. Also, the molybdenum complexes employed in experiments are even larger than the model used in this study, with the metal center being less accessible to solvent molecules, i.e., the *real* effect of solvent molecules is expected to be even smaller than that observed in our calculations.

As reported in Table 1.1, once solvation effects were included in the estimation of reaction energies, the discrepancy between experimental free energies and calculated reaction energies was drastically reduced to less than ~5 kcal mol^{-1},[126] leading to reasonable agreement between experimental and calculated reaction energies. Moreover, our calculations suggested that reactions in heptane are expected to be only slightly exothermic, in line with experimental measurements carried out in benzene.[126]

The reasons behind such a strong solvent effect were revealed by analysis of the solvation energies of the compounds involved in the reactions and by the atoms-in-molecules approach. These clearly showed that LutH$^+$ undergoes a larger stabilization than Lut in the presence of explicit solvent. In particular, Figure 1.2 displays a schematic view of C—H...π interactions between heptane molecules and both forms of proton source (Figure 1.2A'-B'). Lut(heptane)$_4$ has seven BCPs corresponding to C—H...π interactions with a total electron density (ρ_{TOT}) equal to 0.0497 au, where LutH$^+$(heptane)$_4$ shows nine BCPs with ρ_{TOT}=0.0550 au. Similarly, π...π interactions bind the benzene molecules around the proton source. These interactions are stronger in LutH$^+$(benzene)$_4$ (with seven BCPs and ρ_{TOT}=0.160 au) than in Lut(benzene)$_4$ (with six BCPs and ρ_{TOT}=0.137 au), confirming once more that LutH$^+$ can interact much more strongly with solvent molecules than Lut.

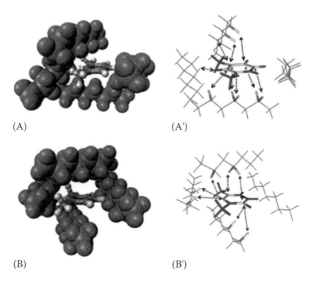

(A) (A')

(B) (B')

FIGURE 1.2 Views of solvation models with four heptane molecules of (A) LutH$^+$ and (B) Lut; schematic view of C—H...π interactions (black arrows) in A' and B', respectively. (Reprinted with permission from Alessandra Magistrato, et al., *J. Chem. Theory Comput.*, 3(5), 1708. Copyright 2007 American Chemical Society.)

In summary, inclusion of explicit apolar solvent molecules improved the energetic description of reactions **I** and **II**. Notably, reaction energies corrected by taking into account explicit solvent molecules were found within the standard DFT error[94] from experimental results.[27] Although this work is clearly a starting point for an extensive study of the artificial nitrogen fixation, it has dual importance. First, it confirmed that the theoretical study of the Yandulov and Schorck cycle is not a trivial task as explicit solvation cannot be neglected. Thus, if one's aim is to describe rigorously the energetics of this process, improvements to the proposed solvation model should be considered such as (i) a drastic increase of the number of solvent molecules, (ii) inclusion of dynamics of solvent molecules, and (iii) temperature effects. This could be achieved by using DFT-based QM/MM calculations, which will allow studying such large systems including these crucial features. Second, this study suggests that widely used continuum models cannot always be employed, especially for those systems where specific solute/solvent interactions occur. Since such interactions can play a crucial role in the energetics of many reactions in organic solvents, explicit solvent models may be required in other systems previously investigated either *in vacuo* or with continuum models.

1.3.2 PLATINUM ANTICANCER COMPLEXES

The most used metal-containing anticancer agent is cisplatin (Figure 1.3), which is particularly effective in solid tumors of testicular, ovarian, head and neck cancers.[133,134] Due to its wide range of applicability in cancer treatment cisplatin is widely studied both experimentally[134–136] and theoretically.[67,137–140] Hydrolysis and alkylation reaction rates as well as the structural consequences of cisplatin binding to DNA are well understood and documented.[134–136,141,142] It is well known, for instance, that cisplatin binds to two adjacent guanines, forming preferentially N7(G)-N7(G) intrastrand crosslinks and that the formation of cisplatin–DNA adducts induces a large kink toward the major groove, a local unwinding at the platinated lesion and an increase of the minor groove width.[142–144] These marked structural deformations are recognized by a series of proteins, which bind to the distorted cisplatin–DNA adduct with high affinity.[67] The consequent formation of cisplatin–DNA/protein adducts may inhibit replication and transcription processes, leading eventually to cell death.[134] However, repair enzymes, in particular nucleotide excision repair enzymes, have also been shown to interact with platinated DNA with high affinity. Thus, the marked structural modifications induced in double strand DNA (dsDNA) by cisplatin may be at the center of cellular resistance toward these anticancer drugs.[135] In fact, the use of cisplatin is characterized by severe drawbacks such as intrinsic and acquired resistance which may be caused by the recognition of the cisplatin lesion by several DNA-binding proteins. In addition, severe side effects (nausea, ear damage and vomiting) occurring upon administration limit the administered dose.[133] All these issues have prompted the design of new anticancer drugs ranging from simple cisplatin derivatives, to more complex polynuclear Pt species, to drugs containing different transition metals.[66,133,141]

Focusing on the problem of intrinsic and acquired resistance to cisplatin (Figure 1.3, **1**) and on a new generation of dinuclear Pt drugs (Figure 1.3, **2** and **3**) which

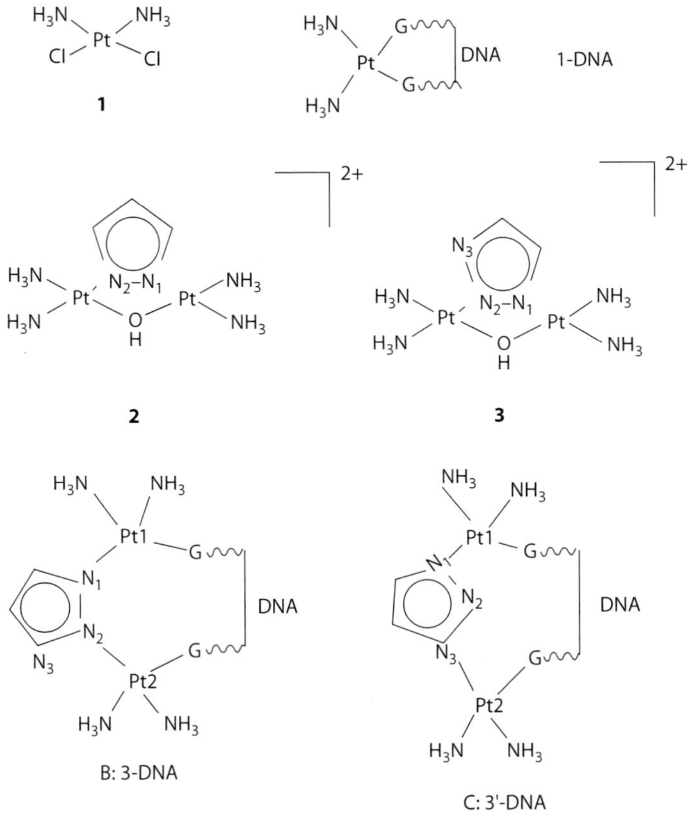

FIGURE 1.3 Cisplatin (1) and its adduct with dsDNA (1-DNA). The azole-bridged diplatinum compounds [Pt2(*i*-OH)(*i*-pz)]²⁺ (2), [Pt2(*i*-OH)(*i*-1,2,3ta)]²⁺ (3), and the atom numbering in the rings. Two different binding modes of 3 to DNA yield to adducts B and C. The 2-DNA adduct (A) is equivalent to B. (Reprinted with permission from Katrin Spiegel et al., *J. Phys. Chem. B*, 111(41), 11873. Copyright Oct 2007 American Chemical Society.)

have been designed to bind DNA has led to a reduction in the structural distortions induced by cisplatin, possibly reducing the risk of cross and intrinsic resistance.[133,145–147] Three hybrid QM/MM simulations (in which the Pt moiety and the guanine are treated at the QM level, while the rest is treated at the MM level) were performed: (i) starting from the x-ray structures of platinated DNA,[142] (ii) the cisplatin–DNA adduct in complex with High Mobility Group (HMG) protein,[144] and (iii) from cisplatin docked to the same oligomer in the canonical B-DNA conformation.[67] All simulations reproduced the relevant experimental and structural features with good accuracy, even though the DNA duplex is rather flexible in all three models. During the simulation, the helical parameters asymptotically approached the values of the simulation based on the x-ray structure with the rise increasing from 4 to 7 Å, the roll angle ranging from 28° to 61° and the global axis curvature ranging from 48° to 57°.[67] However, a complete structural agreement among the three simulations is prevented by the puckering of the sugars: conformational

TABLE 1.2

Selected Helical Parameters at the N7(G)-N7(G) Crosslink Formed by Pt-Drugs Binding to DNA: $[Pt_2(\mu OH)(\tilde{\mu}\text{-pz})](NO_3)_2$ (pz = Pyrazolate) (1) and $[\{cis\text{-Pt}(NH_3)_2\}_2(\mu\text{-OH})(\mu\text{-1,2,3-ta-N1,N2})](NO_3)_2$ (ta = 1,2,3 Triazolate) (2) Bound to 5′-d(CpTpCpTpG*pG*pTpCpTpCp)-3′, Resulting in Complexes A, B, and C (Figure 1.3), Compared to NMR structure,[149] and Reference Simulation of the Unbound Decamer with the Same Sequence (DNA MD)

	A QM/MM	B QM/MM	C QM/MM	A CMD	B CMD	C CMD	DNA MD	NMR
Rise	3.6±0.2	3.6±0.2	4.1±0.3	3.5±0.3	3.4±0.3	4.1±0.5	3.4±0.2	3.3
Roll	9±4	4±4	−5±5	6±7	6±5	−14±9	−3±7	5
Global axis curvature	19±5	10±3	8±4	18±9	18±3	14±7	19±8	5

Rise, major and minor groove width (W) and depth (D) are given in [Å], Roll and global axis curvature are in degrees. The minor and major groove parameters refer to the largest value measured at the platinated site (G–G step). Structural parameters resulting from QM/MM MD and classical MD are labeled as QM/MM and CMD, respectively.

changes of the sugar occur on a time scale of hundreds of ps and thus are not accessible on our simulation time scale. Nevertheless, the extent to which DNA can rearrange in a few ps suggests that our method may qualitatively predict structural changes of drug–DNA adducts for which limited structural information is available.

Azole-bridged dinuclear platinum(II) compounds (**2**, **3**, Figure 1.3)[146,147] have been specifically designed to bind DNA, inducing small structural changes in order to render the platinated lesion less recognizable by repair enzymes.[133] This hypothesis was confirmed by the NMR-structure of the **2**-DNA (the only available structural information of a diplatinated drug–DNA complex) showing structural parameters very similar to that of canonical B-DNA (Table 1.2)[133,148,149] and by an improved cytotoxic behavior of these drugs an relative to cisplatin in several tumor cisplatin resistant cell lines.[147]

Experiments show that for **3**, the mechanism of binding to dsDNA is quite complex.[147] For instance, after the first alkylation step, a nucleophilic attack of the second guanine can occur or, alternatively, Pt2 (and its coordination sphere) can migrate from N2 to N3, followed by the second alkylation step. Therefore, **3** can alkylate two adjacent guanines of dsDNA in both an N1,N2 and an N1,N3 mode. The N1,N3 isomer presents a larger intermetal distance and can lead to the formation of a variety of inter- and intrastrand crosslinks which may be a key factor for the high cytotoxicity of these drugs.[147]

Since the QM/MM approach has been of help in characterizing the structural properties of the cisplatin–DNA adduct[67] we have also used it to characterize and predict the structural features of **2** and **3** in complex with the dsDNA decamer (**A–C**).[66,68]

The NMR structure of the **1**-DNA complex has been used as a template to construct models of **2**-DNA adducts, considering both the N1,N2 and the N1,N3 binding modes (**B** and **C**, respectively) for which no structural information was available.[68]

The overall agreement between the average QM/MM structure of **A** and the NMR structure validates our computational setup. As a general feature, the drug–DNA complexes **A** and **B** display almost the same structural properties, whereas a slightly different structure is observed for **C**, due to its larger intermetal distance.[68]

As displayed in Table 1.2, the binding of the dinuclear drugs is characterized by: (i) a decrease in the roll angle of the platinated G5-G6 bases when going from **A** to **C**, which in **C** becomes negative and markedly smaller than in cisplatin–DNA adducts, (ii) an increase of the rise at the platinated G5-G6 base step that in **C** becomes comparable to the one observed in cisplatin–DNA adducts, (iii) a wider major groove with respect to canonical B-DNA, and (iv) a small overall axis bend for all three complexes.[68] Interestingly, cisplatin causes exactly the opposite effects on the helical parameters.[67]

Although our results do provide a detailed picture of local distortions at the platinated site and give some qualitative trends for the global distortions in DNA, the convergence of these parameters is hampered by the ps simulation time-scale accessible in QM/MM simulations.[66] To confirm the trend observed in global DNA parameters, accurate force field parameters were developed for the diplatinum moiety embedded in its biomolecular environment. This was done using the recently proposed force-matching approach of classical forces to ab initio forces extracted from QM/MM trajectories.[150,151] This approach has an advantage compared to conventional parameterization methods in that the exact local structural properties of the DNA duplex at the drug-binding site need not be known a priori since the QM/MM method can predict small structural changes occurring upon DNA binding. Furthermore, temperature effects and the influence of the DNA environment are automatically taken into account.[150] The accuracy of the newly developed force field parameters is validated by comparison of structural properties from classical MD and hybrid QM/MM simulations.[151] The structural characteristics of the Pt-lesion are well reproduced during classical MD compared with QM/MM simulations and available experimental data (Table 1.2).[151]

These simulations confirmed that upon binding the DNA duplex undergoes minor global distortions. Interestingly, the analysis of local and global DNA parameters revealed that **A** and **B** are almost identical, while the N2, N3 isomerization of the Pt2 coordination sphere imposes different structural properties in **C**. Thus, these simulations show that the azole-bridged diplatinum drugs do not provoke distortions in the DNA duplex (in **A** and **B**), while modest local and global distortions, which are exactly opposite to the ones induced by cisplatin, occur in **C**.[68,151] This may give a first rationale for the lack of cross-resistance of these drugs in cell lines resistant to cisplatin.

Local helical parameters as well as overall curvature of dsDNA have been related to protein recognition processes,[152,153] suggesting that even subtle differences may play a role in cellular events associated to the recognition of platinated DNA[126,152] and be directly related to observed differences in cytotoxic activity.[147] A detailed understanding of how these structural differences in cisplatin- and diplatin–DNA adducts affect protein recognition and other cellular processes is a highly challenging task.[126]

Groove flexibility plays an important role in many biological functions involving DNA; in particular, groove deformability can influence DNA recognition by many DNA-binding proteins, including those proteins involved in repair mechanisms.[152] Generally, the binding of proteins to the DNA minor groove resulted in a significant deformation from the standard B-DNA conformation, e.g., an opening of the minor groove and a bending of DNA toward the major groove.[153,154] To date this aspect has not been considered in the case of Pt drugs, therefore we have investigated the effect of cisplatin and dinuclear Pt drugs on minor groove flexibility by performing classical MD simulations in combination with an adaptive bias force (ABF)[155] method to calculate free-energy profiles. Simulations were carried out on **A**, **B**, **C** and cisplatin–DNA (**D**) considering a hexadecamer with the same sequence as the x-ray structure of the platined DNA in complex with HMG with the sequence (5′-d(CpCp UpCpTpCpTpG*pG*pACpCpTpTpCpCp)-3′). The free oligonucleotide (**E**) was also considered for comparison.[126]

By choosing the reaction coordinate as the distance between the center of mass of the backbone atoms of G9 and A10 on the first strand and to C25 and A26 on the second strand (Figures 1.4 and 1.5), the calculated free-energy profiles reveal that

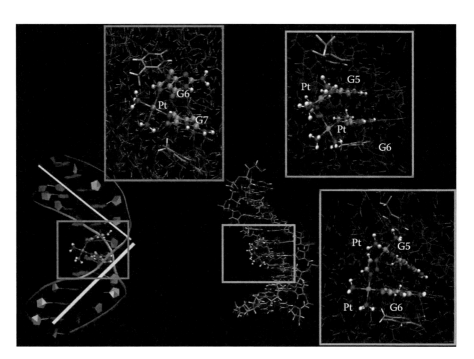

FIGURE 1.4 (See color insert.) Cisplatin–DNA adduct and (in the orange rectangle) a close view of the cisplatin binding region; yellow lines highlight the kink induced by cisPt. Diplatin–DNA adduct and (in the blue rectangle) a close view of the N1, N2 and N1, N3 binding mode in **B** and **C** complexes.

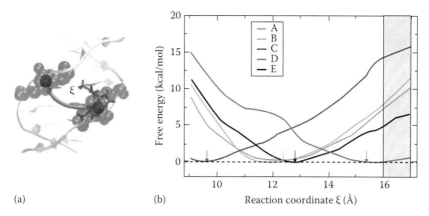

(a)   (b)   Reaction coordinate ξ (Å)

FIGURE 1.5 (See color insert.) (a) Backbone atoms and their centers of mass (transparent and solid blue spheres, respectively) defining the reaction coordinate ξ. (b) Free-energy profiles (kcal/mol) for **A–E**. Small arrows indicate the free-energy minima, and the yellow bar shows the minor groove width of DNA-HMG.

opening the minor groove incurs a high free-energy cost ($\Delta G \sim 7$–$8\,$kcal mol^{-1}) for **A** and **B** which substantially increases for **C** ($\Delta G \sim 15\,$kcal mol^{-1}), which is higher for dinuclear drugs than for canonical B DNA ($\Delta G \sim 5\,$kcal mol^{-1}) and for the corresponding cisplatin–DNA adduct ($\Delta G \sim 0\,$kcal mol^{-1}).[126]

Thus, the formation of diplatinum drug–DNA adducts may significantly affect the binding of excision repair enzymes and other proteins involved in the cytotoxic activity of common Pt drugs. Subsequently, this process could lead to a different cellular response and, in turn, to a lower resistance and cross-resistance with respect to cisplatin.[126] Although the development of drug resistance is a highly complex mechanism, our findings provide an additional rationale for the improved cytotoxic activity of these compounds in cell lines resistant to cisplatin.

1.3.3 Catalytic Mechanism of Metallo β-Lactamases

β-Lactams are the most widespread antibiotics on the market. They have exerted a large evolutionary pressure on bacteria, triggering sophisticated resistance mechanisms. Among them, the most used is the expression of β-lactamases, hydrolases which use different protein scaffolds and catalytic architectures to inactivate β-lactam drugs.[156]

β-Lactamases from classes A, C, and D are the serine hydrolases, whereas class B β-lactamases are characterized by the presence of Zn ions bound to their active sites.[157] Metallo β-lactamases (MβLs), despite not being as ubiquitous as serine β-lactamases, hydrolyze all kinds of β-lactam antibiotics, including the latest generation of carbapenems (Scheme 1.1). MβLs are increasingly spreading among pathogenic bacteria in the clinical setting and are resistant to all current clinical inhibitors on the market.[158–161] Thus, understanding their function at the molecular level is of paramount importance for designing effective drugs.

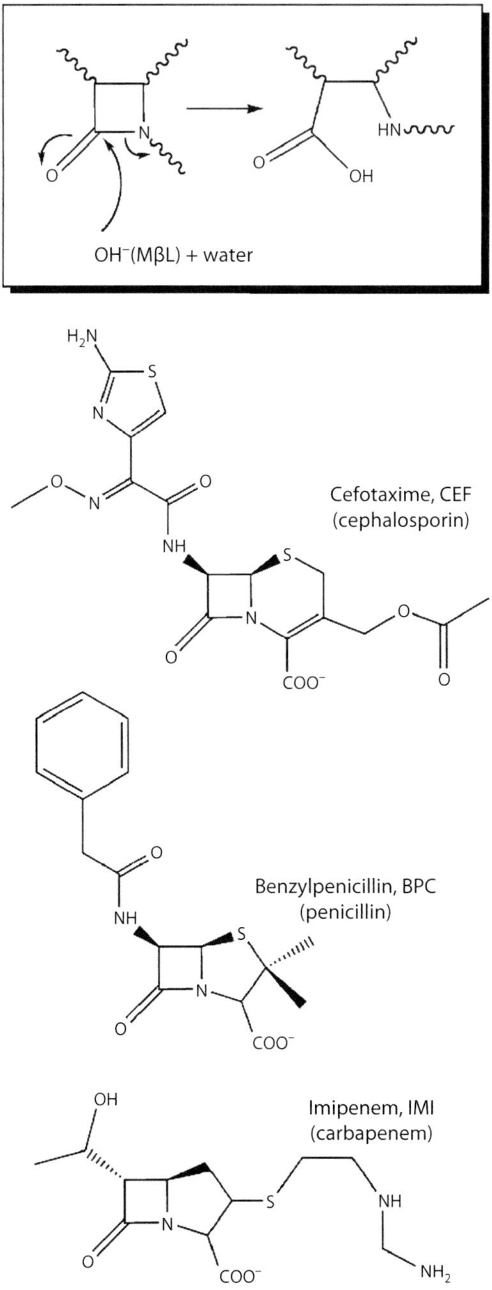

SCHEME 1.1 Metallo β-lactamase hydrolysis of a general β-lactam and β-lactam substrates representative of major antibiotic families.

MβLs are classified by sequence homology in three subclasses: B1, B2, and B3.[159–162] We focus here on the B1 subclass, which includes several chromosomally encoded enzymes such as the ones from *Bacillus cereus* (BcII),[163–168] *Bacteroides fragilis* (CcrA)[169,170] and *Elizabethkingia meningoseptica* (BlaB)[171,172] as well as the transferable VIM, IMP, SPM, and GIM-type enzymes.[173–179] Subclass B1 represents, to date, the best structurally and functionally characterized MβL species[161,163,165–170,180–186] based on its clinical relevance.[187] The MβL folding frame is characterized by a compact αβ/βα sandwich, accommodating an active site that can allocate one or two Zn ions.[158,188] At the active pocket, the first metal site (Zn1) is tetrahedrally coordinated by three histidine ligands (His116, His118, and His196) and the nucleophilic hydroxide, called the 3H site (Figure 1.6). The coordination of the second metal site (Zn2) is provided by the nucleophile, one water molecule and a ligand triad of protein residues which includes Asp120, His263, and Cys221. The metal ion occupancy in B1 enzymes has been also a matter of debate. The first crystal structure of BcII (at 2.5 Å, PDB code 1bmc)[189] showed one zinc ion bound at the 3H site. Subsequent structures of BcII and CcrA revealed a conserved dinuclear metal center in all B1 enzymes[163,190,191] (Figure 1.6). The metal ions are essential for hydrolysis—fast mixing techniques coupled to Trp fluorescence have shown that apo-BcII, despite being properly folded, is unable to bind any substrate.[168] Unfortunately, the characterization of catalytic mechanism in Zn-enzymes by means of experimental techniques is difficult as Zn is invisible to most spectroscopic techniques. Thus, computational quantum mechanical methods may provide a full structural and energetics description of the reactive mechanism, providing a valuable instrument to dissect enzymatic catalysis.

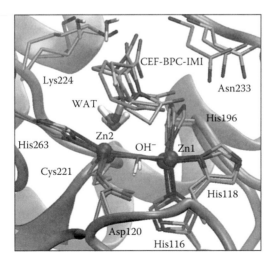

FIGURE 1.6 (See color insert.) MD structural insights into the binding mode of di-Zn MβL CcrA in complex with benzylpenicillin, cefotaxime, and imipenem β-lactam substrates (only substrate bi-cyclic cores and residue heavy atoms are shown for sake of clarity, see also Schemes 1.1 and 1.2). (Adapted from Dal Peraro, M., Vila, A.J., Carloni, P., and Klein, M.L., *J. Am. Chem. Soc.*, 129, 2808, 2007.)

Molecular dynamics simulations of both the mono-Zn and di-Zn B1 MβLs in complex with different types of β-lactams (e.g., benzylpenicillin, imipenem, and cefotaxime) pointed to a few crucial interactions for binding recognition across subclass B1 (Scheme 1.1, Figure 1.6), regardless of the metal content (i.e., in both mono-Zn BcII, and di-Zn CcrA).[192–195] (1) A water bridges the β-lactam carboxylate group and the metal center (WAT in ES in Scheme 1.2 and Figures 1.6 and 1.7). WAT H-bonds to the β-lactam carboxylate moiety it is highly persistent within the catalytic pocket and is spatially conserved in mono and dizinc systems and functional to binding recognition pattern found for different β-lactams in the pocket (Figure 1.6). When two Zn metals are bound at the active site (like in CcrA), WAT is bound to the second Zn (Zn2), completing its coordination shell. (2) A water-mediated salt bridge is maintained between the β-lactam carboxylate moiety and Lys224 (a residue which is conserved in most B1 enzymes) (Scheme 1.2 and Figure 1.6).

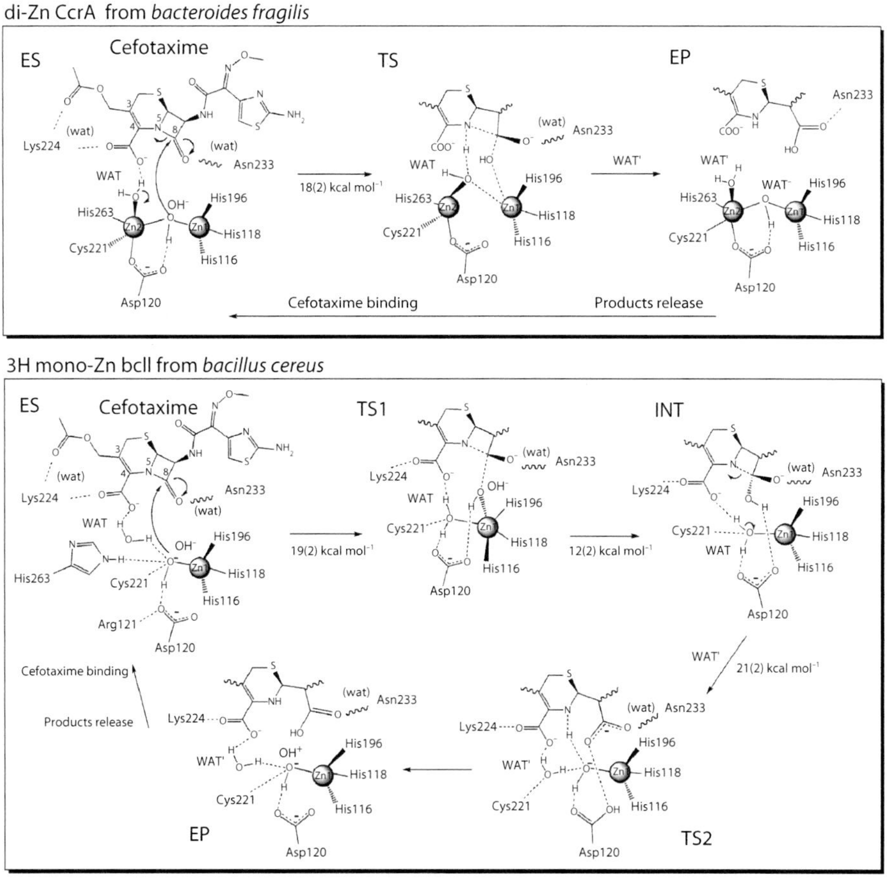

SCHEME 1.2 Di-Zn and mono-Zn reaction mechanisms in B1 MβLs as found in QM/MM studies (see also Figures 1.6 and 1.7). (Adapted from Dal Peraro, M., Vila, A.J., Carloni, P., and Klein, M.L., *J. Am. Chem. Soc.*, 129, 2808, 2007.)

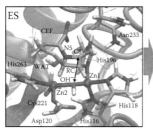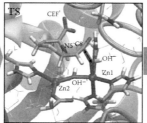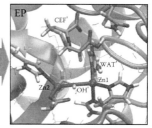

FIGURE 1.7 (See color insert.)　QM/MM structural insights into the reaction mechanism of di-Zn MβL CcrA from *B. fragilis* (atoms shown in licorice are included in the QM cell, see also Schemes 1.1 and 1.2). (Adapted from Dal Peraro, M., Vila, A.J., Carloni, P., and Klein, M.L., *J. Am. Chem. Soc.*, 129, 2808, 2007.)

It is striking that these common minimal features are sufficient to accommodate different β-lactams creating productive conformations for the enzymatic reaction in all complexes; indeed, the putative reaction coordinate (i.e., the distance between the nucleophilic carbonyl oxygen and β-lactam carbonyl carbon) ranges from 3.2 to 3.5 Å, giving plausible models for the Michaelis complexes.[194,195]

We used classical MD and hybrid QM/MM calculations within the DFT/BLYP framework to investigate the hydrolysis of a commonly used cephalosporin (cefotaxime), which is actively degraded by mono-Zn and di-Zn species.[194,195] BcII from *B. cereus* shows that the metal ion is bound in the 3H zinc site, whereas CcrA from *B. fragilis* has both the 3H and DCH sites occupied (Scheme 1.2). Calculations show that both enzymes are able (1) to promote the nucleophilic attack by a metal-bound hydroxide and (2) to catalytically activate a water (WAT) which eventually triggers the C–N bond cleavage of the β-lactam ring. Nonetheless, the chemistry and kinetics of the two reactions strongly depends on the Zn architecture and content (Figure 1.7, Scheme 1.2).

In di-zinc CcrA MβL, the nucleophile OH⁻ approaches the β-lactam carbonyl carbon maintaining its coordination with the metal ions. When the transition state (TS) is reached (reaction coordinate RC=2.0 Å, where RC is the distance between the β-lactam carbonyl carbon and the nucleophilic oxygen), a cascade of almost simultaneous events occurs (Figures 1.6, 1.7 and Scheme 1.2). (i) OH⁻ moves out from the Zn1–Zn2 plane attacking C8 while the Zn1–OH⁻ bond is elongated (2.2 Å). (ii) The OH⁻–Zn2 bond is lost upon nucleophilic attack and the Zn1–Zn2 distance increases (3.8 Å). A hydroxide simultaneously bound to Zn1 and Zn2 is expected to be a poor nucleophile and thus it is reasonable that OH⁻ attacks while detaching from at least one Zn ion. As a consequence, the Zn2 coordination number changes from 5 to 4. (iii) The WAT ligand bonds tightly to Zn2 forming a bipyramidal polyhedron. The WAT–Zn2 distance gradually decreases, thus lowering the WAT pK_a. (iv) The Zn2-bound WAT gets closer to the β-lactam ring, consequently increasing the C–N distance. The partial negative charge on N together with the enhanced nucleophilicity of WAT produces a proton shuttle from WAT to the N5 atom that ultimately triggers the C8–N5 bond cleavage. (v) Deprotonated water binds Zn1, completing the tetrahedral coordination sphere of Zn2 and replacing the position of the OH⁻ nucleophile in ES state. Zn1 then switches to a penta-coordinated bipyramidal coordination from the initial tetrahedral coordination. When the constraint on RC is released, the system evolves into the product state (EP, Figure 1.7, Scheme 1.2), where OH⁻ reorients H-bonding to Asp120 and bridging the Zn1 and Zn2 ions as in the ES state. The CEF substrate is now completely hydrolyzed, the β-lactam ring is open and C8 acquires planar sp^2 hybridization and is finally detached from the metal center. The estimated free energy of activation (ΔF) for the single-step mechanism is $\Delta F = 18(\pm 2)$ kcal mol⁻¹, which is consistent with experimental evidence that suggests a single-step reaction for cephalosporin hydrolysis and a free energy of activation of about 17 kcal mol⁻¹ for similar reactions.

In mono-Zn BcII, Zn1 undergoes a large rearrangement of its coordination shell upon nucleophilic attack. When the hydroxyl nucleophile attacks the β-lactam carbonyl carbon, Zn1 distorts its polyhedron and coordinates WAT as a fifth ligand. This leads to the formation of a stable high-energy intermediate state (INT, $12(\pm 2)$ kcal mol⁻¹, Scheme 1.2). Thus, the presence of only one Zn equivalent in BcII species leads to a two-step mechanism so that WAT replaces the nucleophile during the first step upon entering the Zn1 coordination shell and then is activated in the second step as proton donor for β-lactam N5 (the latter being the rate limiting step of the reaction). In CcrA, the presence of Zn2 merges these two movements in a concerted single step—Zn2 already activates WAT as soon as the OH–Zn1 bond is broken upon nucleophilic attack. This also explains the improved catalytic efficiency of di-Zn versus mono-Zn variants, as found in the calculation of the free energy of activation for di-Zn ($\Delta F = 18(\pm 2)$ kcal mol⁻¹) and mono-Zn ($\Delta F = 21(\pm 3)$ kcal mol⁻¹) species, respectively.

Di-Zn CcrA showed a highly concerted single-step mechanism in which the role of the two metal ion is crucial (Figure 1.7, Scheme 1.2). In particular, the Zn2 site promotes the protonation of the bridging nitrogen that kinetic studies indicate to be the rate-determining step.[165,180,186,196] A series of mutagenesis studies on di-Zn BcII confirms this view, since engineering a more buried position of the Zn2 site gives rise to an inactive β-lactamase even though it is dinuclear.[197] Instead, an optimized BcII obtained by in vitro evolution has been shown to be more efficient through the action of a second-shell ligand, which optimizes the position of Zn2 for the protonation step.[198] These works provide an excellent example of close cooperation between theoretical and experimental studies in a metalloenzyme, and point to an important and more flexible role of the second metal site in B1 species. Moreover, recent mechanistic studies on BcII have shown the accumulation during turnover of an active mono-Zn variant with the metal ion localized at the DCH site, in contrast to what was observed in the first reported x-ray structure of BcII.[199,200] Unfortunately, the lack of structural data for such conformations hampers QM/MM investigations of this alternative pathway, but points consistently to the crucial role of the Zn2 metal-binding site (Zn1).

Because the zinc ligands and most active site residues are highly conserved among subclass B1 MβLs, the binding and mechanism found for CcrA may be a template for the entire B1 subclass, where β-lactams might follow similar catalytic pathways as all the groups involved in the catalysis do not depend on the substrate chemical diversity. Of course, slightly different activation free energies are expected because of the substrate binding modes generated by different β-lactam substituents and differing long-range electrostatics of the MβL scaffold. This is also indirectly confirmed by the results obtained from classical MD simulations of different β-lactam antibiotics docked at the CcrA binding pocket (Scheme 1.1, Figure 1.6).[192,194,195] The reaction coordinates for these additional adducts are always relatively short (3.2(±2) Å for IMI binding and 3.5(±2) Å for BPC). This is consistent with the fact that both are efficiently hydrolyzed by CcrA and BcII. The key interactions of the bicyclic core at the catalytic cleft are still preserved. More importantly, the catalytic water molecule (WAT) is buried at the cleavage site in an equivalent position, as found for the adducts complexed with cefotaxime.

These common elements are flexible enough to accommodate different substrates and elicit a broad spectrum activity in MβLs based on a similar water-assisted hydrolysis mechanism generally plausible for penicillins, cephalosporins, and carbapenems. They might also provide a rationale as to why monobactams such as aztreonam that lack the common β-lactam bicyclic core and carboxylate are not efficiently hydrolyzed by MβLs.

In summary, our studies of MβL catalysis allow us to suggest that a Zn-bound water (WAT in Figures 1.6, 1.7, and Scheme 1.2) turns out to be a common and crucial chemical feature across B1 MβLs. The carboxylate group present in all β-lactam antibiotics stabilizes WAT at the active site upon binding so that the water/β-lactam entity should be considered as the favorite template for the design of new inhibitors. These results also point to the crucial role of the Zn in the second metal-binding site (Zn2), which stabilizes the negative charge developed at β-lactam nitrogen upon nucleophilic attack. This functional advantage is evident in the mechanism of di-Zn B1 MβLs and is completely missing in mono-Zn species which are indeed characterized by low-efficiency turnover and step wise mechanisms. This may suggest, along with recent kinetic findings,[199] that at low concentrations of zinc equally efficient catalysis is achieved when the metal ion is accommodated in the second metal site.

1.4   GENERAL CONCLUSIONS

In this contribution, we have briefly discussed selected examples of first principles DFT and QM/MM studies of biosystems for which metal centers play a pivotal role. First, we showed that even a relatively simple system such as a molybdenum-based catalyst of nitrogen fixation may represent a tremendous scientific challenge. Many DFT functionals failed in reproducing reaction energies derived from experiment. A careful choice of the model and inclusion of solvation effects led us to a drastic improvement of the theoretical description of the entire reaction.

Second, we used extensive first principles calculations to characterize the effect of DNA platination with two specific platinum-based anticancer drugs, i.e., cisplatin and diplatin. Our findings provide a clear rationale for the improved cytotoxic activity of these compounds in cell lines that are instead resistant to cisplatin. Finally, we provided insights into the role of zinc for the mechanism of action of β-lactamases by a combination of classical MD and QM/MM approaches. In summary, we conclude that first principles calculations, despite their limitations (the timescale problem being the most difficult to address), are central for the investigation of biological systems in which metal centers often play a pivotal role.

REFERENCES

1. Orvig, C.; Abrams, M. J. *Chem. Rev*. 1999, *99*, 2201–2204.
2. Spiegel, K.; Magistrato, A.; Maurer, P.; Ruggerone, P.; Carloni, P.; Rothlisberger, U.; Reedijk, J.; Klein, M. L. *J. Comput. Chem*. 2008, *29(1)*, 38–49.
3. Frausto da Silva, J. J. R.; Williams, R. J. P. *The Biological Chemistry of the Elements. The Inorganic Chemistry of Life*. Clarendon Press: Oxford, 1994.
4. Hanzlik, R. P. *Inorganic Aspects of Biological and Organic Chemistry*. Academic Press: New York, 1976.
5. Voet, D.; Voet, J. G.; Pratt, C. W. *Fundamentals of Biochemistry*. Wiley & Sons, Inc: New York, 1999.
6. Ragsdale, S. W. *Chem. Rev*. 2006, *106*, 3317–3337.
7. Andrews, N. C. *Curr. Opin. Chem. Biol*. 2002, *6*, 181–186.
8. Cowan, J. A. *Chem. Rev*. 1998, *98*, 1067–1088.
9. Vasák, M.; Hasler, D. W. *Curr. Opin. Chem. Biol*. 2000, *4*, 177–183.
10. Cox, E. H.; McLendon, G. L. *Curr. Opin. Chem. Biol*. 2000, *4*, 162–165.
11. Siegbahn, P. E. M. *Curr. Opin. Chem. Biol*. 2002, *6*, 227–235.
12. Rosenberg, B.; VanCamp, L.; Krigas, T. *Nature (London)* 1965, *205*, 698.
13. Gordon, M.; Hollander, S. *J. Med*. 1993, *24*, 209–265.
14. Kostova, I. *Curr. Med. Chem*. 2006, *13*, 1085–1107.
15. Alessio, E.; Mestroni, G.; Bergamo, A.; Sava, G. *Met. Ions Biol. Syst*. 2004, *42*, 323–351.
16. Kapitza, S.; Pongratz, M.; Jakupec, M. A.; Heffeter, P.; Berger, W.; Lackinger, L.; Keppler, B. K.; Marian, B. *J. Cancer Res. Clin. Oncol*. 2005, *131*, 101–110.
17. Chen, H.; Parkinson, J. A.; Parsons, S.; Coxall, R. A.; Gould, R. O.; Sadler, P. J. *J. Am. Chem. Soc*. 2002, *124*, 3064–3082.
18. McNae, I. W.; Fishburne, K.; Habtemariam, A.; Hunter, T. M.; Melchart, M.; Wang, F. Y.; Walkinshaw, M. D.; Sadler, P. J. *Chem. Commun*. 2004, *16*, 1786–1787.
19. Ghosh, P.; D'Cruz, O. J.; Narla, R. K.; Uckun, F. M. *Clin. Cancer Res*. 2000, *6*, 1536–1545.
20. Calamai, P.; Carotti, S.; Guerri, A.; Mazzei, T.; Messori, L.; Mini, E.; Orioli, P.; Speroni, G. P. *Anti-Cancer Drug Des*. 1998, *13*, 67–80.
21. Katsaros, N.; Anagnostopoulou, A. *Crit. Rev. Oncol./Hematol*. 2002, *42*, 297–308.
22. Sava, G.; Giraldi, T.; Mestroni, G.; Zassinovich, G. *Chem. Biol. Interact*. 1983, *45*, 1–6.
23. de Hoog, P.; Boldron, C.; Gamez, P.; Sliedregt-Bol, K.; Roland, I.; Pitie, M.; Kiss, R.; Meunier, B.; Reedijk, J. *J. Med. Chem*. 2007, *50*, 3148–3152.
24. Neese, F. *Angew. Chem. Int. Ed*. 2006, *45*, 196–199.
25. Yandulov, D. V.; Schrock, R. R. *Science* 2003, *301*, 76–78.
26. Yandulov, D. V.; Schrock, R. R. *Inorg. Chem*. 2005, *44*, 5542–5542.
27. Yandulov, D. V.; Schrock, R. R. *Inorg. Chem*. 2005, *44*, 1103–1117.
28. Ritleng, V.; Yandulov, D. V.; Weare, W. W.; Schrock, R. R.; Hock, A. S.; Davis, W. M. *J. Am. Chem. Soc*. 2004, *126*, 6150–6163.
29. Yandulov, D. V.; Schrock, R. R.; Rheingold, A. L.; Ceccarelli, C.; Davis, W. M. *Inorg. Chem*. 2003, *42*, 796–813.
30. Yandulov, D. V.; Schrock, R. R. *J. Am. Chem. Soc*. 2002, *124*, 6252–6253.
31. van Mourik, T. *Phil. Trans. R. Soc. A* 2004, *362*, 2653–2670.
32. Siegbahn, P. E. M.; Blomberg, M. R. A. *Chem. Rev*. 2000, *100*, 421–438.
33. Davidson, E. R. *Chem. Rev*. 2000, *100*, 351–352.
34. Rotzinger, F. P. *Chem. Rev*. 2005, *105*, 2003–2038.
35. Carloni, P.; Rothlisberger, U.; Parrinello, P. *Acc. Chem. Res*. 2002, *35*, 455–464.
36. Colombo, M. C.; Guidoni, L.; Laio, A.; Magistrato, A.; Maurer, P.; Piana, S.; Rohrig, U. et al. *Chimia* 2002, *56*, 13–19.
37. Klahn, M.; Schlitter, J.; Gerwert, K. *Biophys. J*. 2005, *88*, 3829–3844.

38. Ludger, W.; Michele, L.; Francesco, M.; Angel, R. *Phys. Rev. B* 2005, *71*, 241402.
39. Piana, S.; Sebastiani, D.; Carloni, P.; Parrinello, M. *J. Am. Chem. Soc.* 2001, *123*, 8730–8737.
40. Simone Raugei, F. L. G. P. C. *Phys. Stat. Sol. (b)* 2006, *243*, 2500–2515.
41. Rodziewicz, P.; Melikova, S. M.; Rutkowski, K. S.; Buda, F. *ChemPhysChem* 2005, *6*, 1719–1724.
42. Hobza, P.; Sponer, J. *Chem. Rev.* 1999, *99*, 3247–3276.
43. Himo, F.; Siegbahn, P. E. M. *Chem. Rev.* 2003, *103*, 2421–2456.
44. Laio, A.; Gervasio, F. L.; VandeVondele, J.; Sulpizi, M.; Rothlisberger, U. *J. Phys. Chem. B* 2004, *108*, 7963–7968.
45. Gervasio, F. L.; Laio, A.; Parrinello, M.; Boero, M. *Phys. Rev. Lett.* 2005, *94*, 158103.
46. Tomasi, J.; Mennucci, B.; Cammi, R. *Chem. Rev.* 2005, *105*, 2999–3094.
47. Cramer, C. J.; Truhlar, D. G. *Chem. Rev.* 1999, *99*, 2161–2200.
48. Rode, B. M.; Schwenk, C. F.; Hofer, T. S.; Randolf, B. R. *Coord. Chem. Rev.* 2005, *249*, 2993–3006.
49. Friesner, R. A.; Guallar, V. *Annu. Rev. Phys. Chem.* 2005, *56*, 389–427.
50. Lin, H.; Truhlar, D. *Theor. Chem. Acc.* 2007, *117*, 185–199.
51. Dal Peraro, M.; Spiegel, K.; Lamoureux, G.; De Vivo, M.; DeGrado, W. F.; Klein, M. L. *J. Struct. Biol.* 2007, *157*, 444.
52. Cavalli, A.; Carloni, P.; Recanatini, M. *Chem. Rev.* 2006, *106*, 3497–3519.
53. Jensen, F. *Introduction to Computational Chemistry.* Wiley & Sons: Chichester, 1999.
54. Leach, A. R. *Molecular Modelling. Principles and Applications.* Prentice Hall: Harlow, 2001.
55. Svensson, M.; Humbel, S.; Froese, R. D. J.; Matsubara, T.; Sieber, S.; Morokuma, K. *J. Phys. Chem. A* 1996, *100*, 19357–19363.
56. Maseras, F.; Morokuma, K. *J. Comput. Chem.* 1995, *16*, 1170–1179.
57. Matsubara, T.; Sieber, S.; Morokuma, K. *Int. J. Quantum Chem.* 1996, *60*, 1101–1109.
58. Svensson, M.; Humbel, S.; Morokuma, K. *J. Chem. Phys.* 1996, *105*, 3654–3661.
59. Komin, S.; Gossens, C.; Tavernelli, I.; Rothlisberger, U.; Sebastiani, D. *J. Phys. Chem. B* 2007, *111*, 5225–5232.
60. Sulpizi, M.; Laio, A.; VandeVondele, J.; Cattaneo, A.; Rothlisberger, U.; Carloni, P. *Proteins* 2003, *52*, 212–224.
61. Biarnes, X.; Ardevol, A.; Planas, A.; Rovirat, C.; Laio, A.; Parrinello, M. *J. Am. Chem. Soc.* 2007, *129*, 10686–10693.
62. Stirling, A.; Iannuzzi, M.; Parrinello, M.; Molnar, F.; Bernhart, V.; Luinstra, G. A. *Organometallics* 2005, *24*, 2533–2537.
63. Kastner, J.; Senn, H. M.; Thiel, S.; Otte, N.; Thiel, W. *J. Chem. Theory Comput.* 2006, *2*, 452–461.
64. Jensen, V. R.; Koley, D.; Jagadeesh, M. N.; Thiel, W. *Macromolecules* 2005, *38*, 10266–10278.
65. Zhu, X.; Yethiraj, A.; Cui, Q. *J. Chem. Theory Comput.* 2007, *3*, 1538–1549.
66. Spiegel, K.; Magistrato, A. *Org. Biomol. Chem.* 2006, *4*, 2507–2517.
67. Spiegel, K.; Rothlisberger, U.; Carloni, P. *J. Phys. Chem. B* 2004, *108*, 2699–2707.
68. Magistrato, A.; Ruggerone, P.; Spiegel, K.; Carloni, P.; Reedijk, J. *J. Phys. Chem. B* 2006, *110*, 3604–3613.
69. Atkins, P. W.; Friedman, R. S. *Molecular Quantum Mechanics.* Oxford University Press: Oxford, 1997.
70. Szabo, A.; Ostlund, N. S. *Modern Quantum Chemistry: Introduction to Advanced Electronic Structure Theory.* Macmillan Publishing Co. Inc.: New York, 1982.
71. Thomas., L. H. *Cambridge Phil. Soc.* 1926, *23*, 542.
72. Fermi, E. *Rend. Lincei 6,* 1927.

73. Hohenberg, P.; Kohn, W. *Phys. Rev.* 1964, *136*, B864–B871.
74. Kohn, W.; Sham, L. J. *Phys. Rev.* 1965, *140*, A1133.
75. Parr, R. G.; Yang, W. *Density-Functional Theory of Atoms and Molecules.* Oxford University Press: New York, 1989.
76. Kohanoff, J. *Electronic Structure Calculations for Solids and Molecules: Theory and Computational Methods.* Cambridge University Press: Cambridge, 2006.
77. Becke, A. D. *Phys. Rev. A* 1998, *38*, 3098–3100.
78. Lee, C. T.; Yang, W. T.; Parr, R. G. *Phys. Rev. B* 1988, *37*, 785–789.
79. Becke, A. D. *Phys. Rev. A* 1988, *38*, 3098.
80. Perdew, J. P. *Phys. Rev. B* 1986, *33*, 8822–8824.
81. Perdew, J. P.; Burke, K.; Ernzerhof, M. *Phys. Rev. Lett.* 1996, *77*, 3865.
82. Perdew, J. P.; Burke, K.; Ernzerhof, M. *Phys. Rev. Lett.* 1997, *78*, 1396.
83. Perdew, J. P.; Chevary, J. A.; Vosko S. H.; Jackson, K. A.; Pederson, M. R.; Singh, D. J.; Fiolhais, C. *Phys. Rev. B* 1993, *48*.
84. Burke, K.; Perdew, J. P.; Wang, Y. In: *Electronic Density Functional Theory: Recent Progress and New Directions*; Eds. J. F. Dobson, G. Vignale, and M. P. Das, Plenum Press: New York, 1998.
85. Perdew, J. P. In: *Electronic Structure of Solids '91*. Akademie Verlag: Berlin, 1991.
86. Sherwood, P. *Hybrid Quantum Mechanics/Molecular Mechanics Approaches*, in *Modern Methods and Algorithms of Quantum Chemistry*; *Proceedings*, 2nd edition, 2000.
87. Garcia-Viloca, M.; Gao, J. *Theor. Chem. Acc.* 2004, *111*, 280–286.
88. Konig, P. H.; Hoffmann, M.; Frauenheim, T.; Cui, Q. *J. Phys. Chem. B* 2005, *109*, 9082–9095.
89. Laio, A.; VandeVondele, J.; Rothlisberger, U. *J. Phys. Chem. B* 2002, *106*, 7300–7307.
90. Gao, J.; Amara, P.; Alhambra, C.; Field, M. J. *J. Phys. Chem. A* 1998, *102*, 4714–4721.
91. Car, R.; Parrinello, M. *Phys. Rev. Lett.* 1985, *55*, 2471.
92. Latajka, Z.; Bouteiller, Y.; Scheiner, S. *Chem. Phys. Lett.* 1995, *234*, 159–164.
93. Baker, J.; Andzelm, J.; Muir, M.; Taylor, P. R. *Chem. Phys. Lett.* 1995, *237*, 53–60.
94. Friesner, R. A. *Proc. Natl. Acad. Sci.* 2005, *102*, 6648–6653.
95. Hunter, C. A.; Sanders, J. K. M. *J. Am. Chem. Soc.* 1990, *112*, 5525–5534.
96. Burley, S. K.; Petsko, G. *A Science* 1985, *229*, 23–28.
97. Hunter, C. A.; Singh, J.; Thornton, J. M. *J. Mol. Biol.* 1991, *218*, 837–846.
98. McGaughey, G. B.; Gagne, M.; Rappe, A. K. *J. Biol. Chem.* 1998, *273*, 15458–15463.
99. Watson, J. D.; Crick, F. H. C. *Nature* 1953, *171*, 737–738.
100. Waller, P.; Robertazzi, A.; Platts, J. A.; Hibbs, D. E.; Williams, P. A. *J. Comput. Chem.* 2006, *27*, 491–504.
101. Wang, W. Z.; Pitonak, M.; Hobza, P. *Chemphyschem* 2007, *8*, 2107–2111.
102. Grimme, S. *J. Comput. Chem.* 2004, *25*, 1463–1473.
103. Cerny, J.; Hobza, P. *Phys. Chem. Chem. Phys.* 2007, *9*, 5291–5303.
104. Riley, K. E.; Vondrasek, J.; Hobza, P. *Phys. Chem. Chem. Phys.* 2007, *9*, 5555–5560.
105. Swart, M.; van der Wijst, T.; Guerra, C. F.; Bickelhaupt, F. M. *J. Mol. Model.* 2007, *13*, 1245–1257.
106. Overgaard, J.; Waller, M. P.; Piltz, R.; Platts, J. A.; Emseis, P.; Leverett, P.; Williams, P. A.; Hibbs, D. E. *J. Phys. Chem. A* 2007, *111*, 10123–10133.
107. Grimme, S.; Antony, J.; Schwabe, T.; Muck-Lichtenfeld, C. *Org. Biomol. Chem.* 2007, *5*, 741–758.
108. Zhao, Y.; Truhlar, D. G. *J. Chem. Theory Comput.* 2007, *3*, 289–300.
109. Meyer, M.; Steinke, T.; Suhnel, J. *J. Mol. Model.* 2007, *13*, 335–345.
110. Antony, J.; Grimme, S. *Phys. Chem. Chem. Phys.* 2006, *8*, 5287–5293.
111. Stein, M.; Lubitz, W. *Phys. Chem. Chem. Phys.* 2001, *3*, 2668–2675.
112. Holland, J.; Green, J. C.; Dilworth, J. R. *Dalton Trans.* 2006, 783–794.

113. Wang, X. J.; Wang, W.; Koyama, M.; Kubo, M.; Miyamoto, A. *J. Photochem. Photobiol. A* 2006, *179*, 149–155.

114. Schlogl, R. *Angew. Chem. Int. Ed.* 2003, *42*, 2004–2008.

115. Schrock, R. R. *Acc. Chem. Res.* 2005, *38*, 955–962.

116. Triplett, E. W. *Prokaryotic Nitrogen Fixation.* Horizon Scientific: Wymondham, 2000.

117. Selvaraj, S.; Kono, H.; Sarai, A. *J. Mol. Biol.* 2002, *322*, 907–15.

118. MacKay, B. A.; Fryzuk, M. D. *Chem. Rev.* 2004, *104*, 385–401.

119. Hidai, M. *Coord. Chem. Rev.* 1999, *186*, 99–108.

120. Burgess, B. K.; Lowe, D. J. *Chem. Rev.* 1996, *96*, 2983–3012.

121. Rees, D. C.; Howard, J. B. *Curr. Opin. Chem. Biol.* 2000, *4*, 559–566.

122. Le Guennic, B.; Kirchner, B.; Reiher, M. *Chem. Eur. J.* 2005, *11*, 7448–7460.

123. Cammi, R.; Mennucci, B.; Tomasi, J. *J. Phys. Chem. A* 2000, 9100.

124. Zexing Cao; Zhou, Z.; Wan, H.; Zhang, Q. *Int. J. Quantum Chem.* 2005, *103*, 344–353.

125. Studt, F.; Tuczek, F. *Angew. Chem. Int. Ed.* 2005, *44*, 5639–5642.

126. Spiegel, K.; Magistrato, A.; Carloni, P.; Reedijk, J.; Klein, M. L. *J. Phys. Chem. B* 2007, *111*, 11873–11876.

127. Hatfield, M. P. D.; Palermo, N. Y.; Jozsef, C. Y.; Murphy, R. F.; Lovas, S. *Int. J. Quantum Chem.* 2008, *108*, 1017–1021.

128. Meyer, M.; Suhnel, J. *J. Phys. Chem. A* 2008, *112*, 4336–4341.

129. Bader, R. F. W. *Chem. Rev.* 1991, *91*, 893–928.

130. Bader, R. F. W. *Atoms in Molecules—A Quantum Theory.* Oxford University Press: Oxford, 1990.

131. Boyd, R. J.; Choi, S. C. *Chem. Phys. Lett.* 1986, *129*, 62–65.

132. Howard, S. T.; Lamarche, O. *J. Phys. Org. Chem.* 2003, *16*, 133–141.

133. Reedijk, J. *Proc. Natl. Acad. Sci. USA* 2003, *100*, 3611–3616.

134. Zorbas, H.; Keppler, B. K. *Chembiochem* 2005, *6*, 1157–1166.

135. Jung, Y. W.; Lippard, S. J. *Chem. Rev.* 2007, *107*, 1387–1407.

136. Rabik, C. A.; Dolan, M. E. *Cancer Treat. Rev.* 2007, *33*, 9–23.

137. Deubel, D. V. *J. Am. Chem. Soc.* 2006, *128*, 1654–1663.

138. Robertazzi, A.; Platts, J. A. *Chem. Eur. J.* 2006, *12*, 5747–5756.

139. Carloni, P.; Andreoni, W.; Hutter, J.; Curioni, A.; Giannozzi, P.; Parrinello, M. *Chem. Phys. Lett.* 1995, *234*, 50–56.

140. Carloni, P.; Sprik, M.; Andreoni, W. *J. Phys. Chem. B* 2000, *104*, 823–835.

141. Wang, D.; Lippard, S. J. *Nat. Rev. Drug Discov.* 2005, *4*, 307–320.

142. Takahara, P. M.; Rosenzweig, A. C.; Frederick, C. A.; Lippard, S. J. *Nature* 1995, *377*, 649–652.

143. Pasheva, E.; Ugrinova, I.; Spassovska, N.; Pashev, I. *Int. J. Biochem. Cell Biol.* 2002, *34*, 87–92.

144. Ohndorf, U.; Rould, M.; He, Q.; Pabo, C.; Lippard, S. J. *Nature* 1999, *399*, 708–712.

145. Komeda, S.; Kalayda, G. V.; Lutz, M.; Spek, A. L.; Yamanaka, Y.; Sato, T.; Chikuma, M.; Reedijk, J. *J. Med. Chem.* 2003, *46*, 1210–1219.

146. Komeda, S.; Bombard, S.; Perrier, S.; Reedijk, J.; Kozelka, J. *J. Inorg. Biochem.* 2003, *96*, 357–366.

147. Komeda, S.; Lutz, M.; Spek, A. L.; Yamanaka, Y.; Sato, T.; Chikuma, M.; Reedijk, J. *J. Am. Chem. Soc.* 2002, *124*, 4738–4746.

148. Komeda, S.; Lutz, M.; Spek, A. L.; Chikuma, M.; Reedijk, J. *Inorg. Chem.* 2000, *39*, 4230–4236.

149. Teletchea, S.; Komeda, S.; Teuben, J. M.; Elizondo-Riojas, M. A.; Reedijk, J.; Kozelka, J. *Chem. Eur. J.* 2006, *12*, 3741–3753.

150. Maurer, P.; Laio, A.; Hugosson, H.; Colombo, M. C.; Rothlisberger, U. *J. Chem. Theory Comput.* 2007, *3*, 628–639.

151. Spiegel, K.; Magistrato, A.; Maurer, P.; Ruggerone, P.; Rothlisberger, U.; Carloni, P.; Reedijk, J.; Klein, M. L. *J. Comput. Chem.* 2008, *29*, 38–49.

152. Zacharias, M. *Biophys. J.* 2006.

153. Olson, W. K.; Gorin, A. A.; Lu, X. J.; Hock, L. M.; Zhurkin, V. B. *Proc. Natl. Acad. Sci.* 1998, *95*, 11163–11168.

154. Deremble, C.; Lavery, R. *Curr. Opin. Struct. Biol.* 2005, *15*, 171–175.

155. Henin, J.; Chipot, C. *J. Chem. Phys.* 2004, *121*, 2904–2914.

156. Fisher, J. F.; Meroueh, S. O.; Mobashery, S. *Chem. Rev.* 2005, *105*, 395–424.

157. Hall, B. G.; Barlow, M. *J. Mol. Evol.* 2003, *57*, 255–260.

158. Galleni, M.; Lamotte-Brasseur, J.; Rossolini, G. M.; Spencer, J.; Dideberg, O.; Frere, J. M.; Grp, M.-B.-L. W.; *Antimicrob. Agents Chemother.* 2001, *45*, 660–663.

159. Garau, G.; Di Guilmi, A. M.; Hall, B. G. *Antimicrob. Agents Chemother.* 2005, *49*, 2778–2784.

160. Walsh, T. R.; Toleman, M. A.; Poirel, L.; Nordmann, P. *Clin. Microbiol. Rev.* 2005, *18*, 306–325.

161. Crowder, M. W.; Spencer, J.; Vila, A. J. *Acc. Chem. Res.* 2006, *39(10)*, 721–728.

162. Hall, B. G.; Barlow, M. *J. Antimicrob. Chemother.* 2005, *55*, 1050–1051.

163. Orellano, E. G.; Girardini, J. E.; Cricco, J. A.; Ceccarelli, E. A.; Vila, A. J. *Biochemistry* 1998, *37*, 10173–10180.

164. Prosperi-Meys, C.; Wouters, J.; Galleni, M.; Lamotte-Brasseur, J. *Cell Mol. Life Sci.* 2001, *58*, 2136–2143.

165. Bounaga, S.; Laws, A. P.; Galleni, M.; Page, M. I. *Biochem. J.* 1998, *331*, 703–711.

166. Rasia, R. M.; Ceolin, M.; Vila, A. J. *Protein Sci.* 2003, *12*, 1538–1546.

167. Rasia, R. M.; Vila, a. J. *Biochemistry* 2002, *41*, 1853–1860.

168. Rasia, R. M.; Vila, a. J. *J. Biol. Chem.* 2004, *279*, 26046–26051.

169. Wang, Z. G.; Fast, W.; Benkovic, S. J. *J. Am. Chem. Soc.* 1998, *120*, 10788–10789.

170. Wang, Z. G.; Fast, W.; Benkovic, S. J. *Biochemistry* 1999, *38*, 10013–10023.

171. Garcia-Saez, I.; Hopkins, J.; Papamicael, C.; Franceschini, N.; Amicosante, G.; Rossolini, G. M.; Galleni, M.; Frere, J. M.; Dideberg, O. *J. Biol. Chem.* 2003, *278*, 23868–23873.

172. Rossolini, G. M.; Franceschini, N.; Riccio, M. L.; Mercuri, P. S.; Perilli, M.; Galleni, M.; Frere, J. M.; Amicosante, G. *Biochem. J.* 1998, *332*, 145–152.

173. Docquier, J. D.; Lamotte-Brasseur, J.; Galleni, M.; Amicosante, G.; Frere, J. M.; Rossolini, G. M. *J. Antimicrob. Chemother.* 2003, *51*, 257–266.

174. Oelschlaeger, P.; Mayo, S. L.; Pleiss, J. *Protein Sci.* 2005, *14*, 765–774.

175. Yamaguchi, Y.; Kuroki, T.; Yasuzawa, H.; Higashi, T.; Jin, W. C.; Kawanami, A.; Yamagata, Y.; Arakawa, Y.; Goto, M.; Kurosaki, H. *J. Biol. Chem.* 2005, *280*, 20824–20832.

176. Materon, I. C.; Beharry, Z.; Huang, W. Z.; Perez, C.; Palzkill, T. *J. Mol. Biol.* 2004, *344*, 653–663.

177. Oelschlaeger, P.; Schmid, R. D.; Pleiss, J. *Protein Eng.* 2003, *16*, 341–350.

178. Moali, C.; Anne, C.; Lamotte-Brasseur, J.; Groslambert, S.; Devreese, B.; Van Beeumen, J.; Galleni, M.; Frere, J. M. *Chem. Biol.* 2003, *10*, 319–329.

179. Toleman, M. A.; Simm, A. M.; Murphy, T. A.; Gales, A. C.; Biedenbach, D. J.; Jones, R. N.; Walsh, T. R. *J. Antimicrob. Chemother.* 2002, *50*, 673–679.

180. Yanchak, M. P.; Taylor, R. A.; Crowder, M. W. *Biochemistry* 2000, *39*, 11330–11339.

181. Garrity, J. D.; Carenbauer, A. L.; Herron, L. R.; Crowder, M. W. *J. Biol. Chem.* 2004, *279*, 920–927.

182. Crawford, P. A.; Yang, K. W.; Sharma, N.; Bennett, B.; Crowder, M. W. *Biochemistry* 2005, *44*, 5168–5176.

183. Davies, A. M.; Rasia, R. M.; Vila, A. J.; Sutton, B. J.; Fabiane, S. M. *Biochemistry* 2005, *44*, 4841–4849.

184. Fast, W.; Wang, Z. G.; Benkovic, S. J. *Biochemistry* 2001, *40*, 1640–1650.
185. Wang, Z. G.; Fast, W.; Valentine, A. M.; Benkovic, S. J. *Curr. Opin. Chem. Biol.* 1999, *3*, 614–622.
186. Wang, Z. G.; Fast, W.; Benkovic, S. J. *Biochemistry* 1999, *38*, 10013–10023.
187. Walsh, T. R.; Hall, L.; Assinder, S. J.; Nichols, W. W.; Cartwright, S. J.; Macgowan, A. P.; Bennett, P. M. *BBA—Gene Struct. Expr.* 1994, *1218*, 199–201.
188. Garau, G.; Bebrone, C.; Anne, C.; Galleni, M.; Frere, J. M.; Dideberg, O. *J. Mol. Biol.* 2005, *345*, 785–795.
189. Carfi, A.; Pares, S.; Duee, E.; Galleni, M.; Duez, C.; Frere, J. M.; Dideberg, O. *EMBO J.* 1995, *14*, 4914–4921.
190. Concha, N. O.; Rasmussen, B. A.; Bush, K.; Herzberg, O. *Structure* 1996, *4*, 823–836.
191. Fabiane, S. M.; Sohi, M. K.; Wan, T.; Payne, D. J.; Bateson, J. H.; Mitchell, T.; Sutton, B. J. *Biochemistry* 1998, *37*, 12404–12411.
192. Dal Peraro, M.; Vila, A. J.; Carloni, P. *Proteins* 2004, *54*, 412–423.
193. Simona, F.; Magistrato, A.; Vera, D. M.; Garau, G.; Vila, A. J.; Carloni, P. *Proteins* 2007, *69*, 595–605.
194. Dal Peraro, M.; Llarrull, L. I.; Rothlisberger, U.; Vila, A. J.; Carloni, P. *J. Am. Chem. Soc.* 2004, *126*, 12661–12668.
195. Dal Peraro, M.; Vila, A. J.; Carloni, P.; Klein, M. L. *J. Am. Chem. Soc.* 2007, *129*, 2808–2816.
196. Llarrull, L. I.; Fabiane, S. M.; Kowalski, J. M.; Bennett, B.; Sutton, B. J.; Vila, A. J. *J. Biol. Chem.* 2007, *282*, 18276–18285.
197. Gonzalez, J. M.; Martin, F. J. M.; Costello, A. L.; Tierney, D. L.; Vila, A. J. *J. Mol. Biol.* 2007, *373*, 1141–1156.
198. Tomatis, P. E.; Fabiane, S. M.; Simona, F.; Carloni, P.; Sutton, B. J.; Vila, A. J. *Proc. Natl. Acad. Sci. USA* 2008, *105*, 20605–20610.
199. Llarrull, L. I.; Tioni, M. F.; Vila, A. J. *J. Am. Chem. Soc.* 2008, *130*, 15842–15851.
200. Tioni, M. F.; Llarrull, L. I.; Poeylaut-Palena, A. A.; Marti, M. A.; Saggu, M.; Periyannan, G. R.; Mata, E. G.; Bennett, B.; Murgida, D. H.; Vila, A. J. *J. Am. Chem. Soc.* 2008, *130*, 15852–15863.

2 Structural and Thermodynamic Studies of α-Synuclein Proteins Related to Parkinson's Disease: Impact of Aqueous Solution and Fe³⁺ Coordination

Olivia M. Wise, Liang Xu, and Orkid Coskuner

CONTENTS

2.1 INTRODUCTION

Parkinson's disease is a severe neurodegenerative disorder that currently affects more than 1.5 million people in the United States. The total health care costs for Parkinson's disease are approximately \$25 billion per year, and this cost is expected to rise unless an effective treatment can be found.[12] For an efficient treatment to be developed, however, the cause (molecular mechanism) of the disease must first be determined and a therapy must be devised. Often the most effective ways to discover the cause of any disease is to start with data and observations from those patients already presenting symptoms and work backward through the progression of the disease until the root cause is exposed. A thorough understanding of what is already known about Parkinson's disease will make clear why the avenue of research discussed in this chapter may aid in the search for the cause of this disease.

Every year, approximately 50,000 new cases of Parkinson's disease are diagnosed. However, it is estimated that many more people are affected by this disease.[12] The average age of onset for Parkinson's disease is 60 years, but many people frequently go years with either an incorrect diagnosis or without realizing they have the disease because the symptoms are often associated with normal aging. There is no definitive test to determine if a patient actually has Parkinson's disease; instead, a diagnosis is obtained from a combination of medical history and a neurological examination.[12,13] Tests such as magnetic resonance imaging (MRI) and computed tomography (CT) scans are performed to rule out other diseases with symptoms similar to Parkinson's disease, as brain scans of those affected often appear normal in the beginning stages of the disease. Unfortunately, the progression and development of the disease varies from person to person, making it even harder to diagnose.[13]

The first recorded case of Parkinson's disease was documented in 1817 by James Parkinson, after whom the disease was later named, in "An Essay on the Shaking Palsy."[14] The symptoms described in his paper are still the symptoms considered to be the hallmark of this debilitating disease. Hand tremors, limb rigidity, bradykinesia (slowness of movement), and postural instability (impaired balance) are the four most common symptoms observed among all Parkinson's disease patients.[12] These symptoms cause those affected with this disease to be unable to perform even the simplest tasks as the disease continues to progress. Patients are not only affected by well-known external symptoms, but may also suffer from pain, dementia, sleep disturbances, depression, fatigue, and memory difficulties.[12] Although the average onset occurs at 60 years, there are several cases of early onset (before the age of 50) and even juvenile parkinsonism (onset at approximately 20 years old).[12] Current statistics also show that on average more men than women are affected by Parkinson's disease.[12]

This disease affects not only the patient, but the caregivers as well. Often family members of patients are the primary caregivers, and in some instances, depending on the severity of the disease, may spend in excess of 40 h per week caring for affected family members.[15] This level of care can have many adverse effects beyond the amount of time required, such as lower income due to time lost from work and increased medical expenses. Some statistics have shown that caregivers suffer from higher levels of stress.[15] As the "baby boomer" generation approaches the average age for onset of this disease, more and more people will be affected, either as a patient or as a caregiver.

Although there are many hypotheses on the cause of Parkinson's disease currently being investigated, the underlying cause remains a mystery. The hypotheses being investigated are varied and it may ultimately prove to be a combination of these hypotheses that results in a full understanding of Parkinson's disease. The current hypotheses being investigated include increased nigral iron,[16] mitochondrial dysfunction,[17] changes in antioxidant systems,[18] neuromelanin/iron interaction,[19] α-synuclein/iron interaction,[16] protein aggregation,[20] Lewy body formation,[4] and oxidative stress.[21] All of these factors, or a combination of several factors, may lead to dopaminergic neuronal cell death in the *substantia nigra* that occurs as a result of Parkinson's disease.

The *substantia nigra* is a compact collection of mostly dopaminergic neurons located in the mid brain and appears to control motor function throughout the body.[19] This region of the brain is easily characterized by its dark color, which is also the origin of the name *substantia nigra* (from the Latin meaning black body), relative to the surrounding neurons due to the presence of a pigment called neuromelanin in the cells.[19] Currently, no active neuromelanin synthesis pathway is known, but it is hypothesized to be a by-product of the autoxidation of dopamine and noradrenaline in neuronal cells, which, until recently, was assumed to be inert.[19] One recent hypothesis is that the neuromelanin, like peripheral melanins in cells outside of the central nervous system, serves a protective role in dopaminergic neuronal cells.[22] Peripheral melanins protect cells from free radical damage by inactivating them.[22] It is hypothesized that neuromelanin may protect the cell from toxic by-products of dopamine metabolism and oxidation and may also play a role in maintaining iron homeostatis in the neuronal cells.[22]

Experimental evidence shows that as the dopaminergic cells degenerate during the progression of Parkinson's disease, the dying cells release stored neuromelanin, causing a marked decrease in the pigmentation of the *substantia nigra*.[19] Although these are not the only cells affected by Parkinson's disease, the decrease in dark color of the *substantia nigra* of the brain is one of the pathological diagnostic criteria for this disease.[19] Some studies suggest that this pattern of discoloration is not random and that neuromelanin may play a role in the pathogenesis. However, whether the discoloration has a positive or negative effect on the cell is still debated.[19] One theory suggests that the presence of neuromelanin may have a negative influence on neuronal cells since the pathway from which neuromelanin is believed to be synthesized also produces other active oxidizers such as quinine, semiquinone, and superoxide radicals.[19] Therefore, increased neuromelanin would indicate an increased amount of oxidizers present in the cell, which could lead to cell damage in a variety of ways. Another theory suggests that neuromelanin may in fact have a positive influence on cells during the progression of Parkinson's disease.[19] This theory comes from an observation by Gibb that *substantia nigra* neuronal cells that contain less neuromelanin are more susceptible to degeneration, i.e., they are the first to die, in Parkinson's disease than cells that are more heavily pigmented by neuromelanin.[19] Additionally, there is some evidence that the structure of neuromelanin in the parkinsonian brain may differ from a healthy brain or that parkinsonian brain neurons may contain less neuromelanin than healthy neurons.[19] Despite these studies, the role of neuromelanin in the brain still remains unclear.

Experimental evidence shows that neuromelanin is capable of binding a variety of metal ions, including iron, copper, zinc, and lead. In addition, the *substantia nigra* contains several metal ions, including iron, at concentrations up to four times that of other neuronal cells.[19,22] There is some experimental evidence of iron binding to neuromelanin, but the actual role of this binding remains unclear.[19,22] It is hypothesized that neuromelanin may have a function similar to the protein ferritin, which is to store iron in the cell in an effectively chemically inactive state until it is needed.[22]

However, much research needs to be done in this area to determine the role this might play in the development of Parkinson's disease.

The ubiquitin-proteasome (UPS) pathway has also been implicated in the progression of Parkinson's disease. The UPS pathway is one of the major pathways for degradation of proteins in the cell.[23] In general, a protein is recognized to be ready for destruction because it has either reached the end of its lifetime or is defective in some manner and the pathway is activated to add ubiquitin to the protein.[23] Ubiquitin acts as a label on the protein so that the proteasome system can identify, unfold, and finally destroy it.[23] Malfunction or inhibition of this system causes the cell to be unable to degrade proteins that could potentially have a harmful effect.

Several theories have been suggested for errors in the UPS pathway and their effect on the brain. Postmortem examination of brains of Parkinson's disease patients has shown that proteasome activity is significantly decreased, and recent evidence has linked several of the mutations associated with Parkinson's disease to the UPS pathway.[24] One of the most significant links between the UPS pathway and Parkinson's disease is the protein parkin, which is one of several ubiquitin ligase (E3) proteins.[23] E3 proteins transfer ubiquitin from an ubiquitin-conjugating enzyme (E2) protein, which accepts the ubiquitin from an ubiquitin activating protein (E1), to the protein that is going to be degraded.[23] E3 proteins are vital to the UPS pathway because they not only recognize the proteins that need to be degraded but also transfer the marker for degradation to that protein.[23] An autosomal recessive (both genes must be mutated in order for the effect to be observed) mutation in parkin is found in over 50% of people with juvenile and early onset parkinsonism.[25] Interestingly, α-synuclein and synphilin-1 have been identified as proteins that parkin is responsible for targeting for degradation and these proteins are also the main components, along with ubiquitin, of Lewy bodies.[23]

Mutations in ubiquitin C-terminal hydrolase L1 (UCH-L1) have also been associated with interfering in the UPS pathway in an autosomal dominant (only one gene must be mutated for the effect to be observed) form of hereditary Parkinson's disease.[24] UCH-L1 is another key enzyme in the UPS pathway and the mutation of this enzyme was discovered in a rare case of familial Parkinson's disease.[24] The actual function of this enzyme remains unknown, so the effect of this mutation on the cell is still undetermined.[24] Like many topics in Parkinson's disease, further research is needed to better understand how this mutation may cause the devastating effects associated with this disease.

While parkin and UCH-L1 mutations are associated with hereditary Parkinson's disease, this accounts for less than 10% of those affected. Environmental factors, such as dopaminergic neurotoxins, have also been shown to produce effects on the UPS system.[17] Compounds such as 1-methyl-4-phenyl-1,2,3,4-tetrahydropyridine (MPTP), rotenone, paraquat, dopamine, 6-hydroxydopamine (6-OHDA), iron, manganese, and dieldrin have all shown evidence of affecting the UPS pathway either directly or by causing other effects in the cell, such as oxidative stress and protein aggregation.[25] This strengthens the theory that it is actually a combination of intracellular events that results in the dopaminergic cell death associated with Parkinson's disease.

Oxidative stress is another prominent hypothesis in the development of Parkinson's disease.[21] Oxidative stress results when the cell produces or is exposed to more free radicals or reactive oxygen species (ROS) than it can neutralize before the ROS causes damage to either the DNA, proteins, or lipids.[21] ROS such as $^\bullet OH$, $^\bullet O_2$, $NO^\bullet$, and H_2O_2 can cause DNA mutation and breakage, protein mutation and mis-folding, and membrane damage by reacting with lipids.[26] There are a variety of ways that ROS can be produced in the cell. One of the most common and well-known methods is the Fenton reaction, which converts hydrogen peroxide (H_2O_2) in the presence of Fe^{2+} to form OH^-, and $^\bullet OH$ (the hydroxyl radical) and oxidizes Fe^{2+} to Fe^{3+}.[7] If excess iron is present in the cell, as is the case with Parkinson's disease, it can produce excess ROS, which, left unchecked, will produce extensive damage throughout the cell.

While Parkinson's disease symptoms are typically a result of a lack of dopamine-producing neurons, dopamine itself can also cause oxidative stress in cells. Dopamine is degraded by either monoamine oxidase A or autoxidation.[21] When dopamine undergoes autoxidation, it increases the amount of ROS and precursors to ROS. Coupled with the increase in nigral iron, this can have a devastating effect on the cell. Postmortem analysis of parkinsonian brains has shown that damage from oxidative stress is prevalent throughout all of the neuronal tissue, not only the *substantia nigra*.[21]

Exposure to some pesticides, herbicides, and industrial chemicals are also potential risk factors for Parkinson's disease primarily due to their ability to increase oxidative stress in neuronal cells. One prominent example of this is MPTP, which is a toxic by-product of the production of an opioid similar to meperidine.[17] The effects of MPTP were discovered when a contaminated opioid was injected by drug users who subsequently developed symptoms of Parkinson's disease.[17] MPTP is converted into 1-methyl-4-phenylpyridinium (MPP+), a pyridinium ion, by monoamine oxidase B in the glial cells (non-neuronal brain cells).[18,26] After MPP+ is taken up into the dopaminergic cells via the dopamine transporter, it interferes with mitochondrial respiration by inhibiting complex I, which causes a decrease in mitochondrial ATP production and an increase in ROS.[18,26]

Inhibition of complex I is a common theme in another hypothesis of the development of Parkinson's disease, mitochondrial dysfunction.[17] Complex I is the enzyme that catalyzes the transfer of electrons from nicotinamide adenine dinucleotide (NADH) to ubiquinone in the first step in the electron transport chain in mitochondria.[26] Mitochondrial dysfunction causes an increase in ROS since that is where the majority of them are formed. A variety of chemicals and mutations can result in mitochondrial dysfunction.

Genetic mutations are also commonly observed in the study of Parkinson's disease.[27] By studying the known genetic mutations that result in Parkinson's disease, it is possible that an understanding of the mechanism that results in the clinical symptoms will be determined and assist in identifying the cause of idiopathic forms of Parkinson's disease. A variety of genetic mutations are discussed in the following that are linked to Parkinson's disease.

A mutation in the gene PINK1 causes an autosomal recessive form of early onset parkinsonism by decreasing the function of PINK1.[26,28] The exact function of PINK1 in the cell is unknown, but the effects of the mutation on the function of mitochondria are far reaching.[26] Decreased PINK1 results in a decrease in the mitochondrial membrane potential, inhibition of complex I and IV activity in the electron transport chain, a decrease in adenosine triphosphate (ATP) production, an increase in

ROS production and structural changes to the mitochondria.[26] The effects of PINK1 are just beginning to be investigated and much research remains to be done on the impact of this protein.

A mutation in the LRRK2 gene is the most common autosomal dominant mutation in familial Parkinson's disease.[26] Very little is known about the function of this protein except that it exhibits some kinase activity.[26] It has been shown experimentally that LRRK2 can bind to the outer mitochondrial membrane, but it has not yet been determined if it has any effect on the membrane itself.[26]

Mutations in the protein α-synuclein are also prevalent in familial Parkinson's disease. The common mutations of α-synuclein that have been shown to be present in brains affected by Parkinson's disease are A53T (the 53rd amino acid is changed from alanine to threonine) and A30P (the 30th amino acid is changed from alanine to proline).[13] Alpha-synuclein is a 140 amino acid protein that is extremely flexible and has been shown experimentally to have a natively unfolded structure in solution.[29] It has further been shown experimentally that it has some α-helical conformations, especially in the presence of phospholipid membranes.[29] Currently, the exact function of α-synuclein in the cell remains unknown, but it is the main component of one of the hallmarks (and, it has been theorized, a pathological feature) of the disease, Lewy bodies.[13] Lewy bodies are composed of a wide variety of proteins, the most prevalent being α-synuclein, neurofilaments, ubiquitin, and ubiquinated proteins as well as components of the UPS pathway.[24] Recent studies have suggested that iron is present in Lewy bodies as well, implying that iron concentration may play a role in the aggregation of α-synuclein.[24] Very little is known about the formation of these inclusions and what effect, if any, that they have on the cell.

While Lewy bodies are a hallmark of Parkinson's disease, they are just one form of a class of protein inclusions termed synucleopathies. Synucleopathies is a general term used to refer to any abnormal aggregation of α-synuclein in brain cells and they are associated not only with Parkinson's disease but also with Lewy body dementia, Lewy body variant Alzheimer's disease, multiple system atrophy, and neurodegeneration with brain iron accumulation type I.[5,13] As a result of many diseases being linked to the presence of synucleopathies, aggregation of α-synuclein is theorized to have a pathogenic effect on cells.

Alpha-synuclein has been found experimentally in both small oligomeric and filamentous aggregates.[4] Experimental evidence shows that the filamental structures may be composed of oligomeric structures and that further aggregation of these structures leads to the formation of Lewy bodies.[4] By studying the mechanism and cause of aggregation of α-synuclein, more can be understood about the progression of not only Parkinson's disease, but several other neurodegenerative diseases as well. Factors such as concentration, pH, ionic strength, temperature, protein inhibition and stimulation, lipids, pesticides, phosphorylation, missense mutations, oxidative stress, and the presence of metals are proposed to affect α-synuclein aggregation.[4,30] A triplication of the α-synuclein gene is present in one form of autosomal dominant Parkinson's disease, which results in a dramatic increase in the amount of α-synuclein present in the cells.[4] This dramatic increase, inevitably leading to aggregation and development of Parkinson's disease, shows that the concentration of α-synuclein is an important factor to its aggregation. This has been shown both in laboratory experiments and in the brain. Changes in pH, ionic strength, and temperature can also affect the rate of aggregation of α-synuclein.[4] The aggregation of α-synuclein has been observed experimentally under varying conditions of pH and ionic strength.[4,31] These experiments demonstrated that as the pH decreases, the rate of aggregation increases and that with increasing ionic strength the aggregation rate increases.[4,31] Increasing ionic strength also causes the aggregate to take a more amorphous shape compared to the cross-β conformation normally observed.[4] Temperature has also been shown to increase the rate of aggregation of α-synuclein and as with ionic strength the conformation of the filaments produced is different from the conformation at physiological temperature.[4]

Several proteins are also capable of inhibiting or stimulating α-synuclein aggregation. Both β- and γ-synuclein are homologues, meaning that they contain the same sequences, but may be shorter or longer and perform similar functions compared to α-synuclein. Interestingly, they both have been shown to inhibit in vitro α-synuclein aggregation.[4] On the other hand, proteins such as tau, histones, and brain-specific protein p25α have all been shown to increase the in vitro aggregation rate of α-synuclein.[4] Tau proteins stabilize microtubules by interacting with tubulin in central nervous system neurons and defects in tau proteins are generally associated with Alzheimer's disease. However, tau is also a component of Lewy bodies.[4] Brain-specific protein p25α is a tubulin polymerization promoting protein that has been shown to be a component of Lewy bodies.[4] Histones are proteins that are responsible for ordering DNA in the nucleus of eukaryotic cells.[4] It has been shown experimentally that histones and α-synuclein often associate in the presence of toxic species; however, histones have not been observed in Lewy bodies.[4] The mechanism by which these proteins interact with and affect α-synuclein still remains unknown, but there are theories that suggest that the process occurs through ionic interactions.

Lipids and pesticides can also increase the rate of aggregation of α-synuclein.[4] Experimental evidence has already shown that in the presence of lipids, the conformation of α-synuclein changes from its normally unstructured form to an α-helical structure.[2,29] Additionally, α-synuclein has been shown to fibrilize and form amorphous aggregates in the presence of fatty acids.[4] The mechanism that causes this increase in aggregation remains largely unknown. As discussed previously in the chapter, exposure to pesticides such as paraquat can cause a temporary aggregation of α-synuclein resulting in Parkinson's disease symptoms,[25] but again, the mechanism through which it increases aggregation remains unclear.

Phosphorylation of proteins can also greatly affect their structure. Under normal in vivo conditions, α-synuclein primarily remains unphosphorylated.[4] Upon examination of the protein isolated from synucleopathies, α-synuclein shows phosphorylation at the 129th amino acid (serine).[4] These results have been reproduced in transgenic mice as well as in vitro.[4] Additional in vitro experiments have demonstrated that phosphorylated α-synuclein forms aggregates at a faster rate than the wild-type α-synuclein.[4] The actual effect that this phosphorylation has on the monomer structure remains unknown and is one of the topics under study in our group.

Missense mutations, such as A53T and A30P, have been shown to accelerate α-synuclein aggregation, but in different manners.[4,32] The A30P mutation shows a correlation with the oligomeric aggregation of α-synuclein, whereas the A53T mutation shows a correlation with fibril aggregation of α-synuclein.[32] Another recently discovered mutation is E46K (the 46th amino acid glutamate is mutated to lysine), but the effect of this mutation on α-synuclein aggregation remains unknown.[33] While a correlation between aggregation and α-synuclein mutations exists, the exact reason for their effect on the protein structure is unknown because the effects on the monomer structure and the mechanism for formation of the initial oligomers, which are proposed to be pathogenic, is unknown.

As mentioned earlier, oxidative stress may play an important role in the development of Parkinson's disease. Several reactions in neuronal cells are capable of producing ROS, such as[16]

$$O_2 \rightarrow O_2^- + 2H^+ \rightarrow H_2O_2 \rightarrow OH + OH^- \rightarrow H_2O \quad \text{(Mitochondrial respiration)} \quad \text{(R1)}$$

$$\text{Dopamine} + O_2 \rightarrow \text{DOPAC} + NH_3 + H_2O_2 \quad \text{(Dopamine metabolism)} \quad \text{(R2)}$$

$$H_2O_2 + Fe^{2+} \rightarrow OH + OH^- + Fe^{3+} \quad \text{(Fenton reaction)} \quad \text{(R3)}$$

All of the reactions (R1–R3) produce ROS that are capable of making oxidative modifications to proteins. Experimental evidence has shown that oxidative nitration of α-synuclein (oxidatively adding a dopamine adduct) and peroxide oxidation of α-synuclein increase the rate of aggregation of the protein into Lewy bodies.[4] Oxidative stress is still one of the most prevalent theories for the etiology of Parkinson's disease and is also a prevailing theory in the etiology of Alzheimer's disease, in which α-synuclein may also play a role. However, as with the other theories for the causes of α-synuclein aggregation, the exact mechanism by which these modifications affect the structure–function relationships of α-synuclein is still unclear.

Another characteristic feature in the pathogenesis of Parkinson's disease is increased iron concentration in the *substantia nigra*.[34] Iron concentration is typically high in this region of the brain, but it has been demonstrated experimentally that in patients with Parkinson's disease, the concentration is always abnormally high both throughout the disease and postmortem.[34] The exact reason for this marked increase is unknown. However, many hypotheses explaining the increase and its effect on the *substantia nigra* are being researched.[16]

On average, the human body contains 3–5 g of iron (this concentration is generally lower in women).[34] Iron is essential to many processes in the cell, especially for function and development in the brain, but iron concentration needs to be maintained at a specific level to avoid cytotoxic reactions.[35] The concentration of free iron in the cell is controlled by interactions between transferrin, lactoferrin, divalent metal transporter 1 (also called DMT1), ferritin, and ferroportin.[35] Transferrin, lactoferrin, and DMT1 are proteins involved in the import of iron into cells.[35] Transferrin and lactoferrin are iron carrying proteins that interact with receptors in the plasma membrane of cells and import iron into the cell through endocytosis, whereas DMT1 is an ion channel that transports iron through the cell wall.[35] Ferritin is an iron storage protein that essentially binds free iron in the cell to prevent it from participating in harmful side reactions.[35] Finally, ferroportin is an iron export protein that regulates the flow of iron out of the cell.[35] While there may be more biomolecules involved, these few are known to have a significant impact on the free iron concentration in the cell, and mutations or mis-folding of these proteins could certainly result in excess free iron.[34]

Excess free iron present in the cell can have a significant impact on the intracellular chemistry of neuronal cells and is assumed to play a role in many of the current theories of Parkinson's disease, such as UPS pathway inhibition or malfunction, oxidative stress, α-synuclein aggregation, and Lewy body formation.[34] Experimental evidence has shown that iron chelators actually reduce proteasome inhibition, or rather that excess free iron can inhibit the UPS pathway.[36] As discussed earlier, inhibition of this pathway can have a deleterious effect on the cell.

The presence of Fe^{3+} and other metal ions such as Al^{3+}, Co^{3+}, Mn^{2+}, Ca^{2+}, and Cu^{2+} in solution is also associated with an increase in α-synuclein aggregation.[4] The Fenton reaction, shown earlier, increases the concentration of Fe^{3+} within the cell. Previous simulations using Car–Parrinello molecular dynamics (CPMD) have shown that Fe^{3+} initiates water dissociation and forms Eigen (H_3O^+) and Zundel ($H_5O_2^+$) complexes. The formation of these complexes may have an effect on the mechanism of the Fenton reaction as well as on the pH and chemical reactivity of the solution around it, which (as discussed earlier) affects α-synuclein aggregation. It has been proposed that α-synuclein is also a metal-binding protein and that this binding in turn affects the protein structure and aggregation rates.[37] However, little is known about the mechanism of this binding and current experimental techniques are not capable of measuring the interaction of iron and α-synuclein at a molecular level.

A popular current theory not only in Parkinson's disease, but also in Alzheimer's and other neurodegenerative disorders, is the early aggregation and monomer hypothesis.[31] As already discussed, one of the hallmarks of Parkinson's disease is the formation of synuclein aggregates. Familial forms of Parkinson's disease involve mutations in α-synuclein that result in a higher propensity for aggregation. The early aggregation and monomer hypothesis focuses on inhibiting aggregation at the earliest stage possible so as to prevent the development of the disease before aggregates form. The formation of aggregates is theorized to be the result of decreasing solubility of the α-synuclein monomers, resulting in the formation of partially folded intermediates that are hypothesized to lead to the formation of the observed insoluble amorphous aggregates and fibrils.[31,38]

While some of the factors discussed earlier have been shown experimentally to increase the rate of aggregation, one aspect of protein folding and interaction that is not often investigated is the effect of hydration on protein structure and folding mechanism.[39] The presence of water can have a significant effect on the structure of proteins by interfering with and preventing some intramolecular interactions. The absence of or decrease in the number of water molecules present can also have a considerable effect on protein structure by allowing intramolecular interactions not normally seen, resulting in mis-folded proteins or partially folded intermediates that lead to an increased rate of aggregation. Without fully understanding the impact of water, the structure and dynamics of a protein in vivo cannot be fully understood.

Specifically, solvent effects cannot be ignored in predicting the structure of a biomolecule and its conformation since approximately 60% of the human body consists of water. Experimental studies are usually limited to providing direct information about the local structure of the solvent molecules around the biomolecules and their impact on the structure and conformation of these biomolecules. In general, x-ray and neutron diffraction measurements have been useful in providing structural information. However, these measurements yield radial distribution functions (RDF) that are spherically averaged to reduce the amount of noise in the distribution functions from the experimental data[40] and thus the anisotropic and asymmetric character of the biomolecules is lost. X-ray and neutron diffraction measurements also only provide information on the first solvation shell, which are the water molecules coordinated to the protein.[40] In addition, nuclear magnetic resonance (NMR), Raman, and infrared spectroscopy measurements can provide information about the structure and conformation of aqueous biomolecules; however, of these three techniques, only NMR is capable of detecting the coordination of water molecules to a biomolecule.[41] While the detection of water molecules is possible with NMR, the process is still difficult and only water molecules within 5 Å (the first coordination shell) of the protein can be detected.[41]

One experimental technique that attempts to investigate the impact of water on protein structure involves the use of hydrostatic pressure.[38,39] This technique was developed in an effort to confirm the results obtained by theoretical studies.[39] By subjecting proteins in an aqueous environment to high pressure, water is forced into the normally hydrophobic core of these proteins causing a partial or complete disruption of the native observed structure of the proteins.[38,39] Once the pressure is released, the refolding of the protein can be observed using high-pressure NMR and fluorescence spectroscopies and a folding mechanism can be proposed.[38,39]

In the case of hydrostatic pressure studies of synuclein, limited data is available and the only studies performed thus far have been on α-synuclein fibrils.[39] Alpha-synuclein fibrils subjected to high hydrostatic pressure are dissociated into soluble forms and then monitored as fibrillogenesis occurs so that a fibrillation mechanism can be determined.[39] Through these studies, researchers have found that the mutant forms (A53T and A30P) of the α-synuclein proteins are more inclined to denaturation by high pressure than the wild-type form.[39] It has been proposed that this experimental result relates to the hypothesis that it is actually the oligomers, and not the fibrils and aggregates, that are pathogenic to cells.[39] Therefore, in the case of the A53T and A30P mutants of α-synuclein, the hypothesis is that they are more soluble in oligomer form and thus more toxic to neurons in the *substantia nigra*.[39] Researchers have also found that α-synuclein and its mutants in aggregate and fibril form are highly susceptible to pressure denaturation and, as a result, hydration and dehydration effects may play an important role in α-synuclein aggregation and Lewy body formation.[39]

Although this experimental technique does investigate the impact of water on protein structure, there are no current hydrostatic pressure investigations into the effect of water on aggregation at a monomer/early aggregation level. Additionally, this technique is still not capable of providing precise details of the amino acids of the protein that are the most and least hydrated at the molecular level. Currently, only theoretical studies are capable of determining hydration properties of a protein on such a detailed level. The combination of hydrostatic pressure experiments and computational techniques could, however, provide strong evidence as to the validity of the early aggregation and monomer hypothesis.[39]

Many theories as to etiology of Parkinson's disease have been proposed, but as explained throughout the preceding paragraphs, the mechanisms by which these theoretical processes progress remain mostly unexplored. Current experimental techniques cannot measure at the molecular level and therefore the mechanistic detail from experimental data is minimal. However, a true understanding of the mechanism through which Parkinson's disease progresses can only be gained through understanding reaction mechanisms at a molecular level. Since experiments cannot currently measure reactivity at this level, the most effective means of determining this mechanism is through computational studies.

Through molecular dynamics (MD) simulations of α-synuclein and other factors believed to affect α-synuclein, such as phosphorylation, oxidative modification, and metal binding, a clearer understanding of the path leading to α-synuclein conformational changes, aggregation, and toxicity can be reached. MD simulations provide insight into intra- and intermolecular hydrogen bonding, torsion transitions, solvation effects, and thermodynamic properties. The results from these simulations can be compared to experimental data to confirm that the theoretical studies are in agreement with experimental findings. Experimental techniques are especially deficient in gaining information about the local molecular structure of the solvent around the protein and the impact of the solvent on the protein structure. Knowledge at this level can only be gained through MD simulations.

Theoretical studies have been extremely useful in investigations of the structural and conformational properties of biomolecules in solution. Classical MD simulations have been applied extensively in these studies.[42,43] Even though these simulations ignore quantum effects and many body interactions (usually higher than three body terms) and depend strongly on the quality of the force field parameters, results obtained for proteins and peptides using these tools are largely in agreement with existing experimental data. In addition, ab initio electronic structure calculations have been performed on biomolecules in solution, but they cannot capture dynamic changes in the intermolecular hydrogen bonds or study large size biomolecules.

Besides ab initio electronic structure calculations, ab initio MD simulations using the Born–Oppenheimer or Car–Parrinello approaches study aqueous biomolecules using dynamics based on traditional ab initio (Hartree–Fock based) and density functional theory (DFT) calculations of the potential energy surface, respectively, and thus capture both the impact of dynamical changes in intermolecular hydrogen bonding interactions and of the dynamics. Even though these ab initio simulations are more accurate and powerful, the practical size of the system currently cannot exceed 200 atoms.

Here, we illustrate the importance of the inclusion of solvent effects via comparison of the structures and conformations of α-synuclein and A53T obtained from gas and aqueous-phase simulations as well as investigate metal ion coordination effects. The protein structure and conformation studies have been performed using classical MD simulations as described earlier and the coordination chemistry studies of α-synuclein and A53T with Fe^{3+} ions have been performed using CPMD/MM simulations. The CPMD/MM simulations are an example of a mixed quantum mechanics/molecular mechanics (QM/MM) method.[44] This method enables more accurate studies of systems with many atoms, such as the present study of metal-containing biomolecules in aqueous solution, by partitioning the system into a part which is treated using a quantum mechanics method and a part that is treated using a molecular mechanics method. Typically, the chemically "interesting" part of the system is treated using quantum mechanics methods. In the present example, we treat the metal ion and corresponding amino acid residues with the CPMD method. This enables the study of bond-breaking and bond-forming interactions (something that is not possible using molecular mechanics methods) and provides detailed electronic structure information typical of ab initio and DFT methods along with the dynamical information. The rest of the system is treated with classical MD simulations as described earlier. We find unprecedented insights into the impact of aqueous solution on the protein structure, conformation, and coordination chemistry of Fe^{3+} ions with α-synuclein and A53T, which are presented in the following text.

Other computational studies have also been performed on wild-type and mutant α-synuclein monomers. However, many of these studies were performed in the presence of sodium dodecyl sulfate (SDS) micelles or lipid bilayers due to the theorized role of α-synuclein in vesicle binding.[2] The reported results of these simulations heavily emphasize the interaction of α-synuclein with SDS micelles and lipid bilayers and very little information on the impact of the aqueous solution on the structure of the protein is described or determined.[2]

Other classical MD simulations of the wild-type and mutant monomers in aqueous solution have also been performed, but the water model used has not been specified, which raises questions about the accuracy of the simulation.[3] These previous studies also provide only a minimal description of the percentage of α- and π-helix content of the monomer during the course of the simulations,[3] and attempt to determine probable aggregation sites using a model with questionable accuracy.[45] To date, a detailed analysis of the structural, dynamical, intra- and intermolecular interactions of the protein (with and without the presence of aqueous solution), hydration characteristics, and thermodynamic properties of the monomers is not available in the literature. The present analysis provides a detailed understanding of the wild-type and A53T mutant monomers as well as a possible explanation for the observed experimental differences between the two variants. Additionally, we report the coordination chemistry between the Fe^{3+} ion and the proteins and compare the resulting structures to one another.

2.2   METHODS

Classical MD simulations operate by propagating the positions of a set of nuclei as a function of time on a potential energy surface representing a particular molecular system according to Newton's equations of motion (hence the term "classical").[46] Such simulations permit particles to interact for a period of time and lead naturally to interpretation of results via statistical mechanics. One of the justifications for MD simulations is based on the equality of the statistical ensemble averages and corresponding time averages. This is known as the "ergodic hypothesis." In other words, the time that an atom is located in a phase space of microstates with the same energy is directly proportional to the volume of the phase space, i.e., the probability of each accessible microstate is equivalent over a long period of time. Various ensembles from canonical to grand canonical ensembles can be used in simulations.[47] For studies of biocomplexes, the canonical and isothermal–isobaric ensembles are usually chosen. In our studies, we have used the isothermal–isobaric ensemble (NPT) in which the number of particles, the pressure, and the temperature are kept constant. The partition function can be described using the canonical partition function

$$\Delta(N,P,T) = \int Z(N,V,T)\exp(-\beta PV)C dV \qquad (2.1)$$

where
    N is the number of particles
    P is the pressure
    V is the volume of the system
    C is a normalization constant
    $\beta = k_B^{-1}T^{-1}$, k_B is the Boltzmann constant
    T is the temperature

The Gibbs free energy can be calculated

$$G(N,P,T) = -k_B T \ln\Delta(N,P,T) \qquad (2.2)$$

In order to more accurately mimic the conditions of experimental studies, the temperature and pressure of the system is controlled using thermostats and barostats during the simulations. Widely used thermostats include Berendsen coupling,[48] Nosé–Hoover temperature coupling[49] and Langevin dynamics.[50] In the Berendsen thermostat, the system is weakly coupled to an external heat bath whose temperature T_0 is held constant. The temperature of the system, $T(t)$, is adjusted by a temperature coupling time constant τ according to the formula:

$$\frac{dT(t)}{dt} = \frac{T_0 - T(t)}{\tau} \qquad (2.3)$$

The Berendsen thermostat is an efficient method for a system to reach a desired temperature. The Nosé–Hoover thermostat was introduced by Nosé and developed by Hoover. It features an extra variable s with mass Q added to the system. The motion of each atom with mass m_i and velocity v_i is given by the Nosé equations

$$\frac{d^2 r_i}{dt} = \frac{F_i}{m_i s^2} - \frac{2v_i}{s}\frac{ds}{dt}$$

$$\frac{d^2 s}{dt} = \frac{1}{Qs}\left(\sum_i m_i s^2 v_i^2 - g k_B T_0\right) \qquad (2.4)$$

Here, $g = N_{d.o.f}$ or $g = N_{d.o.f} + 1$, $N_{d.o.f}$ denotes the number of degrees of freedom, k_B is the Boltzmann constant, T_0 is the bath temperature, r is the position, and F is the force. The Nosé equations may also be written as

$$\frac{d\gamma}{dt} = -\frac{1}{\tau_{N-H}}\left(\frac{g}{N_{d.o.f}}\frac{T_0}{T(t)} - 1\right)$$

$$\gamma = \frac{1}{s}\frac{ds}{dt}$$

$$\tau_{N-H} = \sqrt{\frac{Q}{N_{d.o.f}k_B T}} \tag{2.5}$$

where τ_{N-H} is an effective relaxation time that can be estimated during the simulation.

Similarly, pressure is controlled by coupling the system to a pressure bath and relaxing the system to the desired pressure P_0 with the Berendsen barostat

$$\frac{dP(t)}{dt} = \frac{P_0 - P(t)}{\tau_P} \tag{2.6}$$

where τ_P is the pressure coupling time constant.

Another temperature and pressure control method is Langevin dynamics, where additional damping and random forces are introduced into the system. This thermostat and barostat maintain the kinetic energy of the system and therefore maintain the system temperature (or pressure) at a constant value

$$m_i \frac{d^2 r_i(t)}{dt^2} = F_i(r_i(t)) - \gamma_i \frac{dr_i(t)}{dt}m_i + R_i(t) \tag{2.7}$$

where
   F is the force on atom i with mass m_i and position $r_i(t)$ at time t
   $\gamma_i m_i$ is the frictional damping applied to the atom
   R_i are the random forces applied to the atom

Interactions between particles can be described using force fields or quantum chemical models, or via mixing force fields and quantum chemical models in a single simulation. In classical MD simulations, a force field refers to the set of parameters from which the potential energy is calculated as a function of the atomic coordinates. These parameters are derived from fits to experimental measurements and quantum chemical calculations. The most widely used force field parameters in the studies of biocomplexes via classical MD simulations are either "all atom" or "united atom" models. The all atom model provides a parameter for every atom, whereas the united atom model treats groups of atoms, such as the methyl group, as a single particle to reduce the system size and thus the computational time.

The most widely used force fields include parameters for the atomic mass, van der Waals radius, and partial charge of each atom. In addition, force fields may contain equilibrium values of bond lengths, bond angles, and dihedral angles for two, three, and four atoms, respectively, and corresponding force constants. Most force fields have fixed partial charges on the atoms, thus the partial charge of a specific atom does not correlate with a change in the conformation of a given biocomplex. However, the newest force fields incorporate polarizability in their parameterization and are able to describe the change of the atomic partial charge via interactions with surrounding atoms. The use of a polarizable force field model is generally more expensive (in terms of CPU time) than the corresponding fixed partial charge model.

Force field parameters enable the calculation of the potential energy of the system via calculation of bonded (covalent bonds) and nonbonded terms (electrostatic and van der Waals interactions). Classical MD simulations are not capable of bond-breaking and bond-forming interactions due to the nature of the potential energy function. Additionally, dihedral angle potentials and improper torsion potentials, which are needed to represent the planarity of aromatic and conjugated compounds, are added to the potential energy function. The van der Waals term is calculated using the Lennard-Jones potential and the electrostatic term is computed via Coloumb's law.[50] In our classical MD simulations, we use the CHARMM27 force field parameter set.[51] Using this force field, the potential energy is calculated via

$$V(r) = \frac{1}{2} \sum_{bonds} K_b \left(b - b_0\right)^2 + \frac{1}{2} \sum_{angles} K_\theta \left(\theta - \theta_0\right)^2 + \frac{1}{2} \sum_{torsional} K_\phi [1 + \cos(n\phi - \delta)]$$

$$+ \sum_{van\,der\,Waals} 4\varepsilon \left[\left(\frac{\sigma}{r}\right)^{12} - \left(\frac{\sigma}{r}\right)^6 \right] + \sum_{electrostatic} \frac{q_1 q_2}{r_{12}} \tag{2.8}$$

The first three terms of the potential function (Equation 2.8) are bonded interactions, which represent the interaction energy between two (bonds), three (bending angles), and four (torsion angles) atoms joined by one, two, and three bonds, respectively. The last two terms represent the contribution due to nonbonded interactions, i.e., van der Waals and electrostatic interactions, respectively. The parameters in each term are usually obtained via fitting data from quantum chemistry calculations and/or spectroscopic measurements. The interaction energy between two atoms is described by a harmonic potential as a function of the deviation of a bond length b from its equilibrium value b_0. The force constant K_b describes the interaction of the bonds, and is usually fit to experimental infrared stretching frequencies or frequencies calculated using quantum chemistry. The equilibrium bond length b_0 can be inferred from high-resolution crystal structures or determined from spectroscopy. The second term of the potential energy function (Equation 2.8) describes the bending interaction between three bonded atoms. The energy arising due to the deviation of the bond angle θ from its equilibrium angle θ_0 is expressed by a harmonic potential function with a force constant K_θ. The third term describes the torsion energy and is defined as a periodic cosine function, where K is the force constant, n describes the periodicity of the torsion potential, ϕ is the torsion angle, and δ is the phase shift.

Besides these terms, the Urey–Bradley term, which is used to describe the interaction between atoms separated by two bonds (Equation 2.9) and the improper dihedral term, which is used to maintain planarity and chirality (Equation 2.10) are added to the potential energy function (Equation 2.8).

$$E_{U-B} = \frac{1}{2} \sum_{1,4} K_{\phi}[1 - \cos(n\phi)]^2 \tag{2.9}$$

$$E_{improper} = \frac{1}{2} \sum K_{\psi}[\psi - \psi_0]^2 \tag{2.10}$$

The van der Waals interaction term is represented by a Lennard-Jones 6–12 potential (the fourth term in Equation 2.8) that has two components, a repulsive potential represented by the r^{-12} term, preventing atoms from overlapping at short distances, and an attractive potential at long distance, represented by the r^{-6} term. The value of ε controls the depth of the potential well, σ is the distance at which the potential goes to zero and r is the internuclear separation of the two atoms. Coulomb's law is used to calculate the electrostatic interactions of any two atoms (the fifth term in Equation 2.8) with partial charges of q_1 and q_2 separated by a distance r_{12}, where ε_0 is the dielectric constant of the medium. For simulations of the solvent, we use the transferable intermolecular potential 3P (TIP3P) model for water that includes a three-site rigid water molecule with partial charges and a Lennard-Jones–van der Waals potential on each atom.[52]

The classical MD simulations presented in this chapter use cubic box periodic boundary conditions. When an atom leaves the box from one side, another atom enters the box on the opposite side of the box with the same velocity.[50] Under these conditions, angular momentum is not conserved because of the lack of rotational symmetry. However, the linear momentum is conserved. In our simulations, the systems have neutral net charges. If this were not the case, the charge on the infinite periodic system would be infinite. To achieve neutral charge, we add sodium counterions to the system. The minimum image convention requires a spherical cutoff with a length of the half box length for calculating nonbonded interactions.[50] Additionally, the periodic boundary conditions should be applied with extreme care because a small box size relative to the size of the biocomplex could result in interactions of the atoms with their own images in neighboring boxes. Periodic boundary conditions can be used with methods such as Ewald sum method,[53] which treat long-range electrostatic interactions. This method defines these interactions via calculation of the short-range part of the potential in real space and the long-range part of the potential in Fourier space. It is assumed that the short-range term is easier to calculate than the long-range term, and thus the short-range calculation is used as the method to treat long-range interactions. The Fourier summation is applied to the infinite periodic system. The long-range electrostatic energy is computed via integration of the interaction energies between the charges of the unit cell and all the charges of the lattice, with an infinitely large cell that is a collection of the unit cells of the initial periodic boundary conditions cubic box (Equation 2.11):

$$E_{elec} = E^{direct} + E^{reciprocal} + E^{self} + E^{correction} \tag{2.11}$$

where

$$E^{direct} = \sum_{i}^{N} \sum_{j=i+1}^{N} q_i q_j \frac{erfc(\alpha \mathbf{r}_{i,j,\mathbf{n}})}{\mathbf{r}_{i,j,\mathbf{n}}} \tag{2.12}$$

$$E^{reciprocal} = \frac{1}{2\pi V} \sum_{i=1}^{N} \sum_{j=1}^{N} q_i q_j \sum_{\mathbf{m} \neq 0} \frac{\exp\left[-\left(\dfrac{\pi^2 \mathbf{m}^2}{\alpha^2} \right) + i2\pi \mathbf{m} \cdot (\mathbf{r}_j - \mathbf{r}_i) \right]}{\mathbf{m}^2} \tag{2.13}$$

$$E^{self} = -\frac{\alpha}{\sqrt{\pi}} \sum_{i=1}^{N} q_i^2 \tag{2.14}$$

$$E^{correction} = \frac{2\pi}{(1+2\varepsilon)V} \left(\sum_{i=1}^{N} q_i \mathbf{r}_i \right)^2 \tag{2.15}$$

E^{direct} is the energy in real space, N is the number of atoms and $erfc(x)$ is the complimentary error function defined by

$$erfc(x) = 1 - \frac{2}{\sqrt{\pi}} \int_0^x e^{-u^2} du \tag{2.16}$$

$E^{reciprocal}$ is the energy in reciprocal space, $V = a_1 \times a_2 \times a_3$ is the volume of the box with unit vector of a_1, a_2, and a_3, $\mathbf{m} = m_1 a_1^* + m_2 m_2^* + m_3 m_3^*$ is the vector of the reciprocal space, where a_1^*, a_2^*, and a_3^* are the unit vectors of the simulation box in reciprocal space, and m_1, m_2, and m_3 are integers defining the length of the unit vector.

The value of α ($\alpha > 0$) is used to control the convergence of Ewald sum. If α is large, the value of E^{direct} decreases, which means the value of E^{direct} can be neglected when α is large enough. Thus, the minimum image convention can be employed to calculate the value of E^{direct}. E^{self} is the interaction energy of an atom with itself, $E^{correction}$ is the dipole moment correction, and ε is the dielectric constant. Another commonly used method is the Particle-Mesh Ewald (PME) method,[54] which calculates the $E^{reciprocal}$ term using point charges distributed on surrounding grid points, and thus the electrostatic energy is approximated via integration of values on those grid points using interpolation functions. The summation can be efficiently calculated using the fast Fourier transform (FFT), reducing the calculation time dependence from $O(N^2)$ of Ewald summation to $O(N \log N)$ of PME, where N is the number of atoms in the periodic box.[54] The method of PME has been widely applied due to its high efficiency and accuracy.

Specific details about the simulations are now presented. The initial α-synuclein structure was taken from the protein data bank (PDBID 1XQ8) and was placed in a box with dimension $104.9 \times 204.5 \times 90$ Å containing 61,921 water molecules. In order to neutralize the charge of the system, nine sodium ions were randomly placed within the box. For the simulations of A53T, the structure was initially placed in a box containing water molecules of the same dimensions as for α-synuclein simulations, but the protein shifted out of the box during the initial simulations so the dimensions of the box were expanded to $124.9 \times 224.5 \times 110$ Å, which increased the number of water molecules to 98,784. Again, nine sodium ions were randomly placed to neutralize the system. Both systems were minimized for 10,000 steps (20 ps) and then the system was gradually heated to 310 K over 500 ps. The simulations then ran for an additional 3.5 ns. The integration time step was set to 2 fs. The simulation was performed using periodic boundary conditions as described in detail earlier. The cutoff value for the nonbonded interactions was set to 12 Å. The full long-range electrostatic interactions were calculated using the PME method as explained earlier. The simulations were performed using the isobaric-isothermal ensemble using Langevin dynamics to control the temperature and the Langevin piston method to maintain the pressure at 1 bar. The RATTLE algorithm was used to restrain the bonds between the hydrogen atoms and heavy atoms.[55] For simulations in the gas phase, both systems were minimized for 1000 steps and then the system was heated to 50 K for 200 ps. The simulations ran for an additional 19.8 ns. The gas-phase simulations were performed using the canonical ensemble.

Classical MD simulations are extremely valuable in studies of biocomplexes and biometallic species in water, though their accuracy depends on the quality of the force field parameters as described above. The classical nature of the simulations means that quantum effects that are not accounted for in the parameterization are ignored, which can lead to errors in predicting structure–function relationships of biometallic species. Despite the success achieved in simulations of biomolecules, force field parameters cannot give an accurate representation of the electronic structure of metals that are necessary for accurate studies of the coordination chemistry between biomolecules and transition metal ions. In addition, prediction of the types of bonding, especially π-bonding, between biomolecules and transition metal ions is difficult to derive from classical MD simulations.

Recently, multiscale simulations have become an extensively researched area in which a chemically active region is represented using quantum mechanics, a region surrounding the quantum mechanical region is represented using force fields, and the outermost part of the system is represented using coarse-grained models.[47,56] The positions of all atoms in the system are then propagated according to MD. Here we briefly introduce the concept of quantum mechanical (QM) calculations, as well as the combination QM/MM, which is used to investigate the mechanism of aggregation of α-synuclein. The principle and application of CGMD (coarse grained MD) simulations can be found elsewhere.[57]

The total energy for a QM/MM system can be represented as a sum of three contributions:

$$E = E_{QM} + E_{MM} + E_{QM/MM} \tag{2.17}$$

where

E_{MM} represents the energy calculated using a force field (e.g., CHARMM, as described earlier)

E_{QM} represents the energy of the part of the system treated using quantum mechanics, usually described within the Born–Oppenheimer approximation[58] as

$$E_{QM} = -\frac{1}{2} \sum_i^{electrons} \nabla^2 - \sum_i^{electron} \sum_j^{nuclei} \frac{Z_j}{R_{ij}} + \sum_i^{electron} \sum_{j<i}^{electrons} \frac{1}{r_{ij}} + \sum_i^{nuclein} \sum_j^{nuclein} \frac{Z_i Z_j}{R_{ij}} \tag{2.18}$$

The four terms in Equation 2.18 represent contribution from the kinetic energy, the electron–nucleus interaction, the electron–electron interaction, and the nuclear–nuclear repulsion interaction, respectively. When coupled with molecular mechanics, additional van der Waals and electrostatic force terms are added for each QM atom. Thus, the energy of the combined QM/MM system, $E_{QM/MM}$, is written

$$E_{QM/MM} = - \sum_i^{QM\ electrons} \sum_j^{MM\ atoms} \frac{q_j}{r_{ij}} + \sum_i^{QM\ nucleis} \sum_j^{MM\ atoms} \frac{Z_i q_j}{R_{ij}} + \sum_i^{QM\ nucleis} \sum_j^{MM\ atoms} \left(\frac{A_{ij}}{R_{ij}^{12}} - \frac{B_{ij}}{R_{ij}^6} \right) \tag{2.19}$$

where

A_{ij} and B_{ij} are van der Waals parameters
Z_{ij} is the effective charge of nuclei i
q_j is the atomic partial charge of atom j
r_{ij} represents distance between electron i and atom j
R_{ij} represents the distance between nuclei i and atom j

The calculation of E_{MM} is performed as described earlier for empirical force fields. Standard methods in ab initio quantum mechanics (such as the Hartree–Fock method) and in DFT (via the Kohn–Sham equations)[59] are used to calculate the E_{QM}. The so-called "semiempirical" QM methods, which are, in principle, less accurate than QM methods but also less expensive to evaluate, may also be used for the QM region.

One critical consideration in a combined QM/MM method is the means by which the QM and MM parts are joined. Various strategies have been proposed to treat the boundary atoms between the QM and MM part to make the calculation of the coupling, $E_{QM/MM}$, more accurate and efficient. If there is a covalent bond between the QM and MM region, a "link atom," usually a hydrogen atom, is placed between the two bonded atoms and bonds to this atom are computed in both the QM and MM regions. The combination of QM and MM has been widely applied to study reaction mechanisms in biological systems, and especially to biological compounds and metal ion coordination chemistry.

To study the coordination of a transition metal ion to the protein, we performed QM/MM simulations. The gradient corrected Becke–Lee–Yang–Parr (BLYP) functional was used to treat the active part of the protein (histidine, according to experiment) and the Fe^{3+} ion using ab initio MD simulations with the CPMD approach. The CPMD method permits efficient evaluation of the dynamical evolution of the molecular system. The observation of the dynamics of a molecule provides a distinct advantage over the more common use of static (fixed geometry) electronic structure calculations for studies of the type reported here. Nevertheless, static first principles calculations in the gas phase and in water using implicit and explicit solvent molecules have been widely used for studies of biometallic species since these calculations are less expensive than the CPMD simulations. Such calculations may encompass a few points of dynamical interest, but are no substitute for a dynamical trajectory. The remainder of the protein outside the QM region, the water molecules and the counter ions were treated using classical MD.

Recently, we predicted a mechanism for water dissociation in aqueous solution that is induced by the presence of a Fe^{3+} ion and the impact of the dissociated states of water on the hydration structure of the Fe^{3+} in water using CPMD simulations coupled with transition path sampling (TPS) calculations. (TPS is a specialized method for simulating rare events and is critical to the study of metal ions in water.) The initial geometry of the $Fe(OH)_3(H_2O)_6$ complex and of the surrounding water molecules and aqua ions (H_3O^+ and OH^- complexes) were taken from the CPMD/TPS study to investigate the coordination chemistry of the hydrated Fe^{3+} ion with the proteins and the impact of dissociated states of water on the coordination chemistry. For details see Chapter in this book entitled *Metal Ions in Aqueous Solution*.

The details of the QM/MM studies are as follows. Dangling bonds were terminated with hydrogen bonds. A plane-wave basis cutoff of 76 Ry was used in the DFT calculations. Troullier–Martin pseudopotentials for core electrons along with a double zeta quality valence basis set were used on all atoms. The CPMD method was used to perform dynamical calculations of the metal ion binding site of the protein. A time step of 0.2 fs and a fictitious electron mass of 600 au were used. The temperature was controlled using a Nosé–Hoover thermostat. A cutoff value of 12 Å was used for nonbonded interactions and the PME method was used for treating long-range interactions. The temperature was fixed at 310 K.

When simulating proteins in aqueous solution, it is often necessary to ascertain structural properties of the protein by calculating a root-mean-square deviation (RMSD) from a reference set of atomic coordinates

$$RMSD = \sqrt{\frac{\sum_{i=1}^{N} \left(r_i(t) - r_{ref} \right)^2}{N}} \tag{2.20}$$

where
  N is the number of atoms
  $r_i(t)$ represents the coordinates of atom i at time t
  r_{ref} represents the coordinates of the reference structure, which is usually taken from a crystal structure and is sometimes used as the initial structure in a simulation

The RDF for amino acid residues are computed by calculating the center of mass for each residue and then calculating the center of mass and water molecule oxygen atom RDF using

$$g(r) = \frac{V}{N^2} \frac{\langle \Sigma_i n_i(r) \rangle}{V_{cap}\Delta r}$$

(2.21)

where
  r is the radial distance
  $n_i(r)$ is the coordination number of water molecules around center i within a distance of r to $r+\Delta r$
  V is the volume of the system
  V_{cap} is the spherical cap volume, which is defined as $V_{cap} = \frac{1}{3}\pi h^2(3R - h)$, where R is the spherical radius and h is the spherical cap height

The coordination number of water molecules at distances of 4, 5, and 6 Å is calculated via integration of the RDF functions for the corresponding distances from the center of mass of each specific residue.

The solvation Gibbs free energy values are calculated to provide insight into the thermodynamics of the most preferred conformations of the wild-type and mutant α-synuclein proteins in water using the MM/PBSA (Molecular Mechanics/Poisson-Boltzmann surface area) method.

$$G_b = U_{MM} + G_{solvation} - TS$$

(2.22a)

Each term used in the above equation is computed by

$$G_{solvation} = G_{solvation\text{-}electrostatic} + G_{non\text{-}polar}$$

(2.22b)

$$U_{MM} = U_{electrostatic} + U_{vanderWaals} + U_{bond} + U_{angle} + U_{dihedral}$$

(2.22c)

For the $G_{solvation\text{-}electrostatic}$ calculation, the internal and external dielectric constant values were set to 1 and 80, respectively.[60] Dipolar boundary conditions were applied. Separate calculations on the proteins using 3000 linear iterations showed that the calculations were converged.[61]

The nonpolar Gibbs free energy contribution was approximated using the solvent accessible surface area (SASA) based on $\Delta G_{non\text{-}polar} = 0.00542 \times \Delta SASA + 0.92$[62] and entropy calculations based on the Schlitter method were employed for calculating the entropy for each protein using separate simulations.[63] The Schlitter formula yields an upper bound to the true entropy (S_{true})

$$S_{true} < S = \frac{1}{2} k_B \ln \det\left[1 + \frac{k_B T e^2}{h^2} M\sigma\right] \tag{2.22d}$$

where

k_B is the Boltzmann constant

T is the temperature

e is Euler's number

h is Planck's constant

M is a diagonal matrix of masses associated with the atomic degree of freedom

σ is the covariance matrix of atom fluctuations defined as

$$\sigma_{ij} = \left\langle \left(x_i - \left\langle x_i \right\rangle\right)\left(x_j - \left\langle x_j \right\rangle\right)\right\rangle$$

x_i are the coordinates after least squares superposition of the trajectory with respect to all protein backbone atoms. Even though the selection of protein atoms in these calculations might influence the translational and rotational calculations of the entropy, it is shown that this influence is negligible for relative comparisons. The energies and entropies were calculated using all trajectories for gaining insights into the convergence and the impact of simulation time on these thermodynamic properties.

2.3   RESULTS AND DISCUSSION

The structures of aqueous and micelle-bound α-synuclein have been determined experimentally using NMR techniques by Ulmer et al.[29,64] The micelle-bound structure shows anti-parallel α-helical conformation from residues 3 to 37 (which has been termed as the N-helix) and 45 to 92 (which has been termed as the C-helix) with a short extended region from 93 to 97.[29] The rest of the protein is unstructured according to the experimental observations.[29] The structure of the free α-synuclein protein has also been investigated experimentally by Ulmer et al. as well as Elsevier et al. using NMR techniques.[64,65] Both experiments showed that the first 100 residues have a preference toward a helical structure while the 40 C-terminal residues lack any secondary structural conformation.[64,65] The structure of the A53T mutant has also been determined experimentally via NMR techniques by Ulmer et al. and the results show that the structure of the micelle-bound A53T mutant is indistinguishable from the wild-type by NMR techniques.[64] Uversky et al. characterized wild-type α-synuclein and the A53T and A30P mutants in free solution through far-ultraviolet-circular dichroism (far-UV-CD) and Fourier transform infrared (FTIR) measurements and found that at physiological pH, the structures of the three variations of the proteins are virtually indistinguishable.[30,31] However, our simulations show that there are differences in the structures of the wild-type and A53T mutant that can only be observed on an atomic level and yield insight to possible aggregation mechanisms.

Pictorial descriptions of the wild-type α-synuclein and the A53T mutant conformations taken from our MD simulations in the gas phase are depicted in Figures 2.1a and 2.2a, respectively. The corresponding secondary structures are illustrated in Figures 2.1b and 2.2b. Even before an in-depth analysis of structural changes during the course of our simulations, drastic differences between the equilibrated structures of the wild-type and A53T mutant, as well as between the initial and equilibrated structures of each respective mutant, are obvious.

Considering only the wild-type α-synuclein structure in the gas phase (Figure 2.1a), the equilibrated structure forms an ovoid donut shape that looks nothing like the original experimental structure. However, this depiction does not provide an accurate illustration of the changes in the secondary structure of the protein during the course of the simulation. The structure of the protein is completely different from the initial experimental structure within a matter of nanoseconds. As shown in Figure 2.1b, the α-helical structure from residue 1 to 92 has changed from an

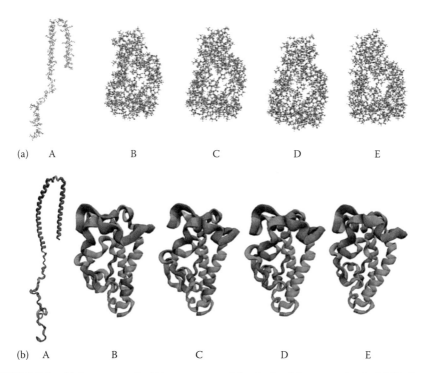

FIGURE 2.1 (a) Structures of wild-type α-synuclein obtained from our classical MD simulations in the gas phase at various times: (A) 0 ns, (B) 5 ns, (C) 10 ns, (D) 15 ns, and (E) 20 ns. (b) (**See color insert.**) Secondary structures of the wild-type α-synuclein protein obtained from classical MD simulations in the gas phase at various times: (A) 0 ns, (B) 5 ns, (C) 10 ns, (D) 15 ns, and (E) 20 ns. The different secondary structures are presented utilizing various colors: α-helix (purple), 3_{10} helix (orange), bridge β (blue), turn (green), coil (red).

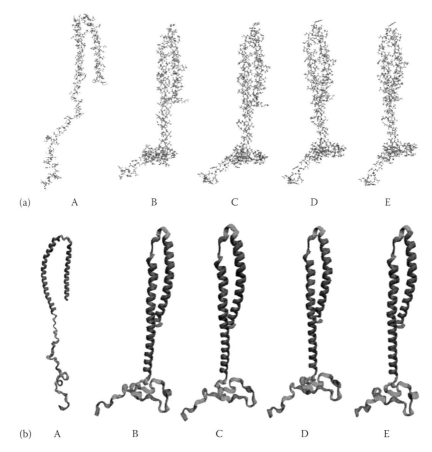

(a)   A   B   C   D   E

(b)   A   B   C   D   E

FIGURE 2.2 (a) Structures of the A53T α-synuclein protein obtained from classical MD simulations in the gas phase at times: (A) 0 ns, (B) 5 ns, (C) 10 ns, (D) 15 ns, and (E) 20 ns. (b) (**See color insert.**) Secondary structures of the A53T α-synuclein protein obtained from classical MD simulations in the gas phase at (A) 0 ns, (B) 5 ns, (C) 10 ns, (D) 15 ns, and (E) 20 ns. The α-helix (purple), bridge β (blue), turn (green), and coil (red) are depicted.

α-helix to a combination of turns and coils with a few select regions of a 3_{10}-helical structure within the first nanosecond. Although the α-helical structure of these initial 92 residues is no longer present, the general U-shape is maintained except that the position and the length of the connecting coil and turn, initially residues 35–44, expands to residues 35–52. Additionally, the N-terminus moves such that it appears to be positioned close to residue 77 converting the U-shape to an almost tear drop shape. The helical region of the protein extends through residue 99 before ending in a long C-terminal tail but after equilibration and without the presence of explicit solvent, the C-terminal tail region begins at residue 77. The C-terminal tail also coils upon itself forming a small clump, leaving only the last 20 residues loose. The last 20 residues then link the U-portion of the protein and the clump of the C-terminal tail, resulting in the final observed structure. These structural changes equilibrate within approximately 1 ns and then the structure does not change significantly for the remainder of the simulation (19 ns).

In contrast to wild-type α-synuclein, our structural analysis reveals that the A53T mutant geometry in the gas phase is completely different from the wild-type geometry in the gas phase. The initial and the equilibrated structures of A53T are more similar than in the case of the wild-type (see Figure 2.2a). Unlike the wild-type, the initial 92 residues of the A53T mutant remain in the α-helical conformation throughout the simulation (see Figure 2.2b). Additionally, the coil/turn connection between the two α-helical portions remains between residues 37 and 45 throughout the course of the simulation. While the actual number of amino acids involved in this connection does not change, the angle at which this connection occurs is obviously different. Also, there does not appear to be any additional unraveling of the α-helical conformation like what is observed in the wild-type simulation. There are similar structural motifs to the changes in the structure of the A53T mutant when compared to wild-type. There is interaction between the N-terminus and the other α-helical portion of the protein (around residue 69) that is similar to the interaction of the N-terminus in the wild-type structure. Also, the C-terminal tail coils upon itself in a similar manner to the wild-type structure except that the remaining loose tail region is approximately 10 residues long instead of 20 residues long in the wild-type structure. The A53T mutant structure equilibrates in about the first nanosecond of the simulation and the structure remains very stable throughout the remaining 19 ns. Thus the effect of changing a single amino acid is seen to have a significant impact on the structural conformation of the protein.

The presence of an aqueous solution environment can also have a significant impact on the structure of a protein due to various solute–solvent interactions, including the intermolecular hydrogen bonds formed between the solute and solvent molecules. The initial structures of the wild-type and mutant A53T were also simulated in aqueous solution using an explicit model for water as described in Section 2.2 to determine the differences, if any, between the structures in the presence of water. The results of the simulations in the presence of explicit solvent molecules show that the presence of water does have a significant effect on the structure. The structures of both the wild-type and mutant A53T α-synuclein have greatly different conformations when compared to the gas-phase structures as well as when the solvated structures of each protein are compared to each other.

The solvated wild-type α-synuclein structure shows the largest structural changes between the initial experimental structure in the gas phase and in aqueous solution. The solvated structure resembles the initial experimental structure more than the structures obtained from our MD simulations in the gas phase (Figure 2.3a). Again, the changes in protein structure are more easily observed by looking at the secondary structures throughout the course of the simulation, as illustrated in Figure 2.3b. Although the overall change in structure is not as significant as in the gas phase, there are still regions of the protein that show significant changes. In agreement with the gas-phase simulations, we observe that the coil/turn connection is between residues 37 and 45. However, this region of the protein appears to change its structural characteristics frequently throughout the course of the simulation. Structural changes between amino acids 68 and 71 in the middle of the C-terminal side of the

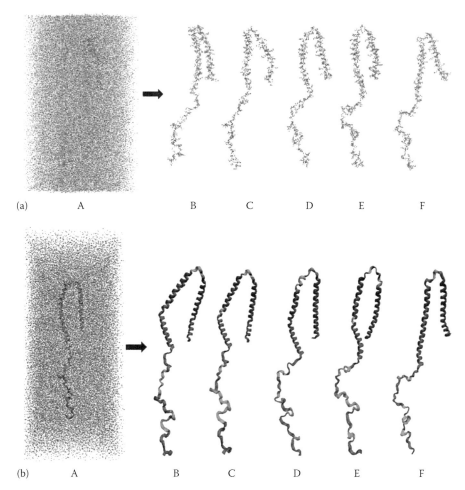

FIGURE 2.3 (a) Structures of the wild-type α-synuclein protein in aqueous solution (A) obtained from classical MD simulations at (B) 0 ns, (C) 1 ns, (D) 2 ns, (E) 3 ns, (F) and 4 ns. (b) (**See color insert.**) Secondary structures of the wild-type α-synuclein in aqueous solution (A) obtained from classical MD simulations at (B) 0 ns, (C) 1 ns, (D) 2 ns, (E) 3 ns, and (F) 4 ns. The α-helix (purple), 3_{10} helix (orange), bridge β (blue), turn (green), and coil (red) are depicted.

α-helix of the protein also occur throughout the course of the simulation. However, the C-terminal tail appears to be the most flexible part of the protein. Unlike the gas phase structure, in the solvated structure, the interaction between the N-terminus and C-terminal side α-helix does not occur and we conclude that this interaction is inhibited by the presence of water molecules.

As with the solvated wild-type simulations, the structural changes of the A53T mutant in aqueous solution resemble the initial experimental structure more than the equilibrated gas-phase structure. By comparing Figure 2.4a through Figure 2.2a, the differences are evident. The differences between the gas-phase and

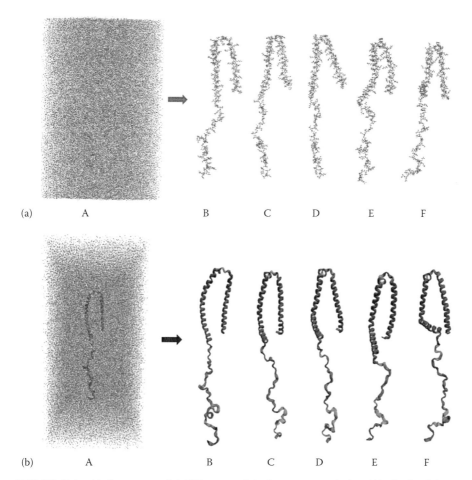

(a)   A   B   C   D   E   F

(b)   A   B   C   D   E   F

FIGURE 2.4 (a) Structures of A53T α-synuclein in aqueous solution (A) obtained from classical MD simulations at (B) 0 ns, (C) 1 ns, (D) 2 ns, (E) 3 ns, and (F) 4 ns. (b) (**See color insert.**) Secondary structures of A53T α-synuclein in aqueous solution (A) obtained from classical MD simulations at (B) 0 ns, (C) 1 ns, (D) 2 ns, (E) 3 ns, and (F) 4 ns. The α-helix (purple), 3_{10} helix (orange), bridge β (blue), turn (green), and coil (red) are depicted.

solvated structures are more pronounced when looking at the secondary structure (Figures 2.4b and 2.2b). The wild-type structure does not exhibit any interaction between the N-terminus and the rest of the protein in aqueous solution, but in the case of the A53T mutant, there does appear to be some interaction between the N-terminus and the C-terminal α-helix. The interaction is not as pronounced as in the gas phase and is partially inhibited by the presence of water molecules, but not as completely inhibited as the wild-type in aqueous solution. The C-terminal tail is again most flexible in the solvated molecular dynamic simulations. The region between residues 68 and 71 undergoes significant structural changes during the course of the simulation as well. The turn and coil section between the two α-helical portions of the mutant protein is also flexible throughout the course of the simulation. However, the changes are not as significant as in the wild-type aqueous structure. In the case of the wild-type, the coil/turn connection grows by unraveling the α-helix on one side, contracting by reforming the α-helix, and then growing again by unraveling on the opposite side before finally contracting back to its original size. The A53T mutant structure in aqueous solution does not exhibit this behavior. Overall, the areas of structural change in the A53T mutant are similar to the wild-type, but they do not change in the same way.

The STRIDE program, which uses computed hydrogen bond energies and torsional angles to determine the secondary structure[66] was used to assign the secondary structure of each residue for each protein in the gas phase and in aqueous solution (Tables 2.1 and 2.2). For the wild-type structure in the gas phase, the

TABLE 2.1

A Quantitative Description of the Secondary Structure Composition of the Wild-Type α-Synuclein Obtained from Classical MD Simulations in the Gas-Phase and in Aqueous Solution

Secondary Structure	Gas Phase					Aqueous Solution				
	0 ns (%)	5 ns (%)	10 ns (%)	15 ns (%)	20 ns (%)	0 ns (%)	1 ns (%)	2 ns (%)	3 ns (%)	4 ns (%)
α-Helix	58.6	0.0	0.0	0.0	0.0	58.6	57.9	55.0	56.4	58.6
Coil	28.6	19.3	20.7	21.4	23.6	26.4	22.9	25.0	17.9	24.3
Turn	11.4	66.4	57.9	66.4	65.7	13.6	19.3	20.0	25.7	13.6
Bridge β	1.4	0.0	0.0	0.0	0.0	1.4	0.0	0.0	0.0	1.4
3_{10}-Helix	0.0	10.7	17.9	12.1	10.7	0.0	0.0	0.0	0.0	2.1
π-Helix	0.0	3.6	3.6	0.0	0.0	0.0	0.0	0.0	0.0	0.0

TABLE 2.2

A Quantitative Description of the Secondary Structure Composition of A53T Mutant α-Synuclein Obtained from Classical MD Simulations in Gas-Phase and in Aqueous Solution

Secondary Structure	Gas Phase					Aqueous Solution				
	0 ns (%)	5 ns (%)	10 ns (%)	15 ns (%)	20 ns (%)	0 ns (%)	1 ns (%)	2 ns (%)	3 ns (%)	4 ns (%)
α-helix	58.6	57.1	57.1	57.1	57.9	58.6	58.6	57.9	57.1	57.9
Coil	27.1	20.7	22.1	20.7	18.6	26.4	26.4	20.0	25.0	28.6
Turn	12.9	22.1	20.7	22.1	23.6	13.6	15.0	22.1	17.9	13.6
Bridge β	1.4	0.0	0.0	0.0	0.0	1.4	0.0	0.0	0.0	0.0
3_{10}-Helix	0.0	0.0	0.0	0.0	0.0	0.0	0.0	0.0	0.0	0.0
π-Helix	0.0	0.0	0.0	0.0	0.0	0.0	0.0	0.0	0.0	0.0

α-helical conformation from residues 3 to 36 and 45 to 92 making up almost 60% of the protein structure completely disappears during the simulation. The majority of the protein converts to a turn conformation, as shown by a 55% increase in content. Some of the α-helix and coil regions of the protein are also converted to a 3_{10}-helix, which increases from 0% to 11%. The bridge β conformation between residues 119 and 123 disappears completely within 1 ns. Four percent of the protein briefly converts to a π-helix conformation between residues 69 and 73, but this conformation does not persist. In aqueous solution, a different trend is observed: approximately 60% of the protein remains in an α-helical conformation while 24% is in a coil structure and 14% is in a turn structure. The bridge β structure between residues 119 and 124 returns after 4 ns. Additionally, a 3_{10}-helix forms between residues 98 and 100, accounting for the 2% decrease in the degree of coil of the protein.

For the A53T mutant simulated in the gas phase, the α-helix structure comprising 60% of the protein between residues 3 and 36 and 45 and 92 remains fairly constant throughout the course of the simulation, decreasing by only 1%. The degree of the coil regions decreases by 8% and is converted along with the bridge β conformation between residues 119 and 123 to a turn conformation. In aqueous solution, the structural conformation remains constant except for conversion of the bridge β conformation to a coil conformation.

Even though the secondary structure analysis indicates that the structures and conformations of both the wild-type and mutant synucleins remain similar to their initial structures in aqueous solution, a more detailed comparison can be made by analyzing the dihedral angle changes of the backbone atoms of these proteins. Table 2.3 provides a summary of the calculated dihedral angles that show significant changes during the course of our simulations in aqueous solution. In the wild-type protein, the dihedral angles that undergo the greatest change (between 100° and 150°) occur in the C-terminal tail except for the dihedral angles from 43N to 44N, which correlate to the observed structural changes. The remaining dihedral angles fluctuate between 70° and 90°, but also occur in the C-terminal tail except for 39N–40N. There are also a few dihedrals (124N–125N, 127αC–128αC, 127N–128N, 97αC–98αC) that are initially very flexible (having a large standard deviation) and end up rigid (having a significantly smaller deviation). Only one bond exhibits the opposite trend (126αC–127αC).

In the case of the mutant protein, dihedrals that undergo the greatest change are also in the C-terminal tail except for 44αC–45αC. The remaining dihedral changes also occur in the C-terminal tail except for the 44C–45C and 39N–40N dihedral angles. In addition to the dihedral angles listed, several regions of the protein were found in the C-terminal tail, the coil/turn region from residues 37 to 44 and the region from residues 67 to 70 that are extremely flexible (also having large values of standard deviation due to their flexibility) throughout the course of our simulation. These flexible regions and regions of greatest change account for the fluctuating structural changes illustrated in Figures 2.1 through 2.4, especially from residues 67 to 70.

Perlmutter et al. have also performed MD simulations of wild-type α-synuclein as well as the A53T mutant, but in the presence of SDS micelles and lipid bilayers because the protein is thought to play a role in synaptic vesicle regulation by

TABLE 2.3

Specific Dihedral Angles That Show Significant Changes in Comparison to Their Values in the Initial Experimental Structure for Both the Wild-Type and A53T Mutant α-Synuclein in Aqueous Solution

	Wild-Type α-Synuclein			A53T Mutant α-Synuclein	
Dihedral Atoms	Initial Dihedral Angle	Final Dihedral Angle	Dihedral Atoms	Initial Dihedral Angle	Final Dihedral Angle
100n-100ca-100c-101n	133.61 (±12.80)	−8.25 (±16.52)	100n-100ca-100c-101n	75.15 (±33.12)	−17.49 (±14.92)
101n-101ca-101c-102n	−18.79 (±24.03)	−135.78 (±48.47)	101c-102n-102ca-102c	59.44 (±10.64)	−69.62 (±12.56)
105c-106n-106ca-106c	−93.52 (±14.76)	69.39 (±11.93)	101n-101ca-101c-102n	−4.51 (±8.09)	−144.13 (±67.17)
107n-107ca-107c-108n	60.58 (±15.79)	144.64 (±32.06)	102n-102ca-102c-103n	55.55 (±8.52)	142.14 (±9.36)
113n-113ca-113c-114n	−115.28 (±39.18)	146.15 (±32.02)	103n-103ca-103c-104n	110.15 (±12.27)	11.80 (±15.80)
124n-124ca-124c-125n	−49.86 (±126.1)	153.29 (±58.40)	104n-104ca-104c-105n	55.99 (±129.33)	149.79 (±68.25)
126c-127n-127ca-127c	60.85 (±15.53)	−71.92 (±22.04)	105c-106n-106ca-106c	−128.99 (±22.35)	−60.15 (±10.77)
126ca-126c-127n-127ca	−155.35 (±32.88)	47.64 (±163.99)	106c-107n-107ca-107c	52.46 (±7.27)	−82.88 (±19.77)
126n-126ca-126c-127n	58.77 (±25.52)	138.98 (±30.87)	107n-107ca-107c-108n	60.80 (±7.80)	152.58 (±9.62)
127ca-127c-128n-128ca	87.30 (±135.5)	160.10 (±33.67)	108n-108ca-108c-109n	71.72 (±45.50)	−5.35 (±26.81)
127n-127ca-127c-128n	0.51 (±165.41)	130.53 (±27.61)	109c-110n-110ca-110c	−82.97 (±11.46)	60.57 (±12.49)
129c-130n-130ca-130c	35.05 (±9.64)	−89.21 (±22.76)	110n-110ca-110c-111n	−21.95 (±26.55)	50.31 (±11.59)
129n-129ca-129c-130n	−38.53 (±14.80)	126.92 (±27.79)	119ca-119c-120n-120ca	−164.29 (±8.60)	−80.84 (±152.48)
130n-130ca-130c-131n	−126.57 (±27.57)	−18.51 (±14.17)	121n-121ca-121c-122n	−42.20 (±11.54)	21.75 (±13.17)
132c-133n-133ca-133c	−1.19 (±65.18)	−106.52 (±24.55)	123n-123ca-123c-124n	−103.91 (±35.89)	42.12 (±16.55)
134c-135n-135ca-135c	−153.8 (±32.65)	−60.16 (±15.22)	129c-130n-130ca-130c	28.25 (±8.68)	−72.88 (±13.47)
39n-39ca-39c-40n	86.75 (±23.42)	−7.77 (±10.84)	132n-132ca-132c-133n	−71.92 (±27.64)	22.86 (±11.16)
43n-43ca-43c-44n	30.28 (±16.59)	146.80 (±47.78)	133n-133ca-133c-134n	−68.02 (±15.41)	42.39 (±10.01)
97ca-97c-98n-98ca	−32.85 (±173.95)	−156.57 (±68.68)	139n-139ca-139c-140n	−39.83 (±13.53)	125.62 (±11.53)
			39n-39ca-39c-40n	97.58 (±19.25)	4.50 (±11.63)
			44n-45ca-45c	−50.39 (±10.77)	47.42 (±9.91)
			44ca-44c-45n-45ca	−164.89 (±6.76)	80.87 (±154.34)
			94n-94ca-94c-95n	132.42 (±27.78)	42.84 (±8.28)

binding to the vesicle membranes.[2] They found that in the presence of SDS micelles and lipid bilayers, the overall structures of both versions of the protein retain the same major structural features observed in the NMR structure in solution and that the C-terminal tail conformation changes the most throughout the course of their simulation.[2] Their simulations show a high degree of flexibility at residues 67 and 68 in both the wild-type and the mutant type and suggest that the flexibility of this section may impact its vesicle binding abilities.[2] They have also reported that the flexibility of these residues is higher in the wild-type than in either the A53T or A30P mutant.[2] From our structural analyses of both the wild-type and the A53T mutant, this region appears to change more frequently in the wild-type simulation than the mutant in aqueous solution. While the A53T mutant kink is clearly present throughout the simulation, it trends in the same direction throughout the simulation whereas the wild-type fluctuates back and forth frequently. Overall, our findings are in excellent agreement with experiments and agree partially with previously reported simulation results. While other simulations have been performed, the hydration characteristics, including both structural and thermodynamics calculations, have not been investigated in detail. In the following paragraphs, we present the change in the hydration characteristics of these two proteins during each nanosecond of our simulations.

The RMSD of the protein backbone was calculated using VMD[67] for each simulation as described in Section 2.2. According to the values in Table 2.4, the wild-type structure shows a difference of 532% between the gas-phase and aqueous solution and the A53T mutant structure differs by 157% between the gas-phase and solution structure. The equilibrated wild-type synuclein changed by 99% more than the equilibrated mutant protein in the gas phase whereas in aqueous solution, the mutant protein changes by 24% more than the wild-type.

The value of R_{ee}, the distance from the nitrogen at the N-terminus to the carboxyl carbon at the C-terminus, was also calculated for each of the structures. The value of R_{ee} for the wild-type α-synuclein changes by 30 Å more than for the A53T mutant during the equilibration period of the simulation. Once equilibrated, the wild-type α-synuclein has the smallest standard deviation of the R_{ee} value in the gas phase.

TABLE 2.4

RMSD, R_{ee}, and Percentage Change in H-Bond Values Calculated Using VMD

	RMSD (nm)	Initial R_{ee} (Å)	Eq R_{ee} (Å)	Intramolecular H-Bonds Disappearing in Water
AS gas phase	4.478 (±0.014)	110	41.89 (±0.49)	—
AS initial in sol'n	0.708 (±0.07)	103.23	109.49 (±5.9)	97.66%
A53T gas phase	2.249 (±0.04)	102.28	70.37 (±2.25)	—
A53T initial in sol'n	0.876 (±0.092)	103.18	106.27 (±4.97)	93.23%

Source: Humphrey, W. et al., *J Mol Graphics.*, 14(1), 33, 1996.

The value of R_{ee} for the equilibrated structures in aqueous solution and in the gas phase changes by 67.6 Å for the wild-type and 35.9 Å for the A53T mutant synucleins. The value of R_{ee} is 3 Å greater in the wild-type than the mutant in aqueous solution. Overall, the differences in the values of R_{ee} between the initial experimental structures and the equilibrated structures is smallest for the wild-type (1 Å). These results also demonstrate the flexibility of the proteins in the absence of solvent molecules and indicate that the initial structures in aqueous solution are stabilized via the interactions of the protein with the surrounding water molecules. Thus, the calculated values of R_{ee} in aqueous solution are similar for both proteins and reflect the similarity between the two structures.

In order to provide a quantitative description of the intramolecular hydrogen bonds that are inhibited by the presence of water, the hydrogen bonding in the gas-phase and aqueous-phase structures was determined for both proteins. The probability (Z) of H-bond formation in the equilibrated structures was calculated by determining the number of frames in which the distance between potentially hydrogen-bonded atoms is less than or equal to 3 Å. The potential of mean force (PMF) values (Tables 2.5 and 2.6) were calculated as described in Section 2.2. The differences in the hydrogen bonding highlight the differences in the dehydrated (gas phase) and solvated structures.

The most common residues involved in hydrogen bonding for the wild-type α-synuclein protein are the lysine and glutamate amino acids. Hydrogen bonds to these two amino acids are also among the most preferred in the gas phase (Table 2.5). The majority of the most stable intramolecular hydrogen bonds are located between the α-helical sections of the protein (residues 3–37 and 45–92) and the C-terminal tail (residues 97–140), explaining the structure of the compactly folded protein in the gas phase. Although there are also a number of lysine and glutamate intramolecular hydrogen-bonds present in aqueous solution, the only common intramolecular hydrogen-bond is the one that is formed between residues 102 and 105, and even so it is not the most preferred of all the intramolecular hydrogen-bonds in aqueous solution (see Table 2.5). These results indicate that intermolecular hydrogen bonding interactions with surrounding water molecules stabilize the protein in a relatively unfolded conformation and prevent the formation of the same intramolecular hydrogen-bonds that resulted in the highly folded structure observed in the gas phase. Threonine residues present in the protein form intramolecular hydrogen bonds in aqueous solution that are not present in the gas phase and highly preferred compared to the other intramolecular hydrogen-bonds of the protein in solution. Looking at these hydrogen-bonds specifically, a pattern may be seen. All of the bonds occur between the hydroxyl hydrogen and a backbone carboxyl oxygen separated by three residues in the α-helical sections of the protein and, therefore, assist in stabilizing the observed α-helical structure. Only between residues 68 and 71 is a deviation from the α-helical conformation observed. This may be due to lack of a stabilizing threonine intramolecular hydrogen bond within the backbone in this region.

The intramolecular hydrogen bonds and corresponding PMF values for the A53T mutant protein in the gas-phase and aqueous solution are presented in Table 2.6. As seen in the wild-type α-synuclein gas-phase structure, the most common and

TABLE 2.5
Intramolecular Hydrogen Bonds Present in α-Synuclein in Gas Phase and in Aqueous Solution

	Gas Phase				Aqueous Solution		
Donor	Acceptor	Z	PMF (kJ/mol)	Donor	Acceptor	Z	PMF (kJ/mol)
Ala107 (HN)	Glu126 (OE1)	0.9998	0.0001 (±0.002)	Asn103 (HD22)	Leu100 (O)	0.1817	6.7149 (±3.9489)
Gln99 (HE21)	Glu104 (OE2)	1.0000	0.0000 (±0)	Asn122 (HD22)	Glu126 (O)	0.0455	9.9208 (±4.0745)
Gly106 (HN)	Glu126 (OE2)	0.9951	0.0000 (±0.0019)	Glu131 (HN)	Ser129 (OG)	0.1987	6.8669 (±5.6709)
Gly47 (HN)	Glu137 (OE2)	1.0000	0.0000 (±0)	Lys23 (HZ2)	Glu20 (OE1)	0.2502	3.7397 (±1.176)
HSE50 (HE2)	Ala140 (OT2)	0.8628	0.0622 (±0.0271)	Lys23 (HZ2)	Glu20 (OE2)	0.2065	4.2317 (±1.8338)
Lys10 (HZ1)	Glu114 (OE2)	1.0000	0.000 (±0)	Lys60 (HZ2)	Glu61 (OE1)	0.2144	2.7151 (±0.8855)
Lys10 (HZ2)	Glu83 (OE1)	0.9843	0.0077 (±0.0337)	Lys80 (HZ1)	Glu83 (OE1)	0.2553	4.9001 (±3.4765)
Lys10 (HZ3)	Ala85 (O)	1.0000	0.0000 (±0)	Lys96 (HZ1)	Asp98 (OD1)	0.1149	7.0568 (±7.6113)
Lys102 (HZ3)	Glu105 (OE2)	1.0000	0.0000 (±0)	Lys96 (HZ1)	Asp98 (OD2)	0.0850	7.2565 (±4.8273)
Lys12 (HZ3)	Gly84 (O)	1.0000	0.0000 (±0)	Lys102 (HZ3)	Glu105 (OE2)	0.3360	3.3385 (±2.1973)
Lys21 (HZ1)	Asp119 (OD1)	1.0000	0.0000 (±0)	Met127 (HN)	Glu126 (OE2)	0.3123	3.3499 (±3.1458)
Lys32 (HZ3)	Glu137 (OE1)	1.0000	0.0000 (±0)	Thr22 (HG1)	Ala18 (O)	0.6385	1.1552 (0.2342)
Lys34 (HZ1/HZ3)	Asp121 (OD2)	0.6914	0.1930 (±0.2784)	Thr33 (HG1)	Ala29 (O)	0.5960	1.3093 (±0.2342)
Lys43 (HZ1)	Tyr133 (OH)	0.9944	0.0023 (±0.0017)	Thr54 (HG1)	Hse50 (O)	0.5637	1.5575 (±0.5326)
Lys58 (HZ3)	Glu139 (OE1)	1.0000	0.0000 (±0)	Thr64 (HG1)	Lys60 (O)	0.7994	0.5986 (±0.2706)
Lys60 (HZ3)	Glu139 (OE2)	1.0000	0.0000 (±0)	Thr75 (HG1)	Val71 (O)	0.2754	2.7947 (±2.9523)
Lys97 (HZ1)	Glu105 (OE1)	1.0000	0.0000 (±0)	Thr81 (HG1)	Val77 (O)	0.7512	0.7548 (±0.2104)
Met1 (HT3)	Ala78 (O)	1.0000	0.0000 (±0)	Thr92 (HG1)	Ile88 (O)	0.8379	0.4637 (±0.1960)
Ser42 (HG1)	Ser129 (O)	1.0000	0.0000 (±0)	Tyr125 (HH)	Pro117 (O)	0.3914	4.0763 (5.6375)
Ser87 (HN)	Glu114 (OE2)	1.0000	0.000 (±0)				
Ser9 (HG1)	Glu83 (O)	0.9825	0.0087 (±0.0364)				
Thr54 (HG1)	Ala140 (OT1)	0.9890	0.0051 (±0.0223)				
Thr81 (HG1)	Asp2 (OD2)	0.9999	0.0010 (±0)				

The H-bond between the Lys102 and Glu105 residues is also observed in water, indicating that the presence of intermolecular H-bonds cannot break the intramolecular H-bond.

TABLE 2.6

Intramolecular Hydrogen Bonds Present in A53T α-Synuclein in Gas-Phase and in Aqueous Solution

Gas Phase				Aqueous Solution			
Donor	Acceptor	Z	PMF (kJ/mol)	Donor	Acceptor	Z	PMF (kJ/mol)
Gln109 (HE21)	Asp115 (OD2)	1.0000	0.0000 (±0)	Gln134 (HE21)	Glu137 (OE1)	0.0615	8.7754 (±4.0539)
Gln134 (HE22)	Glu137 (O)	1.0000	0.0000 (±0)	Lys10 (HZ3)	Glu13 (OE2)	0.3449	2.9905 (±1.4471)
Gln99 (HE22)	Glu126 (OE2)	0.9995	0.0002 (±0.0003)	Lys43 (HZ1)	Glu35 (OE1)	0.3647	2.9946 (±1.7460)
Glu130 (HN)	Glu131 (OE1)	1.0000	0.0000 (±0)	Lys80 (HZ3)	Glu83 (OE1)	0.1374	5.4411 (±4.3100)
Lys102 (HZ1)	Glu105 (OE1)	1.0000	0.0000 (±0)	Lys102 (HZ1)	Glu104 (OE1)	0.0666	9.0837 (±4.4800)
Lys102 (HZ2)	Gln109 (O)	1.0000	0.0000 (±0)	Thr22 (HG1)	Ala18 (O)	0.5546	1.5725 (±0.3242)
Lys102 (HZ3)	Glu110 (OE2)	1.0000	0.0000 (±0)	Thr33 (HG1)	Ala29 (O)	0.5040	1.9060 (±0.9341)
Lys12 (HZ1)	Glu13 (OE2)	1.0000	0.0000 (±0)	Thr44 (HG1)	Val40 (O)	0.6859	1.0400 (±1.1320)
Lys12 (HZ2)	Ser9 (O)	1.0000	0.0000 (±0)	Thr53 (HG1)	Val49 (O)	0.9100	0.2466 (±0.0703)
Lys12 (HZ3)	Ser9 (OG)	0.9970	0.0013 (±0.0031)	Thr54 (HG1)	Hse50 (O)	0.6824	0.9840 (±0.3268)
Lys21 (HZ2)	Gln24 (OE1)	1.0000	0.0000 (±0)	Thr59 (HG1)	Val55 (O)	0.7041	0.9514 (±0.4146)
Lys21 (HZ3)	Glu28 (OE1)	1.0000	0.0000 (±0)	Thr64 (HG1)	Lys60 (O)	0.8860	0.3280 (±0.3679)
Lys23 (HZ1)	Glu20 (OE2)	1.0000	0.0000 (±0)	Thr75 (HG1)	Val71 (O)	0.7121	0.9174 (±0.2481)

Donor	Acceptor			Donor	Acceptor		
Lys23 (HZ2)	Glu20 (OE1)	1.0000	0.0000 (±0)	Thr81 (HG1)	Val77 (O)	0.6359	1.3456 (±0.8714)
Lys32 (HZ1)	Gly41 (O)	1.0000	0.0000 (±0)	Thr92 (HG1)	Ile88 (O)	0.8082	0.5568 (±0.1439)
Lys34 (HZ1)	Val37 (O)	1.0000	0.0000 (±0)				
Lys45 (HZ1)	Lys43 (O)	1.0000	0.0000 (±0)				
Lys45 (HZ3)	Glu46 (OE1)	1.0000	0.0000 (±0)				
Lys58 (HZ2)	Glu61 (OE1)	1.0000	0.0000 (±0)				
Lys6 (HZ1)	Asp2 (OD1)	1.0000	0.0000 (±0)				
Lys6 (HZ3)	Val3 (O)	1.0000	0.0000 (±0)				
Lys60 (HZ3)	Glu57 (OE1/OE2)	1.0000	0.0000 (±0)				
Lys97 (HZ1)	Glu131 (OE2)	1.0000	0.0000 (±0)				
Lys97 (HZ3)	Glu126 (O)	1.0000	0.0000 (±0)				
Ser129 (HG)	Glu131 (OE1)	1.0000	0.0000 (±0)				
Thr53 (HG1)	Val49 (O)	1.0000	0.0000 (±0)				
Tyr125 (HH)	Glu123 (OE2)	1.0000	0.0000 (±0)				
Tyr133 (HH)	Gln134 (OE1)	1.0000	0.0000 (±0)				
Tyr136 (HH)	Ala140 (OT2)	0.5524	0.0050 (±0.0167)				

The H-bonds between Gln134 and Glu137 as well as between Thr53 and Val49 also exist in water.

highly preferred intramolecular hydrogen bonds occur between lysine and glutamate residues. However, the only intramolecular hydrogen bond in common between the wild-type and A53T mutant is between residues 102 and 105. While majority of the intramolecular hydrogen bonds present in the wild-type in the gas phase were between residues that are not normally close to one another, in the A53T mutant a majority of the hydrogen bonds present are between residues that are no more than 10 residues apart. Interestingly, the mutation at residue 53 from an alanine to a threonine results in the formation of an intramolecular hydrogen bond. Intuitively, one would assume that mutation of only one residue would not cause a significant structural change but as seen earlier, the result is vastly different. As discussed earlier, the wild-type protein in the gas phase expands the coil/turn region, linking the N-helix and C-helix from residues 35 to 44 to residues 35–52. In the case of the A53T mutant, the intramolecular hydrogen-bond formation as a result of the mutation plays a similar role to the threonine intramolecular hydrogen-bonds present in the wild-type in aqueous solution by stabilizing α-helix conformation and preventing the movement of the structure into the wild-type conformation.

As seen in the wild-type simulations, the intramolecular hydrogen bonds present in the gas-phase and in aqueous solution are very different. The only intramolecular hydrogen bonds that appear both in the gas phase and in water are between residues 134 and 137 and the one intramolecular hydrogen bond formed due to the mutation between residues 53 and 49. The majority of the intramolecular hydrogen bonds, and the most preferred ones, observed in the A53T mutant structure in aqueous solution are between hydroxyl hydrogens of threonine residues and backbone carboxyl oxygens separated by three residues in the α-helical sections of the protein that further stabilize the α-helical conformation. In addition, many of these structurally stabilizing threonine intramolecular hydrogen bonds are also present in the wild-type structure. Unlike the gas-phase structures, the wild-type and A53T mutant α-synucleins have similar hydrogen bonding and conformation in aqueous solution. Our results are similar to the studies of Perlmutter et al. that showed that the A53T mutation results in an intramolecular hydrogen-bond with the carboxyl oxygen on residue 49 when interacting with an SDS micelle or lipid bilayer.[2] Here we provide more detailed analyses of these configurational changes. As shown in our studies, this hydrogen bond is also found in the aqueous solution simulation of the A53T mutant and is in fact the most preferred hydrogen bond.

As shown earlier, hydration can have a significant impact on the protein structure by affecting the intermolecular interactions. However, there are very few experimental methods that can determine the effects of hydration on protein structure. A recent publication by Silva and Foguel nicely summarizes how hydration characteristics are studied experimentally and what effects can be seen in protein structure.[39] In order to investigate hydration properties experimentally, hydrostatic pressure techniques are used.[39] Essentially, the protein becomes more solvated than it would be natively due to increased water pressure, and this increased solvation denatures the protein by forcing water into the normally hydrophobic cores of the proteins.[39] In the case of α-synuclein, fibrils are subjected to hydrostatic pressure that denatures the fibril structure and as the pressure is reduced, fibrillation reoccurs slowly.[39]

These studies show that the aggregation of α-synuclein that is characteristic of Parkinson's disease may be a result of changes in the hydration characteristics of the monomers.[39] Using this technique in the future may aid in the experimental determination of the mechanism of fibrillation associated with Parkinson's disease. However, the actual coordination of water molecules to the protein is not revealed by this technique.

Other factors have also been shown experimentally to affect the aggregation of synucleins in vitro. Uversky et al. showed that by altering solution conditions, a partially folded intermediate is observed that is thought to be the middle ground between the natively unfolded protein and the fibril aggregates.[31] Through further experimentation, they have demonstrated that several factors including decreased pH, increasing temperature, increased protein concentration, and the presence of metal cations increase the rate of fibrillation of both the wild-type and A53T mutant synucleins, where the rate of fibrillation in the A53T mutant is faster, perhaps by increasing the formation of the partially folded intermediate.[31] While these experiments have shown what conditions increase fibrillation, the mechanism through which it occurs still remains unknown. In relation to our studies, perhaps the changes in solution conditions result in a less-hydrated protein that allows the formation of an intermediate that is similar to the structures observed in our gas phase simulations, and this leads to a greater proclivity for aggregation. Therefore, to fully understand the mechanism of aggregation in vivo, the effects of hydration must be taken into account.

As part of our studies, we have investigated the hydration characteristics of the protein. Table 2.7 summarizes the coordination number of surrounding water molecules for the most and least hydrated residues of the wild-type α-synuclein in aqueous solution. All other residues in the simulation have intermediate coordination numbers (varying between 2.5 and 4.5). This data shows that the N- and C-helix regions contain most of the least hydrated residues while the majority of the most hydrated residues reside in the C-terminal tail. However, there are a few notable exceptions to this general observation. The glycine amino acid in position 41 is one of the most hydrated residues and remains at approximately the same coordination throughout the course of the simulation. This is unexpected considering its nonpolar character. Also, it is unexpected that out of all the polar residues in the coil/turn connection, the most hydrated is a nonpolar residue. Another surprising result is that the least hydrated residue, although nonpolar, is the valine at position 118 in the C-terminal tail. The terminal residues start as the most hydrated residues of the protein but as the simulation progresses, the C-terminal coordination increases and the N-terminal dehydrates to an intermediate coordination, which is unexpected considering the polar character of the N-terminal methionine and the nonpolar character of the C-terminal alanine. The dehydration at the N-terminal residue suggests that perhaps the N-terminal is involved in the aggregation mechanism of the monomers.

Table 2.8 summarizes the hydration characteristics of the A53T mutant. All of the other residues in the simulation have intermediate coordination numbers (varying between 2.5 and 4.5). As in the wild-type protein, the N- and C-helix regions

TABLE 2.7

Calculated Coordination Number of Water Molecules for the Wild-Type α-Synuclein at 4, 5, 6, and 7 Å for Each ns of Our MD Simulations via Integration of the Radial Distribution Functions Obtained for Each Residue

Residue	1 ns				2 ns				3 ns				4 ns			
	4Å CN	5Å CN	6Å CN	7Å CN	4Å CN	5Å CN	6Å CN	7Å CN	4Å CN	5Å CN	6Å CN	7Å CN	4Å CN	5Å CN	6Å CN	7Å CN
1 M	4.9	11.5	22.0	37.2	4.5	10.3	20.0	33.7	4.3	10.0	19.3	32.5	4.4	10.2	19.6	33.0
25 G	2.5	5.7	11.1	20.6	2.6	5.9	11.5	20.9	2.4	5.6	10.9	20.1	2.4	5.6	11.0	20.3
41 G	4.5	10.5	19.0	32.5	4.5	10.5	19.3	33.0	4.6	10.8	19.7	33.7	4.6	10.8	19.7	33.8
51 G	2.2	5.3	10.6	20.3	2.5	5.9	11.5	21.6	2.3	5.5	11.0	20.9	2.3	5.5	10.8	20.8
53 A/T	2.6	6.4	12.6	22.5	2.9	7.1	13.8	24.4	2.9	7.1	13.8	24.3	2.8	6.9	13.5	23.9
54 T	2.3	5.9	12.3	23.0	2.5	6.3	12.9	23.7	2.6	6.5	13.2	24.1	2.6	6.5	13.2	23.9
59 T	2.3	5.7	11.9	21.6	2.5	6.4	13.1	23.4	2.3	5.9	12.4	22.5	2.3	5.9	12.2	22.2
63 V	2.2	5.8	12.2	22.9	2.3	6.0	12.5	23.4	2.3	5.9	12.4	23.3	2.2	5.9	12.4	23.4
98 D	4.5	10.8	20.2	33.4	4.4	10.7	20.4	34.4	4.5	10.8	20.6	35.1	4.1	9.6	18.1	30.4
106 G	4.1	10.1	19.3	33.8	4.3	10.4	19.5	33.8	4.4	10.6	20.1	34.7	4.6	11.4	21.5	36.2
110 E	4.5	11.2	21.8	36.2	4.5	11.2	21.7	36.1	4.5	11.3	22.2	37.4	4.6	11.6	22.6	37.6
118 V	2.0	5.3	11.6	22.8	1.8	4.8	10.8	21.7	1.8	4.8	10.6	21.3	1.9	5.2	11.4	22.4
130 E	4.2	10.2	19.9	34.4	4.5	11.3	22.1	37.4	4.6	11.5	22.5	37.8	4.7	11.6	22.2	37.4
135 D	4.5	10.8	20.9	35.2	4.4	10.8	20.8	35.2	4.2	10.2	20.0	33.9	4.4	10.7	20.5	34.7
139 E	4.5	11.2	21.8	36.7	4.4	11.0	21.5	36.5	4.6	11.4	22.2	37.3	4.7	11.6	22.6	38.0
140 A	4.9	12.0	23.2	38.7	5.1	12.3	23.6	39.4	5.3	12.6	24.0	39.8	5.0	12.1	23.3	39.1

The underlined values are the most hydrated residues.

TABLE 2.8

Water Molecule Coordination for the A53T Mutant at 4, 5, 6, and 7 Å for Each ns of Our MD Simulations for the Most and Least Hydrated Residues

Residue	1ns				2ns				3ns				4ns			
	4Å CN	5Å CN	6Å CN	7Å CN	4Å CN	5Å CN	6Å CN	7Å CN	4Å CN	5Å CN	6Å CN	7Å CN	4Å CN	5Å CN	6Å CN	7Å CN
1 M	4.5	10.6	20.6	34.8	4.3	9.9	19.3	32.7	4.5	10.1	19.4	32.6	4.4	10.1	19.6	33.0
9 S	3.0	6.7	12.6	21.7	2.5	5.6	10.7	19.0	2.2	5.2	10.2	18.5	2.2	5.2	10.1	18.4
31 G	2.6	6.5	12.8	23.2	2.4	5.6	11.1	21.3	2.4	5.7	11.6	21.8	2.2	5.5	11.2	21.2
35 E	3.3	8.3	16.2	27.6	2.6	6.7	13.5	23.5	2.5	6.3	12.8	22.3	2.3	5.8	12.1	21.7
38 L	3.0	7.6	15.4	27.2	2.3	5.7	11.9	21.9	2.3	5.8	12.2	22.5	2.3	5.9	12.3	22.5
47 G	2.2	5.3	10.7	20.4	2.1	5.0	10.0	18.8	2.0	4.9	10.0	18.8	2.1	4.7	9.4	18.3
48 V	2.4	6.5	13.4	23.8	2.1	5.8	12.3	22.2	2.4	6.4	13.1	23.1	2.1	5.8	12.0	21.3
53 A	2.7	6.6	13.3	23.7	2.8	6.9	13.7	24.2	2.8	6.8	13.6	24.2	2.8	6.9	13.7	24.2
59 T	2.2	5.5	11.5	21.1	2.4	6.0	12.5	22.6	2.4	5.9	12.1	22.2	2.4	6.0	12.3	22.5
63 T	2.3	6.0	12.6	23.6	2.3	6.2	12.9	24.0	2.3	6.3	13.1	24.2	2.3	6.1	12.6	23.6
67 G	2.4	5.9	11.8	21.9	2.5	6.1	12.1	22.1	2.5	6.2	12.2	22.5	2.3	5.7	11.5	21.5
90 A	3.0	7.7	15.5	27.4	2.1	5.7	12.3	23.6	2.0	5.5	12.0	23.2	2.1	5.7	12.1	23.0
96 K	4.7	11.4	22.1	36.8	4.8	11.7	22.6	37.5	4.6	11.1	21.6	35.8	4.5	10.9	21.0	34.7
101 G	4.3	10.4	19.6	33.6	4.5	10.9	20.4	34.1	4.5	10.8	20.1	33.2	4.5	10.6	19.5	32.5
102 K	4.5	11.0	21.7	36.6	4.5	10.9	21.6	36.7	4.6	11.2	21.9	36.9	4.5	11.1	21.6	36.4
105 E	4.4	11.2	21.9	37.1	4.6	11.4	22.2	37.6	4.6	11.7	22.8	38.2	4.6	11.6	22.6	37.7
106 G	4.1	10.1	19.5	34.4	4.6	11.0	20.8	36.1	4.7	11.5	21.5	36.5	4.6	11.1	20.7	35.2
115 D	4.3	10.3	19.6	32.9	4.6	11.0	21.4	35.7	4.8	11.4	21.8	36.0	4.6	10.8	21.0	35.1
129 S	4.6	11.3	21.4	35.8	4.5	11.2	21.8	36.8	4.6	11.3	22.0	37.1	4.8	11.7	22.3	37.2
139 E	4.5	11.3	21.9	37.2	4.6	11.6	22.6	38.0	4.7	11.7	22.6	37.8	4.8	11.8	23.0	38.6
140 A	5.0	12.0	23.1	38.7	4.7	11.5	22.4	38.0	4.6	11.5	22.3	37.8	5.0	12.1	23.1	38.6

The underlined values are the most hydrated residues.

contain the least hydrated residues while the most hydrated residues reside in the C-terminal tail. Unlike the wild-type, there are no apparent exceptions to this trend, even in the coil/turn connection. In fact, residue 38 is still one of the least hydrated even though it is in the coil/turn connection. This may explain the relative lack of change in this region of the protein in the mutant variant. The least hydrated residue in the mutant protein is the glycine at residue 47 as well as the alanine at residue 90, as expected due to its nonpolar character. The most hydrated residue is again the C-terminal alanine. While it is the most hydrated residue out of all the residues throughout the course of the simulation in both proteins, the average first shell coordination numbers of the wild-type and A53T mutant synucleins at residue 140 are 5.1 and 4.8, respectively. The first shell coordination numbers of the residues identify which amino acid residues in the protein are the most soluble. The first shell coordination number differences for residue 140 indicate that the wild-type is in fact more soluble than the A53T mutant. Not only is the C-terminal less soluble, but overall the mutant has more less-hydrated residues than the wild-type indicating that the mutant-type protein is less soluble than the wild-type. However, this decrease in solubility is not evident at the point of mutation because at residue 53, the coordination is similar between the two variants. Instead, the decrease in solubility is a result of the effect that the mutation has on the structure and dynamics of the protein. Additionally, our results show that the relative hydrophobic character of the N- and C-helices and the hydrophilic character of the C-terminal tail relates to the observed interactions between α-synuclein and the SDS micelles and lipid bilayers from the simulations performed by Perlmutter et al. that showed an increased affinity for SDS micelles and lipid bilayers for the helical regions of the protein, where in fact these regions imbedded themselves within the hydrophobic core of the micelles and the C-terminal tail remained in solution.[2]

As demonstrated, the presence of water molecules can have a significant impact on protein structure and thus its in vitro behavior. Currently, x-ray and neutron diffraction techniques are employed to obtain structural information about biomolecules and NMR, IR, and Raman spectroscopies are used to investigate dynamic structural properties of biomolecules in aqueous solution. While these experimental techniques have been used to elucidate many structural properties, they still cannot capture the impact of water beyond the first coordination shell on the protein structure with the atomistic details obtained via dynamics calculations. The experimental techniques mentioned earlier are not able to capture the dynamic impact of an aqueous solution on the protein structure, especially for water molecules in the second and third solvation shells around the proteins. One of the advantages of MD and CPMD simulations is that these properties can be investigated dynamically and important information associated with the structure, thermodynamics, and molecular mechanisms of the protein folding and unfolding can be determined simultaneously.

The second shell coordination of water molecules around the protein can impact protein structure as well, by interacting with the first shell water molecules. The second shell coordination can also depict the relative polarity of the amino acids within the protein. Residues are identified as polar or nonpolar based on their side chains, but often when combined into a large biomolecule, the relative polarity of the amino acid can change depending on its location and surrounding amino acids. The second shell coordination characteristics identify which residues are the most and least hydrated and therefore the most and least polar, respectively, within that specific protein. The second shell coordination values are obtained by taking the difference between the coordination number at 7 and 4 Å. The second shell coordination numbers for these proteins exhibit different values than in the first shell. The least coordinated residues (with a coordination number of ~18) in the second shell are two glycine residues, 25 and 51, in the N- and C-helices of the wild-type protein. The most hydrated residue is again the C-terminal alanine, with a coordination of 34 water molecules. In the case of the mutant-type protein, the least hydrated residues are the residue 9 cysteine and the residue 47 glycine (~16 coordinated water molecules) and the most hydrated is again the C-terminal alanine (~33 coordinated water molecules). The differences between the wild-type and the mutant are not only evident in the most and least hydrated residues, but the A53T mutant has on average a lower coordination number throughout the protein. This difference further emphasizes that the presence of aqueous solution in the first and second coordination shells can yield insight into the in vivo behavior of the protein and that the impact of water cannot be ignored.

In some experimental techniques, such as electrospray ionization mass spectrometry (ESI-MS), the impact of water is not observed because the protein is converted to an ionized gaseous state for a short time before it enters the solution and is analyzed. Several papers have been published on the use of ESI-MS to characterize the structure of proteins. In a 1993 paper by Mirza et al., they described a method of studying conformational changes in bovine ubiquitin and bovine cytochrome c by ESI-MS using heat to denature the proteins from their natively folded state as they pass through the ESI needle into the mass spectrometer.[68] Studies have also been performed by Kaitashov and Eyles on biomolecular conformations and conformational dynamics of protein folding via ESI-MS.[69] A recent article by Ashcroft describes ESI-MS studies coupled with ion mobility spectroscopy to study fibril formation by amyloids.[11] While these techniques do provide information about the intermolecular (in the case of multiple protein systems) and intramolecular hydrogen bonding for proteins, the impact of water on these structures cannot be observed because the measurements occur while the proteins are in the gas phase.[11] Although techniques are available to denature proteins by heat or pH, the structure observed may still differ from the native in vivo conformation. According to these papers, the ESI-MS and native structures of the proteins were found to be comparable; however, the proteins studied had highly folded structures in vitro.[68,69] Proteins that are highly folded may not have drastic structural changes from aqueous solution to the gas phase, but in the case of natively unfolded proteins such as synucleins, the structures are very different, as shown earlier. Mass spectrometry techniques have been very useful in determining the structure of proteins and their aggregates in the gas phase and also discerning the relative amounts of the components, in the case of aggregates, that are present. However, in order to understand the mechanism of in vivo aggregation, the impact of water should be considered because it can have a significant effect on the formation of aggregates. Transporting a highly folded gas phase structure into solution may not result in reformation of the wild-type structure since the folded structure may be extremely stable and structurally different than the wild-type. Thus, to understand the monomer and early aggregation mechanism, proteins should remain in solution at all times.

The early aggregation and monomer hypothesis is currently being tested in the case of another closely related neurodegenerative disease, Alzheimer's Eli Lilly has developed a drug called Solunezumab that is presently in Phase III clinical trials, and is one of the few Alzheimer's drugs that has successfully continued on to this phase of clinical trials.[8] This drug has an interesting mechanism when compared to other Alzheimer's drugs that are being developed and/or tested. Most other Alzheimer's drugs work by inhibiting enzymes that are responsible for cleaving the amyloid protein into the amyloid-β state.[70] The new drug is actually a monoclonal antibody that specifically binds only to the mid-domain of the monomeric form of the amyloid-β protein that is believed to play a pathogenic role in the progression of Alzheimer's disease through its overproduction and subsequent aggregation into amyloid plaques.[10] The antibody then causes the bound protein to be relocated outside of the neurons into the cerebral spinal fluid and blood plasma, effectively reducing the amyloid-β concentration in the brain.[9] By selectively binding to the monomer and reducing its concentration, the drug effectively prevents aggregation. If Solunezumab successfully completes Phase III clinical trials, then it could potentially change the approach of treatment development for diseases resulting from protein aggregation.

Similar to drug studies regarding Alzheimer's disease, information gained about the structure and conformation of synuclein monomers in solution, their solubility, and how they dimerize in solution (reaction mechanism) could help to design more efficient drugs that increase the solubility of these monomers and prevent oligomerization via blocking pathways to oligomerization of these proteins. Our studies are directly linked to these properties and we provide detailed insights into the hydration characteristics of these monomers, which in turn are directly related to their behavior in solution and solubility. The structural and thermodynamic properties that we report herein may be used in the design of more effective drugs or treatments for Parkinson's disease.

Additionally, the RDFs of the backbone carboxyl carbon and nitrogen of residues 1, 53, and 140 were calculated via MD simulations in aqueous solution and are shown in Figure 2.5. By looking at the atomic RDFs for these specific residues, the differences in hydration can be more accurately determined and attributed to specific amino acids. For residue 1, both variants equilibrate to very similar minima and first shell coordination numbers in both the C and N residues. In the case of residue 53, however, the first minimum (first shell) occurs at a shorter distance and a much lower value for the mutant than in the case of the wild-type synuclein. The result of this mutation makes the protein much more hydrophobic in this area of the protein than the wild-type. Also, the carboxyl side of the amino acid is more hydrated than the amine terminus. The hydroxyl group on the side of the threonine amino acid is hydrogen bonded to another residue on the N-terminal side so the dehydration of the amine terminus from the wild to mutant type is expected. For residue 140, the RDFs show that the majority of the coordination at this residue occurs at the carboxyl on the C-terminus. The presence of two oxygens, versus one in comparison to the nonterminal residues, explains the increased water coordination of this residue.

In addition to the effect of hydration on the proteins, the average Gibbs free energy changes of both the wild-type and A53T mutant α-synuclein have been calculated as described in the methods section. For the calculations of the overall Gibbs free energy (G) of each protein for each nanosecond of the simulation, the standard deviation is too high to make a direct comparison of these values. However, a trend can be observed from these values. The enthalpic contribution ($E_{MM} + G_{PBSA}$) to the overall Gibbs free energy of the proteins is much greater than the entropic

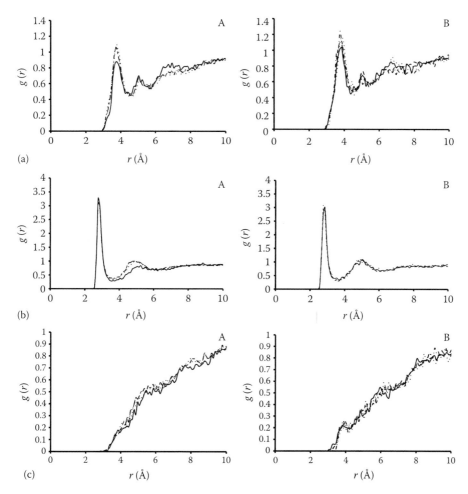

FIGURE 2.5   Radial distribution functions of water molecules around (a) the backbone carboxyl carbon of residue 1, (b) the backbone nitrogen of residue 1, (c) the backbone carboxyl carbon of residue 53,

(continued)

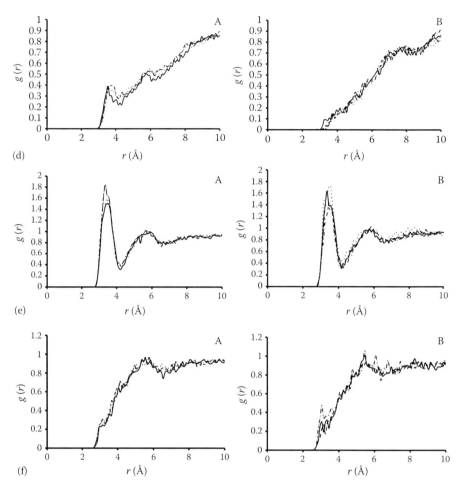

FIGURE 2.5 (continued) (d) the backbone nitrogen of residue 53, (e) the backbone carboxyl carbone of residue 140, and (f) the backbone nitrogen of residue 140 for the wild-type (A) and mutant (B) synucleins at 1 ns ( ), 2 ns (- -), 3 ns (---) and 4 ns (····).

contribution ($-TS$). Also, the overall enthalpic contribution decreases throughout the course of the simulations (Table 2.9).

We now turn from purely classical studies to mixed QM/MM studies. In the following, we present coordination chemistry studies of Fe^{3+} with the tyrosine residue of the wild- and mutant-type α-synuclein proteins in aqueous solution, using the trajectories that we obtained from our classical MD studies presented earlier. For studying the coordination of the transition metal ion to the protein, we performed QM/MM simulations (25 ps). Recently, we predicted the water dissociation mechanism in aqueous solution that is induced by the presence of the Fe^{3+} ion and the impact of the dissociated states of water on the hydration structure of

TABLE 2.9

Molecular Mechanics Potential Energy, Gibbs Free Energy of Solvation and Entropic Contributions to Calculation of the Gibbs Free Energy of the Monomers in Solution for Each ns of Our Simulations

Protein	Time	E_{MM} (kcal/mol)	G_{PBSA} (kcal/mol)	$-TS$ (kcal/mol)	G (kcal/mol)
Wild-type	1 ns	1511.911	−4466.61	−372	−3326.698
		(±211.867)	(±64.47)		(±221.459)
	2 ns	1312.136	−4161.64	−372.93	−3222.434
		(±199.188)	(±124.65)		(±234.975)
	3 ns	1325.735	−4152.66	−372.62	−3199.545
		(±83.658)	(±81.81)		(±117.011)
	4 ns	1374.319	−4249.12	−375.41	−3250.211
		(±116.706)	(±121.6)		(±168.543)
A53T	1 ns	1494.363	−4442.37	−376.03	−3324.037
		(±214.044)	(±69.51)		(±225.048)
	2 ns	1306.162	−4201.75	−367.04	−3262.628
		(±107.126)	(±85.31)		(±136.944)
	3 ns	1300.654	−4239.32	−366.73	−3305.396
		(±62.364)	(±58.73)		(±85.665)
	4 ns	1322.308	−4230.01	−371.07	−3278.772
		(±129.356)	(±86.97)		(±155.874)

Fe^{3+} in water using CPMD simulations coupled with TPS calculations.[71] The geometry of the resulting $Fe(OH)_3(H_2O)_6$ ion with the surrounding water molecules and aqua ions (H_3O^+ and OH^- complexes) was obtained from the trajectories computed in those studies. This geometry has been used in the present studies as the basis for the QM part of the simulation to investigate the coordination chemistry of the hydrated Fe^{3+} ion with α-synuclein proteins and the impact of dissociated states of water on this coordination chemistry. For details of the Fe^{3+} work see the Chapter entitled *Metal Ions in Aqueous Solution*. Dangling bonds in our QM/MM studies were terminated with hydrogen bonds. For the portion of the system treated with molecular mechanics, the same procedure presented in the Methods section was used except that a velocity Verlet propagation algorithm was applied with a time step of 0.12 fs.

The results of several experiments suggest that the C-terminal region of α-synucleins has a low affinity for binding metal ions whereas the N-terminal region has a high affinity for binding metal ions. The molecular mechanism of metal binding in PD and the binding affinities of various metal ions to α-synuclein are unknown. Various spectroscopic measurements have proposed His50 and the residues between 110 and 140 to be active toward Fe(III) binding. The reason for the chemical reactivity of the region between residues 110 and 140 might be

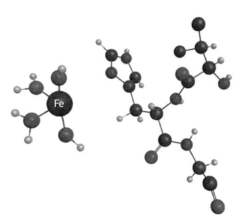

FIGURE 2.6 (See color insert.)   Pictorial representation of the Fe^{3+} ion coordination to the His50 residue of the wild-type α-synuclein protein in water at room temperature obtained from QM/MM simulations.

due to the presence of Asp119, Asp121, Tyr125, Tyr133, Asp135, and Tyr136, or simply the presence of all these Asp and Tyr residues might be sufficient to attract metal ions.

We report results of a study of the binding of Fe(III) with the His50 residue of the wild-type and mutant α-synuclein proteins in water. The initial simulation starts with the Fe^{3+} in an octahedral $Fe(OH)_3(H_2O)_3$ structure (taken from previous studies) at a distance of 11 Å from the His50 residue of both proteins. We find that the Fe^{3+} ion coordinates to the imidazole ring N atoms of the His50 residue within 12 and 19 ps for the A53T and wild-type α-synuclein proteins in water, respectively. This result supports previous experimental measurements. For the wild-type α-synuclein in aqueous solution, we find that the Fe^{3+} ion coordinates to two hydroxyl groups and two water molecules and to the two imidazole N atoms with average distances of 3.6 ± 0.5 and 4.8 ± 0.3 Å, respectively (see Figure 2.6).

For the A53T protein, our results suggest a totally different coordination mechanism from that of the wild-type α-synuclein. The Fe^{3+} ion coordinates to one imidazole ring N atom, one hydroxyl group, and one water molecule. One imidazole ring N atom loses its hydrogen atom via hydrogen bonding interactions with the free hydroxyl group in water. The distance between the Fe^{3+} ion and the coordinating imidazole N atom is shorter with an average distance of 2.9 ± 0.3 Å in comparison to the distance obtained for the Fe^{3+} ion coordination with the wild-type α-synuclein His50 residue (see above). The Fe^{3+} ion also coordinates to the peptide N atom located between the Val49 and His50 residues with a distance of 4.7 ± 0.4 Å (see Figure 2.7). Our findings indicate that the Fe^{3+} coordination mechanism and the structures and conformations obtained for the resulting metalloproteins are different from each other for the wild-type and A53T mutant-type α-synuclein proteins in water. We also note that the water chemistry around these peptides is notably different for the mutant and wild-type α-synuclein proteins. These differences might be attributed to the different folding and thermodynamic preference of these two proteins, which were described in detail earlier.

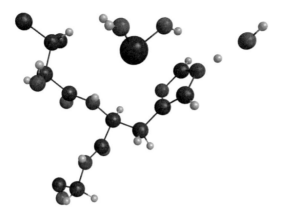

FIGURE 2.7 (See color insert.) Pictorial representation of the Fe^{3+} ion coordination to the His50 residue of the mutant-type A53T α-synuclein protein in water at room temperature obtained from QM/MM simulations.

2.4 CONCLUSION

Although the structural properties for α-synuclein have been determined experimentally, the differences between the dynamical structure, hydration characteristics, and thermodynamic properties of the two variants at a molecular level had not yet been determined. Our studies have revealed that there are noticeable dynamic structural differences between the wild-type and the A53T mutant both in the gas phase and in aqueous solution.

As discussed in this chapter, the structures of both variants are vastly different in the gas phase. Without the presence of solvent molecules, the protein structure is altered from the observed experimental structures to the structures shown in this chapter. These structures are extremely stable and the original solvated configuration is not regained *on the timescale of the simulation* when the molecules are again solvated. This result indicates that for further experimental and theoretical studies of the aggregation of these proteins, the impact of the solution environment cannot be ignored.

One of the hypotheses for the progression of Parkinson's disease (as well as other neurodegenerative disorders) is that the brains of affected patients become dehydrated compared to unaffected brains. In the case of a dehydrated brain, the inhibition of the intramolecular interactions by water molecules observed in the fully solvated simulations would not be as complete. As a result, the intramolecular interactions may cause the protein to fold into the partially folded intermediates that are proposed to be the intermediate structure between the free monomer and the aggregate state present in Lewy bodies and synuclein fibrils. In future studies, the impact of these interactions on aggregation should be investigated.

Not only are differences evident between the gas phase and aqueous solution structures, but the wild-type and A53T mutant synucleins exhibit structural and dynamical differences throughout the course of the simulations. The most notable structural difference between the two variants is that the N-terminus of the A53T mutant interacts with the C-helix of the protein, which may explain its increased rate of aggregation in comparison to the wild-type α-synuclein. This result is similar to a discovery in Alzheimer's disease that led to the development of the pharmaceutical Solanezumab. By determining what region of the protein is most likely involved in the aggregation mechanism, treatments can be designed to specifically target this region. A full understanding of the molecular mechanism of aggregation permits identification of the most advantageous points of attack for potential treatments to prevent synuclein aggregation.

Although it may seem odd to think that even one intramolecular interaction could significantly impact protein structure, the evidence of the effect of one intramolecular interaction is apparent in the differences between the wild-type and mutant structures. Changing one amino acid between the two protein variants results in the formation of one hydrogen bond that completely changes the dynamical structure of the protein throughout the course of the simulations. Any intramolecular interaction can have a significant effect on the overall structure and, thus, the function of the protein in the intracellular environment. Therefore, if the delicate balance of the cell environment is altered in such a way as to allow more intramolecular interactions, the structural and chemical properties of that protein are altered, perhaps permanently.

However, the structural and dynamical differences in the protein are not the only significant findings in this study. The hydration characteristics of both the wild-type and mutant synucleins have also been determined. The hydration trend between the two proteins is similar, but the exact coordination of water molecules around the proteins is different. In both proteins, the N- and C-helix regions are the least hydrated and the C-terminal tail is the most hydrated. However, in the wild-type there are some notable exceptions. In the coil/turn connection between the two helical regions of the protein, residue 41 is highly coordinated compared to the rest of the residues. The hydration at this point explains the higher flexibility of the wild-type protein in the coil/turn region. This region is much more flexible in the wild-type than in the mutant protein and this flexibility could prevent the interaction of the N-terminus with the rest of the protein, affecting the solubility and aggregation of the protein. Additionally, the least hydrated residue in the wild-type is actually in the C-terminal tail. The dehydration in this region may increase the stability and decrease the flexibility of the tail. Neither of these trends is observed in the dynamics of the A53T mutant. The lack of hydration at the coil/turn region allows for greater interaction of the N-terminus with the rest of the protein, perhaps leading to decreased solubility and increased aggregation.

In looking at the overall differences in the coordination numbers of water molecules for both proteins, the A53T mutant is less hydrated than the wild-type protein. This difference indicates that A53T is less soluble than the wild-type protein, which explains its higher propensity for aggregation into insoluble aggregates as compared to the wild-type. If the mutant is already less soluble than the wild-type, any changes in the environment of the cell could further decrease its solubility and thus increase its rate of aggregation. Small changes in the intracellular environment can have a significant impact on the intracellular components.

The thermodynamics characteristics of the synuclein variants, which have not been investigated previously, have also been calculated in our studies. While the deviation of the average Gibbs free energy of the proteins during these simulations does not allow for a direct comparison, the general trend of the data can be taken into account. The most significant contribution to the average free energy of each of the synucleins is the enthalpic contribution. Of the two components of the enthalpic contribution (according to the method of calculation used in this work), the solvation free energy contributes the most to the average free energy of the protein. However, the trend in both monomers is that the solvation free energy is increasing throughout the simulation time and therefore the monomers are less stable in solution. By calculating and analyzing the thermodynamic values of wild-type and A53T mutant α-synuclein, the overall stability of these proteins and the factors that have the greatest effect on their overall stability are ascertained. From this data, it is evident that changing the solvation characteristics of the protein will have a significant impact on the overall stability of the protein in a solution environment.

Changes in the intracellular environment due to changes in pH, protein concentration, temperature, phosphorylation, mutations, oxidative stress, and metal ions have been shown experimentally to affect the aggregation of α-synuclein and its mutants. However, the exact mechanism by which these changes affect the monomer structure and the interaction of the monomers with the environment and each other remains unknown. Further investigation of the impacts of these factors on the structure, dynamics, hydration characteristics, and thermodynamic properties of the variants of α-synuclein is needed to fully understand the aggregation mechanism in the in vivo environment.

After further investigation and the determination of an aggregation mechanism for these proteins, it is possible that a new treatment can be developed. A similar approach to that of the drug Solunezumab (Eli Lilly) might yield a pharmaceutical capable of inhibiting the aggregation of α-synuclein. Or, upon further investigation, an even better method of inhibiting aggregation might be determined once the full mechanism is understood. While other factors still need to be investigated to arrive at a final mechanism, this study represents a significant initial step in elucidating the properties of these proteins at a molecular level.

As discussed earlier in this chapter, the presence of metal ions such as Fe^{3+} has been implicated in the aggregation of synucleins in Parkinson's disease. Therefore, determining where and in what manner the Fe^{3+} ion binds to the wild type and A53T mutant may lead to a better understanding of the aggregation mechanism of these proteins. From our studies, the binding of Fe^{3+} to the His50 residue was studied due to its hypothesized involvement from spectroscopic measurements. Our studies demonstrate that the Fe^{3+} ion does in fact bind to the His50 residue in both variants, but in completely different manners. The binding differences may be due to the different structures and thermodynamic characteristics of the proteins as described earlier. These studies confirm that the wild type and mutant synucleins have different characteristics in solution.

REFERENCES

1. Licker, V.; Kovari, E.; Hochstrasser, D. F.; Burkhard, P. R. Proteomics in human Parkinson's disease research. *J Proteomics* 2009, *73* (1), 10–29.
2. Perlmutter, J. D.; Braun, A. R.; Sachs, J. N. Curvature dynamics of alpha-synuclein familial parkinson disease mutants molecular simulations of the micelle- and bilayer-bound forms. *J Biol Chem* 2009, *284* (11), 7177–7189.

3. Tsigelny, I. F.; Sharikov, Y.; Miller, M. A.; Masliah, E. Mechanism of alpha-synuclein oligomerization and membrane interaction: Theoretical approach to unstructured proteins studies. *Nanomedicine: Nanotechnol Biol Med* 2008, *4* (4), 350–357.

4. Lundvig, D.; Lindersson, E.; Jensen, P. H. Pathogenic effects of alpha-synuclein aggregation. *Mol Brain Res* 2005, *134* (1), 3–17.

5. Baba, M.; Nakajo, S.; Tu, P. H.; Tomita, T.; Nakaya, K.; Lee, V. M. Y.; Trojanowski, J. Q. et al. Aggregation of alpha-synuclein in Lewy bodies of sporadic Parkinson's disease and dementia with lewy bodies. *Am J Pathol* 1998, *152* (4), 879–884.

6. Giasson, B. I.; Duda, J. E.; Quinn, S. M.; Zhang, B.; Trojanowski, J. Q.; Lee, V. M. Y. Neuronal alpha-synucleinopathy with severe movement disorder in mice expressing A53T human alpha-synuclein. *Neuron* 2002, *34* (4), 521–533.

7. Tabner, B. J.; Turnbull, S.; El-Agnaf, O. M. A.; Allsop, D. Formation of hydrogen peroxide and hydroxyl radicals from A beta and alpha-synuclein as a possible mechanism of cell death in Alzheimer's disease and Parkinson's disease. *Free Radical Bio Med* 2002, *32* (11), 1076–1083.

8. Jarvis, L. M. The amyloid question. *Chem Eng News* 2010, *88* (14), 12–17.

9. Dean, R. A.; DeMattos, R.; Gelfanova, V.; Hale, J.; Lachno, R.; Racke, M.; Talbot, J. et al. ELISA-based measurement of plasma total A[beta]1–40 and A[beta]1–42, and CSF total and free A[beta]1–40 and A[beta]1–42 in the presence of solanezumab, a mid-domain anti-A beta antibody. *Alzheimer's Dementia* 2009, *5* (4, Supplement 1), P133–P133.

10. DeMattos, R. B.; Racke, M. M.; Gelfanova, V.; Forster, B.; Knierman, M. D.; Bryan, M. T.; Hale, J. E. et al. Identification, characterization, and comparison of amino-terminally truncated A[beta]42 peptides in Alzheimer's disease brain tissue and in plasma from Alzheimer's patients receiving solanezumab immunotherapy treatment. *Alzheimer's Dementia* 2009, *5* (4, Supplement 1), P156–P157.

11. Ashcroft, A. E. Mass spectrometry and the amyloid problem—How far can we go in the gas phase? *J. Am. Soc. Mass. Spectrom.* 2010, 21, 1087–1096.

12. Parkinson's Disease Foundation. http://www.pdf.org/

13. Bennett, M. C. The role of alpha-synuclein in neurodegenerative diseases. *Pharmacol Therapeut* 2005, *105* (3), 311–331.

14. Parkinson, J. An essay on the shaking palsy (Reprinted). *J Neuropsych Clin N* 2002, *14* (2), 223–236.

15. National Family Caregivers Association. http://www.nfcacares.org/

16. Kaur, D.; Andersen, J. Does cellular iron dysregulation play a causative role in Parkinson's disease? *Ageing Res Rev* 2004, *3* (3), 327–343.

17. Bueler, H. Impaired mitochondrial dynamics and function in the pathogenesis of Parkinson's disease. *Exp Neurol* 2009, *218* (2), 235–246.

18. Fukae, J.; Mizuno, Y.; Hattori, N. Mitochondrial dysfunction in Parkinson's disease. *Mitochondrion* 2007, *7* (1–2), 58–62.

19. Double, K. L.; Ben-Shachar, D.; Youdim, M. B.; Zecca, L.; Riederer, P.; Gerlach, M. Influence of neuromelanin on oxidative pathways within the human substantia nigra. *Neurotoxicol Teratol* 2002, *24* (5), 621–628.

20. Forloni, G.; Terreni, L.; Bertani, I.; Fogliarino, S.; Invernizzi, R.; Assini, A.; Ribizzi, G. et al. Protein misfolding in Alzheimer's and Parkinson's disease: Genetics and molecular mechanisms. *Neurobiol Aging* 2002, *23* (5), 957–976.

21. Hald, A.; Lotharius, J. Oxidative stress and inflammation in Parkinson's disease: Is there a causal link? *Exp Neurol* 2005, *193* (2), 279–290.

22. Fedorow, H.; Tribl, F.; Halliday, G.; Gerlach, M.; Riederer, P.; Double, K. L. Neuromelanin in human dopamine neurons: Comparison with peripheral melanins and relevance to Parkinson's disease. *Prog Neurobiol* 2005, *75* (2), 109–124.

23. Tanaka, K.; Suzuki, T.; Hattori, N.; Mizuno, Y. Ubiquitin, proteasome and parkin. *Biochim Biophys Acta* 2004, *1695* (1–3), 235–247.
24. McNaught, K. S. P.; Olanow, C. W. Protein aggregation in the pathogenesis of familial and sporadic Parkinson's disease. *Neurobiol Aging* 2006, *27* (4), 530–545.
25. Sun, F.; Kanthasamy, A.; Anantharam, V.; Kanthasamy, A. G. Environmental neurotoxic chemicals-induced ubiquitin proteasome system dysfunction in the pathogenesis and progression of Parkinson's disease. *Pharmacol Therapeut* 2007, *114* (3), 327–344.
26. Winklhofer, K. F.; Haass, C. Mitochondrial dysfunction in Parkinson's disease. *Biochim Biophys Acta* 2010, *1802* (1), 29–44.
27. Xiromerisiou, G.; Dardiotis, E.; Tsimourtou, V.; Kountra, P. M.; Paterakis, K. N.; Kapsalaki, E. Z.; Fountas, K. N. et al. Genetic basis of Parkinson disease. *Neurosurg Focus* 2010, *28* (1), E7.
28. Maguire-Zeiss, K. A.; Hoff, S. A.; Su, X.; Yehling, E.; Federoff, H. J. Synuclein, dopainine, and oxidative stress in Parkinson's disease. *Movement Disord* 2006, *21*, S65–S65.
29. Ulmer, T. S.; Bax, A.; Cole, N. B.; Nussbaum, R. L. Structure and dynamics of micelle-bound human alpha-synuclein. *J Biol Chem* 2005, *280* (10), 9595–9603.
30. Li, J.; Uversky, V. N.; Fink, A. L. Conformational behavior of human [alpha]-synuclein is modulated by familial Parkinson's disease point mutations A30P and A53T. *Neurotoxicology* 2002, *23* (4–5), 553–567.
31. Uversky, V. N., Fink, A. L. Folding and Misfolding of α-Synuclein. http://www.chemistry.ucsc.edu/faculty/Fink/2005/ReviewFolding-misfoldingRevd.pdf
32. El-Agnaf, O. M.; Jakes, R.; Curran, M. D.; Wallace, A. Effects of the mutations Ala30 to Pro and Ala53 to Thr on the physical and morphological properties of alpha-synuclein protein implicated in Parkinson's disease. *FEBS Lett* 1998, *440* (1–2), 67–70.
33. Conway, K. A.; Harper, J. D.; Lansbury, P. T. Fibrils formed in vitro from alpha-synuclein and two mutant forms linked to Parkinson's disease are typical amyloid. *Biochemistry-Us* 2000, *39* (10), 2552–2563.
34. Rhodes, S. L.; Ritz, B. Genetics of iron regulation and the possible role of iron in Parkinson's disease. *Neurobiol Dis* 2008, *32* (2), 183–195.
35. Hirsch, E. C. Iron transport in Parkinson's disease. *Parkinsonism Relat Disord* 2009, *15* (Supplement 3), S209–S211.
36. Zhang, X.; Xie, W. J.; Qu, S.; Pan, T. H.; Wang, X. T.; Le, W. D. Neuroprotection by iron chelator against proteasome inhibitor-induced nigral degeneration. *Biochem Bioph Res Co* 2005, *333* (2), 544–549.
37. Peng, Y.; Wang, C.; Xu, H. H.; Liu, Y.-N.; Zhou, F. Binding of [alpha]-synuclein with Fe(III) and with Fe(II) and biological implications of the resultant complexes. *J Inorg Biochem* 2010, *104* (4), 365–370.
38. Silva, J. L.; Cordeiro, Y.; Foguel, D. Protein folding and aggregation: Two sides of the same coin in the condensation of proteins revealed by pressure studies. *Biochim Biophys Acta (BBA)—Proteins Proteomics* 2006, *1764* (3), 443–451.
39. Silva, J. L.; Foguel, D. Hydration, cavities and volume in protein folding, aggregation and amyloid assembly. *Phys Biol* 2009, *6* (1), 015002-12.
40. Jiang, J. S.; Brunger, A. T. Protein hydration observed by x-ray-diffraction—Solvation properties of penicillopepsin and neuraminidase crystal-structures. *J Mol Biol* 1994, *243* (1), 100–115.
41. Armstrong, B. D.; Han, S. G. Overhauser dynamic nuclear polarization to study local water dynamics. *J Am Chem Soc* 2009, *131* (13), 4641–4647.
42. Karplus, M. Molecular dynamics of biological macromolecules: A brief history and perspective. *Biopolymers* 2003, *68* (3), 350–358.
43. Karplus, M. Molecular-dynamics—Applications to proteins. *Stud Phys Theo Chem* 1990, *71*, 427–461, 788.

44. Senn, H. M.; Thiel, W. QM/MM methods for biomolecular systems. *Angew Chem Int Ed* 2009, *48* (7), 1198–1229.

45. Kumar, S.; Sarkar, A.; Sundar, D. Controlling aggregation propensity in A53T mutant of alpha-synuclein causing Parkinson's disease. *Biochem Bioph Res Co* 2009, *387* (2), 305–309.

46. Frenkel, D.; Smit, B. *Understanding Molecular Simulation: From Algorithms to Applications*. 2nd edn., Academic Press: San Diego, 2002; p xxii, 638 p.

47. van Gunsteren, W. F.; Bakowies, D.; Baron, R.; Chandrasekhar, I.; Christen, M.; Daura, X.; Gee, P. et al. Biomolecular modeling: Goals, problems, perspectives. *Angew Chem Int Ed* 2006, *45* (25), 4064–4092.

48. Berendsen, H. J. C.; Postma, J. P. M.; Vangunsteren, W. F.; Dinola, A.; Haak, J. R. Molecular-dynamics with coupling to an external bath. *J Chem Phys* 1984, *81* (8), 3684–3690.

49. Evans, D. J.; Holian, B. L. The Nose–Hoover thermostat. *J Chem Phys* 1985, *83* (8), 4069–4074.

50. Allen, M. P.; Tildesley, D. J. *Computer Simulation of Liquids*. Clarendon Press; Oxford University Press: Oxford, England, New York, 1987; p xix, 385 p.

51. MacKerell, A. D.; Bashford, D.; Bellott, M.; Dunbrack, R. L.; Evanseck, J. D.; Field, M. J.; Fischer, S. et al. All-atom empirical potential for molecular modeling and dynamics studies of proteins. *J Phys Chem B* 1998, *102* (18), 3586–3616.

52. (a) Jorgensen, W. L. Quantum and statistical mechanical studies of liquids.10. Transferable intermolecular potential functions for water, alcohols, and ethers—Application to liquid water. *J Am Chem Soc* 1981, *103* (2), 335–340; (b) Jorgensen, W. L. Quantum and statistical mechanical studies of liquids.24. Revised tips for simulations of liquid water and aqueous-solutions. *J Chem Phys* 1982, *77* (8), 4156–4163.

53. Toukmaji, A. Y.; Board, J. A. Ewald summation techniques in perspective: A survey. *Comput Phys Commun* 1996, *95* (2–3), 73–92.

54. Darden, T.; York, D.; Pedersen, L. Particle mesh Ewald—an N.Log(N) method for Ewald sums in large systems. *J Chem Phys* 1993, *98* (12), 10089–10092.

55. Andersen, H. C. Rattle—A velocity version of the shake algorithm for molecular-dynamics calculations. *J Comput Phys* 1983, *52* (1), 24–34.

56. Tuckerman, M. E.; Martyna, G. J. Understanding modern molecular dynamics: Techniques and applications. *J Phys Chem B* 2000, *104* (2), 159–178.

57. (a) Zhang, Z. Y.; Lu, L. Y.; Noid, W. G.; Krishna, V.; Pfaendtner, J.; Voth, G. A. A systematic methodology for defining coarse-grained sites in large biomolecules. *Biophys J* 2008, *95* (11), 5073–5083; (b) Noid, W. G.; Chu, J. W.; Ayton, G. S.; Krishna, V.; Izvekov, S.; Voth, G. A.; Das, A., Andersen, H. C. The multiscale coarse-graining method. I. A rigorous bridge between atomistic and coarse-grained models. *J Chem Phys* 2008, *128* (24), 244114-11.

58. Szabo, A.; Ostlund, N. S. *Modern Quantum Chemistry: Introduction to Advanced Electronic Structure Theory*. McGraw-Hill, New York, 1989.

59. Kohn, W.; Sham, L. J. Self-consistent equations including exchange and correlation effects. *Phys Rev* 1965, *140* (4A), A1133.

60. Massova, I.; Kollman, P. A. Combined molecular mechanical and continuum solvent approach (MM-PBSA/GBSA) to predict ligand binding. *Perspect Drug Discovery Des* 2000, *18*, 113–135.

61. (a) Rocchia, W.; Alexov, E.; Honig, B. Extending the applicability of the nonlinear Poisson–Boltzmann equation: Multiple dielectric constants and multivalent ions. *J Phys Chem B* 2001, *105* (28), 6507–6514; (b) Rocchia, W.; Sridharan, S.; Nicholls, A.; Alexov, E.; Chiabrera, A.; Honig, B. Rapid grid-based construction of the molecular

surface and the use of induced surface charge to calculate reaction field energies: Applications to the molecular systems and geometric objects. *J Comput Chem* 2002, *23* (1), 128–137.

62. (a) Srinivasan, J.; Cheatham, T. E.; Cieplak, P.; Kollman, P. A.; Case, D. A. Continuum solvent studies of the stability of DNA, RNA, and phosphoramidate—DNA helices. *J Am Chem Soc* 1998, *120* (37), 9401–9409; (b) Massova, I.; Kollman, P. A. Computational alanine scanning to probe protein–protein interactions: A novel approach to evaluate binding free energies. *J Am Chem Soc* 1999, *121* (36), 8133–8143.

63. Andricioaei, I.; Karplus, M. On the calculation of entropy from covariance matrices of the atomic fluctuations. *J Chem Phys* 2001, *115* (14), 6289–6292.

64. Ulmer, T. S.; Bax, A. Comparison of structure and dynamics of micelle-bound human alpha-synuclein and Parkinson disease variants. *J Biol Chem* 2005, *280* (52), 43179–43187.

65. (a) Eliezer, D.; Kutluay, E.; Bussell, R.; Browne, G. Conformational properties of alpha-synuclein in its free and lipid-associated states. *J Mol Biol* 2001, *307* (4), 1061–1073; (b) Sung, Y.-H.; Eliezer, D., Residual structure, backbone dynamics, and interactions within the synuclein family. *J Mol Biol* 2007, *372* (3), 689–707.

66. Frishman, D.; Argos, P. Knowledge-based protein secondary structure assignment. *Proteins* 1995, *23* (4), 566–579.

67. Humphrey, W.; Dalke, A.; Schulten, K. VMD: Visual molecular dynamics. *J Mol Graphics* 1996, *14* (1), 33–38.

68. Mirza, U. A.; Cohen, S. L.; Chait, B. T. Heat-induced conformational-changes in proteins studied by electrospray ionization mass-spectrometry. *Anal Chem* 1993, *65* (1), 1–6.

69. Kaltashov, I. A.; Eyles, S. J. Studies of biomolecular conformations and conformational dynamics by mass spectrometry. *Mass Spectrom Rev* 2002, *21* (1), 37–71.

70. Sabbagh, M. N. Drug development for Alzheimer's disease: Where are we now and where are we headed? *Am J Geriatr Pharmacother* 2009, *7* (3), 167–185.

71. Coskuner, O.; Jarvis, E. A. A.; Allison, T. C. Water dissociation in the presence of metal ions. *Angew Chem Int Ed* 2007, *46* (41), 7853–7855.

3 Carbohydrate and Trivalent Iron Ion Interactions in the Gas Phase and in Aqueous Solution

*Orkid Coskuner and Carlos A. González**

CONTENTS

3.1 INTRODUCTION

Carbohydrates play important roles in many diverse areas such as medicine, nano-biotechnology, and energy. For example, the selectivity of the biological recognition process of a bacterial cell wall depends on the carbohydrate sequence and conformation since these molecules are the primary components of the secondary cell wall [1–3]. Carbohydrates also play crucial roles in physiological processes, i.e., energy storage, growth of cellular components, and control of water in drought and

* Contribution of the National Institute of Standards and Technology.

cold-resistant organisms [4]. In general, understanding the factors that lead to the conformational stability of a carbohydrate is extremely challenging due to the many torsional degrees of freedom present in these molecules [5–7].

Mannopyranoside is important in bacterial pathogenesis since it is among the most common biomolecules in the lectins of enterobacteria that regulates the adherence of the bacteria to host epithelial cells [8]. Furthermore, high mannopyranoside precursors are found on the glycosylation sites of glycoproteins [9]. The structural and conformational changes of carbohydrates (including mannopyranosides), which are the building blocks of eukaryotic cell surfaces, are used in the design of biomarkers that detect structural and conformational changes associated with the onset of cancer [8,10,17–20]. The structural change of mannopyranoside complexes is therefore a code for cellular physiology. An understanding of this code at both molecular and functional levels is emerging. Although it has been appreciated that carbohydrate expression changes with cellular condition, progress toward delineating the molecular basis of carbohydrate function has been slow relative to the studies of proteins and nucleic acids. This slow progress is partly due to the complex structure and high flexibility of carbohydrates, unlike other biomolecules such as proteins. Furthermore, studies on the mannopyranoside-type carbohydrates on the gp120 protein, which is on the surface of HIV-1, show that an understanding of the structure–function relationship of mannopyranoside-containing glycan complexes is essential for designing carbohydrate-based HIV vaccines and antiviral agents [19–23].

Not only the carbohydrates themselves but also their interactions and bindings to metal ions, such as iron ions, are crucial in biological and environmental processes. For instance, iron nanoparticles and their complexes with carbohydrates have been used in new treatment strategies in regenerative medicine, i.e., stem cell labeling [11–14]. Moreover, iron deficiency causes bacterial pathogens, which involve carbohydrates attached to the cell surface (as described above), to produce greater tissue damage because many virulence factors are determined by the iron supply [15–18]. In environmental chemistry, carbohydrates and iron ion interactions play unique roles in diverse processes. For example, microbial bacteria are known to scavenge iron particles and to induce crystallization of unexpected phases called nontronite, which is an iron-rich clay mineral [9]. There are hundreds of additional processes in biology and environmental chemistry beyond the several examples provided here in which carbohydrate and iron interactions play crucial roles. In general, studies of structure–function relationships in biometallic molecular systems are of critical importance in basic and applied sciences.

Both experimental and theoretical studies of the structure of carbohydrate and iron complexes pose a challenge due to the possibility of multi-dentate coordination of the iron ion and the many degrees of freedom present in carbohydrates. In addition, the fast dynamics associated with the small size of solvent molecules and short timescales further challenge the carbohydrate studies in aqueous solution. Since the experimental studies of Saltman in the early 1960s on carbohydrates and iron complexes [24–26] and the studies on carbohydrate chelating ability by Davis and Deller [27], there has been a large gap in the scientific literature regarding the understanding of these biometallic molecular systems. A recent exception is the studies of Rao et al., who proposed synthetic strategies for the isolation, characterization, and stabilization of carbohydrate and iron complexes [28].

Solvent effects present further difficulties in studying biometallic compounds. However, these effects must be taken into account, given that a significant fraction of the human body is composed of water and that most relevant biomolecular and biometallic reactions of physiological and technological relevance are (to some extent) solvent mediated. Water gives rise to hydrophobic interactions that stabilize the core of globular proteins, is thought to weaken and diminish electrostatic interactions in many biomolecules, and fills the internal cavities of proteins, oligosaccharides, and nucleic acids. Moreover, water molecules can be associated with the surfaces of biomolecules, occupying certain more or less well-defined positions and interacting with surrounding water molecules through hydrogen bonds [29–38]. Even though much effort has been expended on studies of liquid water and aqueous-phase biomolecules, the role of water in the coordination chemistry of biometallic compounds remains poorly understood.

Experimental measurements utilizing extended x-ray absorption fine structure, x-ray and neutron diffraction yield structural information, and nuclear magnetic resonance, infrared and Raman spectroscopies are used for investigating the structural and dynamical properties of biomolecules in aqueous solution [36]. Interpreting the results from these experimental techniques can be immensely challenging, and, in particular, it is difficult to produce direct measurements of the local electronic structure of the reaction medium associated with the fast dynamics of biometallic systems. Limited information can be found in x-ray analysis, which has been used to determine pair correlation functions. However, these functions are usually spherically averaged, and thus, details of structure in aqueous biomolecule solutions are difficult to predict accurately since important information linked to the anisotropic and asymmetric character of the system is lost [37,38].

Challenges to theoretical and computational studies of biometallic systems are relatively large molecular sizes, long timescales, and a large number of degrees of freedom (as mentioned above). Static first principles calculations in the gas phase and in aqueous solution using continuum models for water have been applied extensively for predicting the structure and function of aqueous biomolecules and biometallic compounds. Even though these ab initio calculations are relatively efficient in terms of computational time, static gas-phase calculations can capture neither the impact of the dynamics nor the effect of water on the predicted structures of the compounds of interest. Continuum water calculations ignore intermolecular hydrogen bonding interactions, which can have a large impact on the predicted structure and conformation of most biomolecules and biometallic compounds. Classical molecular dynamics (CMD) and Monte Carlo simulations are the most widely used computational tools to study biomolecules and biometallic systems in aqueous solution. Even though classical molecular simulations have been useful to a great extent, the accuracy of these simulations depends on the quality of the force field parameters, and these simulations ignore quantum effects in predicting the structure and function of biomolecules and biometallic systems. In addition, CMD simulations neither capture bond forming and bond breaking events nor the molecular orbital interactions between the iron ion and the carbohydrate in aqueous solution.

In our recent studies, we presented a novel strategy that employs theoretical calculations, molecular dynamics simulations, and mass spectrometric experiments to identify the reactive sites of methyl-α-mannopyranoside toward the trivalent iron ion (Fe^{3+}) in the gas phase and in aqueous solution [39,40]. Calculations of a chemical reactivity index using density functional theory (DFT) and simulations using Car–Parrinello molecular dynamics (CPMD) show that the Fe^{3+} ion preferentially interacts with specific hydroxyl oxygen atoms of the carbohydrate. These predictions were shown to be in accord with mass spectrometric measurements [39,40]. Furthermore, CPMD simulations indicate that the specific conformational preference of the glycosidic linkage of methyl-α-mannopyranoside in the gas phase is influenced by intramolecular hydrogen bonding interactions [39]. In contrast, CMD simulations using OPLS-AA and UFF force fields for the carbohydrate and the metal ion, respectively, converged to structures that differ from the mass spectrometry data, suggesting that classical molecular simulations are insensitive to these effects [39]. This comparison demonstrated the importance of chemical reactivity calculations and CPMD simulations for predicting the reactive sites of a biomolecule toward metal ions. In addition, we used this integrated approach based on theoretical calculations, molecular simulations, and mass spectrometric measurements to study the active sites of methyl-α-mannopyranoside toward Fe^{3+} ions in aqueous solution [40]. Again, results were compared to those obtained via CMD simulations. The insights obtained from CPMD simulations are consistent with the chemical reactivity calculations, natural partial charge analysis (NPA), and mass spectrometric measurements. Based on our theoretical studies using chemical reactivity calculations, CPMD simulations, NPA, and Gibbs free energy change calculations, we are able to predict the structure, conformation, and coordination chemistry of methyl-α-mannopyranoside and the Fe^{3+} ion in the gas phase and in aqueous solution. Our studies show that the inclusion of the solvent medium impacts the coordination chemistry of the carbohydrate with the Fe^{3+} ion as well as the resulting structure and conformation of the biometallic compound [40]. In addition, we calculated the binding affinities of the Fe^{3+} ion with methyl-α-mannopyranoside in the gas phase and in aqueous solution. Results obtained from our calculations and simulations were compared to experimental measurements, and as mentioned above, we find that the calculations and simulations performed at the ab initio level agree very well with the experiments, whereas the results we obtained using classical molecular simulations deviate both from our first principles calculations and from the experiment. In this chapter, we describe the theoretical methods used in these studies and present the results obtained for methyl-α-mannopyranoside and its complex with the Fe^{3+} ion in the gas phase and in aqueous solution. Our studies indicate that this new theoretical strategy can be used in coordination chemistry studies of any organic compound or biomolecule with metal ions and is not limited to investigations of carbohydrates and metal ions. As mentioned above and presented in Section 3.2, the new strategy yields accurate results for biometallic species not only in the gas phase but also in more complicated environments, such as the dense aqueous solution environment.

3.2   METHODS

Detailed description of the methods used herein is provided in Chapter 2.

3.2.1   CAR–PARRINELLO MOLECULAR DYNAMICS SIMULATIONS

The Becke-Lee-Yang-Parr (BLYP) gradient corrected functional, which was found to be reliable in the studies of glucose, was used along with Troullier–Martins pseudopotentials and a double-ζ basis set in the CPMD simulations [41,42]. The electronic wave functions were expanded in a plane wave basis set with a kinetic energy cutoff of 114 Ry. A homogeneous background charge was applied to compensate for the ionic charge. The time step for the CPMD simulations was 0.1 fs. The electronic mass was 900 a.u. The isotopic mass of deuterium was used for hydrogen. CPMD simulations of 80 ps were performed for methyl-α-D-mannopyranoside (M), M–Fe^{3+}, and M–Fe_2^{3+} in the gas phase. Statistics were collected for 70 ps. Separate CPMD simulations were performed in cubic cells with lattice parameters of 25, 35, and 45 Å using periodic boundary conditions. It was found that these lattice parameters do not have a significant impact on predicted structural and thermodynamic properties in the gas phase. Canonical ensemble CPMD simulations were performed at room temperature using the Nose–Hoover thermostat. Long-range interactions were treated with the Ewald sum method [43].

For simulations in aqueous solution, M was solvated by 58 water molecules in a cubic cell with lattice parameters of 16.6 Å and a homogeneous background charge was applied to compensate for the ionic charge. CPMD simulations of 80 ps were performed for M, M–Fe^{3+}, and M–Fe_2^{3+} in aqueous solution. To assess the effects of the chosen number of water molecules, we performed another set of CPMD simulations with 92 water molecules. The structures and conformations obtained from these simulations showed that the impact of additional water molecules is extremely small with a carbohydrate conformational difference of 1.6%. Canonical ensemble simulations using the periodic boundary conditions at room temperature were performed. The Nose–Hoover thermostat was used to control the temperature. The Ewald sum method was used to treat long-range interactions in aqueous solution simulations of these compounds [43].

3.2.2   CLASSICAL MOLECULAR DYNAMICS SIMULATIONS

The OPLS-AA and UFF force field parameters were used for M and the Fe^{3+} ion, respectively [44,45]. Our recent studies on methyl-α-mannopyranoside in aqueous solution using TIP3P, TIP4P, and TIP5P models for water showed that simulation results with the TIP5P model for water are closer to the results obtained via CPMD simulations [29,30]. Thus, we chose the TIP5P potential function for water in this work to study the impact of water on the predicted conformation of M, M–Fe^{3+}, and M–Fe_2^{3+} complexes in aqueous solution [29,30]. Lorentz–Berthelot mixing rules were used for calculating the cross parameters. Separate simulations of 60 and 50 ns were performed for M and its complexes with the Fe^{3+} ion in the gas phase and in aqueous solution, respectively. Statistics were sampled for 50 and 40 ns in the gas phase and in water, respectively. For the simulations in aqueous solution, M was solvated by 470 water molecules in a cubic box with a box length of 24.4 Å. Periodic boundary conditions were used. A cutoff distance of 12 Å was applied for treating solute and solvent long-range interactions. The Ewald sum method was used for treating the long-range interactions [43]. Canonical ensemble simulations were performed using Langevin dynamics at 300 K. Homogeneous background charge was applied to compensate for the net ionic charge. For the gas-phase simulations, separate simulations with cubic cell lattice parameters of 25, 35, and 45 Å were performed and results showed that these various lattice parameters do not impact predicted structural and thermodynamic properties.

3.2.3 DETERMINATION OF THE PREFERRED CONFORMATION FOR GLYCOSIDIC LINKAGE

Figure 3.1 shows the torsional angle of the glycosidic linkage studied in this work. The glycosidic linkage torsional angle $\varphi = O5–C1–C7–O7$ was varied between $0°$ and $360°$ to study the relative torsional energies in the gas phase and in aqueous solution. The potential of mean force (PMF) method (Equation 3.1) was used for calculating the preferred conformation of the glycosidic linkage of M, $M–Fe^{3+}$, and $M–Fe_2^{3+}$ in the gas phase and in aqueous solution [46–48]:

$$\Delta G = -kT \ln Z \tag{3.1}$$

where Z is the probability of the torsional angle (φ) varying between $0°$ and $360°$ obtained from CPMD and CMD simulations. For the PMF calculations with CPMD simulations, i.e., for $\lambda = 0 \rightarrow \lambda = 1$ ($\lambda = 0$ and $\lambda = 1$ represent the initial and final states of the system) each window was simulated for 80 ps in the gas phase. To understand the adequacy and convergence of PMF, we compared the PMF values calculated for 50, 60, and 70 ps simulations for each window. Standard deviations were computed using results obtained at these different simulation times. For PMF calculations with CMD simulations, the systems were simulated for 800 ps for each window, and the convergence was tested by comparing the PMF results calculated for $\lambda = 0 \rightarrow \lambda = 1$ to

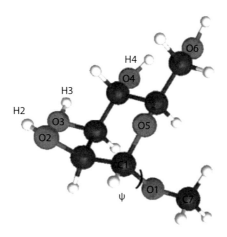

FIGURE 3.1   Methyl-α-mannopyranoside, obtained from CPMD simulations, with conventional numbering of specific atoms.

those obtained for $\lambda=1\rightarrow\lambda=0$. We also compared the PMF results obtained from 500, 600, and 700 ps simulations for each window. The PMF results do not deviate from each other using CMD simulations, indicating that the systems were equilibrated with CPMD as well as with CMD simulations.

The perturbation method was employed and the Gibbs free energy change (ΔG) between states i and $i+1$ with the corresponding Hamiltonians H_i and H_{i+1} were calculated using Equation 3.2 [46–48]:

$$\Delta G\left(\lambda_{i+1} - \lambda_i\right) = -kT \ln \left\langle \exp\left[-(H_{\lambda_{i+1}} - H_{\lambda_i})/kT\right]\right\rangle_{\lambda_i} \qquad (3.2)$$

where
  λ is a continuous coupling parameter that represents the initial ($\lambda=0$) and final ($\lambda=1$) states
  k is the Boltzmann constant
  T is the temperature

Perturbations were broken into 20 windows with a $\Delta\lambda$ value of 0.05. Gibbs free energy values were compared to those calculated using the PMF method.

3.3 CHEMICAL REACTIVITY CALCULATIONS

In order to predict the reactive sites of M with respect to binding a Fe^{3+} ion in the gas phase and in aqueous solution, we performed chemical reactivity analysis of these systems based on Fukui reactivity indices [49]. The Fukui functions (FF), $f(\dot{r})$, provide a robust and efficient way to capture the physics governing chemical reactivity. Within the DFT formalism, FF can be described via Equation 3.3 [49]:

$$f(\dot{r}) = \left|\frac{\delta\mu}{\delta\upsilon(\dot{r})}\right|_N \qquad (3.3)$$

where
  μ is the chemical potential
  $\upsilon(\dot{r})$ is the external potential

and the derivative is taken at a fixed total number of electrons, N. Applying Maxwell's relations and assuming that the total energy of a molecular system is a function of N, Equation 3.3 becomes [50,51]

$$f(\dot{r}) = \left|\frac{\partial\rho(\dot{r})}{\partial N}\right|_\upsilon \qquad (3.4)$$

where $\rho(\dot{r})$ is the electron density. Within the frozen orbital approximation, the FF becomes equal to the Kohn–Sham frontier orbital density, and the following functions can be defined as [50,51]

$$f^{\upsilon} = \left| \phi^{\upsilon}(\dot{r}) \right|^2 \tag{3.5}$$

where

υ is HOMO or LUMO

$\phi^{\upsilon}(\dot{r})$ is the corresponding Kohn–Sham frontier molecular orbital

If υ=HOMO, Equation 3.5 quantifies the susceptibility of the carbohydrate toward electrophilic attack. If υ=LUMO, Equation 3.5 quantifies the susceptibility of the carbohydrate toward nucleophilic attack. A measure of the susceptibility toward radical attacks is given by

$$f^{\text{rad}}(\dot{r}) = \frac{\left(f^{\text{HOMO}}(\dot{r}) + f^{\text{LUMO}}(\dot{r}) \right)}{2} \tag{3.6}$$

A method enabling the calculation of these indices based on condensing the FF to a particular atom i in the molecule was developed by Contreras et al. According to this method, the condensed FF at atom i can be calculated using [52]

$$f_i^{\upsilon} = \sum_{\mu \in i}^{\text{AO}} \left[\left| c_{\mu\upsilon} \right|^2 + c_{\mu\upsilon} \sum_{\kappa \neq \mu}^{\text{AO}} c_{\kappa\upsilon} S_{\kappa\mu} \right] \tag{3.7}$$

where

$c_{\mu\upsilon}$ is the molecular orbital coefficient

AO is the set of atomic orbitals

$S_{\kappa\mu}$ is the atomic orbital overlap matrix

υ is the frontier molecular orbital (HOMO or LUMO)

Analogous to Equation 3.6, the FF for calculating the chemical reactivity of the molecule toward a radical is given by

$$f_i^{\text{rad}} = \frac{\left(f_i^{\text{HOMO}} + f_i^{\text{LUMO}} \right)}{2} \tag{3.8}$$

Given that f_i^{HOMO} provides a quantitative index for the susceptibility of atom i to electrophilic attack, we calculated the FF indices for the chemical reactions of M and $n\text{Fe}^{3+}$ (with $n=0$, 1, and 2). The condensed FF methodology was implemented in a subroutine within a modified version of the Gaussian 03 package [53]. The FFs were calculated using single point calculations on optimized geometries previously obtained from our CPMD simulations with the gradient-corrected DFT PBE exchange-correlation functional using the 6-31G** basis (B3LYP/6_31G**// CPMD).

3.4 RESULTS AND DISCUSSION

3.4.1 COORDINATION CHEMISTRY STUDIES OF M AND Fe^{3+} IN THE GAS PHASE

The chemical reactivity calculations using the condensed FFs indicate that the O2 and O4 atoms (see Figure 3.1) with f^{HOMO} values of 0.38 and 0.32, respectively, possess strong nucleophilic characters (see Table 3.1). In other words, the O2 and O4 atoms of M are the hard bases that react with the hard acid (Fe^{3+}) based on the acid–base concept of Pearson [54]. Furthermore, the calculated natural partial charges show that the hydroxyl oxygen atoms of M have larger partial negative charges compared to the ring oxygen (O5) and methoxy oxygen (O1) atoms. These findings indicate a better Lewis base reactivity for the hydroxyl group oxygen atoms.

CPMD simulations of M and a single Fe^{3+} ion in the gas phase show that the metal ion coordinates to the O2, O3, and O4 atoms of M (Figure 3.2) within 25 ps and stays coordinated for the rest of the simulation (45 ps). These simulation results support

TABLE 3.1

Calculated Chemical Reactivity Indices f^{HOMO} in the Gas Phase and in Water Based on Condensed FF

Atom	M		$M–Fe^{3+}$	
	Gas Phase	In Water	Gas Phase	In Water
O2	0.38106	0.18068	0.50533	0.01602
O3	0.01906	0.00107	0.01135	0.00125
O6	0.00153	0.00004	0.00080	0.00022
O4	0.01999	0.00412	0.07802	0.00309
O5	0.32149	0.00036	0.07753	0.00092
O1	0.01602	0.00165	0.18454	0.00002

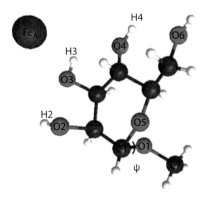

FIGURE 3.2 Methyl-α-D-mannopyranoside and its complex with one Fe^{3+} ion in the gas phase, obtained from CPMD simulations.

the calculated FF indices and the NPA values (see Tables 3.1 and 3.2). The distances between the Fe^{3+} ion and the O2, O3, and O4 atoms vary between 2.7 and 3.8 Å, and the average O2–Fe^{3+}–O4 angle is such that the transition metal ion is symmetrically located between the O2 and O4 atoms (see Table 3.3). In contrast, CMD simulations yield different results regarding the coordination chemistry of the Fe^{3+} ion and M in the gas phase. Based on the CMD simulations, the Fe^{3+} ion is more mobile and less coordinated; its distance to all hydroxyl group oxygen atoms of M varies between 5 and 11 Å. This might be due to the neglect of the partial charge differences of

TABLE 3.2

Calculated Average Natural Partial Charges (NPA) for the Fe Ion and Carbohydrate in the Gas Phase and in Water at the BLYP/6-311+G(2d,p) Level of Theory

Atom	M		M–Fe^{3+}	
	Gas Phase	In Water	Gas Phase	In Water
O1	-0.59 ± 0.06	-0.58 ± 0.04	-0.72 ± 0.03	-0.57 ± 0.03
O2	-0.73 ± 0.04	-0.79 ± 0.02	-0.62 ± 0.05	-0.78 ± 0.02
O3	-0.75 ± 0.04	-0.77 ± 0.02	-0.63 ± 0.04	-0.73 ± 0.01
O4	-0.76 ± 0.05	-0.75 ± 0.04	-0.61 ± 0.06	-0.79 ± 0.02
O5	-0.59 ± 0.04	-0.56 ± 0.02	-0.52 ± 0.03	-0.62 ± 0.05
O6	-0.75 ± 0.03	-0.79 ± 0.03	-0.54 ± 0.07	-0.79 ± 0.02
Fe_I^{III}			1.77 ± 0.08	0.72 ± 0.03

TABLE 3.3

Specific Bond Lengths and Bond Angles of M and Its Complexes with Fe in the Gas Phase Calculated via CPMD and CMD Simulations

	M		M–Fe^{3+}		M–Fe_2^{3+}	
	CPMD	MD	CPMD	MD	CPMD	MD
O3–H2/Å	2.0 ± 0.3	2.3 ± 0.5	2.3 ± 0.2	2.5 ± 0.4	2.7 ± 0.2	2.9 ± 0.4
O4–H3/Å	2.4 ± 0.4	2.7 ± 0.5	2.5 ± 0.2	2.7 ± 0.3	2.4 ± 0.3	2.9 ± 0.2
O6–H4/Å	2.1 ± 0.3	2.9 ± 0.4	2.5 ± 0.2	2.7 ± 0.2	2.4 ± 0.3	3.2 ± 0.4
θ(O2H2O3)/(deg)	146 ± 19	107 ± 23	109 ± 17	116 ± 38	106 ± 18	108 ± 39
θ(O3H3O4)/(deg)	112 ± 28	101 ± 21	103 ± 13	101 ± 25	106 ± 10	98 ± 19
θ(O4H4O6)/(deg)	120 ± 27	115 ± 38	100 ± 22	129 ± 28	105 ± 14	117 ± 31
Fe_I–O2/Å			3.3 ± 0.2	8.4 ± 3.2	3.3 ± 0.2	8.7 ± 2.9
Fe_I–O3/Å			3.0 ± 0.3	7.9 ± 2.2	3.9 ± 0.2	8.8 ± 3.1
Fe_I–O4/Å			3.5 ± 0.3	7.7 ± 3.5	4.1 ± 0.5	8.2 ± 3.6
θ(O2FeO4)/(deg)			69 ± 4		68 ± 3	
Fe_{II}–O1/Å					5.1 ± 1.9	7.3 ± 3.2
Fe_{II}–O2/Å					4.9 ± 1.4	9.8 ± 2.1

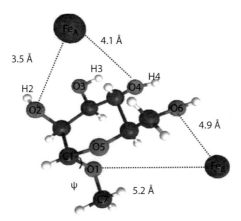

FIGURE 3.3 Methyl-α-D-mannopyranoside and its complex with two Fe^{3+} ions in the gas phase, obtained from CPMD simulations.

the hydroxyl groups when using force field parameters, i.e., failing to distinguish between axial and equatorial orientations of the hydroxyl groups and ignoring the impact of the conformational changes and steric effects on the electrostatics.

To investigate the interaction of M with more than one Fe^{3+} ion in the gas phase, we calculated the chemical reactivity indices of each atom in the vicinity of a Fe^{3+} ion (M–Fe^{3+}) and simulated M with two Fe^{3+} ions (Figure 3.3). Our chemical reactivity index calculations using the condensed FFs indicate that the second Fe^{3+} ion might be coordinated to the O2 atom and to the glycosidic linkage O1 atom (Table 3.1) with f^{HOMO} values of 0.50 and 0.18, respectively. CPMD simulations show that the distances between the second Fe^{3+} ion and the O1 and O2 atoms are shorter than the distance between the second Fe^{3+} ion and the rest of the hydroxyl oxygen atoms of M. In addition, these simulations show that the distances between the second Fe^{3+} ion and the possible coordination sites of M are larger than those obtained for the single Fe^{3+} ion in the gas phase. These results might indicate that the second Fe^{3+} ion coordination is weaker than the coordination of a single Fe^{3+} ion to M in the gas phase. CMD simulations overestimate the mobility of the Fe^{3+} ions in comparison to CPMD simulations.

To further evaluate the binding affinity of two Fe^{3+} ions to M, we computed the binding energies for each Fe^{3+} ion utilizing the trajectories taken from our CPMD simulations using the following relationship:

$$A + B \rightarrow AB$$

$$BE = E_{AB} - E_A - E_B$$

where
    BE represents the binding energy
    A and B represent the Fe^{3+} ion and M

According to these calculations, the coordination of the single Fe^{3+} ion is thermo-dynamically preferred with a BE value of −164.2 kJ mol⁻¹. On the other hand, the coordination of the second Fe^{3+} ion is energetically unfavorable with a BE value of $+328.6\,kJ\,mol^{-1}$. The calculated BEs might be overestimated since we did not include the basis set superposition error in these calculations. Overall, these BE values support the structural properties reported above and indicate that the coordination of a single Fe^{3+} is more likely than the coordination of two Fe^{3+} ions to M in the gas phase.

The relative torsional energies of the glycosidic linkage (φ) of M were calculated for M, $M–Fe^{3+}$, and $M–Fe_2^{3+}$ in the gas phase utilizing both CPMD and CMD simulations. Changes arising in PMF as a function of the torsional angle derived from these simulations are presented in Figure 3.4A. Similar trends for the conformational

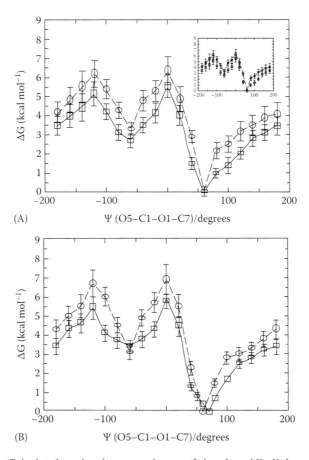

(A) Ψ (O5–C1–O1–C7)/degrees

(B) Ψ (O5–C1–O1–C7)/degrees

FIGURE 3.4 Calculated torsional energy change of the glycosidic linkage of methyl-α-mannopyranoside (solid line) and its energy change in the vicinity of one (dashed line) and two Fe^{3+} ions (dotted line) in the gas phase, with the PMF method: (A) CPMD simulations, (B) CMD simulations. The inset plot (B) presents the calculated energy differences of the glycosidic linkage of methyl-α-mannopyranoside with the perturbation and PMF methods by CMD simulations. Errors were calculated using the time-dependent standard deviation method.

preference of the glycosidic linkage of M were obtained via PMF calculations using both simulation techniques (see Figure 3.4A): $g+ > t > g-$. Furthermore, the difference of the Gibbs free energy values calculated for the prediction of the flexibility of the glycosidic linkage with the PMF and perturbation methods varies between 2% and 9% (Figure 3.4A).

Based on the thermodynamic calculations, the coordination of the Fe^{3+} ion to M impacts the predicted preference of the glycosidic linkage in the gas phase. Namely, the minimum energy torsional angle shifts from 60° to 70°, and the trend for the glycosidic linkage flexibility becomes $g+ > t \approx g-$ (see Figure 3.4B). Interestingly, the CMD simulation results are insensitive to the coordination of the metal ion and do not present a change in the predicted glycosidic linkage flexibility in the vicinity of the metal ion (see Figure 3.4B). Overall, these calculations do not present only the structural properties regarding the coordination chemistry of M and the Fe^{3+} ion but also the thermodynamics obtained via CPMD and CMD simulations.

The observed trends in the glycosidic linkage flexibility can be described via detailed analysis of the structural properties. According to the CPMD simulations, the average intramolecular hydrogen bond distances between O2 and O3, O3 and O4, and O4 and O6 atoms (see Figure 3.1) vary between 1.7 and 2.8 Å, whereas these distances vary between 1.8 and 3.3 Å with the CMD simulations (see Table 3.3). In addition, our studies show intramolecular hydrogen bonds between O2 and O3, and between O4 and O6 atoms for each trajectory (see Table 3.3). These structural parameters could not be demonstrated with the CMD simulations due to the reasons explained above (neglecting quantum effects, fixed partial charges on the hydroxyl groups, and not treating the d orbital interactions of the metal ion).

For the most preferred glycosidic linkage conformation of M (+60°), we find short-lived intramolecular hydrogen bonds between O2 and O3, O3 and O4, and O4 and O6 atoms via CPMD simulations. On the other hand, for the $g-$ and t orientations of the glycosidic linkage of M, we find no intramolecular hydrogen bonds between the O2 and O3 atoms. This finding might indicate that the intramolecular hydrogen bonds observed between O2 and O3 for the $g+$ orientation of φ stabilizes this conformation over its t and $g-$ orientations in the gas phase. For the t orientation, the intramolecular hydrogen bond between O3 and O4 is 23% shorter-lived in comparison to that computed for the $g-$ orientation. This result might be indicative of the role of intramolecular hydrogen bonding between the O3 and O4 atoms in stabilizing the $g-$ orientation of the glycosidic linkage over its t orientation.

In the vicinity of a single Fe^{3+} ion, the intramolecular hydrogen bond between O2 and O3 atoms reported for M (see above) diminished according to our CPMD simulations. Short-lived intramolecular hydrogen bonds were observed between O4 and O6 atoms. The average number of intramolecular hydrogen bonds remained low. For the $g+$ orientation of φ, which is coordinated to a single Fe^{3+} ion, we observe intramolecular hydrogen bonds between O2 and O3, and O4 and O6 atoms. The intramolecular hydrogen bond observed for O2 and O3 atoms diminished for the t and $g-$ orientations of φ. This finding indicates that the hydrogen bond between O2 and O3 atoms stabilizes the $g+$ orientation of the glycosidic linkage over its $g-$ and t orientations.

3.4.2 COORDINATION CHEMISTRY STUDIES OF M AND Fe^{3+} IN AQUEOUS SOLUTION

Chemical reactivity calculations using the condensed FFs (Table 3.1) on a particular geometry of M in water extracted from CPMD simulations indicate that the O2 atom (Figure 3.1) with a f^{HOMO} value of 0.18 possesses a strong nucleophilic character. The NPA shows that the O2 atom has a larger partial negative charge than the ring oxygen (O5) and methoxy oxygen (O6) in aqueous solution (see Table 3.2). This analysis indicates that the hydroxyl oxygen atoms of aqueous M have a better Lewis base reactivity than the ring and methoxy oxygen atoms. This finding is in agreement with our results obtained for M in the gas phase (see above), which show O2, O3, and O4 atoms as being reactive toward Fe coordination.

The CPMD simulations of M in aqueous solution show that a water molecule is coordinated to the O2 atom after 14 ps and stays coordinated for the rest of the simulation (56 ps). This finding supports the FF analysis, which shows that the O2 atom is the most reactive site of M toward Fe coordination. The coordination of a water molecule to the O2 atom and electronic and steric effects seem to decrease the reactivity of the O2 atom toward Fe coordination in water. CPMD simulations show that a single Fe^{3+} ion, which is initially coordinated to six water molecules in an octahedral arrangement, coordinates to the O3 and O4 atoms within 27 ps and stays coordinated for the rest of the simulation (43 ps) as shown in Figure 3.5. CPMD simulations further show that two water molecules bind to the carbohydrate-coordinated single Fe^{3+} ion in aqueous solution within 31 ps, with two water molecules and two hydroxyl atoms (O3 and O4) of M placed in a square-planar arrangement around the metal ion (see Figure 3.5). In other words, these simulations show that the Fe^{3+} ion interacts with two types of ligands; the carbohydrate hydroxyl groups and water molecules.

The partial charges indicate that the O3 atom possesses a slightly smaller partial negative charge in the vicinity of the Fe^{3+} ion (Table 3.2). The distances between the Fe^{3+} ion and O3 and O4 atoms are 3.1 ± 0.1 and 4.0 ± 0.1 Å, respectively. NPA results support this finding and indicate a smaller negative charge for O3 than for O4. This effect is a result of Fe coordination. A comparison with our results for the gas phase, which are presented above, shows that the Fe and O3 and O4 distances are slightly larger (about 10%) in water than those in the gas phase. This result may be due to the coordination of additional water molecules to Fe. The calculated binding energy for the Fe^{3+} ion coordination to the aqueous carbohydrate is -98.7 kJ mol^{-1}, indicating that the single transition metal ion coordination is thermodynamically favorable.

In summary, in comparison to the studies of M–Fe^{3+} in the gas phase, we find that the presence of water impacts the coordination of the Fe^{3+} ion to the carbohydrate. Our gas-phase studies show a coordination of the metal ion to O2, O3, and O4 atoms, but the solution-phase CPMD simulations show that the transition metal ion is coordinated to O3 and O4, and not to O2. It is found that a water molecule is coordinated to O2 instead, with an intermolecular hydrogen bond distance of 2.8 ± 0.2 Å (Figure 3.5 and Table 3.4). Results regarding inter- and intramolecular hydrogen bonding interactions are discussed in the following paragraphs. Our CMD simulations (Figure 3.5) do not seem to detect a coordination between M and Fe in water.

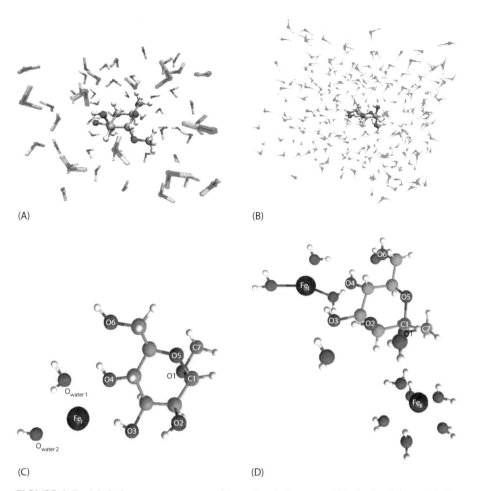

FIGURE 3.5 Methyl-α-D-mannopyranoside molecule in water (A) obtained from CPMD simulations, (B) obtained from CMD simulations, and its complexes with one and two Fe^{3+} ions, and (C, D) obtained from CPMD simulations, with conventional numbering of specific atoms.

To study the coordination of more than one Fe^{3+} ion to M, we performed chemical reactivity index calculations in the vicinity of Fe^{3+} and water molecules and simulated M in water with two Fe^{3+} ions (Figure 3.5). The chemical reactivity calculations on a trajectory of M–Fe^{3+} with the first shell of water molecules obtained from CPMD simulations show that the heavy atoms of the carbohydrate are not reactive toward a second Fe^{3+} ion (Table 3.2). This finding is consistent with the experimental results. CPMD simulations show that the first Fe^{3+} ion is still coordinated to O3 and O4 atoms in the presence of the second Fe^{3+} ion. However, the distances between the Fe^{3+} ion and O3 and O4 atoms are slightly larger in the presence of the second Fe^{3+} ion, with values of 3.4 ± 0.1 and 4.3 ± 0.2 Å, respectively (Figure 3.5 and Table 3.4). Furthermore, the simulations with a single Fe^{3+} ion show the metal ion to be almost

TABLE 3.4

Specific Bond Lengths and Bond Angles of M and Its Complexes with Fe in Water Calculated via CPMD and CMD Simulations

	M		M–Fe^{3+}		M–Fe_2^{3+}	
	CPMD	CMD	CPMD	CMD	CPMD	CMD
O3–H2/Å	2.1±0.3	2.4±0.5	2.1±0.2	2.4±0.4	2.2±0.1	2.5±0.4
O4–H3/Å	2.5±0.4	2.6±0.6	3.0±0.3	2.7±0.5	3.1±0.3	2.8±0.4
O2–O_{water1}/Å	2.8±0.3	3.1±0.3	2.9±0.2	2.9±0.5	3.2±0.2	2.9±0.3
O3–O_{water1}/Å	3.4±0.2	2.9±0.5	4.8±0.3	3.0±0.4	5.1±0.4	3.1±0.4
O4–O_{water1}/Å	3.8±0.3	3.2±0.4	5.0±0.5	3.1±0.5	5.2±0.4	3.2±0.4
O6–H4/Å	4.5±0.4	3.6±0.9	2.5±0.4	3.3±0.6	2.6±0.3	3.5±0.7
θ(O2H2O3)/(deg)	127.9±8.0	105.2±22.8	130.7±4.5	107.7±19.8	128.8±5.3	109.9±13.5
θ(O3H3O4)/(deg)	115.3±11.4	103.7±25.4	117.7±5.6	105.2±20.2	113.6±4.9	113.3±18.4
θ(O4H4O6)/(deg)	123.5±12.3	107.8±22.5	124.6±7.3	106.9±19.9	122.1±8.2	108.0±16.2
Fe_I–O3/Å			3.1±0.2	>5.5	3.4±0.1	> 5.5
Fe_I–O4/Å			4.0±0.2	>6.1	4.4±0.2	> 6.3
Fe_I–O_{water1}/Å			2.1±0.2		2.0±0.3	
Fe_I–O_{water2}/Å			2.2±0.1		2.3±0.1	
θ(O3FeO4)/(deg)			54.4±12.0		44.2±9.7	
Fe_{II}–O_{water1}/Å					2.2±0.3	

symmetrically located between the O3 and O4 atoms with an angle of $54.4 \pm 12°$. In the presence of the second Fe^{3+} ion, this angle becomes $44.2 \pm 9.7°$ and the first Fe^{3+} ion attached to M is coordinated to three water molecules instead of two. The second Fe^{3+} ion prefers to be arranged with six water molecules in an octahedral symmetry in the second solvation shell. The closest distance between the first and second Fe^{3+} ions is 10.2 Å. We also note that the intermolecular hydrogen bond distance between O2 and the coordinated water molecule increases to 3.25 ± 0.24 Å in the presence of the second Fe^{3+} ion. In agreement with our studies in the gas phase, we find that the presence of a second Fe^{3+} ion increases the distances between the first Fe^{3+} ion and O3 and O4 atoms of the carbohydrate. Coordination of the second Fe^{3+} ion to the biomolecule in the gas phase is possible, whereas the inclusion of explicit solvent molecules shows that the presence of water impacts the coordination chemistry since the second Fe^{3+} ion coordinates to water molecules rather than to the carbohydrate (see above).

Our NPA analysis (including all solute and solvent molecules) indicates that the partial positive charge on the first Fe^{3+} ion increases from 0.7 to 1.4 in the presence of the second Fe^{3+} ion (Table 3.2). This finding is consistent with our simulations presented above since the distances of the coordinated hydroxyl group oxygen atoms of M to the first Fe^{3+} ion are larger when a second Fe^{3+} ion is present in the system, indicating less electron donation capability for the carbohydrate-coordinated Fe^{3+} ion.

The present first principles results are in agreement with the experiments, which show only one Fe^{3+} ion coordinated to the carbohydrate in solution. In contrast to the experiment and ab initio simulations, our CMD simulations show that both Fe^{3+} ions form complexes only with water molecules and do not coordinate to the carbohydrate. These discrepancies can be attributed to the quality of the force field parameters for the transition metal ion, usage of fixed partial charge parameters for the different hydroxyl oxygen atoms in the carbohydrate (neglecting charge polarization), failure to distinguish between equatorial and axial orientations of the carbohydrate hydroxyl groups, and ignoring the conformation dependence of steric effects.

The relative torsional energies of the glycosidic linkage (φ) of M (see above) were studied using CPMD and CMD simulations in water. Changes arising in PMF as a function of torsional angle derived from these calculations are presented in Figure 3.6. Figure 3.6A shows the calculated PMF values of the glycosidic linkage for M, M–Fe^{3+}, and based on CPMD simulations (see Section 3.2). Different trends for the conformational preference of the glycosidic linkage of M are observed in CPMD simulations: the $g+ > t > g-$ trend observed in the free M becomes $g+ > g- > t$ and the minimum shifts from 60° to 70° upon coordination of one Fe^{3+} ion to M in water. The presence of the second Fe^{3+} ion does not influence the calculated thermodynamic trend for the glycosidic linkage conformation in the vicinity of a single Fe^{3+} ion, consistent with the fact that a second Fe^{3+} ion does not coordinate to the M–Fe^{3+} complex in water.

According to our CMD simulations, the trend obtained for the glycosidic linkage conformation $g+ > g- > t$ for the free carbohydrate is retained in aqueous solution in the presence of the Fe^{3+} ions, which are coordinated to water molecules rather than to the carbohydrate. Furthermore, in stark contrast to the CPMD results, our CMD simulations do not indicate a significant shift of the minimum energy torsional angle

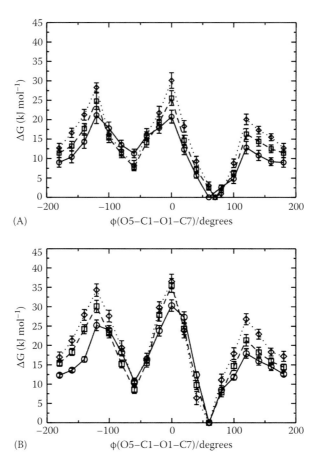

FIGURE 3.6 Calculated torsional energy change of the glycosidic linkage of methyl-α-mannopyranoside (solid line) and its energy change in the vicinity of one (dashed line) and two Fe^{3+} ions (dotted line) in water, with the PMF method: (A) CPMD simulations, (B) CMD simulations. The inset plot (B) presents the calculated energy differences of the glycosidic linkage of methyl-α-mannopyranoside with the perturbation and PMF methods by CMD simulations. Errors were calculated using the time-dependent standard deviation method.

of the glycosidic linkage and predict $+60°$ as the minimum for M, M–Fe^{3+}, and in water. For the gas phase, this trend is $g+ > t \approx g-$ as reported above and indicates that water stabilizes the $g-$ orientation of the glycosidic linkage over its t orientation upon coordination of Fe.

According to the CPMD simulations, the average intramolecular hydrogen bonding distances O2–O3, O3–O4, and O4–O6 of M in water vary between 2.1 and 4.5 Å. In contrast, CMD simulations predict that these intramolecular distances are between 2.4 and 3.7 Å. The intramolecular hydrogen bonds between O2 and the H attached to O3 are shorter than those for O3–O4 and O4–O6 according to the CPMD simulations (Table 3.4). Furthermore, the predicted intermolecular hydrogen bonds between O2–water, O3–water, and O4–water indicate a stronger coordination of the O2 atom to a water molecule (as compared to the O3–water and O4–water hydrogen bonds) via the shorter O2–water distance (see Table 3.4). This finding is in agreement with our chemical reactivity calculations, which identify the O2 atom as the most reactive carbohydrate hydroxyl oxygen toward ligands. CPMD simulations show that the water ligand tends to coordinate to O2 rather than to O3 or O4. CMD simulations do not show a significant difference in intermolecular hydrogen bond distances for these hydroxyl oxygen atoms in water. In addition, a comparison with the results obtained for the gas phase indicates that intermolecular hydrogen bonds between the carbohydrate and solvent molecules increases the intramolecular hydrogen bond distances O2–O3, O3–O4, and O4–O6 by values varying between 0.4 and 1.7 Å.

Figure 3.7 shows the average number of hydrogen bonds between the carbohydrate and surrounding water molecules obtained from three distinct CPMD simulations for the $g+$, $g-$, and t orientations of the glycosidic linkage (each simulation was performed for 60 ps). For these analyses, following the work of Molteni and Parrinello [48] and our previous studies, the criteria for the existence of a hydrogen bond were defined as $O_D–H \leq 1.5$ Å and $H–O_A \leq 2.4$ Å, where O_D and O_A represent donor and acceptor oxygen atoms, respectively. The angular criterion was values larger than $120°$. According to this analysis, the total number of intermolecular hydrogen bonds is highest for the $g+$ conformation, while the t conformation exhibits a slightly larger number than the $g-$ orientation.

CPMD simulations and a comparison with our results obtained from the simulations in the gas phase indicate that the presence of a single Fe^{3+} ion weakens the intramolecular hydrogen bonds between the O3 and O4 atoms since these hydroxyl oxygens are coordinated to the metal ion (see Table 3.4). The average O4–O6 intramolecular hydrogen bond distance becomes shorter in comparison to those obtained for the aqueous carbohydrate without a metal ion (Table 3.4). This result indicates that the intramolecular hydrogen bonds between these hydroxyl groups become

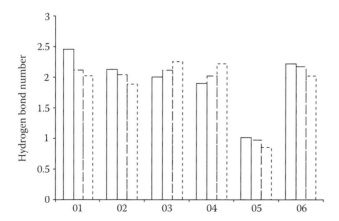

FIGURE 3.7 Average number of hydrogen bonds between specific methyl-α-mannopyranoside oxygens and water molecules; for the (—) $g+$, (---) t, and (...) $g-$ orientations of the glycosidic linkage of methyl-α-mannopyranoside via CPMD simulations.

stronger due to the coordination of Fe to O3 and O4. In addition, we find that two water molecules are coordinated to the carbohydrate-coordinated Fe^{3+} ion with average Fe–water oxygen distances of 2.1 ± 0.2 and 2.2 ± 0.1 Å. In contrast, CMD simulations predict intramolecular hydrogen-bonds between O2 and O3 and between O3 and O4 in the presence of Fe (Table 3.4). The hydroxyl oxygen atoms are also coordinated to water molecules with average intermolecular hydrogen bond distances varying between 2.7 and 3.5 Å.

According to the CPMD simulations, the presence of the second Fe^{3+} ion does not impact the number of intra and intermolecular hydrogen bonds of the carbohydrate. However, both intra and intermolecular hydrogen bond distances are slightly larger in comparison to those calculated for the hydrogen bonds in the vicinity of a single Fe^{3+} ion (Table 3.4). Instead of only two water molecules, three water molecules are coordinated to the carbohydrate-coordinated Fe^{3+} ion (Figure 3.5). This arrangement might be due to the larger distances between Fe and the O3 and O4 atoms of the carbohydrate (see above) since it may allow electron donation from more than two water ligands as shown by our NPA results. The average distance between the second Fe and its six coordinated water oxygen atoms is 2.2 ± 0.3 Å in accord with our recent studies and with diffraction spectroscopy measurements for hydrated Fe^{3+} ion [47]. The first shell coordination number was found to be six in all cases for the second Fe^{3+} ion, indicating that the interaction between Fe^{3+} and the carbohydrate does impact the coordination number of water molecules reported for a hydrated Fe ion. In contrast, CMD simulations suggest that both Fe^{3+} ions are coordinated to water molecules and not to the carbohydrate.

The presence of water molecules affects the chemical reactivity, coordination chemistry, and structural and thermodynamic properties of biometallic complexes and calls for care when one attempts to predict the chemical properties of biometallic complexes in solution from the results of gas-phase calculations.

3.5 CONCLUSION

In this chapter, we have presented a means of identifying sites of carbohydrates reactive toward metal ions in the gas phase and in aqueous solution via study of the coordination chemistry of methyl-α-mannopyranoside and Fe^{3+} ion. Our approach is based on chemical reactivity calculations, NPA, and CPMD simulations. The first principles calculations and simulations show that three hydroxyl groups are coordinated to Fe in the gas phase, whereas two hydroxyl groups and two water molecules are coordinated to Fe in a square-planar arrangement in aqueous solution. The second Fe^{3+} ion is coordinated to six water molecules in an octahedral arrangement in the second solvation shell and does not coordinate to the carbohydrate. We compared the results obtained from CPMD simulations to those obtained from CMD simulations, and we found that the CMD simulations cannot capture the coordination chemistry of biomolecules and transition metal ions in the gas phase and in aqueous solution; both Fe^{3+} ions coordinate to water molecules instead of the carbohydrate in the CMD simulations.

Our theoretical studies using first principles methods show that more than one hydroxyl oxygen atom of methyl-α-mannopyranoside coordinates to the Fe^{3+} ion in the gas phase and in water. Our CPMD simulations and binding energy calculations show that a single Fe^{3+} coordinates to the carbohydrate both in the gas phase and in aqueous solution. The coordination of the transition metal ion impacts the preferred glycosidic linkage conformation in both the phases. Specifically, the glycosidic linkage preference changes from $g+ > g- > t$ to $g+ > g- \approx t$ in the gas phase and CPMD simulations demonstrate that the preference for the $g+$ orientation of the glycosidic linkage is due to intramolecular hydrogen bonding.

In the aqueous solution, we find that the preference of glycosidic linkage changes from $g+ > t > g-$ to $g+ > g- > t$ upon coordination to Fe^{3+}. CPMD simulations clearly indicate that the conformational preference of the glycosidic linkage of the carbohydrate and the coordination of the metal ion are influenced by intra and intermolecular hydrogen bonds. Our results further demonstrate that CMD simulations are insensitive to these effects and suggest a need for better force field parameters for the aqueous Fe^{3+} ion and carbohydrates.

This study illustrates that first principles studies including chemical reactivity calculations, NPA, and CPMD simulations can provide important insights into the structure and coordination chemistry of biometallic complexes in the gas phase and in aqueous solution. Furthermore, we show that the solution environment impacts the reactive sites of a biomolecule toward ligands (metal ions). We are confident that validated first principles approaches in different phases (gas and solution) will help us to predict unknown reaction mechanisms of important physiological processes, such as receptor and ligand interactions that are crucial to drug design, medicine, biochemistry, and nanobiotechnology.

REFERENCES

1. Coskuner, O., *J. Chem. Phys.*, 2007, **127**, 015101.
2. Coskuner, O.; Bergeron, D. E.; Rincon, L.; Hudgens, J. W.; and Gonzalez, C. A., *J. Chem. Phys.*, 2008, **129**, 045102.
3. Ilk, N.; Kausma, P.; Puchberger, M.; Egelseer, E. M.; Mayer, H. F.; Silty, U. B.; and Sara, M., *J. Bacteriol.*, 1999, **24**, 7643.
4. Green, J. L. and Angell, C. A., *J. Phys. Chem.*, 1989, **93**, 2880.
5. Paly, J. P. and Leumieux, R. U., *J. Chem.*, 1987, **65**, 213.
6. Srivastava, O. M.; Hindsgaul, O.; Shoreibah, M.; and Pierce, M., *Carbohydr. Res.*, 1988, **179**, 137.
7. Sabesan, S.; Bock, K.; and Paulson, J. C., *Carbohydr. Res.*, 1991, **218**, 27.
8. Aksoy, M., *Am. J. Clin. Nutr.*, 1972, **25**, 262.
9. Vangen, B. and Hemre, G. I., *Aquaculture*, 2003, **219**, 597.
10. Van Oijen, T.; Van Leeuwe, M. A.; Gieskes, W. W. C.; and De Baar, H. J. W., *Eur. J. Phycol.*, 2004, **39**, 161.
11. Heinrich, H. C., *Arzneimittel-Forschung/Drug Research*, 1975, **25**, 420.
12. Phyner, K. and Ganzoni, A. M., *Schweizerische Medizinische Wochenschrift*, 1972, **102**, 561.
13. Jensen, P. D.; Peterslund, N. A.; Poulsen, J. H.; Jensen, F. T.; Christensen, T.; and Ellegaard, J., *Br. J. Haematol.*, 1994, **88**, 56.
14. Idiman, E.; Ozakbas, S.; Tosun, D.; Sagut, O.; Sahin, O.; and Coskuner, E., *Eur. J. Neurol.*, 2007, **14**, 280.
15. Hemmerich, S., *Drug Discov. Today*, 2001, **6**, 27.

16. Lai, J.; Bernhard, O. K.; Turville, S. G.; Harman, A. N.; Wilkinson, J.; and Cunningham, A. L., *J. Biol. Chem.*, 2009, **284**, 11027.

17. Alexandre, K. B.; Gray, E.; Pantophlet, R.; Chikwamba, R.; McMahon, J.; O'Keefe, B.; and Morris, L., *AIDS Res. Hum. Retrovir.*, 2008, **24**, 151.

18. Koziel, H.; Kruskal, B. A.; Ezekowitz, R. A. B.; and Rose, R. M., *Chest*, 1993, **103**, S111.

19. Lopez-Herrera, A.; Liu, Y.; Rugeles, M. T.; and He, J. J., *Biochim. Biophys. Acta—Mol. Basis Dis.*, 2005, **1741**, 55.

20. Mbemba, E.; Gluckman, J. C.; and Gattegno, L., *Glycobiology*, 1994, **4**, 13.

21. Marzi, A.; Mitchell, D. A.; Chaipan, C.; Fisch, T.; Doms, R. W.; Carrington, M.; Desrosiers, R. C.; and Pohlmann, S., *Virology*, 2007, **368**, 322.

22. Hall, M. and Ricketts, C. R., *J. Pharmacy Pharmacol.*, 1968, **20**, 662.

23. Liang, C. H.; Wang, C. C.; Lin, Y. C.; Chen, C. H.; Wong, C. H.; and Wu, C. Y., *Anal. Chem.*, 2009, **81**, 7750.

24. Saltman, P., *J. Chem. Edu.*, 1965, **42**, 682.

25. Spiro, T. G. and Saltman, P., *Struct. Bond.*, 1969, **6**, 116.

26. Charley, P. J.; Sarkar, B.; Stitt, C. F.; and Saltman, P., *Biochem. Biophys. Acta*, 1963, **69**, 313.

27. Davis, P. S. and Deller, D. J., *Nature*, 1966, **212**, 40.

28. Rao, P. C.; Geetha, K.; Raghavan, M. S. S.; Sreedhara, A.; Tokunaga, K.; Yamaguchi, T.; Jadhav, V.; Ganesh, K. N.; Krishnamoorty, T.; Ramalah, K. V. A.; and Bhattacharyya, R. K., *Inorg. Chem.*, 2000, **297**, 373.

29. Coskuner, O. and Deiters, U. K., *Z. Phys. Chem.*, 2006, **220**, 349.

30. Coskuner, O. and Deiters, U. K., *Z. Phys. Chem.*, 2007, **221**, 785.

31. Stigter, D.; Alonso, D. O.; and Dill, K. A., *Proc. Nat. Acad. Sci.*, 1991, **88**, 4176.

32. Wolfenden, R. and Radzicka, A., *Science*, 1994, **265**, 936.

33. Kirschner, K. N. and Woods, R. J., *Proc. Nat. Acad. Sci.*, 2001, **98**, 10541.

34. Qin, S. and Zhou, H.-X., *Biopolymers*, 2007, **86**, 112.

35. Jung, A.; Berlin, P.; and Wolters, B., *IEE Proc. Nanobiotechnol.*, 2004, **151**, 87.

36. Ohtaki, H. and Radnai, T., *Chem. Rev.* (Washington D.C.), 1993, **93**, 1157.

37. Thanki, N.; Thornton, J. M.; and Goodfellow, J. M., *J. Mol. Biol.*, 1988, **202**, 637.

38. Rossky, P. and Karplus, M., *J. Am. Chem. Soc.*, 1979, **101**, 1913.

39. Coskuner, O.; Bergeron, D. E.; Rincon, L.; Hudgens, J. W.; and Gonzalez, C. A., *J. Phys. Chem. A*, 2008, **112**, 2940.

40. Coskuner, O.; Bergeron, D. E.; Rincon, L.; Hudgens, J. W.; and Gonzalez, C. A., *J. Phys. Chem. A*, 2009, **113**, 2491.

41. Troullier, N. and Martins, J. L., *Phys. Rev. B*, 1991, **43**, 1993.

42. Lee, C.; Yang, W.; and Parr, R. C., *Phys. Rev. B*, 1988, **37**, 785.

43. Rodgers, M. T. and Armentrout, P. B., *Acc. Chem. Res.*, 2004, **37**, 989.

44. Damm, W.; Frontera, A.; Tirado-Rives, J.; and Jorgensen, W. L., *J. Comput. Chem.*, 1997, **18**, 1955.

45. Rappe, A. K.; Casewit, C. J.; Colwell, K. S.; Goddard III, W. A.; and Skiff, W. M., *J. Am. Chem. Soc.*, 1992, **114**, 10024.

46. Coskuner, O. and Jarvis, E. A. A., *J. Phys. Chem. A*, 2008, **119**, 7999.

47. Coskuner, O.; Jarvis, E. A. A.; and Allison, T. C., *Angew. Chem. Int. Ed.*, 2007, **46**, 7853.

48. Molteni, C. and Parrinello, M., *J. Am. Chem. Soc.*, 1998, **120**, 2168.

49. Parr, R. and Yang, W. *Density Functional Theory of Atoms and Molecules*; Oxford University Press, New York, 1989, (b) Parr, R. G. and Yang, W., *J. Am. Chem. Soc.*, 1984, **106**, 4049, (c) Yang, W.; Parr, R. G.; and Pucci, R., *J. Chem. Phys.*, 1984, **81**, 2862, (d) Yang, W. and Parr. R. G., *Proc. Nat. Acad. Sci.*, 1985, **82**, 6723.

50. Ciosloski, J.; Martinov, M.; and Mixon, S. T., *J. Phys. Chem.*, 1993, **97**, 10948.

51. Senet, P., *J. Chem. Phys.*, 1997, **107**, 2516.

52. Contreras, R. R.; Fuentealba, P.; Galvan, M.; and Perez, P., *Chem. Phys. Lett.*, 1999, **304**, 405.
53. Frisch, M. J.; Trucks, G. W.; Schlegel, H. B.; Scuseria, G. E.; Robb, M. A.; Cheeseman, J. R.; Montgomery, J. J. A. et al., *Gaussian* 03, Revision C.02, Gaussian, Inc: Wallingford, CT, 2004.
54. Pearson, R. J., *J. Am. Chem. Soc.*, 1963, **85**, 3583.

4 Aqueous Solutions of Metal Ions

*Orkid Coskuner and Thomas C. Allison**

CONTENTS

4.1 INTRODUCTION

Metal ions play crucial roles in diverse areas such as medicine, nanotechnology, catalysis, environmental sciences, and geochemistry. For example, the chemistry of metal ions with biomolecules and organic compounds is extremely important for a wide range of medicinal and environmental processes that depend on metal ions as active participants [1–3]. Organic complexes that coordinate to metal ions are becoming increasingly prevalent as therapeutic agents for treating a wide variety of metabolic disorders and diseases. For instance, cisplatin has been shown to be toxic to cancer cells (see Chapter 1 for a more detailed discussion of cisplatin) and the platinum-based compound known as "*trans,trans,trans*-$Pt(N_3)_2(OH)_2(NH_3)(py)$" has been shown to be highly toxic and less stable than cisplatin. It is between 13 and 80 times more toxic to cancer cells than cisplatin and kills cancer cells via a different mechanism so it can also kill cisplatin-resistant cancer cells. When we consider that most biological processes occur in solution and that the human body consists of ~60% water, the role of the aqueous solution environment on the structure–function relationships of metal ion–based organometallic drugs cannot be ignored.

In environmental sciences, understanding the interactions of metal ions that are toxic to plants and organisms and that are constituents of natural and wastewaters

---

* Contribution of the National Institute of Standards and Technology.

will enable the design of more efficient techniques for their removal. For example, exposure to arsenic (As) occurs through anthropogenic and natural sources [4–9]. Arsenic has adverse effects on human health through poisoning leading to the formation of cancer [10,11]. Many tissues such as the skin, gut, lungs, blood vessels, and heart are damaged via exposure to As [12]. Arsenic is known to damage chromosomes. Arsenic in the environment may occur naturally in deep-water wells. There are 40,000 cases of arsenicosis in Bangladesh and health experts expect that there will be 2.5 million additional cases within the next 50 years. Drinking As-free water is currently the only way to prevent this disease. A detailed understanding of the As ion interaction with water molecules can help in the design of better and more efficient processes for the removal of this metal ion from natural waters.

From the perspective of catalysis, metal ions and their complexes with organic compounds can be good catalysts because of their ability to change oxidation states or to adsorb other substances on their surfaces and activate them during chemical processes [13–15]. For example, the Haber process uses iron (Fe) as a catalyst for producing ammonia via the reaction between hydrogen and nitrogen: $N_2(g) + 3H_2(g) \rightleftarrows 2NH_3(g)$. Another example is the chemical reaction between persulfate and iodide ions, $S_2O_8^{2-} + 2I^- \rightarrow 2SO_4^{2-} + I_2$, which is extremely slow in aqueous solutions without the Fe catalyst [16,17].

The production of liquid fuels from biomass resources is attractive due to the desire to find a renewable fuel supply and the need to reduce carbon emissions into the atmosphere. Metal catalysts have an important role to play in the conversion of biomass to liquid transportation fuels. Recent experimental studies have shown that sugar alcohols can be converted to lighter alkanes using metal catalysts [18,19]. Processes that produce liquid alkanes via aqueous phase processing of biomass-derived carbohydrates using metal catalysts provide a renewable source for transportation fuel. An improved understanding of metal catalysts for carbohydrate to alkane conversion in aqueous solutions could help in the design of more efficient and economical processes for obtaining biofuels from carbohydrate complexes. Thus, detailed knowledge of carbohydrates and metal catalysts in water is required, which, in turn, requires knowledge of the hydration properties of the metal catalyst and its impact on the solution environment that affects the conformation and structure–function relationships of carbohydrates in aqueous solution [2,3].

As discussed above, metal complexes and metal ions play important roles in many important areas. However, a fundamental understanding of metal ions in aqueous solution and of the nature of the solution environment in the vicinity of metal ions is largely unknown. A detailed molecular-level understanding of metal ion structure–function relationships in aqueous solution and the impact of the nature of the solution environment on the structure, coordination chemistry, and hydration properties of the metals can aid in the design and synthesis of more efficient catalysts, therapeutics, drugs, biomarkers, and nanocomplexes, leading to processes that are beneficial to human health, to the environment, and to the economy.

The title of this chapter is deceptively simple. Our knowledge of the chemistry of these solutions is extremely limited. One reason lies in the small size and fast dynamics of these systems, which makes experimental measurements and theoretical predictions challenging. Another reason lies in the nature of the solution environment itself. Even though water is the most abundant and widely used solvent on our planet, and despite the great amount of experimental and theoretical study on water, there is currently considerable debate over the structure of pure water at room temperature due to differing interpretation of experimental studies such as x-ray absorption and neutron diffraction [20,21]. Quantum chemistry calculations are also employed on both sides of this debate. Direct measurement of the local electronic structure of solvents is immensely difficult. Some information can be obtained from x-ray absorption studies of liquids, which may be used to calculate the radial distribution function, a quantity that is directly comparable to the data that may be extracted from simulations. However, radial distribution functions are usually spherically averaged and information associated with anisotropic and asymmetric characters is lost [22,23].

Another area that has been examined by state-of-the-art theoretical and experimental techniques is the autoionization of water and the dissociation of water in dense systems [24–35]. From a theoretical standpoint, special sampling techniques coupled with ab initio molecular dynamics simulations can provide unprecedented insight into dynamical events in water [24]. One might be surprised to learn that the quest to understand proton transfer in water has been going on for more than 200 years. Grotthuss introduced the concept of "structural diffusion" more than 200 years ago and described the mechanism of proton transfer along a network of hydrogen bonds. Some of the most important aspects of understanding the behavior of the proton in solution, and probably the link to a deeper understanding of acid and base chemistry in aqueous solution, include the structure and energetics of hydrated protons that form during the course of the dynamics in dense aqueous solutions [36–43]. Our modern understanding of hydrated protons is dominated by the well-known Eigen and Zundel complexes. In the middle of the nineteenth century, Eigen proposed a $H_9O_4^+$ proton complex in which the hydronium ion (H_3O^+) is coordinated to surrounding water molecule via hydrogen bonding interactions (Figure 4.1), and Zundel proposed a $H_5O_2^+$ proton complex in which a proton (H^+) is coordinated to two water molecules via hydrogen bonding interactions in dense water (Figure 4.1) [44–46]. The structural properties and associated energetics of these proton complexes in dense aqueous solutions containing various inorganic and organic systems are still being investigated and debated just as in the case of pure water solutions [47–51].

Proton transfer studies—mostly in pure water and in organic solutions—have benefited from the application of ab initio molecular dynamics simulations during the last decade. In particular, Car–Parrinello molecular dynamics (CPMD) simulations have been extremely useful in studies of hydrogen-bonded complexes [52–58]. The electronic structure of the Zundel complex in pure water has been studied using first principles calculations that find the favored position of the proton in the middle of the two water molecules. This arrangement is termed "symmetrically coordinated." It is clear that a quantitative description including microscopic properties with hydrogen bond interactions in the presence of solvent fluctuations is required for the proper description of liquid water. The introduction of AIMD simulations, such as Born–Oppenheimer and CPMD simulations, enable dynamics studies of proton transfer with solvent fluctuations including the

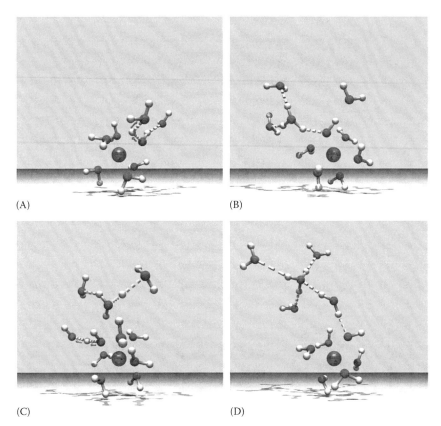

(A)   (B)

(C)   (D)

FIGURE 4.1 (See color insert.)   Illustration of proton transfer in water around Fe^{3+} ions and Cr^{3+} ions from CPMD/TPS simulations. Many water molecules are deleted to clearly show the molecular mechanism. See text for discussion of mechanism. Illustration of the mechanism of water dissociation around metal (Cr^{3+}, Fe^{3+}) ions which begins with a slightly elongated O–H bond (A). Proton migration leads to the hydroxylation of the metal ion (B). Movement of a proton into the second solvation shell leads to the formation of a Zundel complex (C). Further proton transfer leads to the formation of an Eigen complex (D). (Coskuner, O. et al., *Water Dissociation in the Presence of Metal Ions*, 7853, 2007. Copyright Wiley-VCH Verlag GmbH & Co. KGaA. Reproduced with permission.)

detailed electronic structure information afforded by first principles methods such as many body interactions [60,61]. CPMD simulations of Be^{2+} in water have demonstrated the advantages of including many body interactions in the dynamics of metal solutions [41,61–64].

   The migration of a proton in water and the migration of OH^- in water have also been simulated. It was shown that proton migration is a *topological defect* in the hydrogen bond network rather than a single proton that migrates though the network [41]. The notion of "structural diffusion" mentioned above is used to describe this topological defect mechanism that includes the breaking and forming of O–H bonds. AIMD simulations of the proton in water showed that both Eigen and Zundel complexes are observed in the structural diffusion molecular mechanism. These studies have shown that Eigen and Zundel complexes interconvert in pure water by hydrogen-bond fluctuations. Later, it was shown that Warshell's empirical valence bond approach could also be used to obtain a mechanistic description of Grotthuss diffusion in liquid water [41].

Proton transfer via the Grothuss mechanism including the formation of Zundel and Eigen complexes is the currently accepted view. The pH of pure water, which is related to the autoionization of pure liquid water and thus to the proton transfer mechanism, is of considerable interest. Geissler et al. studied water autoinoization using CPMD simulations coupled with transition path sampling (TPS) calculations [24]. The pH of ambient water implies that a water molecule will spontaneously dissociate once every 10 h on the average. Thus, water dissociation is indeed a rare event, and one that would not be feasible to study without TPS. Geissler et al. showed that the spontaneous dissociation of a water molecule leads to the propagation of a proton through the solution along a network of hydrogen bonds. Once the H_3O^+ and OH^- are separated from each other, they may form Zundel or Eigen complexes in the aqueous solution or may recombine. These events occur on a timescale of about 100 fs. Only when a fluctuation is coincident with another hydrogen bond breaking in the hydrogen-bonded network along which the two charged aqua ions separate is fast recombination not possible [24]. Recently, we applied CPMD simulations with TPS calculations on metal ion solutions (aqueous solutions of Fe^{3+}, Cr^{3+}, As^{3+}, and As^{5+}) [65,66]. We implemented classical simulation techniques, such as perturbation theory, for calculating the Gibbs free energy associated with proton transfer in these simulations. We were able to observe the mechanism of water dissociation around these transition metal ions and found both Eigen and Zundel complexes in the solutions of Fe^{3+}, Cr^{3+}, As^{3+}, and As^{5+} ions. We predicted the hydration structures of these metal ions including the impact of the dissociated water on their structures. For the Al^{3+} solution, fast recombination of the H_3O^+ and OH^- ions prevented us from observing the Grotthuss mechanism or the Eigen and Zundel ions in the solution [65,66].

In general, Al is the most abundant metal in the Earth's crust. An understanding of hydrated Al structures is of prime interest in wastewater treatment, neurodegenerative diseases, catalysis, and environmental sciences. For example, it is known that Al is the principal toxic substance that kills the fish in the northern United States. Al affects the function of fish gills by causing ion regulatory and respiratory dysfunctions. Furthermore, it has been hypothesized that the Al ion changes the conformation of amyloid β (Aβ) peptides and is able to trigger the mis-folding mechanism of Aβ peptides in the brain, which has been implicated in Alzheimer's disease. Many studies of aqueous Al ions and their reactions with organic compounds and biomolecules have been performed [1,67–70]. Nevertheless, a thorough explanation of the reaction mechanism of aqueous Al ions or Al polymers with biomolecules, such as Aβ peptide, using ab initio molecular dynamics is still lacking.

The structure of a single, solvated Al^{3+} ion that may exist at high or low pH has been studied extensively [71,72]. Experimental and computational chemistry studies agree that the Al^{3+} ion coordinates with surrounding first shell water molecules in an octahedral and tetrahedral arrangement for pH values lower than 3 and higher than 7, respectively [72–75]. However, the coordination chemistry between a single Al^{3+} ion and first shell water molecules for pH values ranging between 3 and 7 has been the subject of much debate and recent studies such as kinetics experiments and ab initio molecular dynamics simulations show that a single Al^{3+} ion prefers to coordinate to five ligands in this pH range [76]. Furthermore, previous theoretical studies of the dimerization reaction mechanism of Al^{3+} ions in water have been performed using an octahedral geometry, which suggests that these polymerization mechanisms can be proposed for low pH values. However, most relevant chemical and environmental reactions of a single or poly-Al^{3+} species occur at pH values between 3 and 7. Two questions of interest are the mechanism of dimerization and the structure of the resulting dimer in water, that is, the arrangement of the water molecules and ions around the Al dimer.

In addition to studies of the mechanism of water dissociation induced by various metal ions, we present the structures of $Al(H_2O)_6^{3+}$ and $Al(H_2O)_4OH^{2+}$ in water and investigate polymerization in water via the modeling of two $Al(OH)(H_2O)_4^{2+}$ species in aqueous solutions using CPMD simulations and first principles calculations utilizing a continuum model for water (COSMO). These studies yield important insights into the structure, stability, and coordination chemistry of Al^{3+} ions in aqueous solution. We find that the reaction between two $Al(OH)(H_2O)_4^{2+}$ ions in aqueous solution yields the hydroxyl-ligand-bridged $Al_2(OH)_2(H_2O)_8^{4+}$ complex and that solution dynamics is a key factor in dimer formation. This study shows that water exchange and substitution reactions are important in the formation of Al ion polymers in water. Here, we demonstrate the impact of the dynamics and solvent molecules on the formation of the $Al_2(OH)_2(H_2O)_8^{4+}$ complex in aqueous solution (for pH values varying between 3 and 7).

4.2 METHODS

Dynamics calculations on a molecular system involve the time propagation of a set of nuclei according to an equation of motion. In the usual case where the equation of motion is Newton's equation, $F = ma$, the force (F) is required. The force is evaluated as the negative at the first derivative of the potential energy with respect to the atomic coordinates of the molecular system. In classical molecular dynamics calculations, the potential energy surface is usually some sort of classical force field [59,77,78]. In ab initio dynamics calculations, the potential energy surface is computed using some first principles method such as Hartree–Fock or density functional theory (DFT). The dynamics calculations reported in this chapter were performed using CPMD. The CPMD method is one of the most popular methods for simulations of liquids and solution systems. Among the many reasons for its popularity is its efficiency, which permits dynamics calculations on systems containing up to a few hundred atoms over tens of picoseconds. The most popular application of CPMD consists of Kohn–Sham DFT with a plane wave basis set for valence electrons and pseudopotentials for core electrons [37,41]. Thus, CPMD is ideally suited for the present study as it is an efficient method for performing dynamics calculations while maintaining the benefits of the DFT treatment of metal atoms. The CPMD methodology has been validated for use in liquid and solution studies. Simulations of pure water solutions by Geissler et al. [24] and previous simulations of metal ions in water are of particular interest.

All CPMD simulations of Al^{3+}, Cr^{3+}, Fe^{3+}, As^{3+}, and As^{5+} ions in water were performed in cubic cells containing 64 water molecules (the same number of water molecules is used in all simulations studies to facilitate meaningful comparison of the results) with periodic boundary conditions. In each of these simulations, the water density corresponds to 1 g cm^{-3}. We used the White and Bird parameterization (PBE96) of the Perdew–Burke–Ernzerhof functional that has proved to be efficient in describing metal–water interactions and Hamann pseudopotentials [65,66]. The electronic wave functions were expanded in a plane wave basis set with a kinetic energy cutoff of 140 Ry. A time step of 0.1 fs was used and the electronic mass was set to 750 au. A homogeneous background charge was applied to compensate the ionic charge. The simulations were performed at 300 K and at a pressure of 0.1 MPa using the Nose-Hoover thermostat for 40 ps; statistics were collected for the last 30 ps. Long-range interactions were treated with the Ewald mesh method. The tunneling and zero-point motion should not affect the qualitative nature of the coordinates as discussed previously by Geissler et al. and by Marx and coworkers [24,41].

The timescale on which water autoionization or dissociation occurs greatly exceeds the capabilities of state-of-the-art computers and algorithms. Thus, to observe the dissociation of water using CPMD, a sampling method that is able to locate and focus on rare events is needed. TPS is capable of precisely this type of sampling, and we discuss some of the features of this method. TPS calculations can be applied without prior knowledge of the dissociation mechanism, as demonstrated by Chandler and coworkers and the present authors [79–88]. CPMD simulations were used to generate trajectories. Attention was restricted to trajectories on a 200 fs timescale. We took the neutral and charged states of water as two stable states as required by the TPS algorithm. TPS focuses on the transition between these two states in which the system jumps from one state to the other by some pathway. Given the large number of degrees of freedom in a solvated system like we consider here, many pathways that join the two stable states are possible. After assigning a probability to each of the transition pathways, a Monte Carlo method is used along the path to generate an ensemble of all transition paths. All structural and thermodynamic properties can be calculated from these ensembles, such as the Gibbs free energy and rates of chemical reactions [80]. It is also possible to extract reaction mechanisms from these calculations. The TPS method is detailed and specialized and the interested reader is encouraged to refer to the literature for additional details of its implementation and application [65,66,80].

In our studies, the initial state is defined as the neutral state of water and the product is defined as water ions coordinate to a metal ion. Thus, pathways involving a transition from neutral water to separated water ions are accepted during the course of the simulation. Details can be found in Ref. [80]. In this study, 60 trial pathways per time step were used for each metal ion solution [65].

Structural studies of hydrated metal ions in water and of aqua ions (Zundel and Eigen complexes) are performed via bond distance and bond angle calculations using trajectories obtained from CPMD simulations coupled with TPS calculations. Radial distribution functions of metal-water oxygen and metal-water hydrogen were calculated using trajectories obtained from our simulations. The coordination numbers (CNs) of water molecules in the first and second hydration shells are calculated as integrals over correlation functions using [89,90]

$$CN = 4\pi\rho \int_0^{r_{min}} r^2 g(r)dr \qquad (4.1)$$

where

    ρ is the number density
    r is the distance
    $g(r)$ is the correlation function

Analysis of the geometric structures obtained from our simulations enable an under-standing of the hydration characteristics of the metal ion and the structural proper-ties of aqua ions. Results obtained for each metal ion solution are compared to each other in order to understand the effect of the size and charge of the metal ion on the hydration structures of the metal ions and aqua ions.

We now turn to the computation of free energy in our metal ion solutions. In gen-eral, the lack of a corresponding microscopic analogue, that is, a function of configu-ration space variables that may be calculated to obtain the free energy, creates one of the principal challenges in the determination of free energy of dense systems. Much effort has gone into the development of free energy calculation methods. Popular methods include the particle insertion, particle deletion scheme, and perturbation methods [22,23,65,90]. Using our implementation of well-known classical molecular simulation free energy methods with CPMD simulations, the preference for Zundel or Eigen complexes in metallic solutions was calculated via the perturbation method. The perturbation method gives the free energy change between two states i and $i+1$ with Hamiltonians H_i and H_{i+1} as

$$\Delta G\left(\lambda_{i+1} - \lambda_i\right) = -kT \ln \left\langle \exp\left[\frac{-(H_{\lambda_{i+1}} - H_{\lambda_i})}{kT} \right] \right\rangle_{\lambda_i} \qquad (4.2)$$

where λ is a continuous coupling parameter such that $\lambda=0$ at the initial state and $\lambda=1$ at the final state. Perturbations are broken into intermediate steps in which the change in the energy does not exceed the value of kT (k is the Boltzmann constant and T is the temperature). For more details see Ref. [24].

For dimerization studies of Al^{3+} ions in water, the Becke-Lee-Yang-Parr (BLYP) gradient corrected functional was used with Troullier–Martins pseudopotentials. The initial trajectory was taken from an equilibrated classical molecular dynamics simu-lation of a single Al^{3+} ion solvated by 64 water molecules at $300\,K$ and at a pressure of $0.1\,MPa$ [65,66]. A modified TIP5P model for water and the UFF (Universal Force Field) potential function were used for the classical molecular simulations of Al ion in water and Lorentz–Berthelot mixing rules were applied for calculating the cross parameters [1,90]. The system was then equilibrated for 15 ps using CPMD followed by a 35 ps production run using a time step of 0.1 fs and an effective electron mass of 700 au. A homogeneous background charge was applied to compensate for the ionic charge. The temperature was controlled using a Nose-Hoover thermostat, and the Ewald mesh method was used for long-range interactions [47]. The valence electron wave functions were expanded in plane waves with an energy cutoff of 140 Ry. In agreement with the recent studies of Swaddle et al., a hexa-coordinate $AlOH^{2+}$ ion is formed via the removal of a single proton from the $Al(H_2O)_6^{3+}$ structure in water [76]. The penta-coordinate $Al(H_2O)_4OH^{2+}$ complex was obtained from the hexa-coordinate $Al(H_2O)_5OH^{2+}$ structure during the course of our simulation (see Section 4.3 for details). In order to study the polymerization reaction mechanism of two Al^{3+} ions in water in detail, we performed separate CPMD simulations of two $Al(OH)(H_2O)_4^{2+}$ species with 118 water molecules for 60 ps with initial trigonal bipyramidal $Al(OH)$ $(H_2O)_4^{2+}$ structures obtained from our CPMD simulations (see Section 4.3 for details). To study the thermodynamic preference of Al^{3+} polymerization in water, the perturbation method was used with CPMD simulations utilizing Equation 4.2. For determining the free energy difference in the formation of Al_2 species via the distance between the metal centers of two $Al(OH)(H_2O)_4^{2+}$ ions in water using the perturbation method, the perturbations were broken into a series of intermediates with a $\Delta\lambda$ value of 0.05.

Ab initio calculations on geometries extracted from trajectories obtained from our CPMD simulations were performed using the GAMESS program utilizing COSMO. The first phase of our studies focused on the calculations of the hydrated Al^{3+} ion in the gas phase and we computed the gas phase enthalpy of formation, ΔE_f, from the chemical reaction [1]

$$Al^{3+} + 6H_2O \rightarrow Al(H_2O)_6^{3+}$$

The results obtained using various first principles methods utilizing the cc-pVTZ basis set are given in Table 4.1.

Calculations of the gas phase enthalpy were made using

$$\Delta H = \Delta E_f + \Delta E_{ZPE} + T\Delta c_v + \Delta(RT) \tag{4.3}$$

where
ΔE_{ZPE} is the zero point energy
$T\Delta c_v$ is the heat capacity contribution
$\Delta(RT)$ is the work term

TABLE 4.1
Thermodynamic Properties of $Al(H_2O)_6^{3+}$ Ion in the Gas Phase and in Aqueous Solution Using a Continuum Model for Water via Employing the PBE Method with the cc-pVTZ Basis Set

Property (kJ mol^{-1})	Gas Phase	Aqueous Solution
$\Delta E_o = \Delta E_f + \Delta E_{ZPE}$	−2861.7	−4639.7
$\Delta E_{298.15}$	−2886.8	−4655.2
ΔH	−2903.5	−4671.8

ZPE denotes zero point energy.

To calculate the hydration enthalpy for the $Al(H_2O)_6^{3+}$ ion and to compare with the available experimental data, we used the Born–Haber cycle presented in Equation 4.4 [1]

$$
\begin{array}{ccc}
Al^{3+} \text{ (gas)} + 6H_2O \text{ (gas)} & \xrightarrow{\ \Delta H_{gas}\ } & Al(H_2O)_6^{3+} \text{ (gas)} \\[2mm]
\Big\uparrow 6\Delta H_{vap} & & \Big\downarrow \Delta H_{solv} \qquad (4.4) \\[2mm]
Al^{3+} \text{ (gas)} + 6H_2O \text{ (aq)} & \xrightarrow{\ \Delta H_{hyd}\ } & Al(H_2O)_6^{3+} \text{ (aq)}
\end{array}
$$

The values computed using this cycle are presented in Table 4.1. The solvation energy for water obtained in these calculations is −38.5 and −37.9 kJ mol^{-1} using the PBE/cc-pVTZ and MP2/cc-pVTZ methods, respectively. These values are slightly larger than the experimental value (−44.1 kJ mol^{-1}), but are in better agreement with the experimental data than an earlier theoretical study, which reported values of −28.6 kJ mol^{-1} using the BLYP/SVP method and the PCM (Polarizable Continuum Model) model for water. The experimental hydration enthalpy of −4668.3 kJ mol^{-1} agrees well with our calculated value of −4671.8 kJ mol^{-1}. Free energies calculated in aqueous solution can be sensitive to the radius of the cavity surrounding the ions embedded in the continuum. We performed calculations for fitting the van der Waals radius of the Al^{3+} ion to the experimental free energy value of −4619.1 kJ mol^{-1}. In these calculations, the van der Waals radius was fit such that

$$\Delta G\left[Al^{3+}\right] = \Delta G_{water}\left[Al^{3+},r\right] - \Delta G_{gas}\left[Al^{3+},r\right] \qquad (4.5)$$

where r is the van der Waals radius of the Al^{3+} ion. The optimized van der Waals radius is 1.33 Å, in good agreement with the studies of Saukkoriipi et al. [74]. Our benchmark studies show that the PBE method using the cc-pVTZ basis set with an optimized van der Waals radius of 1.33 Å for the Al ion with the COSMO method reproduces the experimental hydration enthalpy (see Equation 4.4) for this system.

4.3   RESULTS AND DISCUSSION

The structures of hydrated metal ions, mechanisms of water dissociation in the vicinity of transition metal ions, and the structural and thermodynamic properties of aqua ion (Zundel and Eigen) complexes formed in the vicinity of the metal ion in aqueous solution were predicted utilizing various CPMD simulations along with TPS calculations.

4.3.1   STUDIES OF THE MECHANISM OF WATER DISSOCIATION IN AQUEOUS SOLUTIONS OF FE^{3+} AND CR^{3+} IONS

The mechanisms of water dissociation around Cr^{3+} and Fe^{3+} ions obtained from our simulations are depicted in Figure 4.1. In these simulations, we observe an elongation of the O–H bond in the first hydration shell (Figure 4.1A). It is this event that initiates the ionization of the water molecule. In the following 80 fs, the proton moves away from the metal ion leaving the ion coordinated to hydroxyl groups (Figure 4.1B). Both Zundel and Eigen complexes are formed once the proton moves to the second solvation shell. Within 30 fs, Zundel complex formation is observed in both metal ion solutions via hydrogen bond fluctuations (Figure 4.1C). The Eigen complex is formed after 50 fs in both solutions via proton transfer between second and third hydration shells (Figure 4.1D). Unlike the Al^{3+} solution, no immediate return path to the undissociated state of water is observed once the Zundel and Eigen complexes are formed around Cr and Fe ions in water during our simulation. We observe fast interconversion between the Zundel and the Eigen complexes in agreement with the theoretical and computational studies of a proton in pure water.

The Zundel complex is formed in the second hydration shell and the Eigen complex is located between the second and third shells. The Zundel complex is 1.5 Å closer to these transition metal ions than the Eigen complex. Unlike earlier calculations, which proposed a symmetric Zundel complex geometry, we note that the Zundel complex in these transition metal ion aqueous solutions is asymmetric. The H^+ is coordinated to two water molecules with average H^+–O distances of 1.16 and 1.19 Å. The slight difference is influenced by the close proximity of the Fe and Cr ions in the aqueous solution. The oxygen atoms that are separated by the hydrogen-bonded proton are separated by 2.2 Å. This value is not significantly different from other computed values reported for the Zundel complex. In addition, the asymmetric binding of the proton to the two water molecules has been reported in recent theoretical studies using state-of-the-art first principles methods. The average H^+–O distance is larger for the Eigen complex than for the Zundel complex, with a value of 1.59 Å in the vicinity of Fe and Cr ions. This trend is in excellent agreement with studies performed for pure water. Overall, the structural parameters that we report here are in excellent agreement with experimental and recent theoretical studies for pure water, with a difference in structural parameters of 1%–3%.

Calculations of the change in the Gibbs free energy associated with proton transfer in aqueous solutions of Fe and Cr using perturbation theory (see Equation 4.2) show that the Zundel complex is thermodynamically more stable than the Eigen complex by 2.6 and 3.2 kJ mol^{-1}, respectively. Recalling that these simulations and calculations were performed at room temperature, we note that these Gibbs free energy difference values are small, similar to kT (2.5 kJ mol^{-1}), which indicate that the aqua ion complexes might coexist.

The structures for Cr^{3+} and Fe^{3+} in water are all in octahedral arrangements. Specifically, we find that two and one hydroxyl groups (OH) are coordinated to the Fe^{3+} and Cr^{3+} ions on the average, respectively. Our results indicate a lower pH value for the Fe^{3+} solution as compared to the pH of the Cr^{3+} solution. In fact, pK_a values and hydrolysis constant values support these findings, which indicate that hydroxylation should be easier for the Fe^{3+} ion than for the Cr^{3+} ion in aqueous solution. Structural parameters for the first and second shell water molecules around these two transition metal ions are presented in Table 4.2. The predicted structural properties for the hydrated Fe^{3+} and Cr^{3+} ions in water agree with spectroscopic measurements within 0.2%.

TABLE 4.2
Specific Structural Parameters for the
Aqueous Solutions of the Cr^{3+} and the
Fe^{3+} Ions

Cr^{3+}		Fe^{3+}	
R_{Cr-O_2} (Å)	3.91 ± 0.03	R_{Fe-O_2} (Å)	3.96 ± 0.04
$R_{O_2-H_2}$ (Å)	0.98 ± 0.01	$R_{O_2-H_2}$ (Å)	0.99 ± 0.01
$R_{O_1-O_2}$ (Å)	2.63 ± 0.03	$R_{O_1-O_2}$ (Å)	2.64 ± 0.02
R_{HB} (Å)	1.65 ± 0.04	R_{HB} (Å)	1.63 ± 0.05

HB denotes the hydrogen bond distance between the first and
   second shell water molecules in the vicinity of these two
   transition metal ions.

4.3.2 SINGLE ION AND DIMERIZATION STUDIES OF Al^{3+} IN WATER

For the studies of the Al^{3+} ion in water, we performed several simulations. To model
Al ions in water at the pH of interest ($3 < pH < 7$), we used the octahedral $Al(H_2O)_6^{3+}$
structure (Figure 4.2A) obtained from our simulations with 58 water molecules,
removing one H atom from a first shell water molecule that is coordinated to the Al^{3+}
ion. The resulting aqueous $Al(H_2O)_5OH^{2+}$ structure loses one water molecule and
becomes a penta-coordinate $Al(H_2O)_4OH^{2+}$ structure (Figure 4.2B) within 0.8 ps.
This finding is in excellent agreement with recent kinetics experiments and CPMD
simulations, which report a penta-coordinated $Al(OH)(H_2O)_4^{2+}$ complex in aqueous
solution. Our studies show that the penta-coordinated complex converts to a trigonal
bipyramidal structure (Figure 4.2C) in water within the following 1.7 ps.

   Additional CPMD simulations involving breaking a HO–H bond of a first shell
water molecule were performed in order to understand the impact of the free proton
on the solution dynamics and coordination chemistry. This simulation shows that
the proton moves to the second shell and forms a Zundel-like structure. However,

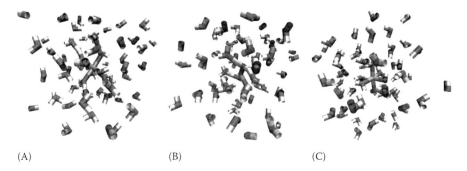

(A)   (B)   (C)

FIGURE 4.2 Octahedral $Al(H_2O)_6^{3+}$(A), square pyramidal $Al(OH)(H_2O)_4^{2+}$ (B), and trigonal bipyramidal $Al(OH)(H_2O)_4^{2+}$ (C) species in aqueous solution, obtained from CPMD
simulations.

the lifetime of the Zundel-like complex is extremely short ($<0.3\,fs$) and the proton migrates back to the OH^- coordinated to the metal ion, forming a neutral water molecule and the initial octahedral $Al(H_2O)_6^{3+}$ geometry in aqueous solution. We do not observe a clear water dissociation mechanism around the Al^{3+} ion, which is in agreement with our previous studies. These findings are in agreement with the work of Lubin et al., who studied the octahedral $Al(H_2O)_6^{3+}$ and the tetrahedral $Al(OH)_4^-$ ionic species in aqueous solution via CPMD simulations and did not observe a clear water dissociation mechanism in the presence of Al ions [73].

Table 4.3 lists several structural parameters for the Al ions in octahedral $Al(H_2O)_6^{3+}$ and trigonal bipyramidal $Al(OH)(H_2O)_4^{2+}$ structures in aqueous solution obtained from our CPMD simulations. The structural parameters are in very good agreement with experiments as well as with previous CPMD simulations for these species in aqueous solution. Based on our CPMD simulation results, the average distance between the oxygen atoms of the first shell water molecules (O_I) and the Al ion in water is 1.95 and 1.98 Å for the octahedral $Al(H_2O)_6^{3+}$ and the trigonal bipyramidal $Al(OH)(H_2O)_4^{2+}$ structures, respectively. The bond distances for the $Al(H_2O)_6^{3+}$ ion in water are slightly larger than the reported x-ray data (1.88–1.90 Å) for the octahedral geometry. However, our values are in excellent agreement with previously reported CPMD simulation results for the octahedral $Al(H_2O)_6^{3+}$ structure in water. As shown in Table 4.3, the $Al–O_I$ bond distance does not change significantly for the optimized octahedral $Al(H_2O)_6^{3+}$ and the trigonal bipyramidal $Al(OH)(H_2O)_4^{2+}$ structures using DFT calculations with a COSMO. This finding indicates that the use of explicit or implicit water models does not significantly change the predicted hydrated Al ion structures.

The average HO–H bond lengths for the first shell water molecules are similar, 0.99 and 1.01 Å for the octahedral $Al(H_2O)_6^{3+}$ and trigonal bipyramidal $Al(OH)(H_2O)_4^{2+}$ species in aqueous solution, respectively, and are only slightly different than results obtained from static DFT calculations utilizing a COSMO (Table 4.3). The second shell water molecule oxygen atoms (O_{II}) to Al^{3+} distance for the octahedral $Al(H_2O)_6^{3+}$ complex in aqueous solution is 4.12 ± 0.02 Å. This value becomes slightly smaller (4.05 ± 0.04 Å) for the trigonal bipyramidal $Al(OH)(H_2O)_4^{2+}$ species in water. These findings are in agreement with x-ray data. In addition, the hydrogen bond lengths between the first and second shell water molecules around the octahedral Al^{3+} ion are 2.69 ± 0.02 Å in our CPMD simulations. This distance is 1.4% smaller than the distance in the aqueous $Al(OH)(H_2O)_4^{2+}$ solution. The second shell CN of water molecules varies between 8 and 12 and 7 and 14 around Al in octahedral $Al(H_2O)_6^{3+}$ and Al in trigonal bipyramidal $Al(OH)(H_2O)_4^{2+}$ arrangements in aqueous solution, respectively. The computed CNs are in agreement with experimental and theoretical studies.

In general, the importance of polymerization reactions of Al is widely accepted as described above. However, a detailed molecular-level understanding of the early stages of the polymerization reactions is lacking. Such information can be useful in the design of more efficient techniques for wastewater treatment or for the removal of metal ions before they form the more stable metal ion polymers in solution, which can be toxic as explained in Section 4.1. In fact, few theoretical studies have successfully studied possible reactions between two hydrated Al ions. Qian et al. studied the reaction mechanism of hydrated Al^{3+} ions with static DFT calculations in the gas phase and using a COSMO [91]. They predicted that the octahedral $Al(H_2O)_6^{3+}$

TABLE 4.3

Structural Parameters for the Octahedral $Al(H_2O)_6^{3+}$ and Trigonal Bipyramidal $Al(OH)(H_2O)_4^{2+}$ Species Obtained from First Principles Calculations Performed in Water Using a COSMO and from CPMD Simulations in Aqueous Solution

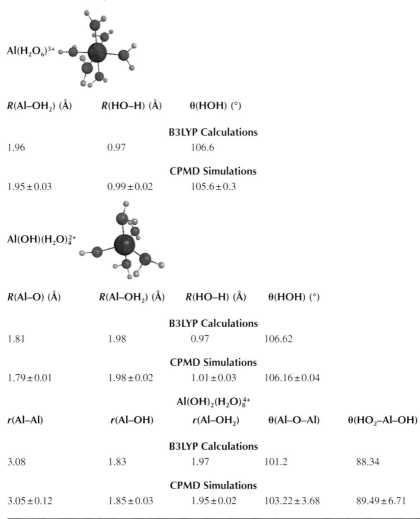

$Al(H_2O_6)^{3+}$

$R(Al-OH_2)$ (Å)	$R(HO-H)$ (Å)	$\theta(HOH)$ (°)
	B3LYP Calculations	
1.96	0.97	106.6
	CPMD Simulations	
1.95 ± 0.03	0.99 ± 0.02	105.6 ± 0.3

$Al(OH)(H_2O)_4^{2+}$

$R(Al-O)$ (Å)	$R(Al-OH_2)$ (Å)	$R(HO-H)$ (Å)	$\theta(HOH)$ (°)
	B3LYP Calculations		
1.81	1.98	0.97	106.62
	CPMD Simulations		
1.79 ± 0.01	1.98 ± 0.02	1.01 ± 0.03	106.16 ± 0.04

		$Al(OH)_2(H_2O)_8^{4+}$		
$r(Al-Al)$	$r(Al-OH)$	$r(Al-OH_2)$	$\theta(Al-O-Al)$	$\theta(HO_2-Al-OH)$
		B3LYP Calculations		
3.08	1.83	1.97	101.2	88.34
		CPMD Simulations		
3.05 ± 0.12	1.85 ± 0.03	1.95 ± 0.02	103.22 ± 3.68	89.49 ± 6.71

structure loses neutral water molecules via substitution and that water exchange reactions occur that drive the formation of $Al_2(OH)_2(H_2O)_8^{4+}$ species. To study the dimerization of Al ions at pH values of 3–7, we performed separate CPMD simulations in water with two $Al(OH)(H_2O)_4^{2+}$ complexes in a trigonal bipyramidal configuration for which the initial geometries were taken from our CPMD simulations as described above.

Initially, the Al ion centers are separated from one another by 7.1 Å. After 5.2 ps, the distance between the Al ions decreases to 5.2 Å. The calculations of the free energy indicate that Al ions separated by two water molecules at a distance of 5.2 Å from one another (Figure 4.3B) are slightly more stable (4.6 kJ mol^{-1}) than the Al ions separated by 7.1 Å (Figure 4.3A). However, this free energy change value is

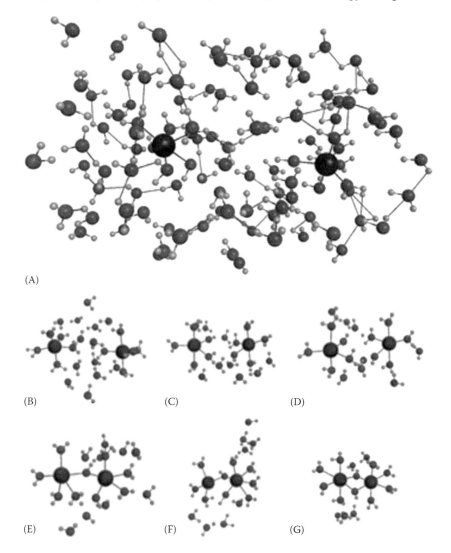

(A)

(B)   (C)   (D)

(E)   (F)   (G)

FIGURE 4.3 Two Al(OH)(H$_2$O)$_4^{2+}$ species in aqueous solution (A), two trigonal bipyramidal Al(OH)$^{2+}$ ions at a distance of 7.1 Å (B), water-separated Al(OH)$^{2+}$ ions at a r_{Al-Al} distance of 5.2 Å (C), formation of a Zundel complex (H$_5$O$_2^+$) and loss of one water molecule from one Al center (D), formation of one OH bridge between the Al centers and formation of a second Zundel-like complex (E), loss of one water molecule coordinated to the Al center and the formation of an Eigen-like aqua complex (F), and formation of OH-bridged Al$_2$(OH)$_2$(H$_2$O)$_8^{4+}$ complex in water (G) obtained from the CPMD simulations.

comparable to kT, which indicates that both configurations are likely to exist in aqueous solution. The hydroxyl oxygen atoms, which are connected to Al ions, are hydrogen bonded to water molecules that are between the Al centers, which in turn interact with the neutral water molecules coordinated to the neighboring Al ion via hydrogen bonding (Figures 4.3B through D). We observe an elongation of the Al–OH_2 bond of 0.2 Å after 1.3 ps (Figure 4.3D), which is in agreement with our previous studies. The proton attached to the water molecule forms a Zundel-like complex with the neighboring water molecule in which the proton is coordinated both to the oxygen atom of the neutral water molecule and to the oxygen atom hydrogen bonded to the Al ion (Figure 4.3D). Next, the Al–OH_2 bond distance for the Al-coordinated water molecule, which participates in a Zundel-like structure, increases by 0.3 Å and the Zundel-like complex moves to the second shell via proton transfer where it forms Eigen-like complexes between the second and third shells within the next 0.2 ps (Figure 4.3E). The system does not interconvert between Zundel and Eigen complexes and an immediate return to the neutral state of water is observed in the next few femtoseconds unlike the proton transfer mechanism found in aqueous solutions of transition metal ions (see above). We observe an elongation of another H–OH bond coordinated to the second Al center, and a second Zundel-like complex formation. The distance between Al ions decreases to 3.6 Å in this structure and the first hydroxyl oxygen atom coordinates to the next Al center (Figure 4.3F). It takes a slightly shorter time (<0.4 ps) for the second hydroxyl oxygen atom to coordinate to the neighboring Al center (Figure 4.3G). Once the aqua ions move to the second and third shells, another water molecule from the second shell coordinates to one Al center within the next 1.4 ps at a distance of 2.12 Å from the Al center (Figure 4.3D). The coordination of the second shell water molecule to the second Al ion takes 0.6 ps longer compared to the coordination of a neutral water molecule to the first Al center (Figure 4.3G). In addition, the distance from this coordinating water molecule oxygen to the Al center is slightly larger with a value of 2.17 Å. After 5.8 ps, the distance between the Al ions becomes smaller (3.05 Å) and the aqueous dimer $Al_2(OH)_2(H_2O)_8^{4+}$ structure is formed with an average Al-(OH)-Al bond angle of 103.2°. Calculations of free energy indicate that the $Al_2(OH)_2(H_2O)_8^{4+}$ structure is thermodynamically preferred (by 10.4 kJ mol^{-1}) over the existence of the water-separated single $Al(OH)(H_2O)_4^{2+}$ species with Al ions separated by 5.2 Å in aqueous solution (Figure 4.4). Overall, these results support previous experimental observations and quantum mechanical calculations, which proposed hydroxyl-bridged poly-Al structures in water. Our simulations indicate that both water exchange and substitution reactions take place in the formation of Al ion dimers in aqueous solution at room temperature and a pressure of 0.1 MPa, supporting previous reaction mechanisms proposed on the basis of static quantum calculations.

For the Zundel-like complex, which we observe in the second shell for a very short time, we note that the proton is asymmetrically coordinated to the neutral water molecule oxygen atom and to the oxygen atom attached to Al with H$^+$–O distances of 1.18 and 1.15 Å, respectively. The O–O distance (water molecule oxygen atom to an oxygen atom attached to Al) is 2.22 Å. The Eigen complex that is formed for a very short time between the second and third shells has H$^+$–O distances of 1.59 and 1.62 Å with an average O–O distance (the distance between the water molecule

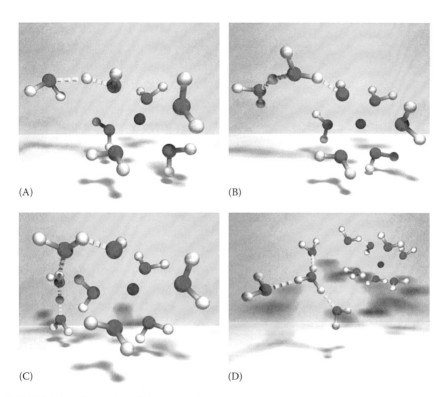

(A) (B)

(C) (D)

FIGURE 4.4 Illustration of the water dissociation mechanism around both As ions in aqueous solution at room temperature obtained from CPMD simulations coupled with TPS calculations. The water molecules were removed to provide a clear presentation. Various steps in the mechanism of water dissociation about As^{3+} are depicted. The mechanism begins with the hydroxylation of the As ion (A) and leads to the formation of a Zundel complex in the first shell (B). A proton transfer leads to the disappearance of the Zundel complex in the first shell and to the formation of another Zundel complex in the second shell (C). Finally, proton migration leads to the formation of an Eigen complex (D). (Coskuner, O. et al., *Dynamic and Structural Properties of Aqueous Arsenic Solutions*, 1187, 2009. Wiley-VCH Verlag GmbH & Co. KGaA. Reproduced with permission.)

oxygen atoms to which the proton is coordinated) of 2.59 Å. These structural parameters do not deviate significantly from previous calculations for pure water and transition metal ion. It is interesting to note that we do not observe a distinct water dissociation event around the Al^{3+} ion because the charged water ions rapidly form the neutral state of water between the second and third shells. This finding does not indicate that water dissociation does not occur in Al solutions. However, it indicates that special sampling methods with extremely large transition pathway sampling may be required to predict the possible pathways of water dissociation around hydrated poly-Al ions.

The structural parameters of the $Al_2(OH)_2(H_2O)_8^{4+}$ complex are depicted in Table 4.3. Based on our CPMD simulation results, the average distance between the two Al ions in $Al_2(OH)_2(H_2O)_8^{4+}$ is 3.05 Å.

TABLE 4.4

Calculated NPA Values for Al and for the Coordination Sphere Atoms of the Octahedral $Al(H_2O)_6^{3+}$, Trigonal Bipyramidal $Al(OH)(H_2O)_4^{2+}$, and $Al_2(OH)_2(H_2O)_8^{4+}$ Structures Using the B3LYP Method and 6-31g** Basis Set

$Al(H_2O)_6^{3+}$		$Al(OH)(H_2O)_4^{2+}$		$Al_2(OH)_2(H_2O)_8^{4+}$	
Al	$+1.99 \pm 0.05$	Al	$+1.93 \pm 0.04$	Al_1	$+1.84 \pm 0.03$
O	-1.02 ± 0.06	$O_{hydroxyl}$	-1.26 ± 0.03	Al_2	$+1.81 \pm 0.04$
		O_{water}	-1.03 ± 0.02	$O_{hydroxyl}$	-0.99 ± 0.09
				O_{water}	-1.02 ± 0.01

The distance between oxygen atoms of the first shell water molecules (O_1) and Al in water is 1.99 Å. This is a slightly larger distance than the corresponding values calculated for the monomers (see above). Furthermore, the distance between one of the coordinating water molecule oxygen atom and one Al center in $Al_2(OH)_2(H_2O)_8^{4+}$ varies between 2.08 and 2.15 Å. These calculations may indicate that the hydrated monomer Al ion structures are more compact in comparison to the hydrated structure of the Al dimer in water. The average Al–O_1 bond distances become slightly smaller when we isolate the hydrated inorganic structure from bulk water and re-optimize the $Al_2(OH)_2(H_2O)_8^{4+}$ structure using a COSMO (Tables 4.2 and 4.3). In addition, the hydrogen bond length between the first and second shells of water molecules around the Al^{3+} ions in $Al_2(OH)_2(H_2O)_8^{4+}$ is 2.71 ± 0.03 Å, indicating that the structural properties of bulk water molecules do not change significantly in comparison to the results obtained for single Al ion solutions (see above). Overall, the calculated structure for $Al_2(OH)_2(H_2O)_8^{4+}$ is in agreement with the previous results [62].

Natural partial charges calculated via the natural population analysis (NPA) method for certain atoms of the aqueous octahedral $Al(H_2O)_6^{3+}$, aqueous trigonal bipyramidal $Al(OH)(H_2O)_4^{2+}$, and aqueous $Al_2(OH)_2(H_2O)_8^{4+}$ structures using the configurations obtained from CPMD simulations are listed in Table 4.4.

Although the metal center possesses a formal oxidation state of +3, the partial charges determined using NPA are lower, indicative of donation of electron density from the ligands to the Al center. We note the larger partial positive charge on Al in $Al(H_2O)_6^{3+}$ relative to the partial charge values for Al in trigonal bipyramidal $Al(OH)$ $(H_2O)_4^{2+}$ and in $Al_2(OH)_2(H_2O)_8^{4+}$ structures. These results indicate that the partial positive charge on Al decreases with increasing number of coordinating hydroxyl groups. Furthermore, the partial negative charge on the oxygen atom in OH is larger than the charge on the oxygen atom in neutral water molecule. These results support the structural parameters and thermodynamics results presented above. The larger electron density donation from coordinated OH groups to Al ions may influence the formation of hydroxyl-bridged Al dimers rather than the formation of water-bridged Al dimers in aqueous solutions.

4.3.3 MECHANISM OF WATER DISSOCIATION IN AQUEOUS SOLUTIONS OF As^{3+} AND As^{5+} IONS

To study the mechanism of water dissociation in As^{3+} and As^{5+} solutions, we employed CPMD simulations coupled with TPS calculations as in our previous investigations of the mechanism of water dissociation induced by Fe^{3+} and Cr^{3+} ions and the impact of dissociated waters on the hydration structure and coordination chemistry of these metal ions. Unlike our Al ion solution studies, but similar to our Fe and Cr solution studies, we observe proton transfer in As solutions. In agreement with our studies of Fe and Cr solutions, in As solutions the dissociation mechanism is initiated via an elongated H–OH bond that subsequently breaks and releases a proton. This leads to hydroxylation of the As ion in water (see Figure 4.4A). After the release of the proton, proton transfer via hydrogen bonding interactions initiates Zundel and Eigen complex formation in water for the aqueous solutions of both As^{3+} and As^{5+} ions. We note that the Zundel complex is formed first via hydrogen bonding of the proton to two surrounding water molecules (see Figure 4.4B). It takes the proton 60 fs to form the Zundel complex in water around the As^{3+} ion, whereas this time is 11 fs longer around the As^{5+} ion. A second Zundel complex is formed between the first and second shells after 38 fs in the vicinity of the As^{3+} ion. The same mechanism is observed around the As^{5+} ion within the following 33 fs (see Figure 4.4C). It takes the system slightly longer (51 fs) in comparison to the dynamics observed in the first and second shells to move the proton between the second and third shells and to form the Eigen complex (see Figure 4.4D). Based on our CPMD/TPS calculations, we observe that the Eigen complex is also formed around the As^{5+} ion, between the second and third shells. The proton transfer that leads to the formation of the Eigen complex takes a slightly shorter time (42 fs) in the vicinity of the As^{5+} ion as compared to the As^{3+} ion. These variations in time might be attributed to differences in ionic charge.

In agreement with our studies on the mechanism of water dissociation induced by Fe^{3+} and Cr^{3+} ions, we find that the Zundel complex is the first water ion complex to be formed and that the Eigen complex is formed afterward in the aqueous solutions of both As ions. We note that the Zundel complex is formed in the first and second shells, whereas the Eigen complex is formed in the second and third hydration shells in the vicinity of both As^{3+} and As^{5+} ions in aqueous solution. Furthermore, the same trends were also obtained from our CPMD/TPS calculations for Fe^{3+} and Cr^{3+} ions in aqueous solution. However, we did not distinguish a clear water dissociation mechanism in the aqueous solutions of Al ions, which might be attributed to the lack of partially or half-filled d orbitals, the ionic radius of the metal ion, or the differences in charge transfer as compared to the transition metal ion solutions.

To understand the thermodynamic preference between the Zundel and Eigen complexes formed in the vicinity of the As^{3+} and As^{5+} ions in aqueous solution, we performed calculations of free energy (Equation 4.2). These calculations show that both Zundel and Eigen complexes can exist in the solution with a free energy difference of less than 0.7 kJ mol^{-1} around the As^{3+} ion and that the Zundel complex is only slightly more stable than the Eigen complex. For the aqueous solution of the As^{5+} ion, we find that the free energy difference becomes larger; the Zundel complex is preferred over the Eigen complex by 2.8 kJ mol^{-1}, which shows the same thermodynamic trend that

TABLE 4.5

Specific Structural Properties and CNs

Obtained for the As^{3+} and As^{5+} Solutions

	r_{H-O}^+ (Å)	R_{O-O} (Å)	r_{M-O} (Å)	CN
		As^{3+}		
Zundel	1.14 ± 0.01	2.19 ± 0.01	3.15 ± 0.39	4.6 ± 0.3
	1.20 ± 0.01			
Eigen	1.60 ± 0.02	2.65 ± 0.02	6.48 ± 0.47	6.8 ± 0.5
		As^{5+}		
Zundel	1.14 ± 0.02	2.26 ± 0.03	2.99 ± 0.28	4.9 ± 0.2
	1.20 ± 0.01			
Eigen	1.61 ± 0.02	2.65 ± 0.02	5.82 ± 0.50	7.2 ± 0.7

we found for the stability of the Zundel and Eigen complexes in the aqueous solutions of the Fe^{3+} and Cr^{3+} ions. We note that the difference in the energy is comparable to kT and that both complexes might exist Co$^-$ in these solutions.

Thermodynamic calculations and structural property calculations of metal ion solutions have led to interesting findings—especially when considered together. For example, the proximity of the Zundel and Eigen complexes is related to trends in the free energy in As ion solutions. Zundel and Eigen complexes are formed closer to the metal ion around As^{5+} as compared to As^{2+}. In addition, the distance between the Zundel and Eigen complexes, which are formed in different hydration shells, is shorter in the As^{5+} solution than in the As^{3+} solution by 21%. We note that the CN of water molecules in the first hydration shell of both As ions is related to the proximity of Zundel and Eigen structures to the metal ion centers. Structural parameters for water ions in both As ion solutions are presented in Table 4.5 and we note that the structural parameters for the Zundel and Eigen complexes are very similar in both As solutions (the largest difference is 3.5%).

The first shell water molecule oxygen atom to As ion distance varies between 1.64 Å (As^{5+}) and 1.76 Å (As^{3+}) and shows agreement with x-ray absorption spectroscopy data [92]. Interestingly, we also find a neutral water molecule located at a distance of 2.65 Å from the As^{3+} ion whose distance is too large to belong to the coordinated water molecules, but is too small to belong to the second shell water molecules. This finding is supported by the most recent x-ray absorption experiments where it was observed that the scattering in the Fourier transforms of EXAFS spectra is long (>2.65 Å) [93]. Our calculations aid in the interpretation of this experiment since the unusually located water molecule could not be observed in the experiments.

Our simulations show that water dissociation affects hydrated metal ion geometries. Specifically, we observe CNs of 4.2 ± 0.6 and 4.8 ± 0.7 Å for the As^{3+} and the As^{5+} ions, respectively, and that they tend to form tetrahedral-like geometries. These findings are in agreement with experimental data [92–95]. Average hydroxylation numbers are 3.2 and 3.5 for the As^{3+} and the As^{5+} ions, respectively, indicating a lower pH for the As^{5+} solution than for the As^{3+} solution.

4.4   CONCLUSION

We report the results of CPMD simulations coupled with TPS calculations for study-ing the mechanism of proton transfer in aqueous solutions of Fe^{3+}, Cr^{3+}, Al^{3+}, As^{3+}, and As^{5+} ions. The CPMD/TPS technique enables studies of bond breaking and bond forming with great accuracy using first principles electronic structure methods for rare events such as water dissociation induced by metal ions. We found that both Zundel and Eigen complexes are formed in the metal ion solutions. However, we were not able to locate a dissociated state of water in the vicinity of Al ions. Our simulations show that a Zundel complex is formed in the first and second hydration shells and the Eigen complex is formed in the second and third hydration shells around the metal ions. We note that the formation of aqua ion complexes and the structural properties of these aqua ions depend on several factors: charge and size of the metal ion and the number of d orbital electrons. In addition, we find that the hydration structure around the metal ions is influenced by the behavior of the solvent. According to our studies, water dissociation affects the solvated metal ion geometries. Fe^{3+} and Cr^{3+} ions form octahedral geometries with average numbers of coordinated hydroxyl groups of 2.1 and 1.7 for the Fe^{3+} and Cr^{3+} ions, respectively. These results indicate a slightly lower pH for the Fe^{3+} ion solution than that for the Cr^{3+} solution. This trend is supported by hydrolysis constant and pK_a values. For the As solutions, we found that these metal ions form tetrahedral-like structures, and that a fifth water molecule becomes coordinated to the As^{5+} ion and then released into the second shell (and vice versa) during our simulation. These simulations show that the average number of hydroxyl groups coordinated to the As^{3+} and As^{5+} ions is 3.2 and 3.5, respectively. These results suggest a lower pH for the As^{5+} solution than for the As^{3+} solution. Atomic force spectroscopy, dialysis, and kinetics experiments support these findings [96,97]. We calculated the Gibbs free energy associated with proton transfer in metal ion solutions utilizing the perturbation method with ab ini-tio molecular dynamics simulations. These simulations predict the Zundel complex to be preferred over the Eigen complex for the Fe^{3+}, Cr^{3+}, and As^{5+} ion solutions. However, this trend is different in the vicinity of the As^{3+} ion, and both Zundel and Eigen complexes may exist at the same time in this solution since the difference in the Gibbs free energy difference is very small ($0.7\,kJ\,mol^{-1}$).

To address the effect of pH on metal ion solution properties and to study the dimerization molecular mechanism of Al^{3+} ions in water, we employed CPMD simu-lations and DFT calculations for pH values between 3 and 7. The insights gained from this study lend additional support for the trigonal bipyramidal $Al(OH)(H_2O)_4^{2+}$ structure, which was reported by recent experimental and CPMD simulation studies. We do not observe a clear water dissociation mechanism around single and poly-Al ions. Our simulations show that (short-lived) charged states for water convert to the neutral state of water around the aqueous $Al_2(OH)_2(H_2O)_8^{4+}$ structure. The structural parameters obtained for the short-live Zundel and Eigen complexes are similar to those reported for pure water and for transition metal ion solutions.

The hydrated single Al ion structures are in good agreement with the available experiments and support the kinetics experiments that report a penta-coordinated single Al ion for solutions with pH between 3 and 7 [98]. Our simulations show formation of hydroxyl-ligand-bridged Al polymers via reaction between $Al(OH)^{2+}$ ions in water, which may be found in solutions with pH values between 3 and 7. In addition, NPA reveals that the coordination of hydroxyl groups donates more electron density to Al than that to the water molecules, which favors the hydroxyl-ligand-bridged Al_2 dimer over the water-molecule-bridged metal dimer in aqueous solution. Another key finding from the simulations is that the hydroxyl-bridged Al_2 dimer is more stable than the water-separated single Al ions in water. Furthermore, a comparison of our CPMD simulation results with DFT calculations utilizing a COSMO shows that the structures of hydrated octahedral $Al(H_2O)_6^{3+}$, trigonal bipyramidal $Al(OH)(H_2O)_4^{2+}$, and $Al_2(OH)_2(H_2O)_8^{4+}$ do not change significantly, indicating that the use of implicit or explicit models for water does not significantly change the hydration structures for hydrated metal ions. Overall, our predicted $Al_2(OH)_2(H_2O)_8^{4+}$ structure agrees well with x-ray measurements. We find that one water molecule coordinated to Al in $Al_2(OH)_2(H_2O)_8^{4+}$ possesses a slightly longer distance to the metal center in comparison to the rest of the water molecules coordinated to both Al centers. Structural and thermodynamic studies for $Al_2(OH)_2(H_2O)_8^{4+}$ support previous experiments and theoretical investigations.

We have demonstrated that theoretical studies performed with the appropriate methods and at an appropriate level of theory can provide complementary insights into structural and thermodynamic studies including dynamics of hydrated metal ions and their polymers in aqueous solution. Such theoretical studies coupled with kinetics and spectroscopic measurements can help to identify the mechanisms of the reactions of metal ions in solution, which may lead to efficient processes for removing toxic metal ions from drinking water or for treating wastewater. In addition, metal ion solutions are important in organic and medicinal chemistry since they can act as catalysts or ligands to organic molecules and biomolecules. An understanding of the solution properties including the acid–base chemistry (pH) of these solutions along with the geometry of the hydrated metal ion can help in the design of more efficient syntheses and lead to a more detailed understanding of coordination chemistry.

In summary, studies of the mechanism of water dissociation remain challenging with experimental techniques due to the small size and short timescale associated with this process. Ab initio molecular dynamics simulations coupled with special sampling techniques enable computational study of these mechanisms as shown in this chapter.

REFERENCES

1. Coskuner, O.; Jarvis, E. A. A., Coordination studies of Al-EDTA in aqueous solution, *J. Phys. Chem. A* 2008, *112* (12), 2628–2633.
2. Coskuner, O.; Bergeron, D. E.; Rincon, L.; Hudgens, J. W.; Gonzalez, C. A., Identification of active sites of biomolecules. 1. Methyl-alpha-mannopyranoside and Fe-III, *J. Phys. Chem. A* 2008, *112* (13), 2940–2947.
3. Coskuner, O.; Bergeron, D. E.; Rincon, L.; Hudgens, J. W.; Gonzalez, C. A., Identification of active sites of biomolecules. II: Saccharide and transition metal ion in aqueous solution, *J. Phys. Chem. A* 2009, *113* (11), 2491–2499.
4. Das, D.; Samanta, G.; Mandal, B. K.; Chowdhury, T. R.; Chanda, C. R.; Chowdhury, P. P.; Basu, G. K.; Chakraborti, D., Arsenic in groundwater in six districts of West Bengal, India, *Environ. Geochem. Health* 1996, *18* (1), 5–15.

5. Piamphongsant, T., Chronic environmental arsenic poisoning, *Int. J. Dermatol.* 1999, *38* (6), 401–410.

6. Karim, M., Arsenic in groundwater and health problems in Bangladesh, *Water Res.* 2000, *34* (1), 304–310.

7. Paul, B. K.; De, S., Arsenic poisoning in Bangladesh: A geographic analysis, *J. Am. Water Resour. Assoc.* 2000, *36* (4), 799–809.

8. Chatterjee, A.; Mukherjee, A., Hydrogeological investigation of ground water arsenic contamination in South Calcutta, *Sci. Total Environ.* 1999, *225* (3), 249–262.

9. Brown, K. G.; Ross, G. L., Arsenic, drinking water, and health: A position paper of the American Council on Science and Health, *Regul. Toxicol. Pharm.* 2002, *36* (2), 162–174.

10. Nakadaira, H.; Endoh, K.; Katagiri, M.; Yamamoto, M., Elevated mortality from lung cancer associated with arsenic exposure for a limited duration, *Occup. Environ. Med.* 2002, *44* (3), 291–299.

11. Wang, J. P.; Qi, L. X.; Moore, M. R.; Ng, J. C., A review of animal models for the study of arsenic carcinogenesis, *Toxicol. Lett.* 2002, *133* (1), 17–31.

12. Saha, J. C.; Dikshit, A. K.; Bandyopadhyay, M.; Saha, K. C., A review of arsenic poisoning and its effects on human health, *Crit. Rev. Environ. Sci. Technol.* 1999, *29* (3), 281–313.

13. Holme, B.; Tafto, J., Mechanism of transformation from Fe(3−x)A1(x)O(4) to porous Fe by exposure to H-2 gas, *Phil. Mag. A* 2000, *80* (2), 373–387.

14. Hill, H. G. M.; Nuth, J. A., The catalytic potential of cosmic dust: Implications for pre-biotic chemistry in the solar nebula and other protoplanetary systems, *Astrobiol.* 2003, *3* (2), 291–304.

15. Matsumoto, O., Plasma catalytic reaction in ammonia synthesis in the microwave discharge, *J. Phys. Iv* 1998, *8* (P7), 411–420.

16. Yiokari, C. G.; Pitselis, G. E.; Polydoros, D. G.; Katsaounis, A. D.; Vayenas, C. G., High-pressure electrochemical promotion of ammonia synthesis over an industrial iron catalyst, *J. Phys. Chem. A* 2000, *104* (46), 10600–10602.

17. Mahapatra, H.; Kalyuzhnaya, E. S.; Yunusov, S. M.; Shur, V. B., Synergism of iron and ruthenium in ammonia synthesis catalyst at low temperature, *Recent Adv. Basic App. Aspects Ind. Catal.* 1998, *113*, 267–270.

18. Huber, G. W.; Chheda, J. N.; Barrett, C. J.; Dumesic, J. A., Production of liquid alkanes by aqueous-phase processing of biomass-derived carbohydrates, *Science* 2005, *308* (5727), 1446–1450.

19. Laggoun-Defarge, F.; Mitchell, E.; Gilbert, D.; Disnar, J. R.; Comont, L.; Warner, B. G.; Buttler, A., Cut-over peatland regeneration assessment using organic matter and microbial indicators (bacteria and testate amoebae), *J. Appl. Ecol.* 2008, *45* (2), 716–727.

20. Wernet, P.; Nordlund, D.; Bergmann, U.; Cavalleri, M.; Odelius, M.; Ogasawara, H.; Naslund, L. A. et al., The structure of the first coordination shell in liquid water, *Science* 2004, *304* (5673), 995–999.

21. Nilsson, A.; Wernet, P.; Nordlund, D.; Bergmann, U.; Cavalleri, M.; Odelius, M.; Ogasawara, H. et al., Comment on "Energetics of hydrogen bond network: Rearrangements in liquid water", *Science* 2005, *308* (5723), 793A–793A.

22. Coskuner, O., Preferred conformation of the glycosidic linkage of methyl-beta-mannose, *J. Chem. Phys.* 2007, *127* (1), 015101.

23. Coskuner, O.; Bergeron, D. E.; Rincon, L.; Hudgens, J. W.; Gonzalez, C. A., Glycosidic linkage conformation of methyl-alpha-mannopyranoside, *J. Chem. Phys.* 2008, *129* (4), 045102.

24. Geissler, P. L.; Dellago, C.; Chandler, D.; Hutter, J.; Parrinello, M., Autoionization in liquid water, *Science* 2001, *291* (5511), 2121–2124.

25. Mucke, M.; Braune, M.; Barth, S.; Forstel, M.; Lischke, T.; Ulrich, V.; Arion, T.; Becker, U.; Bradshaw, A.; Hergenhahn, U., A hitherto unrecognized source of low-energy electrons in water, *Nat. Phys.* 2010, *6* (2), 78–81.

26. Sander, M. U.; Luther, K.; Troe, J., On the photoionization mechanism of liquid water, *Ber. Bunsen-Ges. Phys. Chem. Chem. Phys.* 1993, *97* (8), 953–961.

27. Sukhonosov, V. Y., Geminate recombination of quasi-free electrons in liquid water, *High Energy Chem.* 1995, *29* (1), 5–10.

28. Dunn, K. F.; ONeill, P. F.; Browning, R.; Browne, C. R.; Latimer, C. J., The dissociative photoionization of H_2O/D_2O between 20 and 40 eV, *J. Electron. Spectrosc. Relat. Phenom.* 1996, *79*, 475–478.

29. Sato, H.; Hirata, F., Theoretical study for autoionization of liquid water: Temperature dependence of the ionic product (pK(w)), *J. Phys. Chem. A* 1998, *102* (15), 2603–2608.

30. Buch, V.; Milet, A.; Vacha, R.; Jungwirth, P.; Devlin, J. P., Water surface is acidic, *PNAS* 2007, *104* (18), 7342–7347.

31. Vacha, R.; Buch, V.; Milet, A.; Devlin, J. P.; Jungwirth, P., Response to Comment on autoionization at the surface of neat water: Is the top layer pH neutral, basic, or acidic? by J. K. Beattie, *Phys. Chem. Chem. Phys.*, 2007, 9, DOI: 10.1039/b713702h, *Phys. Chem. Chem. Phys.* 2008, *10* (2), 332–333.

32. Beattie, J. K., Comment on autoionization at the surface of neat water: is the top layer pH neutral, basic, or acidic? by R. Vacha, V. Buch, A. Milet, J. P. Devlin, and P. Jungwirth, *Phys. Chem. Chem. Phys.*, 2007, 9, 4736, *Phys. Chem. Chem. Phys.* 2008, *10* (2), 330–331.

33. Aziz, E. F.; Ottosson, N.; Faubel, M.; Hertel, I. V.; Winter, B., Interaction between liquid water and hydroxide revealed by core-hole de-excitation, *Nature* 2008, *455* (7209), 89–91.

34. Mundy, C. J.; Rousseau, R.; Curioni, A.; Kathmann, S. M.; Schenter, G. K., A molecular approach to understanding complex systems: Computational statistical mechanics using state-of-the-art algorithms on terascale computational platforms, *J. Phys. Conf. Ser.* 2008, 125, 012014.

35. Beattie, J. K.; Djerdjev, A. N.; Warr, G. G., The surface of neat water is basic, *Farad. Discuss.* 2009, *141*, 31–39.

36. Agmon, N., The Grotthuss mechanism, *Chem. Phys. Lett.* 1995, *244* (5–6), 456–462.

37. Marx, D.; Tuckerman, M. E.; Hutter, J.; Parrinello, M., The nature of the hydrated excess proton in water, *Nature* 1999, *397* (6720), 601–604.

38. Cui, Q.; Karplus, M., Is a "proton wire" concerted or stepwise? A model study of proton transfer in carbonic anhydrase, *J. Phys. Chem. B* 2003, *107* (4), 1071–1078.

39. Tuckerman, M. E.; Chandra, A.; Marx, D., Structure and dynamics of OH-(aq), *Acc. Chem. Res.* 2006, *39* (2), 151–158.

40. Morrone, J. A.; Hasllinger, K. E.; Tuckerman, M. E., Ab initio molecular dynamics simulation of the structure and proton transport dynamics of methanol-water solutions, *J. Phys. Chem. B* 2006, *110* (8), 3712–3720.

41. Marx, D., Proton transfer 200 years after von Grotthuss: Insights from ab initio simulations, *ChemPhyschem* 2006, *7* (9), 1848–1870.

42. Han, J. H.; Liu, H. T., Ab initio simulation on Grotthuss mechanism, *Proceedings of the ASME Advanced Energy Systems Division* 2005, *45*, 449–453.

43. Glans, P.; Nordgren, J.; Agren, H.; Cesar, A., Experimental and theoretical investigation of the soft-x-ray emission-spectrum of molecular-oxygen, *J. Phys. B: At. Mol. Opt. Phys.* 1993, *26* (4), 663–673.

44. Eigen, M., General introduction: Kinetics of proton transfer processes, *Discuss. Farad. Soc.* 1965, (39), 7–15.

45. Weideman, Eg; Zundel, G., Influence of environment on proton-transfer in symmetrical hydrogen-bonds, *Z. Naturforsch., A: Phys. Sci.* 1973, *A 28* (2), 236–245.

46. Lindeman, R.; Zundel, G., Symmetry of hydrogen-bonds, infrared continuous absorption and proton-transfer, *J. Chem. Soc., Faraday Trans.* 1972, *68* (2), 979–990.

47. Pankiewicz, R.; Wojciechowski, G.; Schroeder, G.; Brzezinski, B.; Bartl, F.; Zundel, G., FT-IR study of the nature of K+, Rb+ and Cs+ cation motions in gramicidin A, *J. Mol. Struct.* 2001, *565*, 213–217.

48. Pietraszko, A.; Hilczer, B.; Pawlowski, A., Structural aspects of fast proton transport in $(NH_4)_3H(SeO_4)_2$ single crystals, *Solid State Ionics* 1999, *119* (1–4), 281–288.

49. Cukierman, S., Proton mobilities in water and in different stereoisomers of covalently linked gramicidin A channels, *Biophys. J.* 2000, *78* (4), 1825–1834.

50. Islam, M. S.; Davies, R. A.; Gales, J. D., Proton migration and defect interactions in the $CaZrO_3$ orthorhombic perovskite: A quantum mechanical study, *Chem. Mater.* 2001, *13* (6), 2049–2055.

51. Jude, K. M.; Wright, S. K.; Tu, C.; Silverman, D. N.; Viola, R. E.; Christianson, D. W., Crystal structure of F65A/Y131C-methylimidazole carbonic anhydrase V reveals architectural features of an engineered proton shuttle, *Biochemistry* 2002, *41* (8), 2485–2491.

52. Laasonen, K.; Sprik, M.; Parrinello, M.; Car, R., Ab initio liquid water, *J. Chem. Phys.* 1993, *99* (11), 9080–9089.

53. Selloni, A.; Car, R.; Parrinello, M.; Carnevali, P., Electron pairing in dilute liquid-metal metal halide solutions, *J. Phys. Chem.* 1987, *91* (19), 4947–4949.

54. Fois, E. S.; Selloni, A.; Parrinello, M.; Car, R., Bipolarons in metal metal halide solutions, *J. Phys. Chem.* 1988, *92* (11), 3268–3273.

55. Selloni, A.; Fois, E. S.; Parrinello, M.; Car, R., Simulation of electrons in molten-salts, *Phys. Scr.* 1989, *T25*, 261–267.

56. Sorella, S.; Baroni, S.; Car, R.; Parrinello, M., A novel technique for the simulation of interacting fermion systems, *Europhys. Lett.* 1989, *8* (7), 663–668.

57. Galli, G.; Martin, R. M.; Car, R.; Parrinello, M., Carbon—The nature of the liquid-state, *Phys. Rev. Lett.* 1989, *63* (9), 988–991.

58. Buda, F.; Chiarotti, G. L.; Stich, I.; Car, R.; Parrinello, M., Ab initio molecular-dynamics of liquid and amorphous-semiconductors, *J. Non-Cryst. Solids* 1989, *114*, 7–12.

59. Car, R.; Parrinello, M., Unified approach for molecular-dynamics and density-functional theory, *Phys. Rev. Lett.* 1985, *55* (22), 2471–2474.

60. Marx, D.; Sprik, M.; Parrinello, M., Ab initio molecular dynamics of ion solvation. The case of Be^{2+} in water, *Chem. Phys. Lett.* 1997, *273* (5–6), 360–366.

61. Marx, D.; Fois, E. S.; Parrinello, M., Static and dynamic density functional investigation of hydrated beryllium dications, *Int. J. Quant. Chem.* 1996, *57* (4), 655–662.

62. Silvestrelli, P. L.; Bernasconi, M.; Parrinello, M., Ab initio infrared spectrum of liquid water, *Chem. Phys. Lett.* 1997, *277* (5–6), 478–482.

63. Bernasconi, M.; Silvestrelli, P. L.; Parrinello, M., Ab initio infrared absorption study of the hydrogen-bond symmetrization in ice, *Phys. Rev. Lett.* 1998, *81* (6), 1235–1238.

64. Resta, R., Electrical polarization and orbital magnetization: The modern theories, *J. Phys. Condens. Matter* 2010, *22*, 123201–123219.

65. Coskuner, O.; Jarvis, E. A. A.; Allison, T. C., Water dissociation in the presence of metal ions, *Angew. Chem.-Int. Ed.* 2007, *46* (41), 7853–7855.

66. Coskuner, O.; Allison, T. C., Dynamic and structural properties of aqueous arsenic solutions, *ChemPhyschem* 2009, *10* (8), 1187–1189.

67. Kiss, T., Interaction of aluminum with biomolecules—Any relevance to Alzheimers disease, *Arch. Gerontol. Geriatr.* 1995, *21* (1), 99–112.

68. Hu, Y. F.; Xu, R. K.; Dynes, J. J.; Blyth, R. I. R.; Yu, G.; Kozak, L. M.; Huang, P. M., Coordination nature of aluminum (oxy)hydroxides formed under the influence of tannic acid studied by X-ray absorption spectroscopy, *Geochim. Cosmochim. Acta* 2008, *72* (8), 1959–1969.

69. Miu, A. C.; Benga, O., Aluminum and Alzheimer's disease: A new look, *J. Alzheimers Dis.* 2006, *10* (2–3), 179–201.

70. Newman, P. E., Could diet be one of the causal factors of Alzheimers disease, *Med. Hypotheses* 1992, *39* (2), 123–126.

71. Ares, J., Identification of hydrolysis products of aluminum in natural-waters. 2. Alspec, a computerized procedure for quantifying equilibria with inorganic and organic-ligands, *Anal. Chim. Acta* 1986, *187*, 195–211.

72. Martin, R. B., Fe^{3+} and Al^{3+} Hydrolysis equilibria—Cooperativity in Al3+ hydrolysis reactions, *J. Inorg. Biochem.* 1991, *44* (2), 141–147.

73. Lubin, M. I.; Bylaska, E. J.; Weare, J. H., Ab initio molecular dynamics simulations of aluminum ion solvation in water clusters, *2000 International Conference on Modeling and Simulation of Microsystems, Technical Proceedings* 2000, 91–94.

74. Sarpola, A. T.; Saukkoriipi, J. J.; Hietapelto, V. K.; Jalonen, J. E.; Jokela, J. T.; Joensuu, P. H.; Laasonen, K. E.; Ramo, J. H., Identification of hydrolysis products of $AlCl_3 \cdot 6H_2O$ in the presence of sulfate by electrospray ionization time-of-flight mass spectrometry and computational methods, *Phys. Chem. Chem. Phys.* 2007, *9* (3), 377–388.

75. Yang, W. J.; Qian, Z. S.; Miao, Q.; Wang, Y. J.; Bi, S. P., Density functional theory study of the aluminium(III) hydrolysis in aqueous solution, *Phys. Chem. Chem. Phys.* 2009, *11* (14), 2396–2401.

76. Swaddle, T. W.; Rosenqvist, J.; Yu, P.; Bylaska, E.; Philiips, B. L.; Casey, W. H., Kinetic evidence for five-coordination in $AlOH(aq)^{2+}$ ion, *Science* 2005, *308* (5727), 1450–1453.

77. Car, R.; Parrinello, M., The unified approach to density functional and molecular-dynamics in real space, *Solid State Commun.* 1987, *62* (6), 403–405.

78. Car, R.; Parrinello, M.; Payne, M., Error cancellation in the molecular-dynamics method for total energy calculations—Comment, *J. Phys. Condens. Matter* 1991, *3* (47), 9539–9543.

79. Chandler, D.; Lee, J. Y.; Dellago, C., Transition path sampling: Rearrangement of the cage water hexamer, *Abs. Pap. Am. Chem. Soc.* 1999, *218*, U360–U360.

80. Bolhuis, P. G.; Dellago, C.; Geissler, P. L.; Chandler, D., Transition path sampling: throwing ropes over mountains in the dark, *J. Phys. Condens. Matter* 2000, *12* (8A), A147–A152.

81. Geissler, P. L.; Chandler, D., Importance sampling and theory of nonequilibrium solvation dynamics in water, *J. Chem. Phys.* 2000, *113* (21), 9759–9765.

82. Chandler, D.; Dellago, C.; Geissler, P., Transition path sampling and the pathways to auto ionization of a water molecule in liquid water, *Abs. Pap. Am. Chem. Soc.* 2001, *222*, U204–U204.

83. Bolhuis, P. G.; Chandler, D., Transition path sampling of cavitation between molecular scale solvophobic surfaces, *J. Chem. Phys.* 2000, *113* (18), 8154–8160.

84. Marti, J.; Csajka, F. S.; Chandler, D., Stochastic transition pathways in the aqueous sodium chloride dissociation process, *Chem. Phys. Lett.* 2000, *328* (1–2), 169–176.

85. Bolhuis, P. G.; Dellago, C.; Chandler, D., Reaction coordinates of biomolecular isomerization, *PNAS* 2000, *97* (11), 5877–5882.

86. Geissler, P. L.; Dellago, C.; Chandler, D., Kinetic pathways of ion pair dissociation in water, *J. Phys. Chem. B* 1999, *103* (18), 3706–3710.

87. Dellago, C.; Bolhuis, P. G.; Csajka, F. S.; Chandler, D., Transition path sampling and the calculation of rate constants, *J. Chem. Phys.* 1998, *108* (5), 1964–1977.

88. Dellago, C.; Bolhuis, P. G.; Chandler, D., Efficient transition path sampling: Application to Lennard-Jones cluster rearrangements, *J. Chem. Phys.* 1998, *108* (22), 9236–9245.

89. Coskuner, O.; Deiters, U. K., Hydrophobic interactions of xenon by Monte Carlo simulations, *Z. Phys. Chem.* 2007, *221* (6), 785–799.

90. Coskuner, O.; Deiters, U. K., Hydrophobic interactions by Monte Carlo simulations, *Z. Phys. Chem.—Int. J. Res. Phys. Chem. Chem. Phys.* 2006, *220* (3), 349–369.

91. Qian, Z.; Feng, H.; Yang, W.; Zhang, Z.; Wang, Y; Bi, S., Theoretical investigation of dehydration of aquated $Al(OH)_2^+$ species in aqueous solution, *Dalton Trans.* 2009, 1554–1558.

92. Arcon, I.; van Elteren, J. T.; Glass, H. J.; Kodre, A.; Slejkovec, Z., EXAFS and XANES study of Arsenic in contaminated soil, *X-ray Spectrom.* 2005, *34*, 435–438.

93. Ramírez-Solís, A.; Mukopadhyay, R.; Rosen, B. P.; Stemmler, T. L., Experimental and theoretical characterization of arsenite in water: Insights into the coordination environment of As-O, *Inorg. Chem.* 2004, *43*, 2954–2959.

94. Guo, X.; Du, Y.; Chen, F.; Park, H.-S.; Xie, Y., Mechanism of removal of arsenic by bead cellulose loaded with iron oxyhydroxide (β-FeOOH): EXAFS study, *J. Colloid Interface Sci.* 2007, *314*, 427–433.

95. Pokrovski, G. S.; Bény, J. M.; Zotov, A. V., Solubility and Raman spectroscopic study of As(III) speciation in organic compound-water solutions. A hydration approach for aqueous arsenic in complex solutions, *J. Solut. Chem.* 1999, *28*, 1307–1327.

96. Hseu, Y.-C.; Yang, H.-L., The effects of humic acid-arsenate complexes on human red blood cells, *Environ. Res.* 2002, *89*, 131–137.

97. Tongesayi, T.; Smart, R. B., Arsenic speciation: Reduction of Arsenic(V) to Arsenic(III) by fulvic acid, *Environ. Chem.* 2006, *3*, 137–141.

98. Swaddle, T.; Rosenqvist, J.; Yu, P.; Bylaska, E.; Phillips, B. L.; Casey, W. H., Kinetic evidence for five-coordination in $AlOH(aq)^{2+}$ ion, *Science* 1995, *308*, 1450–1453.

5 Structure of Liquid Metal Surfaces: A First Principles Perspective

Brent Walker, Nicola Marzari, and Carla Molteni

CONTENTS

5.1   INTRODUCTION

Liquids or molten metals are important in many industrial refining, smelting, and casting processes, as well as in numerous scientific applications (e.g., [1,2]). From the point of view of research, free surfaces (or equivalently the liquid–vapor transition zones) of liquid metals have provided significant interest for experimental and

135

theoretical study for several years, in addition to the focus on other interfaces incorporating a liquid metal component. A major reason for this is that these surfaces display properties that are more complex than those of the surfaces of ionic and dielectric liquids. The atomic structure of the surface of a simple dielectric liquid is generally believed to be relatively straightforward, exemplified by a monotonic decrease in the transverse density profile from the bulk liquid density to the vapor density. The situation is more complex across the transition zone of a liquid metal; this is particularly apparent in the density profile, which displays oscillations on the liquid side of the interface (see Figure 5.1). A density profile containing such oscillations is indicative of the formation of atomic layers. In the liquid, where the atoms undergo diffusive motion, the layers will exist as preferential regions with higher or lower densities (corresponding respectively to maxima and minima in the density profile). At a fundamental level, the differences seen between the structures of the surfaces of liquid dielectrics and liquid metals reflect the behaviors of the microscopic interactions across the different interfaces. In the case of a dielectric, the nature of the interatomic interactions may not need to change a great deal across the interface, for instance, being van der Waals bonding in both the liquid and vapor phases of the material [3]. In the liquid metal case, there has to be a substantial

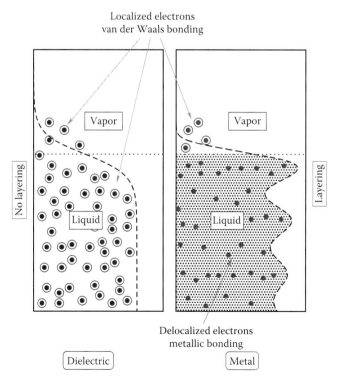

FIGURE 5.1 Comparison between the free surfaces of a liquid dielectric (left) and a liquid metal (right), with the types of interatomic interactions in effect in the different regions indicated.

change in the nature of the interatomic interactions across the surface, as the bonding changes from metallic in the liquid part to dispersive-type interactions between the atoms in the vapor part. These fundamental differences at the microscopic level mean that atomistic simulations, which are able to provide insight at this level, have an extremely useful role to play.

Refining the experimental measurements to clarify the picture of the structure of liquid metal surfaces and the search for a physical explanation for the mechanism of layer formation have been key factors driving research on liquid metal surfaces. Improvements in the quality and scope of the experimental techniques have occurred side by side with advances and extensions in the body of theoretical and simulation research on the subject, although it was only with the availability of high-intensity x-rays from synchrotron sources that layered structures at free liquid surfaces could be seen experimentally with any degree of certainty.

5.2 LIQUID METAL SURFACES: TRANSVERSE SURFACE STRUCTURE

5.2.1 SUMMARY OF EXPERIMENTS ON LIQUID METAL SURFACES

The experimental measurements that have provided major insights into the structures of liquid metal surfaces have been made with the technique of x-ray reflectivity. In this technique, the intensity R of x-rays incident on the surface at angle α and reflected at angle β from the surface is measured. The reflection geometry in an x-ray reflectivity measurement is represented schematically in Figure 5.2a, and a typical experimental measurement setup is illustrated in Figure 5.2b. The reflectivity is a function of the momentum transfer vector (q_z) and the density profile of the surface:

$$\frac{R(q_z)}{R_f(q_z)} \approx \left| \frac{1}{\rho_\infty} \int dz \, \frac{d\langle\rho(z)\rangle}{dz} e^{iq_z z} \right|^2 \tag{5.1}$$

where
   R_f is the Fresnel reflectivity of a perfect step-function density profile
   ρ_∞ is the bulk liquid density (that is, the density far into the bulk, away from the surface)
   $\langle\rho(z)\rangle$ is the density of the liquid averaged in the plane of the surface, which we assume to be perpendicular to the z-axis

To resolve features in the surface structure of the size of a single atomic layer and large momentum transfer vectors q_z are required. This in turn requires very high intensity x-ray sources as the strength of the measurable reflectivity signal decreases rapidly with the magnitude of the momentum transfer ($R(q_z) \sim q_z^{-4}$).

Early x-ray reflectivity measurements made by Lu and Rice [4] on the surface of Hg did not extend to sufficiently high momentum transfers to resolve atomic layers at the surface. It was, however, observed that later measurements made on

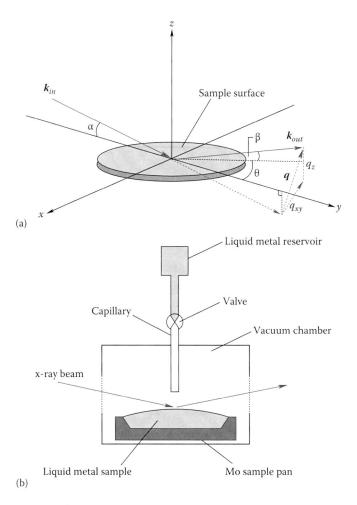

(a)

(b)

FIGURE 5.2 (a) Reflection geometry in experimental x-ray reflectivity measurement. The angles of incidence and reflection are respectively α and β; the incoming and outgoing momenta are k_{in} and k_{out}, and the wave vector transfer is q_z. (b) Illustration of a typical experimental setup for an x-ray reflectivity measurement from a liquid metal surface, as described, for example, in Ref. [16].

liquid Cs [5] were fitted better using a layered surface density profile than a monotonic density profile. Ambiguities in fitting the density profile were resolved with subsequent experiments on liquid Hg [6], which achieved wave vector transfers up to $q_z \sim 2.5$ Å^{-1} and allowed resolutions down to that of a single atomic layer needed to discern definitely the surface structure. Further measurements were performed on various metals, including Ga [7–9], Hg [9,10], and In [11], all of which have relatively high surface tensions, as well as alloys of Ga:Bi [12–14]. Subsequently, measurements were made for a eutectic alloy of potassium and sodium [15], and for the pure alkali metal K [16], both metal systems having low surface tensions. The experimental signature of layering was the appearance of a peak in the reflectivity at high momentum transfers, with the density profile fitted to reproduce the density. In the case of the high surface tension metals, a complete peak in the reflectivity was observed, and thus fitting the surface density profile was unambiguous. In the case of the low surface tension metals (K and eutectic Na:K alloys), only the lead up to a peak from the low q_z side was seen in the experimental measurements; this made the fitting to determine the density profiles more difficult. In addition, the intrinsic structure of the liquid surface is damped by thermally excited capillary waves. Though these are suppressed by higher surface tension, for the low surface tension metals where the damping is not very significant, there is correspondingly an increased difficulty in measuring the intrinsic surface structure. In measurements on liquid Sn [17,18] and Bi [18], a decreased spacing between the surface layer and the second layer from the surface relative to the spacings of the subsequent layers was observed.

Additional studies of liquid metal surfaces have been made using x-ray techniques, including the structures of various nonmetallic layers on liquid metal surfaces [19–23] and the wetting and segregation at the surface of components in alloys, for example, Bi:Sn [24], Au:Si [25,26], Au:Sn [27].

In contrast to the measurements made on the liquid metals, experimental measurements on the surface of water [28] showed no evidence of molecular layer formation at the free surface. Subsequent measurements on the molecular liquid tetrakis(2-ethylhexoxy)silane, an isotropic, dielectric liquid [29] showed that layer formation is possible at dielectric liquid surfaces.

5.2.2 SIMULATIONS VIEWPOINT

We review three possible explanations for the mechanism of layer formation at the free surfaces of liquid metals coming out of simulation research. Systematic examination of these explanations was the focus of simulations of free surfaces of sodium made using ab initio molecular dynamics, which we shall review in Section 5.3, as well as supporting computer "experiments."

5.2.2.1 Layering against Hard Wall

The earliest atomistic simulations of liquid metals were made by Rice and coworkers [30–39] and predicted (in fact, before the experimental techniques were able to ascertain definitely) that the transverse density profiles at liquid metal surfaces would display layering oscillations. Their simulations were made using an "effective Hamiltonian" to describe the electronic structure, together with a Monte Carlo algorithm to sample atomic configurations. A fairly large set of metals were studied using this approach, including Na [30,31,34], Cs [31], Hg [32], Ga [37], and Mg [36]. In addition to these, various alloys were simulated: Na:Cs [33,34], Bi:Ga [38], and Ga:Sn [39]. The physical explanation for the layering mechanism resulting from this work is that the rapid decay of the electronic density across the surface, and the strong dependence of the effective ionic potential on electron density (which accounts for the "phase-transition" from a metallic system with the associated delocalization of the electrons in the liquid, to the atomic state in which the electrons are localized on their corresponding nuclei) acts like a rigid wall that geometrically confines the motion of atoms across the surface. This explanation connects the formation of layers at free liquid surfaces to the situation occurring at the interface between a liquid and a solid [3]; at such an interface, the formation of layers in the liquid part occurs naturally due to geometrical confinement of the liquid atoms by the solid atoms.

5.2.2.2 Geometrical Rearrangement to Increase Density at the Surface

Following the effective Hamiltonian simulations of Rice, studies were made on liquid Au surfaces using glue model potentials by Iarlori et al. [40] and on Pb and Au using the same method by Di Tolla [41] and Celestini et al. [42]. In these simulations, the surface densities showed oscillatory profiles that were explained as occurring so that the undercoordinated atoms near the surface regain some of the favorable coordination they would have in the bulk liquid. This increased density in the outermost part of the surface means that there is an overcoordination in the next layer, which is mitigated by a decrease in density in the next layer. This alternating increase and decrease in density continues in successive layers moving from the surface toward the bulk, and is consistent with the layers seen in the density profile.

The spirit of this proposed mechanism—that there is a rearrangement of the surface to partially recover the bulk coordination environment—is similar to the explanation proposed by Fabricius et al. [43] based on full density functional theory (DFT) simulations of liquid Si surfaces at high temperature (at the temperature considered, 1800 K, Si behaves as a metal). They proposed that there is a rotation of the bonds (considerable covalent character is retained in the high temperature state of Si) at the surface driven by a requirement to recover the bulk density close to the surface that results in the observed oscillating transverse density profiles.

5.2.2.3 Friedel Oscillations in Electronic Density Profile

A third explanation is that Friedel oscillations [44] in the electronic density may drive the atomic layers. Friedel oscillations in the valence electron density would electrostatically drive the atoms toward favorable locations and away from unfavorable ones, giving rise to the appearance of layers in the atomic density. Motivation for this explanation was provided by a set of DFT calculations on solid $Mg(10\bar{1}0)$ surfaces by Cho et al. [45], which indicated that the fine rearrangements in the interlayer relaxations were driven by the changes in the valence electron density upon formation of a surface. The electronic charge density at the surface was subtracted from the density of a bulk system; the difference in charge density, indicating the rearrangement of charge upon formation of the surface, contained definite Friedel oscillations. These damped oscillations are expected to appear in the response of the valence electrons in a metallic system to the addition of a localized defect (e.g., a surface, or inclusion of a point charge) and correspond to a potential $V(r)$ of the form

$$V(r) \sim \frac{\cos(2k_F r)}{r^3} \tag{5.2}$$

where
  r is the distance from the defect (in this case, the surface)
  k_F is the Fermi wave vector

If Friedel oscillations were present at the liquid surface, these might be expected to have an influence on the layer formation there.

5.2.2.4 Surface Layering without Metallic Interactions: Ratio of Melting to Critical Temperatures

In contrast to the layering explanations outlined above, Chacón et al. [46], Tarazona et al. [47], and Velasco et al. [48] proposed that a metallic system is not necessary for a free liquid surface to exhibit layering behavior. Instead they suggested that the criterion determining whether a surface would exhibit layers is that the substance has a low ratio of melting temperature to critical temperature (the temperature above which the system cannot be pressurized to form a liquid). This conclusion was supported with molecular dynamics simulations performed using empirical potentials. These potentials were tuned to give a low ratio of melting temperature (T_M) to critical (T_C) temperature, and though not taking into account the many-body characteristics, they produced layer formation in liquid surface simulations. According to this suggestion, for Hg and Ga, which have low ratios of T_M to T_C ($T_M/T_C < 0.13$), layering should be quite pronounced; for the alkali metals, with intermediate ratios ($T_M/T_C \sim 0.15$), layer formation would be weak and at the limits of experimental detection; for metals with higher T_M/T_C ratios still (such as Al and Mg), layering should not be present.

In considering this conjecture connecting the melting and critical temperatures and its consistency with experimental measurements on metals and nonmetals, we note that measurements on water (which freezes at about $0.42T_C$) showed no evidence of layer formation [28], while measurements on tetrakis(2-ethylhexoxy)silane at $T/T_C \approx 0.25$ did show layer formation [29]. On the other hand, simulations performed using molecular dynamics based on orbital-free DFT techniques of Li, Mg, and Al [49] showed pronounced layer formation, disagreeing with the prediction of Chacón based on the T_M/T_C ratio.

5.3 DENSITY FUNCTIONAL THEORY-BASED MOLECULAR DYNAMICS SIMULATIONS

Recently, simulations of liquid surfaces of metals have been made using molecular dynamics coupled with DFT, with the primary focus on exploring the explanations for surface-induced layering outlined above. The use of quantum mechanical methods is motivated by the need to take account of the changing interactions across the surface, and the difficulties this poses for the use of classical potentials to describe the interactions between atoms. DFT is now widely used for performing computer simulations at the quantum mechanical level as it provides, in practice, a highly favorable ratio of cost to accuracy.

In a full quantum mechanical calculation, the system of interacting nuclei and electrons is described by a wave function that contains all the details of the dynamics of the system. As the nuclei in the system are ~2000 times heavier than the electrons, the motions of the electrons and nuclei can be considered separately by making the Born–Oppenheimer approximation [50]. In most ab initio MD simulations, the dynamical system of interacting electrons and nuclei is considered as two (coupled) subsystems: the systems of nuclei and electrons, which are described at different levels of theory. The system of nuclei is considered as a set of classical point particles (described by appropriate equations of motion), with a component of force coming from the system of electrons, which are treated quantum mechanically. In this formulation, the electrons respond instantaneously (this is termed "abiabaticity") to any changes in the nuclear positions. The motions of the coupled electronic and nuclear systems are determined numerically, with the most time-consuming part of the calculation being computation of the electronic structure. Once calculated, the electronic structure allows determination of the forces on the nuclei (using the Hellman–Feynman theorem [51,52]) and the positions of the nuclei can be evolved using the pertinent equations of motion. The electronic structure is recalculated for this new set of nuclear positions, and the process repeated. At each step in the nuclear motion, we need to solve the problem of the interacting set of electrons in the (fixed at a given time step) field of the nuclei. Though the Born–Oppenheimer approximation is very often made to treat the nuclei as classical particles, effects arising from the quantum mechanics of the nuclei can be included, in particular through the use of path integral techniques (see Section 4.4 of Ref. [50]).

The molecular dynamics technique provides access to the full dynamical behavior of the system, an advantage over, for instance, Monte Carlo techniques that can only provide time-averaged information. This comes at substantial cost in computational resources. Ab initio MD simulations are currently limited to timescales from tens to hundreds of picoseconds. Regardless of whether the computational time required to use molecular dynamics to describe the nuclear configurations is more time-consuming than using Monte Carlo sampling, in an ab initio simulation, the most time-consuming part of the calculation is the evaluation of the forces on the atoms—this is because the electronic structure must be determined at each step. This cost can be mitigated somewhat by using fitted potentials to describe the interactions between the different atoms in the system in performing empirical molecular dynamics, but there is a substantial penalty in transferability (the prime example being that changes in "bond" types and arrangements must be built in beforehand) as such calculations move away from the realm of first principles.

Many methods have been developed to perform the integration of the nuclear equations of motion, including the Gear predictor–corrector schemes [53], the Verlet algorithms [54] (both are so-called finite-difference methods) and the split-step methods [55]. The widely used Verlet algorithms are based on making Taylor series expansions of the positions and velocities of the atoms at time $t + \delta t$ in terms of those at time t and the forces acting on the nuclei, with δt the time interval between successive nuclear configurations.

5.3.1 Density Functional Theory

An immensely popular method of describing a system of interacting electrons that can be used to describe the valence electrons in an ab initio molecular dynamics simulation is density functional theory (DFT) [56,57]. For reviews focusing on DFT, the reader should see, for instance, Refs. [58,59]. Here, we mention only in passing that metallic systems require special care, since there is discontinuity in

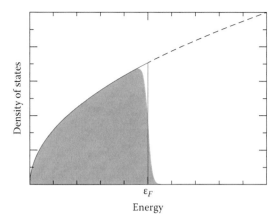

FIGURE 5.3   Smearing of profile of occupation number versus energy in a metal. The solid blue curve shows the occupied states for a metal at zero electronic temperature; the shaded green area shows the occupied states smeared near the Fermi energy at finite electronic temperature. (Adapted from Walker, B.G. et al., *J. Phys. Condens. Matt.*, 16, S2575, 2004. With permission.)

the occupation number of electronic orbitals as the Fermi energy is crossed. This discontinuity makes calculations in metals more challenging (small errors in the charge density give rise to large changes in the electrostatic potential) and also more expensive to converge with respect to the sum over all occupied states (since a discontinuity needs to be integrated carefully). This latter problem can be dealt with by smearing the occupation number profile, adding extra electronic states above the Fermi energy, and allowing noninteger occupancies for those states just above and below the Fermi energy (such an idea is illustrated in Figure 5.3). This is equivalent to the introduction of a finite electronic temperature, and will be discussed further below.

5.3.1.1   Ensemble Density Functional Theory

Introduction of a finite temperature into the description of the electronic system (for instance to deal appropriately with metallic systems) can be achieved through generalization of the energy functional to a free energy functional.

Using an early canonical approach introduced by Mermin [60], the free energy F can be written as a density functional including an entropy term:

$$F[T_{elec};\{\varphi_j\},\{f_j\}] = T_e[\rho] + E_H[\rho] + \int d^3\mathbf{r}\, V_{ext}(\mathbf{r})\rho(\mathbf{r}) + E_{XC,\beta}[\rho] - T_{elec}S$$

$$= T_e[\rho] + E_H[\rho] + \int d^3\mathbf{r}\, V_{ext}(\mathbf{r})\rho(\mathbf{r}) + E_{XC,\beta}[\rho]$$

$$+ \frac{1}{\beta}\sum_j [f_j \ln f_j + (1 - f_j)\ln(1 - f_j)] \tag{5.3}$$

where ρ is the electronic density, the $\{\varphi_j\}$ are a set of Kohn–Sham single-particle orbitals with associated occupancies $\{f_j\}$, $V_{ext}(\mathbf{r})$ is the external potential acting on the electronic system, T_e is the kinetic energy of the non-interacting Kohn–Sham orbitals, E_H is the Hartree energy, $E_{XC,\beta}$ is the exchange correlation energy at finite electronic temperature $T_{elec} = 1/\beta$, and S is the electronic entropy. The last term in the second part of Equation 5.3 gives the entropy in the case that Fermi–Dirac statistics are used to describe the occupancies. In this approach, the occupancies and rotations within the subspace of occupied orbitals are treated separately which, gives rise to ill-conditioned, nonlinear degrees of freedom in the minimization problem.

Marzari et al. [61,62] subsequently introduced a generalization of the Mermin canonical finite-temperature DFT known as ensemble density functional theory (eDFT), which remedied this problem. In this approach, the free energy functional takes the form

$$F[T_{elec};\{\varphi_j\},\{f_{ij}\}] = \sum_{ij} f_{ji}\left\langle\varphi_i\left|\hat{T}_e+V_{ext}\right|\varphi_j\right\rangle + E_H[\rho] + E_{XC,\beta}[\rho] - T_{elec}S[\{f_{ij}\}] \qquad (5.4)$$

$$\rho[\{\varphi_j\},\{f_{ij}\}](\mathbf{r}) = \sum_{ij} f_{ji}\varphi_i^*(\mathbf{r})\,\varphi_j(\mathbf{r}) \qquad (5.5)$$

in which $\{f_{ij}\}$ is the set of "generalized" occupancies. As both the free energy and density are written as traces, the occupancies and rotations within the occupied subspace can both be treated consistently. This allows development of a robust and stable algorithm for solving the minimization problem, the full details of which appear in Refs. [61,62]. Closely associated with the ensemble DFT formulation is the "cold-smearing" technique [61,62]. In this scheme, the smearing function is chosen so that the systematic errors introduced via coupling to the temperature are minimized.

Application of the ensemble density functional theory scheme has been demonstrated in simulations of various metallic systems (including surfaces): studies of the microscopic mechanisms accompanying the sliding of grain boundaries in aluminum [63] and the properties of the Al(110) surface near melting [64].

5.3.2 ORBITAL-FREE MOLECULAR DYNAMICS

Recently, a number of liquid metal surfaces have been studied using "orbital-free" DFT methods (described in detail in Ref. [65]). In this technique, an approximation for the kinetic energy operator in terms of the density is introduced. Instead of calculating the non-interacting kinetic energy as in the usual orbital-based Kohn–Sham formulation, the kinetic energy term is evaluated as

$$T_e = T_W + T_\beta \qquad (5.6)$$

where T_W is the von Weizsäcker term:

$$T_W[\rho(\mathbf{r})] = \frac{1}{8}\int d^3\mathbf{r}\ |\nabla\rho(\mathbf{r})|^2\ \rho^{-1}(\mathbf{r}) \tag{5.7}$$

and

$$T_\beta = \frac{3}{10}\int d^3\mathbf{r}\ \rho(\mathbf{r})^{5/3-2\beta}\tilde{k}(\mathbf{r})^2 \tag{5.8}$$

$$\tilde{k}(\mathbf{r}) = (2k_F^0)^3 \int d\mathbf{s}\ k(\mathbf{s})w_\beta\left(2k_F^0\ |\mathbf{r}-\mathbf{s}|\right) \tag{5.9}$$

$$k(\mathbf{r}) = (3\pi^2)^{1/3}(\rho(\mathbf{r}))^\beta \tag{5.10}$$

where:
   $\rho_e = N/V$ (N being the number of electrons in the system and V the system volume)
      is the mean electronic density
   k_F^0 is the Fermi wave vector for the mean electronic density ρ_e
   w_β is a weight function chosen so that both the linear response theory and
      Thomas–Fermi limits are recovered

As the relatively time-consuming evaluation of the kinetic energy term via orbitals in a full formulation of DFT is avoided, the orbital-free approach can be applied to systems containing greater numbers of atoms, as well as making it possible to collect longer sets of statistics in the simulations. Clearly, questions can be raised regarding the reliability of methods in which the kinetic energy in DFT is approximated. This is expected to be particularly crucial when dealing with the free surface of a liquid metal, where a rapidly varying (i.e., decreasing from the bulk density in the liquid to close to zero in the vacuum) is an important characteristic of the system.

A number of simulations have applied orbital-free DFT to liquid metal surfaces, including applications to Li, Na, K, Rb, Cs, Mg, Ba, Al, Tl, and Si [65,66], Ga [67], as well as the alloys Na:K [68,69], Li:Na [68,69], and Na:Cs [69]. In these simulations, slabs containing ~ 2000 – 3000 atoms were considered. For all the liquid metal surfaces simulated using the orbital-free MD method, oscillations indicating layer formation were observed in the transverse density profiles. With increasing temperature, both the range and amplitude of the layering oscillations in the transverse density profiles across the surface decreased.

5.3.3 ENSEMBLE DFT-BASED MOLECULAR DYNAMICS SIMULATIONS OF LIQUID METAL SURFACES

5.3.3.1 Free Sodium Surfaces

Free liquid surfaces of sodium have been modeled with molecular dynamics with the forces on the atoms determined using ensemble DFT and cold smearing [70–74]. The motivation cited for undertaking simulations on sodium included that, being a prototypical free electron metal, sodium as a system will provide clear insights into the effects of metallic bonding on the surface structure, without complications like partial covalent character present in the bonding of some metals such as Ga. Having access to a system in which the bonding is purely metallic (or close to it) would be especially important when examining the existence and importance of Friedel oscillations. Furthermore, experimental x-ray reflectivity measurements have proven more difficult and less clear-cut in illuminating surface layering for the low surface tension (e.g., alkali) metals than in the high surface tension metals measured earliest, so performing simulations on alkali metals would be expected to provide valuable insights to complement the experimental efforts being made.

These simulations considered the free surfaces of liquid sodium at two temperatures, 400 and 500 K, which correspond to conditions just above the melting temperature of sodium ($T_M = 373$ K), and somewhat above it. As the existence and details of in-plane order (see Section 5.4) was an important item of consideration in the study, the simulations were performed with two cross-sectional geometries in the plane of the surface, the idea being that having simulation cells with different cross-sectional shapes would help to assess the influence of finite-size effects on the existence (if any) of order parallel to the surface. The cell shapes simulated are indicated in Figure 5.4, and the dimensions of the cells in Table 5.1. Slabs of liquid atoms as models of free surfaces were obtained by firstly melting bulk systems (i.e., with no vacuum regions) at high temperatures ($T = 800$ K for 2 ps, followed by $T = 600$ K for 2 ps). The "(001)" cell contained 160 atoms, and the starting bulk simulation cell contained 10 atomic layers along the z-direction (that would later become the direction

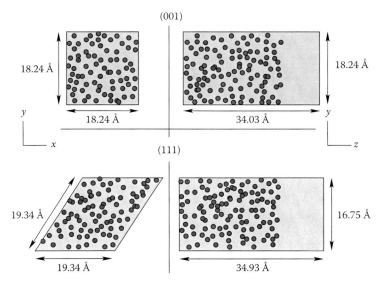

FIGURE 5.4 Unit cell geometries used in simulating the free liquid surfaces of sodium [70–74]. The upper part shows the shape of the "(001)" cells; the lower part shows the shape of the "(111)" cells. The dimensions shown are for temperature $T = 500$ K.

TABLE 5.1

Details of Cell Geometries Used in Ab Initio MD Simulations of Liquid Sodium Surfaces in Refs. [70–74]

Cell	T (K)	a (Å)	b (Å)	c (Å)	α (°)	β (°)	γ (°)
(001)	400	18.06	18.06	34.03	90	90	90
	500	18.24	18.24	34.03	90	90	90
(111)	400	19.16	19.16	34.93	90	90	60
	500	19.34	19.34	34.93	90	90	60

Source: Adapted with permission from Walker, B.G., *Ab Initio Molecular Dynamics Studies of Liquid Metal Surfaces*, PhD thesis, University of Cambridge, copyright 2004; Walker, B.G. et al., *J. Phys. Condens. Matter*, 16, S2575, copyright 2004; Walker, B.G. et al., *J. Chem. Phys.*, 124, 174702, copyright 2006; Walker, B.G. et al., *J. Phys. Condens. Matter*, 18, L269, copyright 2006; Walker, B.G. et al., *J. Chem. Phys.*, 127, 134703, copyright 2007, American Institute of Physics.

a, b, and c are the lengths of the cell vectors (in Å), and α, β, and γ are the Euler angles (in degrees).

normal to the surface); the "(111)" cell contained 162 atoms, with 12 layers in the starting bulk cell. Once mean square displacements showed that the bulk simulation cells had melted satisfactorily, vacuum regions of approximately 11 Å thicknesses were added along the z-direction of each cell, to form liquid slabs. These were then equilibrated, and statistics collected. For each simulation cell and temperature, over 50 ps of molecular dynamics was obtained, with the last 30 ps of each run being used in calculating time-averaged properties. The Verlet algorithm ("leapfrog" version [75]) was used to integrate the ionic equations of motion, with an integration time step $\delta t = 10$ fs. A Gaussian thermostat [76] was used to maintain ionic temperatures at the desired values.

The eDFT calculations were made with plane wave basis sets specified by a cutoff energy $E_{cut} = 100$ eV; all integrations were performed with the Baldereschi k-point (¼, ¼, ¼) [77]. Norm-conserving pseudopotentials of the Troullier–Martins type [78] were used to describe the Na atoms; for the exchange–correlation energy, the gradient-corrected functional of Perdew and Wang (PW91) [79] was used. A nonlinear core-correction [80] with a core radius of 2.0 Bohr was used. In the cold smearing, an electronic temperature equivalent to 0.5 eV was used.

The density profiles for the four simulation cells are shown in Figure 5.5. Atomic and valence electron density profiles for the two simulation cells, labeled "(001)" and "(111)," are shown for the two temperatures at which each was simulated. They show definite oscillations at the liquid surface, indicating the formation of atomic layers. The strength of the layers is seen to be reduced at the higher temperature for each simulation cell. Average layer spacings for the surface and inner layers in each of the free surface simulations are given in Table 5.1. The

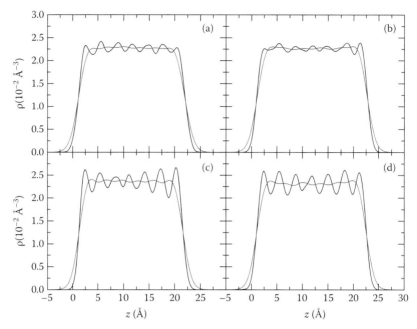

FIGURE 5.5 Density profiles obtained for free liquid sodium surfaces [70–74]. Black lines give the atomic density profiles, red lines the valence electron densities. The simulations are (a) $T = 500\,K$, (001) cell; (b) $T = 500\,K$, (111) cell; (c) $T = 400\,K$, (001) cell; (d) $T = 400\,K$, (001) cell. (Adapted with permission from Walker, B.G., *Ab Initio Molecular Dynamics Studies of Liquid Metal Surfaces*, PhD thesis, University of Cambridge, copyright 2004; Walker, B.G. et al., *J. Phys. Condens. Matter*, 16, S2575, copyright 2004; Walker, B.G. et al., *J. Chem. Phys.*, 124, 174702, copyright 2006; Walker, B.G. et al., *J. Phys. Condens. Matter*, 18, L269, copyright 2006; Walker, B.G. et al., *J. Chem. Phys.*, 127, 134703, copyright 2007, American Institute of Physics.)

average layer spacings extracted from the density profiles are slightly smaller in the surface layer compared to the inner layers (suggesting confinement in the outermost layers); within the quoted uncertainties, the average layer spacings for the two different simulation cell cross-sections are consistent (indicating that no finite-size effects are playing an important role). The layer spacings at the liquid surfaces are uncorrelated with the layer positions that would be seen at a solid surface (i.e., in the surface simulations, the numbers of atomic layers and their spacings are independent of those in the initial bulk simulation cells before they were melted); furthermore, the layer positions are not greatly affected by the temperature.

5.3.3.1.1   Assessment of Finite-Size Effects in Ab Initio DFT MD Free Surface Simulations

Careful examination of the independence of the surfaces on either side of the slabs was made. The density–density autocorrelation functions (see Equation 5.1 of

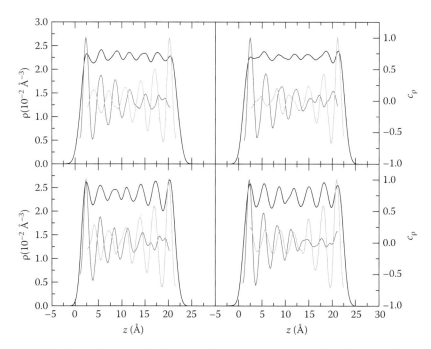

FIGURE 5.6 (See color insert.) Density–density autocorrelation functions (c_ρ) for the Na free surface simulations. In each case, the red and green curves give the density–density autocorrelation functions referenced respectively to the left and right peaks in the atomic densities (which are shown as the black curves). The labeling of the panels is as in Figure 5.5. (Adapted with permission from Walker, B.G. et al., *J. Chem. Phys.*, 124, 174702, copyright 2006, American Institute of Physics.)

Ref. [43]) for the oscillations propagating from the two surfaces were calculated and are displayed in Figure 5.6. As the autocorrelation functions become uncorrelated moving from the respective surfaces into the bulk and each one shows no structure near the opposite surface, the structures of the two surfaces are independent of one another.

The importance of finite-size effects on the simulation cells used in the MD was examined using simulation cells with different thicknesses and cross sections, which were described using the empirical potential used for Na by Chacón et al. [46]. Because the computational requirements of making simulations are much smaller than when using DFT, slab simulations containing far greater numbers of atoms could be studied. In the upper panels of Figure 5.7, the surface profiles for classical simulations are compared for a cell containing the same number of atoms as the (001) cell used in the ab initio MD simulations, and for one having a slab of atoms twice as thick. The classical density profiles show that the effect on the surface structure of having a slab of a larger thickness is minimal. The lower panel of Figure 5.7 compares the density profiles obtained from classical MD simulations of slabs having the same thickness as the (001) ab initio simulation, but with various

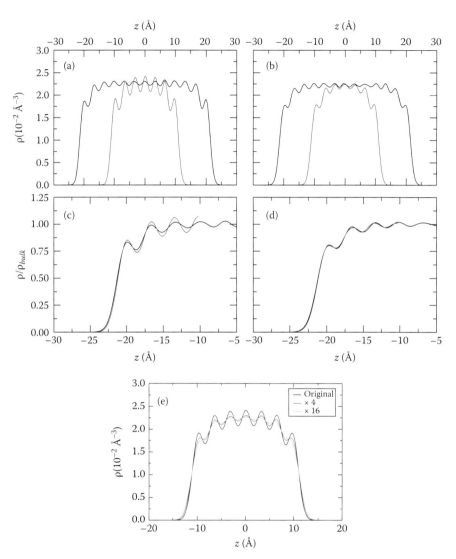

FIGURE 5.7 (See color insert.) (a) Density profiles obtained for classical MD simulations of Na, for a cell having the same number of atoms as in the (001) ab initio simulations (red curves), and for a slab of twice the thickness (black curves), at temperature $T = 650\,K$. (b) As for (a), but at temperature $T = 750\,K$. (c) Surface parts of the density profiles for the slabs in (a) overlaid. (d) The surface parts of the density profiles for the slabs in (b) overlaid. (e) Density profiles for classical MD simulations of a slab having the same thickness as the (001) ab initio simulations (black curve), and for cells having 4 and 16 times the cross-sectional area (red and green curves, respectively). (Adapted with permission from Walker, B.G. et al., *J. Chem. Phys.*, 124, 174702, copyright 2006, American Institute of Physics.)

cross-sectional areas (4 times the area and 16 times the area). These simulations show that although the strengths of the peaks in the density profile are damped when larger cross-sectional areas are considered, their positions are unaffected. The relative heights of the peaks in the classical simulations differ from those in the ab initio simulations, in that in the classical simulations the outermost peaks are lower in height than the inner peaks, while in the ab initio simulations the outermost peaks are of the same or greater height than the inner peaks. The larger height of the outermost peaks in the ab initio simulations is consistent with the idea that the surface is more strongly confined in the eDFT treatment.

5.3.3.2   Introducing a Thin Rigid Wall into Liquid Sodium

As we have noted, at the free surface, there is a rapid decay of the electronic density from the bulk liquid part to the vapor region (where the electronic density is close to zero). In order to probe the importance of this on the layering, and to avoid the possibility of Friedel oscillations in the electronic density, a computational experiment that contained a rigid layer of atoms, but no vacuum regions was performed. In addition to considering the free liquid surface, molecular dynamics has been used to consider the properties of liquid sodium interacting with a fixed layer of sodium atoms [73]. In this case, the charge density is continuous at the fixed layer, meaning that there should be no Friedel oscillations, though the motion of the liquid atoms is restricted by the presence of the fixed layer of sodium atoms. The fixed layer of atoms does not represent a defect for the electronic system in the same way as the presence of a surface as there is no vacuum region and no requirement of a rapid decrease in the electronic density. Simulations were made for liquid sodium in contact with a fixed layer of sodium atoms for temperatures of 400 and 800 K. The density profiles for simulations performed at these two temperatures are shown in Figure 5.8. At both of these simulation temperatures, strong layer formation in the liquid part is shown by the oscillations in the density profiles; the layer formation is much stronger than for the free surfaces, suggesting a strong degree of geometrical confinement of the liquid part in these systems. It is useful to note that there are no undercoordinated liquid atoms at the interface with

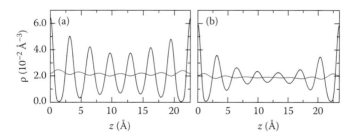

FIGURE 5.8   Density profiles for simulations of liquid sodium in contact with a rigid fixed layer of atoms [73]: (a) $T=400\,K$ and (b) $T=800\,K$. The black curves show the atomic densities, and the red curves show the electronic densities. (Adapted from Walker, B.G. et al., *J. Chem. Phys.*, 124, 174702, 2006; Walker, B.G. et al., *J. Phys. Condens. Matter*, 18, L269, 2006. With permission.)

the fixed layer, precluding the possibility of such atoms driving rearrangements leading to layer formation.

5.3.3.3 Friedel Oscillations in Sodium

To ascertain whether the Friedel oscillations that were observed at the solid $Mg(10\overline{1}0)$ surface in Ref. [45] persist at liquid metal surfaces, the density subtraction procedure used by Cho et al. for solid surfaces was adapted to consider the formation of a liquid surface.

The method of calculating the charge redistribution upon surface formation is illustrated in Figure 5.9 for a crystalline system: the valence electron density for the system in which a surface has been created is subtracted from the density for the bulk system in which there is no vacuum region. Valence charge density redistributions upon surface formation for solid surfaces of sodium are shown in Figure 5.10. Strong Friedel oscillations are found to exist in the redistribution of charge upon surface formation for solid sodium surfaces.

A similar procedure was followed to extend the study of Friedel oscillations at the liquid surface. Though it has been shown (for instance by the work of Chacón et al. [46], Tarazona et al. [47], and Velasco et al. [48]) that layered atomic surface density profiles can be obtained without taking into account the quantum mechanics of the electrons, computation of the response of the electronic system to surface formation requires an accurate quantum mechanical treatment. First, a series of snapshots were taken from a simulation of bulk liquid sodium and a vacuum region was added to

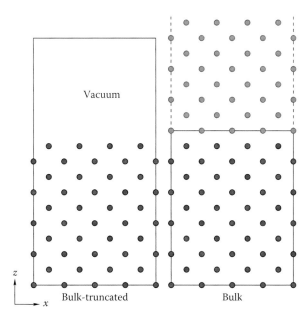

FIGURE 5.9 Unit cells used to calculate the charge redistribution upon surface formation. The unit cell on the left is the "bulk-truncated" cell that contains a vacuum region and that on the right is the bulk cell that contains no vacuum region, and so no surfaces.

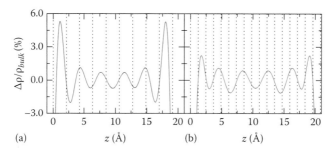

FIGURE 5.10 Friedel oscillations in valence charge density redistributions at solid sodium surfaces [70,73]. Panel (a) is for the (001) surface and panel (b) the (111) surface. The dotted vertical lines show the positions of the underlying layers of atoms. (Adapted from Walker, B.G. et al., *J. Chem. Phys.*, 124, 174702, 2006; Walker, B.G. et al., *J. Phys. Condens. Matter*, 18, L269, 2006; Walker, B.G., *Ab Initio Molecular Dynamics Studies of Liquid Metal Surfaces*, PhD thesis, University of Cambridge, 2004. With permission.)

each to form a surface. Then, the charge density was calculated for this newly formed surface. The density determined for this newly created surface was subtracted from the density for the original bulk snapshot, to determine the redistribution of charge that would occur upon surface formation (no relaxation of the atomic positions in the snapshot is made when the surface is formed by addition of the vacuum region). The charge densities for selected snapshots from the bulk liquid simulation are plotted in Figure 5.11, as well as the charge density differences between the bulk and surface charge density profiles for the liquid snapshots. It is clear that although the valence charge densities in different snapshots are rather different (reflecting the fact that the atoms in the liquid move considerably), when the redistribution profiles of the valence charge upon surface formation are determined they are remarkably uniform for the different snapshots. The charge redistributions are relatively uncoupled from the underlying ionic positions. Furthermore, the redistribution profiles indicate the presence of strong Friedel oscillations upon surface formation, even for the liquid snapshots, where there are no atomic layers. The difference in the forces between the atoms in the bulk and surface snapshots are given in the bottom panel of Figure 5.11. The atoms that become the surface atoms when vacuum regions are added to the bulk simulation snapshots have larger forces on them than those that are near the center of the slabs of atoms created. The forces induced upon surface formation do not correlate with the Friedel oscillations observed in the valence charge redistribution profiles.

The Friedel oscillation wavelengths obtained from these calculations are approximately 3 Å (see Table 6.7 of Ref. [70]), which agree well with the expected Friedel oscillation wavelength for sodium (3.4 Å) [44]. Comparing these numbers with the layer spacings calculated for the free Na liquid surfaces in Table 5.2 shows them to be quite similar. This means that the importance of Friedel oscillations cannot be assessed solely on the basis of compatibility of the layer spacings and the Friedel oscillation wavelength. However, based on the combination of the persistence of Friedel oscillations upon formation of liquid surfaces and the strong layer formation

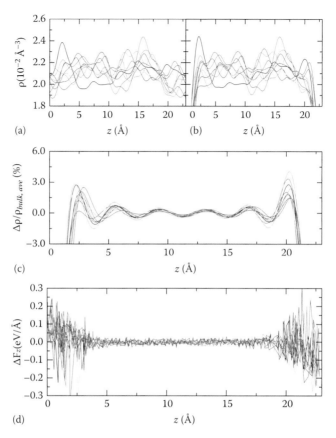

FIGURE 5.11 (See color insert.) Friedel oscillations observed in snapshots from a bulk liquid simulation [73]. Panel (a) shows the valence charge densities for the various MD time steps taken from the bulk liquid simulation. Panel (b) shows the valence charge densities for the same MD snapshots after a vacuum region has been added to each of the unit cells. Panel (c) shows the valence charge density redistributions upon surface formation for the set of bulk MD time steps. Panel (d) shows the differences in the forces acting on atoms (i.e., the forces induced by formation of a surface). (Adapted from Walker, B.G. et al., *J. Chem. Phys.*, 124, 174702, copyright 2006; Walker, B.G. et al., *J. Phys. Condens. Matter*, 18, L269, copyright 2006. With permission.)

in the liquid + fixed-layer simulations in the absence of Friedel oscillations [73], it is very unlikely that Friedel oscillations are responsible for the formation of layers at liquid metal surfaces.

5.4 IN-PLANE ORDERING

In addition to focusing on the arrangement of the atoms transverse to the surface, there is the important issue of ordering in the plane of the surface. Simulations based on glue model potentials of supercooled liquid Au [42] indicated that there should be formation of a quasihexatic phase within the layers at the surface. In such an

TABLE 5.2

Layer Spacings Determined from Sodium Liquid Surface Simulations in Refs. [70–74]

	$d_{surface}$ (Å)		d_{inner} (Å)	
$T(K)$	(001)	(111)	(001)	(111)
400	2.98 ± 0.03	2.97 ± 0.05	2.98 ± 0.24	3.20 ± 0.15
500	2.91 ± 0.15	2.75 ± 0.12	3.04 ± 0.24	3.34 ± 0.43

Source: Adapted with permission from Walker, B.G., *Ab Initio Molecular Dynamics Studies of Liquid Metal Surfaces*, PhD thesis, University of Cambridge, copyright 2004; Walker, B.G. et al., *J. Phys. Condens. Matter*, 16, S2575, copyright 2004; Walker, B.G. et al., *J. Chem. Phys.*, 124, 174702, copyright 2006; Walker, B.G. et al., *J. Phys. Condens. Matter*, 18, L269, copyright 2006; Walker, B.G. et al., *J. Chem. Phys.*, 127, 134703, copyright 2007, American Institute of Physics.

The labels "(001)" and "(111)" refer to the simulation cells having square- and diamond-shaped cross-sectional geometries, respectively.

ordering, the atoms are on average sixfold-coordinated, although there are significant numbers of atoms with coordination numbers of 5 and 7.

Experimental work on liquid lead in contact with a solid silicon surface [81] found evidence for the existence of partial icosahedral fragments (corresponding to fivefold local ordering) at the interface of the lead with the Si surface. In considering free liquid surfaces and liquid–solid interfaces, one can ask whether this type of local ordering might exist at those interfaces in other types of metal.

The DFT-based molecular dynamics simulations of liquid Na discussed in Section 5.3 were analyzed in relation to this question [74]. The simulations yielded indications, although weak, that there is a preference for fivefold- and sixfold-coordinated ordering within the planes at the surface, with the tendency toward sixfold ordering increasing as the temperature is lowered.

The regions of the free surface considered in examining the in-plane structure are indicated in Figure 5.12. The numbers of nearest neighbors in the various regions are shown in Figure 5.13. The dominant number of neighbors is 5, though there are significant weights at 4 and 6. The proportion of atoms having 4 neighbors decreases as the temperature is lowered from 500 to 400 K. The in-plane structure can be further examined with the calculation of bond angle distributions; the bond angle distributions obtained from the molecular dynamics simulations appear in Figure 5.14. The bond angle distributions do not vary much in the different regions of the surface at the same temperature, though in comparing the bond angle distributions at $T = 400$ K with those at $T = 500$ K, a movement of weight to higher bond angles can be seen at the lower temperature. It was suggested [74] that this is consistent with

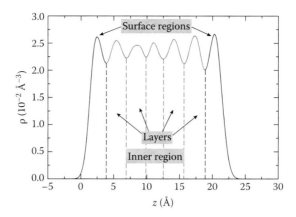

FIGURE 5.12 Division of the free surface simulations (modeled as slabs of atoms, with vacuum on either side) into surface, and inner regions, for examination of structure within the plane of the surface. The extents of the different regions are determined from the stable time-averaged transverse density profiles (that illustrated is for the (001) simulation at $T = 400\,\mathrm{K}$). (Adapted with permission from Walker, B.G. et al., *J. Chem. Phys.*, 127, 134703, copyright 2007, American Institute of Physics.)

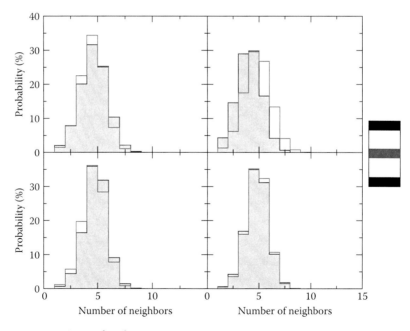

FIGURE 5.13 (See color insert.) Nearest neighbor distributions within the different regions of the liquid surfaces simulated with ab initio molecular dynamics from Refs. [70,74]. As indicated by the bar to the right of the plots: black lines refer to the surface layers and red lines to the inner layers. The labeling of the panels is as in Figure 5.5. (Adapted with permission from Walker, B.G., *Ab Initio Molecular Dynamics Studies of Liquid Metal Surfaces*, PhD thesis, University of Cambridge, copyright 2004; Walker, B.G. et al., *J. Chem. Phys.*, 127, 134703, copyright 2007, American Institute of Physics.)

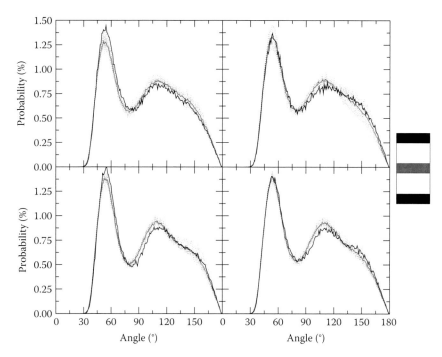

FIGURE 5.14 (See color insert.) Distributions of bond angles determined in the different regions of the simulated liquid sodium surfaces from Refs. [70,74]. The regions referred to by the different curves are as in Figure 5.13. The labeling of the panels is as in Figure 5.5. (Adapted with permission from Walker, B.G., *Ab Initio Molecular Dynamics Studies of Liquid Metal Surfaces*, PhD thesis, University of Cambridge, copyright 2004; Walker, B.G. et al., *J. Chem. Phys.*, 127, 134703, copyright 2007, American Institute of Physics.)

an enhancement of the angles 60°, 120°, and 180°, the bond angles corresponding to hexagonal ordering within the surface layers.

The indication that there is some tendency toward fivefold coordination at the surface was rationalized by considering the introduction of relatively small amounts of disorder to a hexagonal lattice in Ref. [74]. The resulting structure displays regions of mostly fivefold- and sixfold-coordinated order. This type of surface structure can be modeled rather simply starting with perfect hexagonal packing of the atoms into the cross-sections of the cells. As the numbers of atoms in the surface layers (as indicated by integration of the density profiles) are slightly lower than needed for perfect hexagonal packing, some atoms should be removed. Adding a bit of thermal disorder gives a structure having mostly five- and sixfold ordering. This procedure is illustrated in Figure 5.15.

Experimentally, the technique of grazing-incidence x-ray diffraction may be able to explore the existence and form of in-plane order at liquid metal surfaces, though no conclusive measurements relating to this aspect of liquid metal surface structure have appeared.

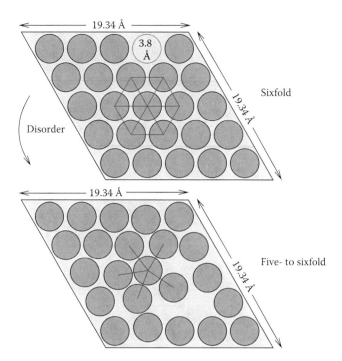

FIGURE 5.15 Illustration of removal of atoms from perfectly ordered hexagonal arrangement in 2D, followed by disordering to obtain an arrangement with remnant order, as a model for the in-plane structure of the surface layers seen in the ab initio MD simulations of Na. (Adapted with permission from Walker, B.G. et al., *J. Chem. Phys.*, 127, 134703, 2007, American Institute of Physics.)

5.5 SUMMARY

In this chapter, we have discussed the properties of liquid metal surfaces, focusing particularly on characterization and understanding of surface structure from computer simulations and experiments. The liquid surfaces of metals demonstrate the interesting phenomenon of surface-induced layering; this has been firmly established on the basis of simulations—the most accurate being quantum mechanical simulations based on density functional theory—as well as extensive x-ray reflectivity measurements. At least three formation mechanisms were suggested on the basis of earlier simulation work. These were assessed by more recent simulations on liquid sodium surfaces using the rigorous ensemble density functional theory technique. Combining simulations of the free liquid surfaces of sodium with computer "experiments" (e.g., simulation of the liquid in contact with a fixed layer of sodium atoms) favors the explanation that the rapid decay of the electronic density across the free surface or liquid–vapor interface acts like a hard wall, geometrically confining the electrons and resulting in the formation of atomic layers. Friedel oscillations as potentially driving layer formation was ruled out by the observation of layers when a rigid layer is inserted into the bulk liquid. Friedel oscillations were seen at solid surfaces of sodium (an observation similar to that seen for $Mg(10\bar{1}0)$ earlier [45]);

more surprising is that strong Friedel oscillations were seen to be present for *liquid* surfaces. Though the atomic arrangements used in the formation of liquid surfaces differed greatly, the response in the valence electron density—displaying Friedel oscillations—was remarkably uniform.

Atomistic simulations have also indicated that there should be ordering *within* the planes of the layers formed at the surfaces of liquid metals. Early simulations indicated that there should be hexatic ordering within the plane of the surface for supercooled Au surfaces. The ensemble DFT/MD simulations on sodium that have been discussed in this chapter favor a predominantly fivefold-coordinated ordering, interspersed with significant numbers of atoms having fewer neighbors (fourfold-coordinated) as well as more neighbors (sixfold-coordinated ordering). The proportion of atoms with five- and sixfold coordination increases as the temperature decreases (an observation that correlates with the expected reduction in disorder at lower temperature). The presence of fivefold ordering may connect with the observation of fivefold local symmetry (due to icosahedral fragments of liquid atoms) at the interface between a liquid metal and a solid substrate [81]. From the experimental point of view, the existence and nature of any order within the surface layers is as yet unresolved.

REFERENCES

1. Magnussen, O.M. et al., Self-assembly of organic films on a liquid metal, *Nature*, 384, 250, 1996.
2. Kraack, H. et al., Structure of a Langmuir film on a liquid metal surface, *Science*, 298, 1404, 2002.
3. Rowlinson, J.S. and Widom, B., *Molecular Theory of Capillarity*, Dover Publications Inc., New York, 2002.
4. Lu, B.C. and Rice, S.A., Determination of the density profile in the liquid–vapor interface near the triple point, *J. Chem. Phys.*, 68, 5558, 1978.
5. Sluis, D., D'Evelyn, M.P., and Rice, S.A., Experimental and theoretical studies of the density profile in the liquid–vapor interface of Cs, *J. Chem. Phys.*, 78, 1611, 1983.
6. Magnussen, O.M. et al., X-ray reflectivity measurements of surface layering in liquid mercury, *Phys. Rev. Lett.*, 74, 4444, 1995.
7. Regan, M.J. et al., Surface layering in liquid gallium: An x-ray reflectivity study, *Phys. Rev. Lett.*, 75, 2498, 1995.
8. Regan, M.J. et al., Capillary-wave roughening of surface-induced layering in liquid gallium, *Phys. Rev. B*, 54, 9730, 1996.
9. Regan, M.J. et al., X-ray studies of atomic layering at liquid metal surfaces, *J. Non-Cryst. Solids*, 205–207, 762, 1996.
10. Di Masi, E. et al., X-ray reflectivity study of temperature-dependent surface layering in liquid Hg, *Phys. Rev. B*, 58, R13419, 1998.
11. Tostmann, H. et al., Surface structure of liquid metals and the effect of capillary waves: X-ray studies on liquid indium, *Phys. Rev. B*, 59, 783, 1999.
12. Tostmann, H. et al., Microscopic structure of the wetting film at the surface of liquid Ga-Bi alloys, *Phys. Rev. Lett.*, 84, 4385, 2000.
13. Huber, P. et al., Wetting at the free surface of a liquid gallium-bismuth alloy: An x-ray reflectivity study close to the bulk monotectic point, *Colloids & Surfaces A*, 206, 515, 2002.
14. Huber, P. et al., Short-range wetting at liquid gallium-bismuth alloy surfaces: X-ray measurements and square-gradient theory, *Phys. Rev. B*, 68, 085401, 2003.

15. Tostmann, H. et al., Microscopic surface structure of liquid alkali metals, *Phys. Rev. B*, 61, 7284, 2000.
16. Shpyrko, O.G. et al., X-ray study of the liquid potassium surface: Structure and capillary wave excitations, *Phys. Rev. B*, 67, 115405, 2003.
17. Shpyrko, O.G. et al., Anomalous layering at the liquid Sn surface, *Phys. Rev. B*, 70, 224206, 2004.
18. Pershan, P.S. et al., Surface structure of liquid Bi and Sn: An x-ray reflectivity study, *Phys. Rev. B*, 79, 115417, 2009.
19. Kraack, H. et al., Fatty acid Langmuir films on liquid mercury: X-ray and surface tension studies, *J. Chem. Langmuir*, 20, 5375, 2004.
20. Tamam, L. et al., The structure of mercapto-biphenyl monolayers on mercury, *J. Phys. Chem. B*, 109, 12534, 2005.
21. Kraack, H. et al., Langmuir films of normal-alkanes on the surface of liquid mercury, *J. Chem. Phys.*, 119, 10339, 2003.
22. Kraack, H. et al., Temperature dependence of the structure of Langmuir films of normal-alkanes on liquid mercury, *J. Chem. Phys.*, 121, 8003, 2004.
23. Tamam, L. et al., Langmuir films of polycyclic molecules on mercury, *Thin Solid Films*, 515, 5631, 2007.
24. Shpyrko, O.G. et al., Atomic-scale surface demixing in a eutectic liquid BiSn alloy, *Phys. Rev. Lett.*, 95, 106103, 2005.
25. Shpyrko, O.G. et al., Surface freezing in liquid AuSi alloy, *Science*, 313, 77, 2006.
26. Shpyrko, O.G. et al., Crystalline surface phases of the liquid Au-Si eutectic alloy, *Phys. Rev. B*, 76, 245436, 2007.
27. Balagurusamy, V.S.K. et al., X-ray reflectivity studies of atomic-level surface-segregation in a liquid eutectic alloy of AuSn, *Phys. Rev. B*, 75, 104209, 2007.
28. Shpyrko, O.G. et al., Surface layering of liquids: the role of surface tension, *Phys. Rev. B*, 69, 245423 2004.
29. Mo, H. et al., Observation of surface layering in a nonmetallic liquid, *Phys. Rev. Lett.*, 96, 096107, 2006.
30. D'Evelyn, M.P. and Rice, S.A., Structure in the density profile at the liquid–metal–vapor interface, *Phys. Rev. Lett.*, 47, 1844, 1981.
31. D'Evelyn, M.P. and Rice, S.A., A pseudoatom theory for the liquid–vapor interface of simple metals: Computer simulation studies of sodium and cesium, *J. Chem. Phys.*, 78, 5225, 1983.
32. D'Evelyn, M.P. and Rice, S.A., A study of the liquid–vapor interface of mercury: Computer simulation results, *J. Chem. Phys.*, 78, 5081, 1983.
33. Gryko, J. and Rice, S.A., The liquid–vapor interface density profiles and the surface pair correlation functions of sodium and a sodium–cesium alloy, *J. Non. Cryst. Solids*, 61/62, 703, 1984.
34. Gryko, J. and Rice, S.A., Comment on the structures of the liquid–vapor interfaces of Na and Na–Cs alloys, *J. Chem. Phys.*, 80, 6318, 1984.
35. Harris, J.G., Gryko, J., and Rice, S.A., Self-consistent Monte Carlo simulations of the electron and ion distributions of inhomogeneous liquid alkali metals. I. Longitudinal and transverse density distributions in the liquid–vapor interface of a one-component system, *J. Chem. Phys.*, 87, 3069, 1987.
36. Gomez, M.A. and Rice, S.A., Self-consistent Monte Carlo simulation of the electron and ion distributions in the liquid–vapor interface of magnesium, *J. Chem. Phys.*, 101, 8094, 1994.
37. Zhao, M. et al., Structure of liquid Ga and the liquid–vapor interface of Ga, *Phys. Rev. E*, 56, 7033, 1997.
38. Zhao, M., Chekmarev, D.S., and Rice, S.A., Quantum Monte Carlo simulations of the structure in the liquid–vapor interface of BiGa binary alloys, *J. Chem. Phys.*, 108, 5055, 1998.

39. Zhao, M. and Rice, S.A., The structure of the liquid–vapor interface of a gallium-tin binary alloy, *J. Chem. Phys.*, 111, 2181, 1999.

40. Iarlori, S. et al., Structure and correlations of a liquid metal surface: Gold, *Surf. Sci.*, 211/212, 55, 1989.

41. Di Tolla, F.D., *Interplay of Melting, Wetting, Overheating and Faceting on Metal Surfaces: Theory and Simulation*, PhD thesis, SISSA, 1996.

42. Celestini, F., Ercolessi, F., and Tosatti, E., Can liquid metal surfaces have hexatic order? *Phys. Rev. Lett.*, 78, 3153, 1997.

43. Fabricius, G. et al., Atomic layering at the liquid silicon surface: A first-principles simulation, *Phys. Rev. B*, 60, R16283, 1999.

44. Ashcroft, N.W. and Mermin, N.D., *Solid State Physics*, Harcout College Publishers, Orlando, FL, 1979.

45. Cho, J.H. et al., Oscillatory lattice relaxation at metal surfaces, *Phys. Rev. B*, 59, 1677, 1999.

46. Chacón, E. et al., Layering at free liquid surfaces, *Phys. Rev. Lett.*, 87, 166101, 2001.

47. Tarazona, P. et al., Layering structures at free liquid surfaces: The Fisher–Widom line and the capillary waves, *J. Chem. Phys.*, 117, 3941, 2002.

48. Velasco, E. et al., Low melting temperature and liquid surface layering for pair potential models, *J. Chem. Phys.*, 117, 10777, 2002.

49. González, L.E., González, D.J., and Stott, M.J., Interplay between the ionic and electronic density profiles in liquid metal surfaces, *J. Chem. Phys.*, 123, 201101, 2005.

50. Marx, D. and Hutter, J., Ab initio molecular dynamics: Theory and implementation, in *Modern Methods and Algorithms of Quantum Chemistry*, NIC Series, vol. 1, Grotendorst, J., Ed., John von Neumann Institute for Computing, Jülich, Germany, 2000, p. 301.

51. Hellmann, H., *Einführung in die Quantenchemie*, Deuticke, Leipzig, 1937.

52. Feynman, R.P., Forces in molecules, *Phys. Rev.*, 56, 340, 1939.

53. Gear, C.W., *Numerical Initial Value Problems in Ordinary Differential Equations*, Prentice-Hall, Englewood Cliffs, NJ, 1971.

54. Verlet, L., Computer "experiments" on classical fluids. I. Thermodynamical properties of Lennard-Jones molecules, *Phys. Rev.*, 159, 98, 1967.

55. Tuckerman, M.E. and Langel, M., Multiple time scale simulation of a flexible model of CO_2, *J. Chem. Phys.*, 100, 6368, 1994.

56. Hohenberg, P. and Kohn, W., Inhomogeneous electron gas, *Phys. Rev.*, 136, B864, 1964.

57. Kohn, W. and Sham, L.J., Self-consistent equations including exchange and correlation effects, *Phys. Rev.*, 140, A1133, 1965.

58. Payne, M.C. et al., Iterative minimization techniques for ab initio total-energy calculations: Molecular dynamics and conjugate gradients, *Rev. Mod. Phys.*, 64, 1045, 1992.

59. Argaman, N. and Makov, G., Density functional theory: An introduction, *Am. J. Phys.*, 68, 69, 2000.

60. Mermin, N.D., Thermal properties of the inhomogeneous electron gas, *Phys. Rev.*, 137, A1441, 1965.

61. Marzari, N., *Ab Initio Molecular Dynamics for Metallic Systems*, PhD thesis, University of Cambridge, 1996.

62. Marzari, N., Vanderbilt, D., and Payne, M.C., Ensemble density-functional theory for ab initio molecular dynamics of metals and finite-temperature insulators, *Phys. Rev. Lett.*, 79, 1337, 1997.

63. Molteni, C. et al., Sliding mechanisms in aluminum grain boundaries, *Phys. Rev. Lett.*, 79, 869, 1997.

64. Marzari, N. et al., Thermal contraction and disordering of the Al(110) surface, *Phys. Rev. Lett.*, 82, 3296, 1999.

65. González, D.J., González, L.E., and Stott, M.J., Surface structure in simple liquid metals: An orbital-free first-principles study, *Phys. Rev. B*, 74, 014207, 2006.

66. González, D.J., González, L.E., and Stott, M.J., Surface structure in simple liquid metals. A first principles simulation, *J. Non-Cryst. Sol.*, 353, 3555, 2007.

67. González, L.E. and González, D.J., Structure and dynamics of bulk liquid Ga and the liquid–vapor interface: An ab initio study, *Phys. Rev. B*, 77, 064202, 2008.

68. González, D.J., González, L.E., and Stott, M.J., Liquid–vapor interface in liquid binary alloys: An ab initio molecular dynamics study, *Phys. Rev. Lett.*, 94, 077801, 2005.

69. González, D.J. and González, L.E., Structure and motion at the liquid–vapor interface of some interalkali binary alloys: An orbital-free ab initio study, *J. Chem. Phys.*, 130, 114703, 2009.

70. Walker, B.G., *Ab Initio Molecular Dynamics Studies of Liquid Metal Surfaces*, PhD thesis, University of Cambridge, 2004.

71. Walker, B.G., Molteni, C., and Marzari, N., Ab initio molecular dynamics of metal surfaces, *J. Phys. Condens. Matter*, 16, S2575, 2004.

72. Walker, B.G., Marzari, N., and Molteni, C., Ab initio studies of layering behavior of liquid sodium surfaces and interfaces, *J. Chem. Phys.*, 124, 174702, 2006.

73. Walker, B.G., Marzari, N., and Molteni, C., Layering at liquid metal surfaces: Friedel oscillations and confinement effects, *J. Phys.: Cond. Matter*, 18, L269, 2006.

74. Walker, B.G., Marzari, N., and Molteni, C., In-plane structure and ordering at liquid sodium surfaces and interfaces from ab initio molecular dynamics, *J. Chem. Phys.*, 127, 134703, 2007.

75. Hockney, R.W., The potential calculation and some applications, in *Methods in Computational Physics*, vol. 9, Plasma Physics, Alder, B., Fernbach, S., and Rotenberg M., Eds., Academic Press, New York/London, 1970.

76. Hoover, W.G. et al., Nonequilibrium molecular dynamics via Gauss's principle of least constraint, *Phys. Rev. A*, 28, 1016, 1983.

77. Baldereschi, A., Mean-value point in the Brillouin zone, *Phys. Rev. B*, 7, 5212, 1973.

78. Troullier, N. and Martins, J.L., Efficient pseudopotentials of plane-wave calculations, *Phys. Rev. B*, 43, 1993, 1991.

79. Perdew, J.P. and Wang, Y., Accurate and simple analytic representation of the electron-gas correlation energy, *Phys. Rev. B*, 45, 13244, 1992.

80. Louie, S.G., Froyen, S., and Cohen, M.L., Nonlinear ionic pseudopotentials in spin-density-functional calculations, *Phys. Rev. B*, 26, 1738, 1982.

81. Reichert, H. et al., Observation of five-fold local symmetry in liquid lead, *Nature*, 408, 839, 2000.

6 Some Practical Considerations for Density Functional Theory Studies of Chemistry at Metal Surfaces

Rudolph J. Magyar, Ann E. Mattsson,
and Peter A. Schultz

CONTENTS

6.1   INTRODUCTION

The aim of this chapter is to introduce the basic challenges and tools involved in modeling chemistry on metallic surfaces, and to provide practical guidance in the use of these methods and tools to design quantitative simulations of surface processes. Calculations involving a metal surface introduce many additional complications relative to quantum chemistry studies of molecular chemistry. These not only include new methods and unfamiliar tools but also novel issues and unexpected pitfalls. Thanks to continuing advances in methods and computational capabilities, first principles simulations of surface chemistry are becoming generally accessible. These advances are founded upon the success of density functional theory (DFT) and the development of large-scale solid-state codes implementing DFT for periodic systems. The size and complexity of first principles simulations has become staggeringly large compared to what was feasible decades ago, yet the codes can be deceptively simple to use. The advanced capabilities of modern computational tools rely on many layers of complexity with many hidden options and assumptions, and it is possible that concealed approximations may ultimately limit the accuracy of a simulation. Obtaining quantitatively meaningful calculations for surface chemistry requires awareness of the physical approximations inherent in the methods and careful design of a viable model. Improper use of these tools can result in drastically unrealistic conclusions. Some general considerations along these lines for solid-state calculations in materials science were considered in a recent review article [1]. Here, we describe several fundamental issues that arise in applying DFT methods to surface chemistry, calculations that explore the intersection between physics and chemistry, the realm of metallic clusters and surface catalysis.

This chapter is divided into three sections. The first details important conceptual differences between finite molecular systems and bulk metals, and the implications these differences have on the computational description of the chemistry at metal surfaces. The second gives a brief overview of DFT, focusing on the importance of the choice of functional can have on results, and how the proper interpretation of DFT may limit the scope of conclusions that can be drawn from the results. The third section describes the design of a viable computational model for surface chemistry: the use of a slab model to describe the surface, construction of an appropriate simulation cell, testing and verifying the evaluation of surface properties, and use of the model to study chemistry. This process is illustrated through detailed examination of two surface chemistry examples: erbium surfaces and CO on Pt.

6.2   CHALLENGES ARISING IN THE SIMULATION OF METALS

The defining difference between metals and nonmetals is the lack of an energy gap between the ground and excited states. In the molecular orbital picture, this transition from nonmetals to metals can be thought of as a vanishing energy difference between the highest occupied and lowest unoccupied molecular orbital, the HOMO-LUMO gap. Once this gap is closed, there is no energy barrier to prevent electrons from exciting to the unoccupied states that are more delocalized than the occupied ones. This ability for electrons to freely excite allows for great chemical and structural flexibility but adds an extra degree of complexity to the numerical simulation. The onset of metallic behavior varies greatly depending on system size and constituents, and we concern ourselves primarily in this chapter with systems that are large and dense enough to exhibit delocalized electronic behavior.

The closing of the energy gap has many practical implications on numerical simulations. The first is that the outer shell or valence electrons can spread out over macroscopic dimensions without an energy barrier. Transient electrons can act collectively as a liquid according to classical electrostatics. For example, in a bulk charged metal, the electrons repel each other forcing any excess charge to the surface, leaving the typical bulk unit cell with zero net change.

The transition from a system with discrete energy levels to another with discrete bands can be imagined using a highly simplified Hückel model of a conjugated polymer. The Hückel Hamiltonian [2] is

$$H_{ij} = \alpha \delta_{ij} + \beta \left(\delta_{i,j+1} + \delta_{i,j-1} \right),$$

where
  α is the on-site Coulomb energy
  β is a transfer term that describes the kinetic energy overlap between atoms
  δ_{ij} is Kronecker delta indexed according to atom sites

We will ignore the on-site Coulomb energy by setting $\alpha = 0$ and retain only the transfer term β. A single isolated dimer has a discrete HOMO-LUMO gap. If we were to bond this dimer with another dimer, we would find a slight decrease in the magnitude of the HOMO-LUMO gap. This decrease arises due to hybridization of the various atomic HOMO orbitals and separately to the LUMO orbitals. Figure 6.1 shows the gap as the length of the polymer is extended. The HOMO-LUMO gap

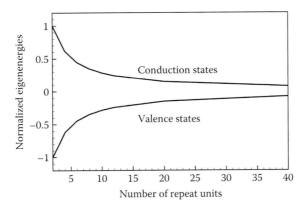

FIGURE 6.1 Closing of the energy gap in model polymer.

decreases with increasing polymer length. In fact, for any finite length polymer within this model, the gap will never become zero. Nevertheless, real polymers at finite temperature conduct electrical current. One of the reasons for this paradox is that at finite temperature, the effective band gap is somewhat less than the zero temperature one and could in fact vanish where a zero temperature one might not. The analogue of the HOMO-LUMO gap in a bulk material is the valence-conduction or band gap. A physical example of the transition from van der Waals to metallic behavior can be seen in experiments on Mercury clusters [3]. For Hg clusters smaller than 20 atoms, the electronic structure, as measured by photo-ionization spectroscopy, appears mostly molecular in character, revealing discrete optical transitions. For larger clusters (20–70 atoms in size), there is a gradual transition in optical response to free-electron behavior. We note that these experiments were performed at high temperatures where a significant fraction of valence electrons could be excited to the conduction level, eventually resulting in an effectively gapless cluster.

Although a large number of atoms are involved, metallic crystals often possess of high degree of symmetry and can be well described by codes with periodic boundary conditions. When switching from the finite system paradigm to a periodic system, several quantities are interpreted differently. The spin per unit cell is now noninte-gral and given as an effective spin per atom—this is because a unit cell is not to be thought of as an isolated molecule. The spin polarization is an effective average over many atoms with one majority spin site out of a host of spinless sites.

It is often shown in the literature that the HOMO is directly related to the Fermi-energy for a finite system at zero temperature. This is not true for a solid or at finite temperature [4]. For a finite system, there is a countable number of electrons and a code can fill quantum orbitals until all electrons are accounted for. In a bulk solid, the number of electrons is finite with an average number of electrons per unit cell. The Fermi level corresponds to the upper energy level required for the integrated density of states to give the number of electrons per unit cell (Figure 6.2).

The implications of the closing of the band gap beyond the effects on the interpretation of electron numbers. The closing of the band gap affects the density of states given by Hartree-Fock (HF) calculations. While for many finite systems HF can be expected to give reasonable if not accurate results, for bulk metallic systems, it provides a qualitatively wrong vanishing of the density of states near the Fermi level [4]. This is because in bulk metallic systems, long-range HF effects are physically compensated with long-range correlation effects not included in HF theory. A corollary is that any quantum chemical method based on a perturbation from a HF reference will be unreliable for bulk metallic metals.

6.3   DENSITY FUNCTIONAL THEORY

The total energy and other non-dynamical physical properties of any material can be calculated, at least in principle, from an exact ground-state solution of the Schrödinger equation, given the information about the nuclei and number of electrons. We restrict ourselves to the Born-Oppenheimer approximation in which the

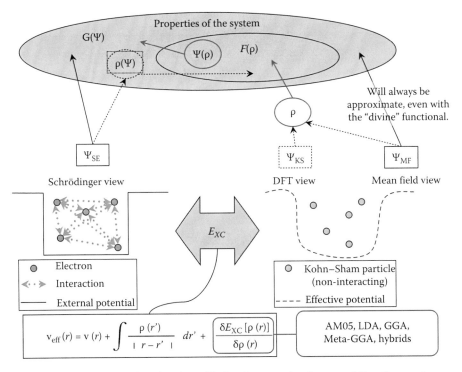

FIGURE 6.2 The exact functional would give the exact density, ρ, and thus the exact properties via density functionals, $F[\rho]$. The unknown wave function, density functional, $\psi[\rho]$, would in this case give all properties that can be calculated from the Schrödinger equation exactly. However, the mean field view of using the KS orbitals as approximate electron wave functions would be approximate. Only the density and properties calculated via density functionals are guaranteed to be exact. The quality of a functional can thus not be judged by how well it reproduces wave function derived properties except where properties of the exact KS wave functions are known.

nuclear positions are independent of the electrons. Even this simplified problem is exponentially difficult mainly due to the difficulty of accounting for many mutually interacting electrons explicitly with a wave function. DFT is a tool to calculate the many-electron density that results from a full solution of the quantum problem. In its most common form, DFT relies on two ideas. First, the Hohenberg-Kohn (HK) theorem provides a one-to-one correspondence between potentials, $v(r)$, and electronic densities, $n(r)$, implying that the density alone contains the same information as the potential [5]. Since the potential gives a formal route to the exact wave function through the Schrödinger equation, so must the density. Second, the Kohn-Sham (KS) scheme provides a route to density by using an auxiliary Schrödinger-like system [6] to represent the kinetic energy. A small but important contribution to the total energy, the exchange-correlation energy $E_{XC}[n]$, is all that remains to be approximated. Over the years, increasing sophistication and experience have been built into ever more complicated functionals to approximate this exchange-correlation contribution to the energy, E_{XC}. Our agenda here is to give the reader a taste of the rationale behind various classes of functionals for E_{XC} and when to trust or distrust them.

Before discussing the practical approximations to exchange-correlation, it is instructive to consider what an exact KS solution using the exact E_{XC} would provide. In practice, the KS equations are solved instead of the full many-electron Schrödinger equation. The KS equations have the familiar form of the independent-electron Schrödinger equation with an effective potential and this has led to a great deal of misappropriate interpretation of the auxiliary quantities of the theory. On one hand, an exact solution of the KS equations using the exact E_{XC} could provide the same accurate prediction of observables that a solution of the electron Schrödinger equation could provide, albeit not always in a direct fashion. Thus, DFT has the extraordinary promise to replace a much more cumbersome problem with a much simpler one. On the other hand, a great deal of work has established understanding regarding the solution of the noninteracting Schrödinger equations. Even prior to the formal introduction of DFT, researchers were developing effective potential theories to provide ever more accurate treatments of many-electron systems, for example, X-α theory [7]. However, the similarities to the noninteracting Schrödinger problem cause confusion. Given the exact functional, DFT is not a mean-field theory (MFT) of electrons but rather an exact theory of the electronic density. There are two fundamental ideas underlying DFT in practice. First is that the proof of a theory can be constructed to reproduce the same density of electrons as quantum mechanics can. The second is that, given this density, useful functionals can be created to compute observables in the same way the expectation values of the many-body wave function can be used to extract observables. In practical KS calculations, these two ideas are merged through the postulated energy functional used to find the density. However, in principle, the two ideas are discrete. Keeping this principle in mind will allow users to better determine, in particular applications, which properties are meaningful and why certain functionals appear to fail or succeed.

In an effective potential theory or MFT, the derivation of the theory is physically motivated but approximate. Take for example the HF theory in which the electrons interact with each other via a Coulomb interaction. The antisymmetry of the many-body wave function with respect to particle interchanges has energetic consequences. The resulting wave function solution is accurate when one Slater determinant dominates and Hartree and exchange account for the majority of the relevant physics beyond the external potential. Since HF is based on a physically motivated model of the structure of the many-electron wave function, it is perfectly sensible to interpret the results by analyzing properties of the HF wave function solution. For example, ionization potentials can formally be related to HF eigenvalues via Koopmans' theorem. Additionally, it would make sense to apply perturbation theory to the resulting HF wave function. The point is that in the MFT, the resulting wave function has formal significance.

The crucial difference between an effective potential theory, like HF theory, and exact KS DFT is that using the exact KS potential would give the exact electronic density rather than the exact electronic wave function. In the KS paradigm, the whole machinery of a Schrödinger-like equation is auxiliary. The KS orbitals in

TABLE 6.1

Physical Properties and the Estimated Level of Difficulty in Extracting them from an Exact Solution Density for a Coulomb Interacting Many-Electron Problem

Easy	Intermediate	Difficult
Finite systems		
Number of electrons	KS orbitals	Exact wave function
Homo eigenvalue	KS eigenspectrum	Total energy
Hartree energy, nuclear energy	Noninteracting kinetic energy, T_s	Exchange–correlation energy, E_{XC}
Moments of the density	KS linear response functions, χ_s	Linear response functions
Nuclear positions and charges		

general have no physical significance. They are not electron orbitals, but rather they are simply mathematical objects that reproduce the exact density and noninteracting kinetic energy. Knowledge of an exact density permits calculation of useful physical quantities like the total electronic energy. In practice, we use the minimization of an energy functional to find the density, but here, we stress the discreteness of the two steps in the KS scheme as a way of illustrating the strengths and shortcomings of the theory.

As a thought exercise, imagine that you are given an exact electronic density distribution either as a highly resolved data set on a grid or a complicated analytic function. In principle, you could extract all physical quantities and even nonmeasurable objects such as the Schrödinger many-particle wave function. Table 6.1 lists the relative work in finding various physical and auxiliary properties from the density. *Easy* means an exact, explicit formula is known as a functional of the density. *Intermediate* means that an explicit numeric scheme can be written down and if cutoff errors and other practicalities do not get in the way, the quantity could be extracted. *Difficult* means that no exact scheme is known to obtain the quantity as a function of the density and perhaps even as a functional of the potential! Most of the KS quantities are of the intermediate type. Thus, KS gives a bridge between difficult to extract properties like the wave function and total energy, and more directly extractable quantities like density moments and total numbers of electrons.

There are many similarities between the machinery of DFT and an effective potential theory, so KS DFT is often misidentified as a MFT (See Figure 6.2). We identify the following two main sources of this confusion:

1. *Limitations on the accuracy of approximate functionals.* DFT may be an exact theory, but all practical KS calculations rely on approximate exchange-correlation density functionals and give approximate densities. More accurate and sophisticated functionals must be designed. As the complexity of functionals increase, so does the expense of the calculation, eventually establishing a practical limit.

2. *Electomorphism of KS particles.* By electomorphism, we mean interpreting KS particles as electrons, and calculating properties with formulae that invoke the KS orbitals as independent electron orbitals. This interpretation, strictly speaking, lies outside of DFT, and can yield misleading results. A prime example of this is the use of the eigenvalue band gap between occupied and unoccupied KS eigenstates from DFT calculations to predict the fundamental band gap in solids. The eigenvalue gap, HOMO-LUMO in DFT is usually (much) smaller than the fundamental gap measured in materials. This difference is often described as a failing of DFT, the so-called *band gap problem.* However, local density approximation (LDA) (and Perdew-Burke-Ernzerhof functional [PBE]) give excellent structural properties for silicon, despite a KS gap that is much smaller than the experimental band gap [8]. Defect energy levels in silicon, charge transition energies computed as differences of properly constructed total energy calculations of different charge, indicate a range of ionization potentials that span an experimental band gap and are not limited by the KS gap [9]. This seeming contradiction simply reflects that DFT is a ground-state theory, and quantities computed from ground-state energies—structures, defect levels—are generally quite accurate, while the electron orbital interpretation of KS—KS gap—is faulty. The band structure and band gap defined by the ideal DFT eigenvalue spectrum from the exact E_{XC} functional may be significantly different from the band structure and band gap observed in experiment. A special density functional approach would be needed to extricate a band structure from a density in a reliable way; it would need to explicitly take into account the nature of the excitations involved in promoting an electron from the valence band into the conduction band.

Because of its utility and computational economy, DFT is a common practical tool and researchers often need to glean as much information from a calculation as possible. Where an explicit density functional is not known, or hard to construct, a researcher might have to appeal to noninteracting particle theory for results. A good example is the use of the Kubo-Greenwood formula to find the conductivity of a material based on KS orbitals [10]. This approach is acceptable with the following caveat: the results extracted in this fashion are no longer true density functional results but rather are based on an MFT interpretation of DFT.

An MFT approach to DFT lies outside the scope of the original HK and KS formalisms. The main drawback of this MFT perspective is that it is unclear, which many-body effects have already been accounted for in the DFT potential. Improvement of such a theory can, at best, be made on empirical evidence and on analogies to wave function theory, while moving further away from the formal exactness based on a rigorous KS picture. Improving the extraction scheme without improving the physics in E_{XC} yields less predictive results.

Response theory illustrates the conceptual difference between interpreting DFT as an exact theory versus an MFT. In the exact theory picture, the excited states of a quantum system can be extracted by examining the structure of the response functional. Formally in DFT for a finite system, finding excited states amounts to identifying the resonances of the physical density-density response function in the time-dependent DFT [11]. This quantity can be related to integrals over pairs of KS excitations including a many-body exchange-correlation kernel and noninteracting auxiliary KS response functions. Within the MFT picture, the KS orbital eigenvalues are equated to actual excited states of the system and KS noninteracting response functions are interpreted as physical response functions. In certain contexts, the mean-field and density functional representations may not differ much. The advantage of the former is that given a highly accurate kernel and KS orbitals, the excited states would tend to the fully interacting exact solutions. The advantage of the latter is that excitations can be understood in terms of familiar orbital objects. This orbital picture allows plotting of electrons and holes as well as useful descriptions in terms of single and double excitations. These distinctions have ambiguous meaning in the exact KS picture.

The exact KS functional as a pure functional of the density may never be known. Since knowledge of the exact functional implies knowledge of the exact solution of the many-particle interacting Schrödinger equation, the exact functional would likely be intractable. The power and beauty in DFT is that tractable approximate functionals for the energy do exist and can provide accuracy comparable to much more expensive solutions of the Schrödinger equation for many systems. Future improvements in exchange-correlation energy functionals will likely extend this trend. However, reliable functionals for quantities such as band gap are still unknown. It may be speculated that a band gap functional would post-process the exact KS band gap, providing the proper band gap. In cases where the density functional for a physical property is unknown, it makes practical sense to interpret the DFT results as MFT.

In what follows, we restrict ourselves to the most important approximate ingredient in practical KS calculations, the exchange-correlation functionals. This is the critical ingredient in finding the KS system. We will give the reader a notion of the origins of several classes of functionals.

In general, DFT functionals need to incorporate information obtained from *outside* the density functional framework, for example, by analysis of the Scrödinger equation [12].

6.3.1 LOCAL DENSITY APPROXIMATION

The LDA was one of the earliest functionals introduced [6]. It is constructed by reference to a solvable reference system through solution of the many-body Schrödinger equation, the uniform many-electron system. The idea is that the energy density of an arbitrary electron density distribution should be the same locally as a uniform reference system with the same density. Despite being completely local, the LDA succeeds for many applications and especially in solid-state applications. It is often used as a benchmark functional expected to be highly predictive and reliable if not always accurate.

Improving upon the LDA, in fact, has proven difficult. The built-in compatibility between exchange and correlation has advantages. When the exchange and correlation are independently constructed, as is often done in practice, and then assembled into a functional, the result will often lack this compatibility. The LDA's exactness when dealing with at least one physical, albeit ideal, system ensures that its predictions are typically at least reasonable.

A natural extension to the LDA is to include gradients of density, $n(\mathbf{r})$, into the functional form, resulting in a gradient expansion approximation (GEA) form,

$$E_{XC}[n] = \int n(\mathbf{r}) \varepsilon_{XC}^{GEA}(n(\mathbf{r}), |\nabla n(\mathbf{r})|) d\mathbf{r} \qquad (6.1)$$

Lamentably, the GEA does not provide systematically better results than the LDA, and a more sophisticated generalized gradient approximation (GGA [14]) needs to be employed. This is because density gradients in real solids and molecules can vary strongly and the expansion must be controlled. Several approaches were developed to overcome these challenges and an assortment of these will be discussed shortly. GGAs allowed sufficient accuracy for DFT to be applied to molecules as well as solids.

6.3.2 FUNCTIONALS FROM MODEL SYSTEMS, THE AM05 FUNCTIONAL

Functional approximations based on the method of reference systems builds upon the strengths of the LDA. Namely, the exchange and correlation energy expressions come from a solvable *model system*, the uniform electron gas. The resulting approximation is therefore internally consistent in its modeling of exchange and correlation, making the combined quantity more widely transferrable and accurate than the individual components. We refer to this feature as *compatible* exchange and correlation. Compatibility is sometimes dismissed as a strong cancellation of errors between the exchange and correlation parts of contributions, but this fails to recognize the important physics behind treating the exchange-correlation energy consistently.

Kohn and Mattsson discussed the creation of an XC functional from a surface-oriented model system and its possible combination with other treatments [15,16]. The approach was formalized and generalized in the subsystem functional scheme by Armiento and Mattsson [17]. The idea is to create separate functionals suitable for different model systems and to merge them using a density functional index [18] that locally determines the nature of the system. These ideas culminated in the AM05 functional [14,19]. The functional invokes two model systems: for regions that are locally bulk-like, the uniform electron gas; and for regions that are locally surface-like, the Airy gas [15] and jellium surfaces [20].

For bulk-like electron regions, LDA with local density correlation according to the Perdew-Wang parameterization [21] based on Ceperly-Alder data [22] is used. Thus, the exchange–correlation energy per particle in the interior is

$$\varepsilon_{XC}^{interior}[n] = \varepsilon_X^{LDA}(n) + \varepsilon_C^{LDA}(n), \qquad (6.2)$$

where

n is the electronic density

ε_C is the correlation contribution

ε_X is the exchange contribution

At edges or regions of space where the density abruptly changes, an exchange energy per particle derived from the Airy gas system [14,15] (the local Airy approximation) is used together with a scaled PW LDA correlation:

$$\varepsilon_{XC}^{edge}[n] = \varepsilon_X^{LAA}(n,s) + \gamma \varepsilon_C^{LDA}(n) \quad \gamma = 0.8098, \tag{6.3}$$

where $s = s(n,|\nabla n|) = \frac{|\nabla n|}{2k_F(n)n}$ is a dimensionless scaled gradient also used in many other GGA-type functionals with $k_F(n) = (3\pi^2 n)^{1/3}$, the local Thomas Fermi wave number.

The interpolation index, $X(s) = \frac{1}{1+\alpha s^2}$ with $\alpha = 2.804$ depends only on s and determines the ratio of edge versus interior functional to use at each point. Thus, the AM05 local energy per particle is

$$\varepsilon_{XC}^{AM05}(n,s) = \varepsilon_{XC}^{interior}(n)X(s) + \varepsilon_{XC}^{edge}(n,s)(1 - X(s)). \tag{6.4}$$

The AM05 functional can be seen as a consistent improvement over LDA in the sense that it reproduces the exact XC energy for *two* types of model systems, the uniform electron gas and the jellium surfaces; two situations with fundamentally different physics. The subsystem functional approach suggests that a sequence of improvements are possible by maintaining proper descriptions of these exact XC model systems, and incorporating additional model systems into a systematically improved functional. An additional functional based on an exactly solvable model system should be introduced to treat qualitatively unique aspects of the density, just as AM05 introduces gradients paired with exact results for edges.

6.3.3 CONSTRAINT-BASED CONSTRUCTION, PBE AND TPSS FUNCTIONALS

When LDA was analyzed to understand its success for nonuniform electron densities, it was observed that the LDA fulfills a number of exact constraints that the exact XC functional must satisfy. Several works, most prominently of Perdew and coworkers, focus on this observation and argue that improved functionals can be constructed by retaining the constraints LDA fulfills and imposing additional physically motivated constraints. A framework for the development of functions has been given by Perdew et al. [13] and termed Jacob's Ladder [13].

One of the most prominent examples of a GGA XC functional is the popular and successful PBE [23]. For atoms, molecules, and transition metals, PBE offers a significant improvement over LDA. The construction explicitly enforces seven constraints. Some of these are argued to be energetically significant and satisfied by the LDA. However, despite PBE being constructed to satisfy the important constraints of LDA, it does not always improve the results with respect to LDA for solids. For example, PBE is less accurate than LDA for a jellium surface [24], a traditional solid-state model system. In general, PBE gives lattice constants in better agreement with experiments than LDA does, but the bulk moduli are often not well reproduced [25]. A newly constructed functional, PBEsol [26] is mostly based on fulfilling constraints, even though a fit to jellium surfaces as in the AM05 construction is used [14]. PBEsol and AM05 both improve upon LDA and PBE for solid-state applications.

By allowing additional degrees of freedom beyond the density and its gradient, further constraints can be satisfied. This leads to the concept of meta-GGAs that include, for example, the kinetic energy density of the KS orbitals. The TPSS (Tao, Perdew, Staroverov, Scuseria) [27] meta-GGA functional is reported to give improved results over PBE. A recent comparison of many GGAs for solids has shown that in many contexts, alternative GGAs outperform PBE, but not universally [28].

6.3.4 EMPIRICAL CONSTRUCTION, THE **BLYP** AND **RPBE** FUNCTIONALS

An alternative to constraint-based functional construction is the more pragmatic approach of empirical functionals. The governing principle is that good exchange and correlation functionals can be obtained from suitable generic expressions fitted to known energies. Traditionally, this approach has been more prevalent in chemistry than in solid-state physics; perhaps due to the more readily available energy data on atoms and molecules used for such fittings.

One of the most widely used functionals created using this concept is the BLYP (Becke, Lee, Yang, Parr) functional. It is comprised of the B88 exchange functional [29] and the LYP correlation functional [30]. Both of these functionals use fits to atomic data to fix free parameters. The BLYP functional has proven very successful for various chemical applications [31]. However, from a theoretical standpoint, LYP has been criticized for a number of shortcomings: (i) it does not reproduce LDA in the limit of uniform density and (ii) it gives zero correlation energy for any fully spin-polarized system.

There have also been attempts to turn constraint-based functionals into a semiempirical functional. Zhang and Yang created revPBE by refitting one of PBE's parameters to atoms ranging from Helium to Argon [32]. Hammer, Hansen, and Nørskov observed that the refit gave improved chemisorption energies [33] but also that this improvement could still be achieved without giving up any of the original PBE constraints by changing the form of exchange. The result, the RPBE functional [33], is not overtly empirical in the sense that it does not directly involve any fits. However, the new exchange functional form was chosen because it reproduces relevant behavior of the exchange of revPBE, which was fitted to atoms. Because of this, RPBE does not improve on PBE for solid-state systems. The performance of BLYP and RPBE for solid-state systems is reported in Ref. [25]. Very recently, inspired by the fact that PBEsol is based on the same functional form as PBE with refit parameters, a study of several possible functionals based on the PBE form has been explored [34].

6.3.5 HYBRID FUNCTIONALS

A hybrid functional combines local or semi-local density-based exchange with some HF exchange. This approach is pragmatically motivated by the benefits of adding DFT correlation to HF and nonlocal exchange in DFT. A more formal motivation has also been given by Becke [35] and further discussed by Perdew et al. [36]. Since many quantum chemistry codes had been designed to do both HF and DFT calculations, combining the two into a hybrid form was readily implemented. The reader is cautioned that hybrid functionals are often coded as a mixture of DFT and HF rather than a proper formulation involving a local potential. It is often argued that the energies differ by a small amount in the two approaches; however, the unoccupied KS spectrum may differ substantially.

A comprehensive analysis of the performance of PBE0 [36,37] and HSE06 [38] for solids was recently given by Paier et al. [38]. A major limitation of hybrid schemes is a steep increase in computational cost, especially if the KS local potential is found. At present, their applicability to large systems and/or long molecular dynamics simulations is unfortunately limited by computational resources.

B3LYPs poor performance on solids is discussed in Ref. [39]. Since any amount of Fock-like exchange will be unscreened, many of the benefits of using DFT on metals are lost when using hybrid functionals.

6.3.6 FUNCTIONAL BASED ON THE ADIABATIC CONNECTION RELATIONSHIP

We have mentioned that the exact KS exchange–correlation functional is unknown, but this is not entirely true. The exchange–correlation functional is not explicitly known as a functional of the density. However, the exact functional is equivalent to the exact many-body solution of the problem and this is known to the extent that the exact solution can be formally written as a perturbative expansion. There are subtleties about the convergence of a truncated version of the full many-body expansion, but formally, the interacting ground-state energy can be found through infinite order perturbation theory in the electron interaction. A similar approach can be applied in DFT [40]. However, the density is fixed to be the density of the interacting system and the resulting curve is called the adiabatic connection relationship. In this case, the energies and orbitals used in the expansion are not the HF ones but rather the KS ones. These are obtained as the orbitals from a single local potential. It turns out that the local potential requirement constrains the expansion so that terms that depend on the local potential contribute in second and higher order. The resulting series is called the Goerling-Levy expansion and provides a useful representation of the exact exchange–correlation functional. The exact exchange–correlation energy is obtained by an integration of this expansion along the adiabatic connection curve. To first order, the functional is the exact exchange functional or a Fock-like exchange term using KS orbitals. This explains why approximate functionals including some fraction of exact exchange tend to be more accurate. The second-order term looks much like a second-order Moller-Plesset (MP2) or an RPA term with a correction involving the first-order local exchange potential.

We have seen that the choice of functional is wrought with uncertainty and, typically, greater accuracy requires a greater computational cost and can result in a less reliable calculation. A general review of the quantitative performance of many functions was published recently [41]. Now, as we embark on the perilous application of DFT to a practical calculation, the choice of functional is only one in a myriad of decisions a researcher must make.

6.4 SIMULATING CHEMISTRY AT METAL SURFACES

Simulations of chemical processes occurring at metal surfaces lie at the crossroads of quantum chemistry and solid-state physics, presenting the challenges of two domains. On one hand, one is interested in the chemical behavior of a molecular system with short-ranged electronic interactions, the focus of computational chemistry. On the other hand, behavior below a surface is dictated by the bulk properties of the substrate. For the highly mobile electrons of a metal, the focus of solid-state physics, the range of interaction can be rather long. While nominally concerned with electronic structure through the solution of the Schrödinger equation, the two branches are built upon rather independent theoretical frameworks and mostly distinct sets of computational tool and techniques owing to the different approximations that are reasonable to make in each discipline.

There exists a vast computational infrastructure developed over the past several decades that enables highly accurate calculations of molecular properties. Semiempirical methods, such as Hückel theory [2], tight-binding [4], and neglected overlap methods [42], are useful for quick estimates of many properties, including excited state properties. More expensive methods, such as HF, have been the mainstay of quantum chemistry for many years. More costly, multiconfigurational methods, such as configuration interaction and coupled cluster treatments, enable almost arbitrarily accurate predictions of molecular electronic structure. Additionally, today's hybrid density functionals, such as B3LYP, have proven to be effective and efficient computational workhorses for predictions for molecular chemistry.

In recent years, there has been an analogous, almost completely independent computational infrastructure developed for simulations of condensed matter systems, with DFT being the workhorse method. The computational demands of extended systems limit the accuracy of the methods that can be applied in tractable simulations.

The challenge of modeling surface chemistry lies in successfully negotiating the compromises of simultaneously satisfying the constraints of achieving sufficient accuracy required for molecular chemistry and the demands of incorporating extended bulk metal effects on that chemistry. The *perfect* calculation is not possible; the Schrödinger equation in its unapproximated form being unsolvable for bulk systems. A deliberate judgment must be made as to what compromises will lead to acceptable errors in the simulation, with a rigorous assessment of the magnitude of those errors and their impact on the conclusions drawn from the simulations.

Smaller computational models, such as modeling a bulk material as a cluster, permit more accurate quantum chemistry methods to be applied at the expense of neglecting bulk effects. With semiconductor or oxide substrates, where the electronic structure can be well approximated as localized bond pairs, a substrate model frequently can be truncated using terminators to tie off the dangling bonds at the edge of a cluster model, such as modified hydrogen atoms ("siligens") for silicon. For many years in surface science, calculations of adsorption on metal surfaces were performed with quantum chemistry codes that simulated molecular adsorption on a cluster, using a molecule bonded to a small number (as few as one!) of metal atoms to model adsorption to a metal surface. Metals, however, possess delocalized electrons; thus, it is difficult to truncate a cluster model of a surface that does not introduce quantitatively intolerable boundary effects. Indeed, it can be exceedingly difficult, if not impossible, to assess the magnitude of edge effects on the computed chemistry for a given cluster model for metals. This assessment must be done on a case-by-case basis for every cluster model.

In recent years, advances in density-functional-based computational methods and improved computational capabilities, particularly the emergence of massively parallel computing, have enabled relatively routine DFT calculations of bulk systems consisting of several hundreds of atoms. With this scale of calculation, it becomes practical to represent chemistry at surfaces within a super-cell approximation.

The local process of interest at a surface (such as adsorption of a carbon monoxide molecule on a platinum surface) is represented as periodically replicated copies of the chemical system on a metal slab. With a larger number of atoms in the super-cell model, a molecule at the surface can be made progressively better isolated from its periodic copies. This is important for using super-cell calculations to model isolated adsorbates or defects at surfaces.

It is the purpose of this section to describe issues that need to be confronted in order to translate this conceptual Ansatz into a useful computational model for quantitatively describing surface chemistry. Modeling chemistry of a molecule on a metal surface using periodic boundary conditions in a super-cell approximation entails a number of complications and approximations that do not arise in quantum chemistry calculations for finite molecular systems. These complications unavoidably include needing to understand how to perform sufficiently accurate calculations of bulk metal properties. But the true complicating factor is adapting these computational tools, fundamentally designed for periodic bulk crystalline systems, to treat problems that lack that periodicity. In the following, we will go through a discussion of a variety of issues that must be considered when constructing meaningful calculations of metal surface properties and chemistry at metals surfaces.

6.4.1 SLAB MODEL OF A SURFACE

As a practical matter in any three-dimensional (3D) periodic code, a computational model of a surface is constructed as a 3D set of periodically repeated two-dimensional (2D) slabs of bulk metal, as illustrated in Figure 6.3. We will describe how the positions and types of atoms are chosen in modeling a surface of choice.

The model of a surface as replicated slabs has several immediate important consequences for the simulation. First, there is not one surface in the slab model, but two, one on the top of the slab and a second on the bottom of the slab above the cell. This complicates any analysis as a second surface is inextricably entwined into the computational model, and it can be difficult to determine the properties of a given

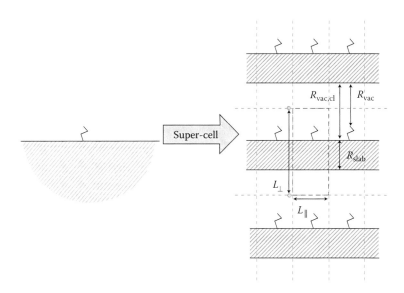

FIGURE 6.3 A schematic depiction of a super-cell approximation to isolated surface adsorption. The idealized isolated molecule is mapped into a super-cell defined by the parallelepiped enclosed by dashed lines and periodically replicated in three dimensions. The adsorbed molecules are translated by the repeat vector parallel to the surface, $L_{\parallel}$. The slab repeat vector $L_{\perp}$ is a sum of the slab thickness R_{slab} and the separation between the slab surfaces, $R_{vac,cl}$, which is the amount of vacuum between slabs for a clean surface. To isolate the slabs from one another, the minimum distance between an atom on the surface and an atom in the slab must be made large enough that they do not significantly interact.

surface in isolation. For example, one property of frequent interest in bulk surface simulation is the surface formation energy. Metals are often high-symmetry crystals where the two surfaces can be constructed to be structurally equivalent so that the surface energy can be computed as half the formation energy of the slab in the limit of an infinitely thick slab. There exist crystal structures and surfaces for which this equivalence between surfaces is not possible in a slab model, particularly in a binary material. A well-known example is the surface energy of the polar (0001) surfaces of a wurtzite crystal. The two sides of the slab are different surfaces with different polar terminations and an individual surface energy cannot be extracted from the slab formation energy, only the sum of the two surface energies. Thus, one can obtain the cleavage energy for this surface, but not the surface formation energy.

Second, the two surfaces can interact with one another across the slab. In metals, with mobile, delocalized electrons, a surface state may extend sufficiently deeply into the slab to interact with the opposite surface. Efficient screening in metals can act to shorten the range of significant interactions, but for some materials, Friedel density oscillations extend long distances [43], reducing the effect of screening. With sufficiently thick slabs, one can, in principle, reduce the interaction between the two surfaces to any desired tolerance. In practice, the thickness of a slab is limited by the available computational resources and complete isolation is never perfectly achieved. As we will describe below, the degree of isolation can and should be assessed to ensure that the surfaces are sufficiently far apart to support the conclusions of the simulation.

Third, periodically replicated slabs can interact with one another, either directly through overlap of the densities, or through long-range Coulombic potentials. To remove direct overlap, the slabs need to be separated by sufficient distance so that the surface densities die off to near-vacuum and have little overlap. Computational cost considerations intrude here, the size of the basis set and, therefore, the computational cost, increases with increasing volume, with the additional cost mostly devoted to reproducing a vacuum region. Hence, any slab calculations should test that the slabs are sufficiently separated that they do not quantitatively affect surface processes. Simulations of surface chemistry mostly involve adsorbates and interactions between them, so the construction of the intervening vacuum in a slab model should anticipate sufficient slab separation to accommodate adsorbates and ensure those adsorbates will not interact with adjacent periodic images of the slab. The separation distance between the top layer of atoms on one surface of slab and the bottom surface of the next slab should be longer than the sum of the range of the basis sets on the two surfaces for a local orbital basis. Our experience is that a 10 Å separation between atoms in different slabs is the minimum required to adequately isolate metallic slabs; however, more than 15 Å is almost certainly wasteful.

If the slab is polar, convergence of computed properties with slab separation is very slow. A dipole across a polar slab interacts with all of the slab periodic images. For clean metal surfaces, slab dipoles are not a significant issue, as metals do not often sustain the localized charge polarizations necessary to generate a significant dipole in a slab. However, adsorption of electronegative species, such as oxygen, on surfaces induces significant charge transfer, creating a surface dipole on the metal surface. Furthermore, there is a lot of interest in the chemistry in the presence of electric and magnetic fields recently, and computational studies that investigate the effects of external fields in slab calculations are becoming more common. A nonpolar metal slab becomes polar in the presence of an applied external field.

An external electric field is typically incorporated into a DFT slab calculation in the form of a sawtooth potential normal to the slab, as illustrated in Figure 6.4. The discontinuity in the potential is placed in the middle of vacuum, and the gradient in the potential across the slab imposes a static external field. This potential gradient induces the electrons in the slab to polarize, creating a slab dipole. For plane-wave slab methods, this application of an external electric field has a second unwanted consequence. With sufficient vacuum inserted into the slab model to isolate the periodic slabs from one another, the potential at the vacuum boundary between the slabs can become deep enough that it is deeper than the binding energy of the electrons in the slab. With local orbital DFT methods, this is not a great concern, as there is no basis set support in the vacuum region that would allow this vacuum occupation. With plane-wave calculations, the basis set treats all regions in the super-cell equally, allowing this artificial well to become populated. Applying external electric fields needs to be done with extra care and monitored carefully to ensure this model artifact does not corrupt the calculation [44].

One common trick to remove an adsorption-induced dipole is to take advantage of the symmetry that often exists in a metal slab model and do equivalent

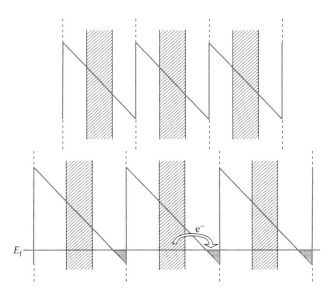

FIGURE 6.4   A schematic depiction of how an electric field is applied to a slab super-cell and the associated hazard. An electric field is applied to a slab super-cell in the form of a "sawtooth" potential from one end of the super-cell, across the slab and to the other end of the super-cell. The thicker the super-cell, the deeper the sawtooth potential at the boundary between the cells, in the middle of the vacuum. If the super-cell is thick enough, the attractive potential of the edge of the sawtooth will artificially extract electrons from the slab. This artifact must be avoided for a meaningful calculation.

adsorption on both sides of the slab. The adsorption-induced dipole on one side of the slab then exactly cancels the adsorption-induced dipole on the other side of the slab. This is often effective in removing the dipole problem but risks exacerbating the interaction between the two sides of the same slab. An adsorption strong enough to generate a significant dipole will more deeply perturb the slab. A slab thick enough to keep that perturbation from reaching the far side of the slab may not be thick enough to prevent interaction with duplicated adsorption on the other surface of the slab. Hence, while adsorption on two sides removes a slab dipole, a thicker slab might be needed to remove the extended interaction, requiring a more expensive calculation.

This duplicated adsorption is an unnecessary expense. A better approach is to explicitly remove the periodic effects of the slab dipole layer, either by introducing a counter-dipole-layer in the middle of vacuum between the slabs [45], or a mixed-dimension Poisson solver that explicitly solves the 2D Poisson equation for the dipole layer [46]. These kinds of fixes for the slab dipole problem are implemented in many DFT codes, but it should not be assumed that they are, or that they are implemented correctly. Any simulation that involves a polar surface should be checked for this dipole correction by assessing the convergence of the computed quantities with respect to the amount of vacuum separation between slabs. It should be noted that the original theory [45] for correcting the slab dipole had a subtle error that was only identified and repaired several years later [47]. Hence, even if the code in question asserts a proper treatment of periodic slab dipole layers, the correctness and effectiveness of the implementation should be verified in test calculations.

In summary, the use of a super-cell slab model in periodic DFT calculations for surface processes complicates the study of surfaces. Reconstructing the appropriate boundary conditions needed for the reduced-dimensional surface simulation from codes that are designed with 3D periodicity and juggling the compromises that must be made as a consequence, are issues that constrain how to approach a given simulation. These are issues that do not arise in typical molecular quantum chemistry calculations, but become crucial for simulations of chemistry on surfaces. These broader issues go into the design of a slab model and dictate the boundary conditions. Construction of a specific slab model within a given DFT code involves manipulating a wide array of input options, subject to the constraints given by the boundary conditions imposed by the slab model. This is the subject of the following sections.

6.4.2   Bulk Lattice Parameters and Slab Models

The initial step in fashioning a viable slab model for a surface is to obtain the properties of the bulk metal within the same computational methodology. The immediate need for constructing a slab model is the optimized bulk structure—the computed lattice parameters and positions of the atoms in the unit cell of a perfect crystal. This step is also important to assess the insensitivity of the results to increases in the quality of computational parameters.

From the perfect crystal, one generates an initial ideal model by truncating a bulk crystal along crystal symmetry planes to create a slab. The lateral dimensions of the slab—those along the surface—are fixed in any slab calculation. The surface lattice parameter should be set to its bulk value to satisfy the requirement that, as the thickness of the slab increases, the center of the slab approaches the perfect crystal bulk material. If this condition is not satisfied, the *ideal* positions of the atoms will not correspond to the perfect truncated bulk atoms and this flaw in the model construction will lead to an artificially reconfigured surface—a reconstruction dictated by improperly applied boundary conditions rather than the creation of a surface.

It is plausible but questionable to use experimental lattice parameters to place the atoms for a model surface. The problem with this approach is that theoretical lattice parameters rarely match experiment. The typical computed DFT bulk lattice parameter differs by 2%–3% from experiment: LDA usually too small and GGA usually too large. Using an experimental lattice parameter for an LDA slab calculation places the atoms on the surface farther apart than the optimal LDA distances. This expansion in the surface area then causes an overall contraction in the lattice in the direction normal to the surface, as the system attempts to reestablish a volume per atom closer to the calculated equilibrium volume. Because the surface dimension is fixed, this contraction is expressed as a shrinking of the distances between the interatomic layers at the surface. It is not possible to separate the motion of atoms responding to this boundary condition from the relaxation of the atoms caused by the creation of the surface or the chemical interaction with an adsorbing molecule.

Using extended x-ray absorption fine structure (EXAFS) it is possible to measure with very high accuracy the interlayer spacings of atoms at a surface, or more precisely, the change of those spacings from their bulk values. A finely tuned surface probe, EXAFS can detect changes in interlayer spacings of less than 0.1%. A typical interlayer spacing at a surface changes by ~5% for a close-packed surface and more for open surfaces. Assuming a 2% deviation of lattice parameter, a slab calculation using the experimental lattice parameter would have a surface area ~4% larger than the LDA ideal surface area. With the lateral dimensions of the slab fixed, the only means that the atoms within the slab have to recover the LDA equilibrium volume is to expand the slab normal to the surface ~4% to compensate for the error in the surface lattice parameter. Note that the resulting error normal to the surface is roughly double the error imposed parallel to the surface and that typical DFT deviations from experimental lattice parameters cause interlayer compensations as large as typically observed surface relaxations.

However, unless the same calculation can be performed for the bulk metal, the theoretical interatomic spacings between the metal substrates are unknown, and the choice of an arbitrary spacing can lead to spurious results.

If substrate atoms participate notably in the chemistry, it is imperative to use interatomic spacings in the substrate that are consistent with theoretical lattice parameters obtained using the same method.

6.4.3 COMPUTATIONAL METHODS

A prerequisite for any slab calculation is, of course, to choose the computational method. The most common choice for periodic calculations is plane-wave basis, pseudopotential (PP) codes, and numerous codes are widely available, from commercial codes such as VASP or CASTEP to open source codes such as ABINIT or PWSCF.

Full-potential, all-electron codes such as WIEN2K, RSPT, or the newly emerging open source EXCITING enable explicit treatment of core electrons with a corresponding increase in computational cost. Linear combination of atomic orbitals (LCAO) approaches are most commonly used for molecular calculations. Periodic LCAO codes, such as CRYSTAL, use contracted Gaussian basis sets, and SIESTA uses with a numerical atomic orbital basis set, may be familiar to practicing computational chemists. LCAO codes often have significant computational cost advantages for periodic calculations over the more common plane-wave codes at the expense of having a less flexible and less complete basis sets. In bulk calculations, the requirement that the LCAO basis sets avoid linear dependence limits which basis sets can be used. Different codes have different features that are particularly useful for a given application, desired computational performance, or availability of needed atomic potentials.

For purposes of illustrating the construction of slab models below, we use SEQQUEST [48], a general purpose DFT code using PPs and a contracted Gaussian basis set in an LCAO approach to periodic DFT calculations. SEQQUEST is based on a computational method specifically developed for slab calculations of metal surfaces [40,41] and has capabilities specifically designed to properly treat reduced-symmetry problems such as molecules, slabs [49], and defects in bulk periodic systems [42] in a periodic super-cell approximation. Hence, it is particularly well-suited to slab-surface calculations. It uses standard, norm-conserving PPs [50,51] taken from an extensive library of atomic potentials accompanied by highly optimized contracted Gaussian basis sets tuned to the particular PP and functional for bulk periodic systems.

6.4.4 PSEUDOPOTENTIALS

PPs, perhaps better known in the molecular quantum chemistry community as effective core potentials (ECPs), are commonly used in periodic DFT codes to remove the explicit treatment of core electrons. The focus on the chemically active valence electrons reduces the number of electrons and the size of the basis sets required in a calculation, dramatically reducing the computational cost. While nominally performing the same role in chemistry and solid-state physics, the practical generation and use of PPs is quite different in the two communities. The ECPs of quantum chemistry are typically constructed from HF atoms and are accompanied by Gaussian-based contracted basis sets to accommodate their use in the prevalent codes in chemistry. A good deal of chemistry concerns itself with first-row atoms and does not require or use ECPs. Solid-state codes are generally concerned with heavier atoms in the periodic table, requiring PPs to make any but the smallest calculations computationally tractable.

One surprising fact for chemists using solid-state codes is that it is commonplace to use PPs for atoms without core electrons—even for a hydrogen atom! This practice is dictated by computational considerations. The cusp of the 1s orbital of the hydrogen atom at the nucleus cannot be reproduced with any reasonably sized plane-wave basis, and the pseudized orbitals are needed to make the computation tractable. The goal of PP construction is to create a smooth nodeless pseudo-orbital that quantitatively reproduces the electronic behavior of the full-potential orbital. Such a construction can be made from any valence orbital, even if it lacks core electrons. With carefully converged calculations that treat the cusp accurately, we were unable to identify any meaningful differences in results comparing a hydrogen PP to a bare-core hydrogen.

One should not use HF PPs in DFT calculations and, indeed, it is not advisable to use a PP generated using one functional in DFT calculations with a different functional [1]. The results can be unpredictable. Mismatching functionals in the PP and calculation, just as mismatching experimental lattice parameters and theoretical lattice parameters in slab models, can produce unphysical results. A further complication is that the functionals that are used in solid-state calculations vary and preferred functionals change over time, and, therefore, PPs need to be reconstructed frequently. The practice of PPs use in solid-state systems is not as systematized as in quantum chemistry. One does not have libraries of accepted PP. Instead of being distributed as collated sets of ECP+basis over atoms, PPs used in solid-state calculations are distributed as codes that implement specific methods for generating PPs and it is often left to the user to fashion their own PP for a specific calculation.

Two well-established methods for PP construction include the norm-conserving PP due to Hamann [52], considered by many, including us, to be the *gold standard* of PP in DFT, and the somewhat softer (smoother), therefore, more computationally tractable PP of Troullier and Martins (TM) [53]. PPs using the Hamann construction are generally *hard*, involving more rapid variations in the resulting valence pseudo-orbitals, thereby achieving better transferability. The softer TM PP compromise some transferability to obtain faster calculations although with suitable settings to make the PP harder, the TM PP can be made almost perfectly equivalent to the hard Hamann PP [54]. Libraries of PP are often provided with a solid-state DFT code, but those libraries are frequently incomplete, or a PP for an element is found to be inadequate. In those cases, the user must construct his own PP, which can be done using available code packages such as fhi98PP [55].

In solid-state calculations, to achieve yet smoother orbitals conducive to more efficient calculations, non-norm-conserving methods such as the ultra-soft potentials due to Vanderbilt [56] are used. Potentials based on the projector-augmented wave method [57] further extend this concept to include some treatment of core states. Non-norm-conserving potentials are more challenging to develop and add significant complexity to code implementation. Users of these methods are often dependent on atomic potential libraries provided with the code, and these libraries are often incomplete or inadequate.

An additional complication in many PPs used in solid-states physics, and almost universally in plane-wave-based methods, is the need to use a separable approximation such as the Kleinman-Bylander (KB) form [58], to make the calculations tractable. With a spatially local PP and a local basis set, the number of matrix elements that need to be evaluated scale linearly with the number of atoms. In contrast for delocalized plane waves, the computational cost of evaluating matrix elements over a standard PP scales badly with the size of the basis. The KB form recovers the linear scaling for a plane-wave basis. However, the use of KB often complicates the construction of the PP, particularly avoiding *ghost states*—artificial states of the PP that intrude into the energy span of the intended valence orbitals and thereby corrupt a DFT calculation. The PP generation codes have tools to help identify the presence of such states in a candidate PP, but might not provide an alternative that eliminates the ghost state and leaves the candidate PP otherwise intact.

The art of PP generation is not yet mature and the creation of a PP almost always involves some compromises. There is no general recipe that guarantees success, although some guidelines for how to design and assess PP can be found in the literature [1]. It is possible to create your own PPs if existing ones are lacking, but this is not a simple exercise. The ultimate figure of merit is how well a PP reproduces an all-electron calculation, but practical considerations and expediency sometimes rule the day. This is true even for the authors of atomic PP libraries and it is prudent to take a skeptical approach to the use of any PP. The important message is that the accuracy of any slab-surface calculation is dependent on the quality of the PP used in the calculation and that the quality of the PP should never be taken for granted. We will illustrate this principle below with an example taken from our own experience.

6.4.5 Computational Model for an Erbium Surface

To illustrate the full process of generating a surface model for a new system, we investigate the surface properties of erbium metal, specifically the potential surface reconstructions and surface adsorption/absorption of oxygen and hydrogen.

The erbium system immediately dictates certain choices in the simulation approach. Erbium is a rare earth metal, occurring late in the lanthanide series, the

Er atom has a $4f^{12} 5d^0 6s^2$ ground-state configuration. Erbium metal has a hexagonal close-packed (hcp) structure with lattice constants $a = 3.56$ Å and $c = 5.59$ Å.

Lanthanides can be problematic for DFT calculations as they possess a partially filled 4f-shell. This is particularly true for PP codes. Many codes might not have PPs available for Er or the default potential might never have been validated. Hence, PP availability may limit the choice of computational method.

If an Er PP does not exist, one must be constructed. The 4f-shell is the first f-shell to be occupied and, therefore, the orbitals have very sharp spatial variations that violate the underlying motivation of PP codes, namely, that all the resulting valence orbitals are smooth. Fortunately, the 4f-orbitals, being the first f-shell and, hence, largely unscreened, are so short ranged that the interaction with the outer valence electrons is quite small. These 4f-electrons are chemically inert, and the oxidation state in chemical environments is mostly fixed; to a good approximation in most chemical environments (with the exception of the bare atom), the erbium atom can be well approximated as having 11 f-electrons, and 3 valence electrons. For our application, we are not interested in magnetic properties, so the presence of the f-electrons in the valence is unnecessary and we use this observation to justify putting the $4f^{11}$ electrons into the core of the PP. Computing bulk properties of Er indicates that the $5p^6$ semi-core states need to be treated as valence electrons and that a partial core correction was needed to incorporate the nonlinear effects of the density from the other core electrons [59]. The tuning of these aspects of the PP closely follows a process described elsewhere for constructing a strontium PP [1]. The basis set is of double-ζ quality with s- and d-orbitals and a single-ζ function for the semi-core $5p^6$ electrons plus a second, single-Gaussian p-function for polarization of the valence electrons [59].

The next steps are to verify that the computational model is sound and validate that it gives reasonable results for the systems of interest. The immediate quantities of interest are the bulk lattice parameters. Using SEQQUEST and the PP described above, the results for the hcp ground-state structure and the fcc closed-packed structure obtained from stress-optimized DFT calculations are presented in Table 6.2. For

TABLE 6.2

Er Crystal Properties from DFT

Bulk Er	Experiment	LDA-CAPZ	GGA-PBE
hcp			
a (Å)	3.559	3.437	3.578
c (Å)	5.587	5.375	5.560
c/a	1.570	1.564	1.554
V (Å^3)	30.64	27.49	30.83
fcc			
a (Å)	n/a	4.769	4.957
V (Å^3)	n/a	27.12	30.45
B (GPa)	n/a	46.5	43.8
ϕE_{fcc} (meV)	>0	+0.8	+25.2

metals, especially, it is necessary to converge the bulk calculations with respect to
k-point sampling [1]. The fcc calculations were performed in the single-atom rhombohedral primitive cell, sampling the Brillouin zone (BZ) with regular grids ranging
from 10^3 to 24^3 points and the hcp calculations in the primitive two-atom cell, using
k-grids ranging from $10^2 \times 6$ to $24^2 \times 15$ points. (Note there is no special significance
to the multiplication symbol in this context, $10^2 = 10 \times 10$.) With BZ sampling of 16^3
(fcc) or $16^2 \times 10$ (hcp), the energy was converged to ~1 meV/Er. The lattice parameter,
a, was converged within 0.001 Å and c to 0.002 Å. The values quoted in Table 6.2
are averages obtained from the BZ samplings from this convergence and better. The
accuracy of the stress-minimized lattice parameters was verified using fits to the
potential energy curve as a function of lattice parameter. The stress-minimized and
energy-minimized lattice parameters for the fcc calculations agreed to better than
0.001 Å.

In an LCAO code like SEQQUEST that implements full Pulay corrections for the
stress formula, this quantitative agreement is expected. With plane-wave calculations,
cell optimizations using stress-elimination must also test convergence of the stress
with respect to the energy cutoff (the size of the plane-wave basis set). The stress
calculation is much more sensitive to incompleteness in the plane-wave basis set than
is the total energy calculation, and to converge the lattice constants using the stress
requires a higher energy cutoff than is necessary to converge the energy [1].

One will do calculations with both the LDA and PBE functionals, simply to get a
sense of the magnitude of the error in a calculation [1]. If LDA and PBE agree, the
calculation is more likely to be accurate, as the different constructions of LDA and
GGA tend to expose aspects of a system that are sensitive to errors. If LDA and PBE
results disagree significantly, then the results should be examined with a skepticism,
as this suggests that the system contains physics that give these functionals trouble.

Here, the preliminary calculations of bulk Er indicate that LDA will likely not be
useful for surface studies. The error in the LDA lattice parameter is not the problem,
deviations of 4% are within the typical errors. But, the nearly equal stability of the
fcc and hcp structures will be a problem. For the surface calculations, we are interested in surface reconstructions and adsorbate-induced relaxation. The fcc and hcp
differ only in the stacking of the close-packed planes and the bulk results indicate
that stacking could easily convert from hcp to fcc. The 1 meV/Er energy difference
between the LDA and PBE is smaller than the variation with BZ sampling for either
crystal, so that there are local environments (such as a surface) where the fcc stacking might be lower (artificially) than the hcp stacking. Hence, for the purposes of the
present surface study, the LDA will not be reliable. The PBE favors the hcp structure
by 25 meV/Er, making it the more suitable choice.

These considerations arise frequently for metal surface calculations and, in addition to providing the structural parameters necessary to construct a slab model, the
bulk calculations are also useful to get a preliminary assessment of the applicability
of a given method. Among the better known examples for the 3d-transition metals,
particularly for iron, the computed description of the bulk material will often limit
the kind of simulation that one can do. While PBE predicts the bulk Fe to be magnetic *bcc*, in agreement with experiment, the LDA ground state Fe is nonmagnetic
fcc, rendering LDA useless for practical surface studies.

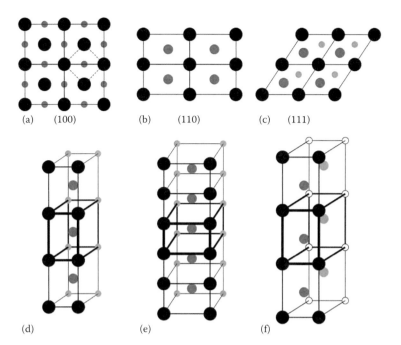

(a)   (100)   (b)   (110)   (c)   (111)

(d)   (e)   (f)

FIGURE 6.5   Low-index surfaces of a simple fcc crystal. The top view of a (100) surface (a), (110) surface (b), and (111) surface (c), and a side view of a segment of the corresponding slab unit cells are shown in (d) through (f).

Construction of the slab requires 2D periodicity within the plane. For the (100) surface, the unit cell is a square and can take two orientations: a larger face-centered square of length a on each side, or the smaller primitive cell rotated by 45° enclosed by the dashed lines of length $a/\!\sqrt{2}$ on a side (see Figure 6.5). The next layer of atoms lies $a/2$ below the top layer, between the top layer atoms, and the third layer is aligned with the top layer. Further layers repeat this pattern, to create a slab as thick as desired. The surface primitive cell for the (110) surface is a rectangle of side a along one edge and $a/\!\sqrt{2}$ along the other, with the next layer of atoms in the center of the rectangle $a/2\!\sqrt{2}$ below the top layer, and the third layer directly below the top layer. This pattern then repeats to fill the desired volume. The (111) surface is a closed-pack triangular lattice having hexagonal lattice vectors of length $a/\!\sqrt{2}$ at 60° to one another. The repeat sequence is three layers, the layers separate by $a/\!\sqrt{3}$. The second layer lies at a distance $a/\!\sqrt{3}$ below the top layer and offset to the center of every second triangle. The third layer is offset further and occupies the other triangle.

The hcp (0001) surface of the Er above is closely related to the (111) surface of the fcc crystal except the repeat sequence is only two layers. The third layer returns to sit directly underneath the top layer, in an $ABABAB$ stacking, instead of an $ABCABC$ stacking of the fcc crystal (the A, B, and C refer to the three unique positions in the 2D hexagonal lattice). Higher index surfaces are similarly constructed, but the surface primitive cells get larger and more complicated.

Using the bulk lattice parameter, the positions of the slab atoms in the truncated bulk crystal are chosen. To complete the definition of the slab calculation, the third lattice vector of the super-cell, the slab repeat vector normal to the surface, is needed. As mentioned above, the slab model of the surface requires 10–15 Å of space between the atoms of one slab and the next to isolate the slabs from one another (remembering that any anticipated adsorbing species, not just the substrate atoms, are included when considering the boundaries of the slab atoms). While, in principle, any spacing between slabs with sufficient vacuum is allowed, it is usually best to set the vacuum spacing to be in increments of the bulk layer spacings normal to the surface, so that the substrate layers are in concordance with the layers of a bulk calculation that would use the same super-cell.

As an example, we return to the Er hcp (0001) slab. For a small slab calculation of a 1×1 primitive cell with five Er layers, the two surfaces layers of the slab are $2c$ (four layer spacings) apart. Using PBE with $c = 5.560$ Å, a distance of at least $2c$ between the atoms of different slabs is needed. To allow some room for adsorbed species, we increase the distance between substrate layers by an additional $2c$ for a total distance of $4c$ between the substrate surface atoms of different slabs. Added to the $2c$ thickness of the slab, the slab repeat lattice vector for the five-layer slab is $6c$. With each two layers added to the slab, this slab repeat vector increases by c.

6.4.6 ERBIUM EXAMPLE: PROPERTIES OF A CLEAN SURFACE

Having constructed the geometry of the slab model of the surface, the first obvious application is to determine the properties of the clean surface. Creating a surface costs energy and causes atoms near the surface to shift from their bulk positions. DFT calculations of the clean surface require assessing many of the same considerations as one would in the bulk.

The k-point sampling now only needs to be done in two dimensions rather than the three dimensions of the bulk, as the third dimension is non-periodic. The same concerns about adequacy of the 2D surface k-sampling apply as in the bulk, and the convergence of results versus k-sampling should be confirmed. A crude rule of thumb is that the linear spacing of the k-grid in the surface BZ should be roughly the same as was needed in the bulk BZ to converge bulk quantities. If bulk hcp Er was converged with a $16^2 \times 10$ sampling of the BZ, a slab calculation should also be converged with the surface-projected 16^2 sampling.

The adequacy of the basis set might also need to be reassessed. While a pure bulk aluminum calculation might be converged with a relatively small plane-wave basis, a surface calculation may introduce sharper variations in the wave functions and require a slightly higher cutoff to achieve similar accuracy. Moreover, adsorbates may require even higher cutoffs, and total energy calculations for molecular chemisorption energies require that all components—the isolated molecule, the clean substrate, and the molecules adsorbed on the surface—are performed consistently, meaning that the bulk and the molecular calculation are done with the same k point sampling, basis and energy cutoffs.

LCAO codes have a different shortcoming in the basis set in that they do not always describe very well the slow fall-off in the density at a metal surface. To reproduce the slow attenuation of the density at the surface, it is particularly important to get an accurate calculation of the work function, which might require very diffuse orbitals on the metal atoms. Very diffuse orbitals overlap with other basis functions in bulk metal and could trigger linear dependence in periodic calculations. To get good work functions, floating orbitals (FOs), basis functions disconnected from any nuclear center, can be used effectively [60]. The specification of FOs for a given system is somewhat arbitrary. We have found that it is sufficient to use a set of FO with just the outermost gaussian of each angular momentum in the metal atom basis, at the bulk position of the first missing layer. For the Er (0001) surface, we add one extra layer of ghost atoms on top of the surface with FO consisting of single-gaussian s-orbitals, p-orbitals, and d-orbitals, a total of nine basis functions per ghost atom. A ghost atom is a set of basis functions, known as floating basis functions, at a position where there is no atom; there is no atomic potential associated with the basis function centers.

The slab formation energy is simply the difference of the total energy of the slab model and the energy of the same number of atoms in the perfect bulk, in the limit of in finitely thick slabs. Assuming the two surfaces are equivalent, the surface energy $E_{surface}$ for a given slab with n_{slab} atoms is half the slab formation energy:

$$E_{surface} = \frac{1}{2}(E_{slab} - n_{slab}E_{bulk}). \tag{6.5}$$

This formula requires a reference energy E_{bulk} of the metal atom in the bulk. This value cannot be simply chosen as the bulk energy computed in the preliminary bulk calculation, as the slab calculation of a bulk atom in the center of the slab and the bulk calculation are not compatible. An incompatibility multiplied by the number of atoms in the slab n_{slab} will not lead to convergence of the surface formation energy as the thickness of the slab is increased [61].

Fiorentini and Methfessel [62] specified a bulk energy from a fit to the incremental change in energy as the slab thickness is increased. With this definition, the surface energy converges by construction. Da Silva et al. [63] showed that if independent bulk and slab calculations were both performed with very high accuracy, converged energy cutoffs, and k-sampling, one could obtain a convergent surface energy. Both of these involve rather expensive calculations, as one needs rather thick slabs or highly converged total energy calculations to demonstrate the necessary numerical convergence.

A different method to obtain a compatible bulk energy for a convergent surface energy calculation is to construct a bulk calculation with a unit cell compatible with the specific slab-surface calculation, using a bulk reference energy derived from a bulk unit cell that is a structural incremental building block of the slab supercell, with equivalent numerics. Construction of this bulk super-cell is illustrated in Figure 6.5 (d)–(f) for the low index surfaces of the fcc crystal. The central cell highlighted in each is a building block that is inserted to make the slab thicker, but

is simultaneously a unit cell for a 3D description of the bulk metal. In the limit of
an infinitely thick slab, this is exactly the bulk cell that is compatible with the slab
calculation. The design of the slab repeat vector described above guarantees that
the real-space grids in the slab and in the bulk have consistent computational algorithms. The k-sampling must also be made consistent and the slab-compatible bulk
calculation would use the 2D slab k-sample multiplied by $N_{k,z}$ in the limit of very
large $N_{k,z}$.

With this definition, the slab-compatible reference bulk energy is obtained from
calculations using a small bulk unit cell. The slab-compatible bulk reference energy
calculations for hcp Er are shown in Figure 6.6 for surface BZ sampling ranging from
8^2 to 16^2. The bulk reference energy is well converged within 1–2 10^{-6} Rydberg, with
$N_{k,z} > 16$ applied to each surface BZ sampling. We average all these larger samplings
to get a slab-compatible reference energy. The effectiveness of this slab-compatible
bulk reference calculation is shown in Figure 6.7 where the hcp(0001) slab formation
energy is plotted as a function of the thickness of the slab with slabs varying from 3
to 27 layers. The atoms in the slab (both sides) are relaxed until the forces on all the
atoms are less than 0.01 eV/Å.

Using the correct slab-consistent bulk reference energy, the slab formation energy
does converge well. Furthermore, the results indicate rapid convergence with respect
to surface BZ sampling. Indeed, the Er(0001) slab formation energy converges very
quickly with slab thickness and is within 0.05 eV of the thick slab limit with even
the smallest, three-layer slab. The resulting surface formation energies (half the slab
formation energy) are within 1 meV of 0.737 eV/1 × 1 with all the k-samples from 10^2

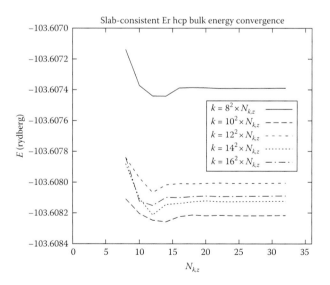

FIGURE 6.6 Convergence of the slab-compatible bulk reference energies as a function of
the dimension of the k-point grid in the normal direction, $N_{k,z}$ for several x–y-plane k-meshes
ranging from 8^2 through 16^2. The energy converges at an $N_{k,z}$ equal to 16 or greater.

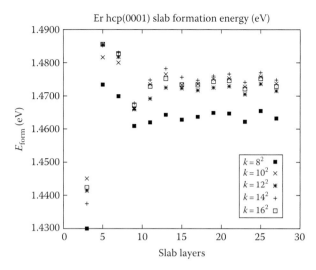

FIGURE 6.7 The convergence of slab formation energy as a function of slab thickness, using the slab-compatible bulk reference energy appropriate to the surface k-grid, which ranges from 8^2 to 16^2. The surface formation energy exhibits greater variation with layers than the bulk reference did with k-sample, indicating that the surfaces are still interacting across the slab, even with slabs as thick as 27 layers. Starting with nine layers and thicker, the formation energy is as well converged as it will be. The 10^2 and greater surface k-grids are fully converged, while the 8^2 k-grid yields a slab formation energy only 0.01 eV less.

through 16^2 and slightly less, 0.732 eV, with the 8^2 k-sample. Hence, the 10^2 sampling looks well converged and the 8^2 is nearly converged.

The virtue of using the approach of slab-compatible bulk reference energies is the reduced computational expense. One does not need highly converged calculations for each piece, either slabs thick enough to extract a reliable bulk limit [62], or basis cutoffs and k-sampling for both slab and bulk calculations sufficient to absolutely converge both the slab and bulk independently [63]. One need only ensure that the slab and bulk calculations are equivalently converged. The computed slab-compatible Er bulk reference energy for the $10^2 \times N_{k,z}$ through $16^2 \times N_{k,z}$ bulk sampling are all within 1 meV of one another (although at 27 layers, this 1 meV error in a single atom would translate into a total 27 meV difference in the corresponding reference energy), but the $8^2 \times N_{k,z}$ reference energy is 4–5 meV less than any of these. Nonetheless, the slab formation energy for the 8^2 BZ sampling is almost perfectly flat (converged) out to at least 27 layers, verifying that the slab-compatible bulk reference procedure gives a valid reference energy for the surface energy calculation. The $8^2 \times N_{k,z}$ is not a highly converged calculation of the total energy for the bulk, nor is the 8^2 sampling a highly converged calculation for the total energy for the slab, but in tandem, these give highly converged calculations of slab formation because their convergence with respect to the computational model setup is compatible.

Other properties of the slab are also well converged with relatively thin slabs and sparse surface k-grids. In Table 6.3, the layer relaxations and work functions for the

TABLE 6.3
Er hcp(0001) Slab Surface Work Function Φ and Surface Relaxation

	Φ (eV)	ϕd_{12} (%)	ϕd_{23} (%)	ϕd_{34}(%)	ϕd_{45} (%)
$k = 12 \times 12$					
3 layers	3.28	−1.83	—	—	—
5 layers	3.25	−2.11	+1.10	—	—
7 layers	3.25	−1.98	+0.90	+0.12	—
9 layers	3.25	−1.90	+0.80	+0.22	−0.08
11 layers	3.25	−1.86	+0.91	+0.30	−0.02
13 layers	3.25	−1.88	+0.91	+0.30	−0.04
15 layers	3.25	−1.88	+0.92	+0.29	−0.04
17 layers	3.25	−1.87	+0.92	+0.29	−0.04
19 layers	3.25	−1.88	+0.92	+0.29	−0.04
21 layers	3.25	−1.88	+0.92	+0.29	−0.04
Bulk limit					
$k = 8 \times 8$	3.26	−1.93	+0.81	+0.46	−0.06
$k = 10 \times 10$	3.25	−1.82	+0.98	+0.24	−0.07
$k = 12 \times 12$	3.25	−1.87	+0.92	+0.30	−0.04
$k = 14 \times 14$	3.25	−1.89	+1.02	+0.27	−0.06
$k = 16 \times 16$	3.25	−1.86	+0.93	+0.29	−0.06

The changes in distances d_{uv} between layers u and v from the bulk separation are given as a percentage of the bulk layer separation $c/2$.

slab calculation are presented. The separation between the first and second layers, d_{12}, shrinks by 1.9% and the separation between the second and third layers, d_{23}, increases by ~0.9%, with deeper layers exhibiting minimal relaxations. This magnitude of relaxation is typical of close-packed surfaces. Somewhat larger relaxations are common on more open surfaces. These geometry relaxations converge quickly with slab thickness for Er(0001) and with surface BZ sample, just as the formation energy did.

The work function, Φ, a commonly measured physical quantity at metal surfaces, is extracted from a DFT slab calculation as the Fermi level of the slab with respect to vacuum. This last caveat is important, SEQQUEST automatically rescales the self-consistent potential so that the vacuum potential is zero, but the reported Fermi level in other codes may need to be adjusted manually by the vacuum level to obtain the correct work function. With a polarized slab with a surface dipole, it is crucial to have appropriately applied boundary conditions, as described earlier, in order to properly identify a vacuum potential.

The work function is very sensitive to the attenuation of the density at the surface, and so the use of FOs in LCAO code is crucial to getting good values. This is a delicate undertaking, as in conventional DFT codes, because the work function is *not* a variational quantity with respect to the basis set. The total energy can vary minimally with changes in the FOs, but the work function will vary strongly with changes in the true FOs. It is possible to converge highly accurate work functions with a LCAO code, even with complex surface chemistry [60], but it is not straight-forward. The work functions quoted in Table 6.3 have not been specifically veri-fied for convergence through augmentation with FOs, but can be used to assess the convergence of the slab model with thickness and BZ sample. We see that the work function converges even more rapidly than the energy or layer relaxation, both with slab thickness and with surface BZ sampling.

This exercise has identified the minimal characteristics necessary in a slab model to serve as an adequate substrate for a surface chemistry simulation. In the case of the Er(0001) surface, the slab thickness should be at least seven layers and it may suffice to use a surface k-sampling as small as 8^2 to get adequately converged results. This is somewhat less than might be deduced from examining the convergence of the bulk properties with k-sampling. This is useful, as the smaller the sampling, the less expensive the calculation and the more atoms can be inserted into a slab model, either in a thicker slab to converge the surface calculation better, or in making the surface super-cell larger and more realistic in the sense that the impurity is further from its copies.

6.4.7   SURFACE SUPER-CELLS AND CHEMISTRY AT SURFACES

These small 1×1 slab models are primitive cells for constructing larger surface super-cells with which to examine longer-ranged surface phenomena. Examples include such phenomena as adsorption of less than one monolayer, investigation of the initial steps of oxidation, or surface reconstructions that entail more than the simple relaxation described above. The most famous of these is perhaps the recon-struction of the silicon (111) surface, which takes a 7×7 reconstruction. To con-struct a 2×2 super-cell, simply double the lattice vectors in the surface and add the atoms enclosed within the expanded cell, replicating the atoms of the 1×1 primitive cell appropriately offset by the primitive lattice vectors. The expanded cell requires a reduced BZ sampling. The required dimensions of the k-grid are inversely pro-portional to the real-space dimensions of the grid. A 2×2 super-cell with 4^2 BZ sampling is formally equivalent to the 1×1 primitive cell with an 8^2 sampling and results for the clean surface (energy, relaxation, and work function) should be the same. A 3×3 super-cell with a 4^2 BZ sampling is equivalent to a 12^2 sampling in the 1×1 primitive cell.

With the definition of the periodic slab substrate in the DFT calculation, the simulations follow the same path as for typical molecular quantum chemistry simu-lation. In order to compute the adsorption of a chemical species, such as carbon monoxide on a platinum surface, for example, one would add a CO to the super-cell. The energy of adsorption is the difference between the total energy of the adsorbed system and the sum of the total energies of the isolated molecule and the clean substrate:

$$E_{ads} = E_{CO/Pt(111)} - (E_{CO} + E_{Pt(111)}). \tag{6.6}$$

These total energies must be consistent with each other, that is, generated with the same functional, method, and computational algorithms. For example, one would not use a CO energy obtained in a molecular quantum chemistry code in the calculation of the adsorption energy with a slab result from a periodic DFT code. The numerical implementation of different codes may be incompatible and even the same DFT functional can differ from code to code. The energy of the isolated CO molecule must be computed in the same code, using the same basis (e.g., energy cutoff in a plane-wave code) and numerical quadratures as in the slab calculation.

Note that an isolated molecule such as CO in a periodic code is represented as a very diffuse 3D crystal of molecules. The periodic slab models need to be separated by sufficient vacuum to isolate the repeating slabs from one another along one dimension, to isolate the molecule requires sufficient vacuum in all three dimensions. In the molecular calculation, dipole, quadrapole, and higher moments between periodically replicated molecules affect the energy, and either the energy must be extrapolated using larger super-cells with more vacuum or the offending multipolar interactions explicitly removed [46,64].

Once this computational infrastructure is in place and the numerical aspects of the calculations are explicitly controlled, one has only to deal with the limitations of DFT in the calculations. This present example, CO/Pt(111) is a classical example of a failure of DFT for surface chemistry [65]. Feibelman recounts multiple computational methods implementing LDA and GGA that attempt to compute adsorption of CO on Pt(111) and fail. Experimentally, it has been observed that CO adsorbs directly atop a surface Pt atom, but many prevalent functionals favor the triangular site midway between three surface Pt atoms. The DFT functionals do very well for the bulk platinum, but do poorly for chemisorption of CO on platinum. The B3LYP functional does quite well for first row chemistry but does poorly for many bulk metals. The present example, CO/Pt(111), remains an unsolved problem for DFT calculations and, therefore, any DFT simulation of chemistry at a Pt surface involving CO is suspect. The CO/Pt "puzzle" [65] is among the most prominent manifestations of this problem, but any molecular chemistry at surfaces should be thoroughly validated against established data before more sophisticated simulations are undertaken.

It is not uncommon for DFT to fail for a crucial aspect of a surface chemistry problem. In the Er example above, the LDA failed to reproduce bulk Er properties, but the GGA-PBE functional succeeded and was able to give reasonable results. For cases like the CO/Pt(111), both the LDA and PBE fail and an alternative approach must be taken.

Among the 3d transition metals and particularly for the oxides of these metals, standard DFT frequently fails. A recent development proposed to circumvent this problem is the DFT+U approach [66] where the interactions among the 3d electrons in the DFT treatment are modified by an empirical U parameter in order to obtain a better description of the bulk system. This method enables applications that would otherwise need to be abandoned with standard DFT approaches. Being outside of formal DFT, this approximation needs to be approached with as much, or more, skepticism than DFT calculations. The LDA+U method can fix some errors of the bulk problem but perhaps at the cost of exacerbating others. Validation of the approach requires following the process described earlier for LDA and PBE to ensure a fully consistent description of a chemical system.

A good example of how to apply LDA+U to a physical problem is described by Mosey and Carter [67] who investigated surface energies of chromia. Conventional DFT calculations fail to describe bulk chromia well, so this study used LDA+U with great success. The distinctive feature of this study was the treatment of the U parameter as yet another tunable computational aspect of the calculation. That the results proved insensitive to the value of U over a large plausible range of values lent the conclusions greater credibility than if the theory had been applied without verifying the calculation.

6.5 CONCLUSIONS

In this chapter, we have discussed several important issues for a researcher to consider when performing quantum mechanical calculations of solid surfaces. We have highlighted the fundamental difference between metals and other materials in terms of the mobility of valence electrons. The most commonly used many-body theory, DFT, is well suited to problems involving metals but is limited by the practical choice of exchange–correlation functionals. Even with highly accurate functionals, a researcher must be careful when extracting results from the auxiliary KS system always being mindful of whether the property is a MFT property or one that is obtained through a proper interpretation of KS results. We outlined a detailed systematic approach to developing a viable computational model of an Er surface to illustrate many of the practical challenges that one experiences when performing surface calculations.

ACKNOWLEDGMENTS

Sandia National Laboratories is a multi-program laboratory operated by Sandia Corporation, a wholly owned subsidiary of Lockheed Martin Corporation, for the U.S. Department of Energy's National Nuclear Security Administration under contract DE-AC04-94AL85000.

REFERENCES

1. A. E. Mattsson, P. A. Schultz, M. P. Desjarlais, T. R. Mattsson, and K. Leung, Designing meaningful density functional calculations in material science–A primer, *Modelling Simul. Mater. Sci. Eng* R1 (2005).
2. E. Hückel, *Zeitschrift fr Physik* **70**, 204, (1931); **72**, 310, (1931); **76**, 628 (1932); **83**, 632, (1933).
3. K. Rademann, B. Kaiser, U. Even, and F. Hensel, Size dependence of the gradual transition to metallic properties in isolated mercury clusters, *Phys. Rev. Lett* 59, 2319–2321 (1987).
4. N. W. Ashcroft. and N. D. Mermin, *Solid State Physics*, Brooks Cole, New York (1976).
5. P. Hohenberg. and W. Kohn, Inhomogeneous electron gas, *Phys. Rev* **136**, B864 (1964).
6. W. Kohn. and L. J. Sham, Self-consistent equations including exchange and correlation effects, *Phys. Rev* **140**, A1133 (1965).

7. J. Slater, A simplification of the Hartree-Fock method, *Phys. Rev* **81**, 385 (1951).

8. C. Fiolhais. and F. Nogueira, *A Primer in Density Functional Theory* Miguel Marques Published by Springer, 2003 ISBN 3540030832, 9783540030836.

9. P. A. Schultz, Theory of defect levels and the "Band Gap Problem" in silicon, *Phys. Rev. Lett* **96**, 246401 (2006).

10. M. P. Desjarlais, J. D. Kress, and L. A. Collins, Electrical conductivity for warm, dense aluminum plasmas and liquids, *Phys. Rev. E* **66**, 025401 (2002).

11. E. Runge. and E. K. U. Gross, Density-functional theory for time-dependent systems, *Phys. Rev. Lett* **52**, 997–1000 (1984).

12. A. E. Mattsson, In pursuit of the "Divine" functional, *Science* **298**, 759 (2002).

13. J. P. Perdew. and K. Schmidt, *Density Functional Theory and its Applications*, Eds. V. Van doren et al. CP577 (2001).

14. R. Armiento. and A. E. Mattsson, Functional designed to include surface effects in self-consistent density functional theory, *Phys. Rev. B* **72**, 085108 (2005).

15. W. Kohn. and A. E. Mattsson, Edge electron gas, *Phys. Rev. Lett* **81**, 3487 (1998).

16. A. E. Mattsson. and W. Kohn, An energy functional for surfaces, *J. Chem. Phys* **115**, 3441 (2001).

17. R. Armiento. and A. E. Mattsson, Subsystem functionals in density-functional theory: Investigating the exchange energy per particle, *Phys. Rev. B* **66**, 165117 (2002).

18. J. P. Perdew, J. Tao, and R. Armiento, How to tell an atom from an electron gas: A semi-local index of density inhomogeneity, *Acta Phys. Chim. Debrecina* **36**, 25 (2003).

19. A. E. Mattsson. and R. Armiento, Implementing and testing the AM05 spin density functional, *Phys. Rev. B* **79**, 155101 (2009).

20. N. D. Lang. and W. Kohn, Theory of metal surfaces: Charge density and surface energy, *Phys. Rev. B* **1**, 4555 (1970).

21. J. P. Perdew. and Y. Wang, Accurate and simple analytic representation of the electron-gas correlation energy, *Phys. Rev. B* **45**, 13244 (1992).

22. D. M. Ceperley. and B. J. Alder, Ground state of the electron gas by a stochastic method, *Phys. Rev. Lett* **45**, 566 (1980).

23. J. P. Perdew, K. Burke, and M. Ernzerhof, Generalized gradient approximation made simple, *Phys. Rev. Lett* **77**, 3865 (1996).

24. S. Kurth, J. P. Perdew, and P. Blaha, Molecular and solid-state tests of density functional approximations: LSD, GGAs, and meta-GGAs, *Int. J. Quantum Chem* **75**, 889 (1999).

25. A. E. Mattsson, R. Armiento, J. Paier, G. Kresse, J. M. Wills, T. R. Mattsson, The AM05 density functional applied to solids, *J. Chem. Phys* **128**, 084714 (2008).

26. Z. Wu. and R. E. Cohen, More accurate generalized gradient approximation for solids, *Phys. Rev. B* **73**, 235116 (2006).

27. J. Tao, J. P. Perdew, V. N. Staroverov, and G. E. Scuseria, Climbing the density functional ladder: Nonempirical meta-generalized gradient approximation designed for molecules and solids, *Phys. Rev. Lett* **91**, 146401 (2003); V. N. Staroverov, G. E. Scuseria, J. T. Tao, and J. P. Perdew, Tests of a ladder of density functionals for bulk solids and surfaces, *Phys. Rev. B* **69**, 075102 (2004).

28. P. Haas, F. Tran, and P. Blaha, Calculation of the lattice constant of solids with semilocal functionals, *Phys. Rev. B* **79**, 085104 (2009).

29. A. D. Becke, Density-functional exchange-energy approximation with correct asymptotic behavior, *Phys. Rev. A* **38**, 3098 (1988).

30. C. Lee, W. Yang, and R. G. Parr, Development of the Colle-Salvetti correlation-energy formula into a functional of the electron density, *Phys. Rev. B* **37**, 785 (1988).

31. J. Tirado-Rives. and W. L. Jorgensen, Performance of B3LYP density functional methods for a large set of organic molecules, *J. Chem. Theory Comput* **4**, 297 (2008).

32. Y. Zhang. and W. Yang, Comment on Generalized Gradient Approximation Made Simple, *Phys. Rev. Lett* **80**, 890 (1998).

33. B. Hammer, L. B. Hansen, and J. K. Nørskov, Improved adsorption energetics within density-functional theory using revised Perdew-Burke-Ernzerhof functionals, *Phys. Rev B* **59**, 7413 (1999).

34. L. S. Pedroza, A. J. R. da Silva, and K. Capelle, Gradient-dependent density functionals of the Perdew-Burke-Ernzerhof type for atoms, molecules, and solids, *Phys. Rev. B* **79**, 201106(R) (2009).

35. A. D. Becke, A new mixing of Hartree-Fock and local density-functional theories, *J. Chem. Phys* **98**, 1372 (1993).

36. J. P. Perdew, M. Ernzerhof, and K. Burke, Rationale for mixing exact exchange with density functional approximations, *J. Chem. Phys* **105**, 9982 (1996).

37. C. Adamo. and V. Barone, Toward reliable density functional methods without adjustable parameters: The PBE0 model, *J. Chem. Phys* 110, 6158 (1999).

38. J. Paier, M. Marsman, K. Hummer, G. Kresse, I. C. Gerber, and J. G. Ángyán, Screened hybrid density functionals applied to solids, *J. Chem. Phys* **124**, 154709 (2006); *ibid* **125**, 249901 (2006).

39. J. Paier, M. Marsman, and G. Kresse, Performance of B3LYP for solids: Why does the B3LYP hybrid functional fail for metals?, *J. Chem. Phys* 127 024103 (2007).

40. A. Goerling. and M. Levy, Exact Kohn-Sham scheme based on perturbation theory, *Phys. Rev. A* **50**, 196–204 (1994).

41. S. F. Sousa, P. A. Fernandes, and M. J. Ramos, General performance of density functionals, *J. Phys. Chem A* **111**, 10439 (2007).

42. C. J. Cramer, *Essentials of Computational Chemistry*, Wiley, New York, 126 (2002).

43. S. H. Vosko, L. Wilk, and M. Nusair, Accurate spin-dependent electron liquid correlation energies for local spin density calculations: A critical analysis, *Can. J. Phys* **58**, 1200 (1980).

44. P. J. Feibelman, Surface diffusion mechanis versus electric field: Pt/Pt001, *Phys. Rev. B* **64**, 125403 (2001).

45. J. Neugebauer. and M. Scheffler, Adsorbate-substrate and adsorbate-adsorbate interactions of Na and K adlayers on Al(111), *Phys. Rev. B* **46**, 16067 (1992).

46. P. A. Schultz, Local electrostatic moments and periodic boundary conditions, *Phys. Rev. B* **60**, 1551 (1999).

47. L. Bengtsson, Dipole correction for surface supercell calculations, *Phys. Rev. B* **59**, 12301 (1999).

48. J. P. Perdew, A. Ruzsinszky, G. I. Csonka, O. A. Vydrov, G. E. Scuseria, L. A. Constantin, X. Zhou, and K. Burke, Restoring the density-gradient expansion for exchange in solids and surfaces, *Phys. Rev. Lett* **100**, 136406 (2008).

49. Z. Yan, J. P. Perdew, and S. Kurth, Density functional for short-range correlation: Accuracy of the random-phase approximation for isoelectronic energy changes, *Phys. Rev. B* **61**, 16430 (2000).

50. P. J. Feibelman, First-principles calculational methods for surface-vacancy formation energies, heats of segregation, and surface core-level shifts, *Phys. Rev. B* **39**, 4866 (1989). *Phys. Rev. B* **38**, 12133 (1988). *Phys. Rev. Lett.* **65**, 729 (1990).

51. P. T. Sprunger, L. Petersen, E. W. Plummer, E. Laegsgaard, and F. Besenbacher, Giant friedel oscillations on the beryllium(001) surface, *Science* **275**, 1764 (1998).

52. D. R. Hamann, Generalized norm-conserving pseudopotentials, *Phys. Rev. B* **40**, 2980 (1989).

53. N. Troullier. and J. L. Martins, Efficient pseudopotentials for plane-wave calculations, *Phys. Rev. B* **43**, 193 (1991).

54. O. A. von Lilienfeld. and P. A. Schultz, Structure and band gaps of Ga-V semiconductors: The challenge of Ga pseudopotentials, *Phys. Rev. B* **77**, 115202 (2008).

55. M. Fuchs. and M. Scheffler, Ab initio pseudopotentials for electron structure calculations of polyatomic systems using density functional theory, *Comput. Phys. Commun* **119**, 67 (1999).

56. D. Vanderbilt, Soft self-consistent pseudopotentials in a generalized eigenvalue formulation, *Phys. Rev. B* **41**, 7892 (1990).

57. G. Kresse. and J. Joubert, From ultrasoft pseudopotentials to the projector augmented wave method, *Phys. Rev. B* **59**, 1758 (1999). L. Kleinman. and D. M. Bylander, Efficacious form for model pseudopotentials, *Phys. Rev. Lett* **48**, 1425 (1982).

58. L. Kleinman. and D. M. Bylander, Efficacious form for model pseudopotentials, *Phys. Rev. Lett* **48**, 1425 (1982).

59. R. R. Wixom, J. F. Browning, C. S. Snow, P. A. Schultz, and D. R. Jennison, First principles site occupation and migration of hydrogen, helium, and oxygen in β-phase erbium hydride, *J. Appl. Phys*, **103**, 123708 (2008).

60. D. R. Jennison, P. A. Schultz, D. B. King, and K. R. Zavedil, BaO/W(111) thermionic emitters and the effectts of Sc, Y. LA, and the density functional used in the computations, *Surf. Sci* **549**, 115 (2004).

61. J. C. Boettger, Nonconvergence of surface energies obtained from thin-film calculations, *Phys. Rev. B* **49**, 16798 (1994).

62. V. Fiorentini. and M. Methfessel, Extracting convergent surface energies from slab calculations, *J. Phys. Condens. Matter* **8**, 6525 (1996).

63. J. L. F. Da Silva, C. Stampfl, and M. Scheffler, Converged properties of clean metal surfaces by all-electron first-principles calculations, *Surf. Sci* **600**, 703–715 (2006).

64. I. Dabo, B. Kozinsky, N. E. Singh-Miller, and N. Marzari, Electrostatics in periodic boundary conditions and real-space corrections, *Phy. Rev. B* **77**, 115139 (2008).

65. P. J. Feibelman, B. Hammer, J. K. Nørskov, F. Wagner, M. Scheffler, R. Stumpf, R. Watwe, and J. Dumesic, The CO/Pt(111) puzzle, *J. Phys. Chem. B* **105**, 4018 (2001).

66. P. A. Schultz, SEQQUEST code, Sandia National Laboratories, see http://dft.sandia.gov/Quest/

67. N. J. Mosey. and E. A. Carter, Ab initio LDA+U prediction of the tensile properties of chromia across multiple length scales, *J. Mech. Phys. Solids* **57**, 287 (2009).

68. P. A. Schultz, Theory of defect levels and the "Band Gap Problem" in silicon, *Phys. Rev. Lett* **96**, 246401 (2006).

69. J. Heyd, G. E. Scuseria, and M. Ernzerhof, Hybrid functionals based on a screened Coulomb potential, *J. Chem. Phys* **118**, 8207 (2003).

70. D. C. Langreth. and M. J. Mehl, *Phys. Rev. B* **28**, 18091834, (1983).

71. A. E. Mattsson, R. Armiento, P. A. Schultz, and T. R. Mattsson, Nonequivalence of the generalized gradient approximations PBE and PW91, *Phys. Rev. B* **73**, 195123 (2006).

72. J. P. Perdew. and A. Zunger, Self-interaction correction to density-functional approximations for many-electron systems, *Phys. Rev. B* **23**, 5048 (1981).

73. L. Vitos, B. Johansson, J. Kollár, and H. L. Skriver, Exchange energy in the local Airy gas approximation, *Phys. Rev. B* **62**, 10046 (2000).

74. G. H. Gonnet, D. E. G. Hare, D. J. Jerey, and D. E. Knuth, On the LambertW Function, *Adv. Comput. Math* **5** 329 (1996).

75. Subroutines and information about AM05 are available at http://dft.sandia.gov/functionals/AM05.html

76. X. Y. Pan, V. Sahni, and L. Massa, Normalization and fermi-coulomb and coulomb hole sum rules for approximate wave functions, *Int. J. Quantum Chem* **107**, 816 (2006).

77. P. A. Schultz, Charged local defects in extended systems, *Phys. Rev. Lett* **84**, 1942 (2000).

78. P. J. Feibelman, Force and total energy calculations for a spatially compact adsorbate on an extended metallic crystal surface, *Phys. Rev. B* **35**, 2626 (1987).

79. P. J. Feibelman, Pulay-type formula for surface stress in a local-density-functional, linear combination of atomic orbitals, electronic-structure calculation, *Phys. Rev. B* **44**, 3916 (1991).

80. H. Kadas, Z. Nabi, S. K. Kwon, L. Vitos, R. Ahuja, B. Johansson, and J. Kollar, Surface relaxation and surface stress of 4d transition metals, *Surf. Sci* **600**, 395–402 (2006).

81. V. I. Anisimov, F. Aryasetiawan, and A. I. Lichtenstein, First-principles calculations of the electronic structure and spectra of strongly correlated systems: The LDA+U method, *J. Phys. Condens. Matter* **9**, 767 (1997).

7 Computational Investigations of Metal Oxide Surfaces

Emily A. A. Jarvis and Cynthia S. Lo

CONTENTS

7.1 INTRODUCTION

The importance of materials in society cannot be overstated. In many respects, a society's potential for success and advancement is fundamentally related to the sophistication of its materials. Ancient civilizations can be distinguished according to the materials they employed. Occasionally, those materials even served as the defining characteristic of that era; accordingly, the "Bronze Age" identifies hundreds of years of history and the thousands of people that lived during that time. To bring this impact closer to home, consider that the industrial revolution was intimately associated with improvements in steel production and that many of the phenomenal engineering and computing feats accomplished over the past few decades were possible only through advances in metal alloys, ceramics, semiconductors, and materials processing. The impact of nanotechnology, and its promise of addressing many of the pressing issues of today has been hampered largely by the difficulty in achieving reproducible and inexpensive production but will appear increasingly in everyday applications as these technical challenges are overcome. New ideas, new techniques, and new materials will be required for continued advancement through the twenty-first century.

Many of our most significant dilemmas concerning materials may include metal oxides in their eventual solution. Computer hardware development predicaments have been highly publicized, with new materials and production techniques seeming necessary to continue the advancement trajectory. Energy produced through fossil fuel combustion has many drawbacks, including the lack of sustainable resources, emissions, and health and safety, which motivate interest in developing alternative and renewable sources. Catalysts are needed in such varied applications to facilitate novel pathways for chemical reactions, to break down environmental pollutants, and to dramatically permit more energy-efficient chemical production. The materials science of thin films, interfaces, and small particles must be advanced. Metal oxides are likely to figure prominently in each of these issues.

The physical understanding required for these advances will be made possible only by experiment and theory via simulations working hand in hand to provide new insight into these complex systems. The qualitative information gleaned from electronic and atomic structure calculations can provide a general understanding of materials behavior and atomic-level explanations for certain trends observed experimentally, in some cases allowing predictions for materials improvements based on theory. A quantitative analysis of calculations permits a more rigorous comparison with experimental findings and furnishes a basis for predicting values that may be difficult to measure via experimental techniques. It is vital to have computational justification, such as density functional calculations, to explain behavior in complex systems, such as heterogeneous catalysts.[1] A recent review by Norskov et al.[2] gives some flavor of the progress that has been made in the computational understanding of catalysis. This review and the references herein describe how recent advances allow strong comparisons with experiment and even aid in design through a predictive understanding gleaned from calculations. The ability to control structure and isolate interactions in calculations permits predictions of material improvements that generally would not be possible to determine simply through experimental measurements.

7.1.1　Metal Oxide Models in Calculations

Frequently, metal oxides have a variety of possible crystalline structures as well as aperiodic structures that satisfy a given stoichiometric ratio. Naturally, the external conditions, usually measured in temperature and pressure, determine the thermodynamically preferred structure under the given thermodynamic conditions, while growth conditions may introduce kinetic limitations to achieving this most favorable structure. In practical applications, a particular phase may be essential to achieve and stabilize over a wider range of physical conditions than the phase diagram would permit. For instance, bulk ZrO_2 exhibits several phases (including cubic, tetragonal, and monoclinic) with distinct material properties and electronic structure. In thermal barrier applications, the tetragonal to monoclinic phase transition must be suppressed due to the unfavorable physical properties of the monoclinic phase in matching to a metal alloy substrate.[3] Accordingly, yttria, Y_2O_3, is added in <10% quantities to inhibit the transition to the monoclinic phase, which thermodynamically would be accessed for pure zirconia within the relevant temperature/pressure cycle.

Metal oxides of interest are often in the form of thin films. Depending on the size, thickness, and other factors, these frequently present electronic band structure differences from the bulk crystalline material stable facets. Generally, for electronic structure calculations, a crystalline facet is chosen that meets one or more of the following criteria:

- It is known to be an experimentally relevant surface.
- It is shown to be the lowest energy facet after comparing the energetics of several low index surfaces.
- It presents the appropriate surface termination for comparison with the experiment, or it avoids models presenting complications, such as polar surfaces, or it presents a small surface area, and thus is computationally more tractable.

Accordingly, determining the appropriate facet choice in a computer model may be influenced by the desired application as well as the computational resources.

Although a variety of techniques could be employed to investigate metal oxides, the primary focus of this chapter will be on applications employing density functional theory (DFT).[4,5] DFT applications have become ubiquitous across both solid state and molecular applications, frequently being used in both descriptive and predictive capacities because such calculations may provide physical insights that complement what can be explored through experiment and allow a level of control in structural modifications that may be challenging or even impossible to explore via experimental techniques alone. Although a few applications of Orbital Free DFT (OFDFT) have been performed,[6,7] this primarily remains an area of ongoing development and research.[8–11] The Kohn-Sham (KS) implementation of DFT is employed in almost all applications since the KS-DFT implementation allows an exact form for the kinetic energy functional in which one-electron orbitals are introduced. The form of the exchange-correlation functional is not known exactly and must be approximated. The primary distinction between these exchange-correlation functionals is local density approximation (LDA) versus generalized gradient approximation (GGA) functionals. Within the GGA functional family, many varieties have been proposed with this being an active area of theoretical development for over two decades.[12] The functional that results in closest agreement with the experiment may depend on the application of interest, with the Becke–Lee–Yang–Parr-type (BLYP) functionals proving popular for many molecular applications, while others (LDA and GGA functionals discussed in more detail in other chapters and in reviews mentioned here) are frequently used in solid-state physics calculations.

One issue that is particularly challenging in the electronic structure calculations of metal oxide surfaces and surface reactions is the choice of a model system. Choosing a model of sufficient complexity to capture the necessary physics must be balanced with computational constraints for system size and complexity. Surface models can be divided into two main categories, namely, cluster and periodic models. Figure 7.1 displays schematics of cluster and periodic models for surface reactions.

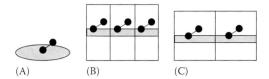

(A)   (B)   (C)

FIGURE 7.1   Possible model surfaces for studying a chemical reaction on a metal or metal oxide surface. (A) shows a "cluster model," (B) displays a periodic model with a small surface area, and (C) displays a periodic model with a larger surface area. Some advantages and disadvantages of each model are discussed in the text.

Cluster models of surfaces may have the advantage of modeling an isolated surface event using a relatively small number of atoms. However, especially for metallic surfaces, terminating the edges of the cluster in a way that maintains the appropriate electronic structure in the region of interest is not trivial. Inserting a metal oxide cluster into a point charge array may provide a means of alleviating some of the difficulties associated with correctly terminating a surface, but careful testing is required to ensure that size effects do not introduce unphysical behavior.[13,14]

Periodic models have the advantage of avoiding artificial surface edge terminations needed in cluster models that can impact the surface band structure. A major concern with periodic models used in simulating surface reactions is the adsorbate interactions with their periodic images if the surface area of the unit cell is too small. Although such a model may capture some features of a high coverage regime, predictions and direct linkage to experiments may be complicated by the enforced periodicity with the small unique surface area of the model, whereas the experimental conditions may relate to very low coverage and depend on defect sites.[15,16] Conversely, increasing the surface area of the unit cell is accompanied by a significantly increased computational cost. Once again, careful testing may be required, although the general "rule of thumb" separations on the order of $10\,\text{Å}$ between periodic images may be used. For applications with periodic interfaces, the issue of lattice misfit between the two materials can be a significant issue in determining the size of the interface cell that must be employed to avoid unphysical strain due to enforced periodicity of commensurate lattices in the xy plane. In catalytic applications, it may be step edges or defect sites that are key to functionality rather than the clean periodic surface. Properly including such sites in a periodic model is challenging since the limiting system size necessitates either a clean periodic surface or the creation of a very high defect density, which may not correspond to the physical system.

A vast number of DFT calculations on metal oxides have been performed to date, and it is certain that this number will continue to increase with corresponding increases in computing power and improvements in techniques allowing much faster theoretical calculations and more complex systems to be addressed. It would not be possible to provide a comprehensive review of all computational metal oxide studies, but we wish to highlight a few applications that are interesting in their own right, and which provide examples of the breadth of applications that exist. Accordingly, in Section 7.2, we present a summary of Al_2O_3 calculations from early studies to current applications, and also detail Fe_2O_3 hematite surface calculations. How experiment and theory have worked together to elucidate the appropriate surface structure will also be described. We then provide, in Section 7.3, an overview of several current applications of density functional calculations to pertinent technological and scientific metal oxide questions. We conclude by emphasizing some of the remaining challenges facing the theoretical and computational studies of metal oxide surfaces.

7.2 APPLICATIONS OF DFT TO AL_2O_3 AND FE_2O_3

7.2.1 ALUMINA CALCULATIONS

In the 1990s, density functional implementations and computing power had reached the stage where a variety of metal oxide surfaces were tractable. Alumina Al_2O_3 is an important material for many applications ranging from use as a high-k dielectric, to a component in thermal barrier coatings, to a support in heterogeneous catalysis. The α-Al_2O_3 phase was studied early due to its stability as well as its symmetry resulting in a reasonable number of atoms in a computational supercell (10 atoms in the rhombohedral unit or 30 atoms in the full hexagonal cell for the space group 167, $R\bar{3}c$, relative to ~100 atoms in the metastable gamma phase.) Figure 7.2 displays a side view of the corundum structure with oxygen positions shown in red and metal ions in blue.

Early calculations of alumina surfaces focused on clean stoichiometric surface terminations for this phase.[17–19] In 1999, a study by Batyrev et al.[20] explored the atomic and electronic structure of α-Al_2O_3 using periodic density functional calculations within a planewave pseudopotential implementation. They explored five different surface terminations with varying excess oxygen terminations in an attempt to explore the impact of oxygen pressure on the favorability of a particular surface termination for the basal plane. They found the stoichiometric aluminum surface termination to be the most energetically favorable of the surfaces studied over a significant range of oxygen partial pressures, further reinforcing the validity of the surface termination chosen in the earlier study by Manassidis et al.[17] Wang et al. showed that stable surface terminations of oxygen were accessible only in the presence of hydrogen and helped resolve differences between theoretically and experimentally predicted surface relaxations.[21] These ab initio thermodynamics studies, whereby surface energies are calculated in DFT and then expressed in terms of

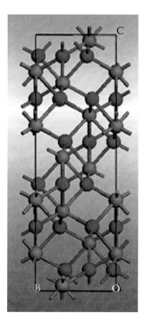

FIGURE 7.2 Side view of the corundum structure for α-Al_2O_3 as well as hematite. The oxygen positions are shown in red with the metal ion positions displayed in blue. Possible cleavage locations for the (0001) basal plane for this crystal will be discussed in the text.

surface phase diagrams dependent on temperature and partial pressure of oxygen and hydrogen for instance, have explored a variety of low index surfaces and their different terminations.[22] Additional work has looked at details of the surface atomic and electronic structure for reconstructions of this plane, which may hold technological significance.[23,24]

As mentioned, it is often thin films rather than bulk crystalline material of metal oxides that are important from an application standpoint, and this may require a good model and understanding of interfacial interactions. A general review of the theory of metal ceramic interfaces was provided by Finnis in 1996.[25] The importance of interfaces is particularly appropriate for alumina, which forms an Angstrom self-terminating coating when formed via Al, NiAl, or Ni_3Al oxidation.[26-28] Jennison and coworkers performed several DFT studies of thin film alumina on various metals and determined that the thin film phase is most similar to $\kappa\text{-}Al_2O_3$.[29-31] A more recent DFT study by Kresse, which included a detailed comparison to scanning tunnel microscope (STM) experiments, determined that the alumina overlayer structure on bulk NiAl (110) had a larger number of simulated atoms than in previous studies and concluded that the film structure differed from all bulk phases of alumina.[32] Numerous additional DFT studies of interfaces between α-alumina and metals have improved our understanding of atomic-level interactions and electronic structure modifications at these interfaces.[33-36]

Many of the most recent DFT studies of α-alumina surfaces are interested in small molecule adsorption with alumina serving as the support rather than on the detailed surface structure of alumina itself. For instance, Casarin et al. investigated chemisorption of CO on the basal plane of alumina ($\alpha\text{-}Al_2O_3$ (0001)) in 2000.[37] Gamallo and Sayos looked at atomic oxygen and nitrogen adsorption on the clean UHV surface (aluminum terminated),[38] and Hernandez and coworkers compared the adsorption of several metals on the clean surface.[39] In 2005, Chatterjee and coworkers explored the deposition of Ag on the basal plane of alumina while attempting to account for the complexity of this surface in non-UHV conditions.[40] Similarly, Yang and Rendell modeled adsorption of Ga on clean alumina[41] and surfaces contaminated through hydrogen chemisorption and physisorption.[42] Recently, DFT studies intending to model larger molecule chemisorption, such as quinizarin,[43] nitromethane, and 1,1-diamino-2,2-dinitroethylene,[44] the fluorination of dichloromethane,[45] uranyl (hydrated),[46] Pd nanoclusters,[47] and ZnO growth[48] on $\alpha\text{-}Al_2O_3$ (0001) have been undertaken.

Calculations of alumina surfaces,[49] adsorption on alumina surfaces, and with alumina surfaces serving as a support have begun to focus on $\gamma\text{-}Al_2O_3$, which is frequently a more relevant phase due to experimental growth conditions, temperatures, and pressures. Because of computational size constraints, it is common to see cluster models employed instead of periodic models, when phases other than $\alpha\text{-}Al_2O_3$ are modeled; however, current studies are able to employ periodic models of $\gamma\text{-}Al_2O_3$. Several recent studies have looked at $\gamma\text{-}Al_2O_3$ with adsorbates such as heavy metals including Pd,[50] Pb (on hydrated surfaces),[51] oxides of Ba[52] and Mo,[53] small molecules including CO and C_2H_4 on supported Pd clusters,[50,54] and MoS_2 sheets.[55,56] Recent studies are able to address adsorption of larger and more complex molecules and reactivity, such as phosphonyl reagents on $\gamma\text{-}Al_2O_3$,[57] chemical warfare agents on a model of hydroxylated γ-Al_2O_3,[58] and alkane metathesis.[59] A recent review by Raybaud et al. describes progress in DFT calculations of fuel hydrotreating and hydrosulfurization catalysts supported on γ-Al_2O_3.[60]

κ-Al_2O_3 is challenging to model due to difficulty in determining structures of complex and metastable materials.[61] Accordingly, DFT studies employing this phase have appeared only within the past few years. Fortrie et al. investigated VO_x/κ-Al_2O_3 (001) using DFT to parameterize Monte Carlo simulations and employed a model with a much larger surface area than could be addressed through a straightforward application of DFT.[62]

Current applications of DFT calculations to alumina are investigating novel mechanistic pathways, low symmetry crystal structures, and comparisons with experiments that were not addressed until recent computational and theoretical advances. For instance, density functional investigations of vacancies,[63–65] interfacial vacancies,[66] and charge defects[67] are now available. Comparisons with chemical shifts in calculated and experimental solid state NMR[68] are also now possible and may assist in creating models that better approximate the electronic structure of alumina in noncrystalline environments, such as thin films, which have traditionally been a challenge for DFT applications. Additionally, calculations of alumina for features typically considered to be associated with length scales beyond the scope of quantum chemistry, and of interest in the materials science and engineering realm are being tackled via DFT, including investigations of mesoporous materials,[69] possible fatigue mechanisms,[70] grain boundary structure,[71] and ion segregation and sliding.[72]

7.2.2 FE₂O₃ CALCULATIONS

Hematite (α-Fe_2O_3) is the mineral form of iron (III) oxide, which is the most commonly mined ore of iron for industrial production.[73] Like alumina, hematite crystallizes in the rhombohedral system, and shares the corundum structure. The bulk structure of hematite has a hexagonal unit cell (space group $R\bar{3}c$), with a distorted close-packed array of oxygen anions and iron cations that occupy two-thirds of the octahedral holes. The atomic layer stacking sequence along the c axis consists of six stoichiometric $-(Fe-O_3-Fe)-$ repeat units per bulk unit cell. The iron cations are staggered along the c_s axis and displaced vertically from their ideal hexagonal close-packed (hcp) positions centered between the oxygen layers. Unlike alumina, hematite is a bulk antiferromagnet with alternating spin layers in the bulk crystal,[73,74] so quantum mechanics methods are of great interest in probing the correlation between the electronic structure and the experimental observations of surface reactivity.

While its crystals do not cleave, hematite's predominant growth faces include the (0001) basal plane[75] and the ($1\bar{1}02$) plane, as shown in Figure 7.3. The clean surfaces of Fe_2O_3 have been well characterized as a function of surface preparation using low-energy-electron-diffraction (LEED); in these studies, the surface was first cleaned by Ar^+ bombardment, and then annealed to determine the effect of temperature and oxygen partial pressure on the surface structure and composition. Several types of LEED patterns, corresponding to different surface reconstructions, were observed upon annealing, but these patterns were found to depend on both the O_2 partial

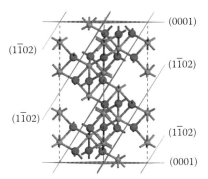

FIGURE 7.3 Bulk unit cell of hematite (α-Fe$_2$O$_3$), along with its dominant cleavage planes, (0001) and ($1\bar{1}02$). The oxygen positions are shown in red and the metal ion positions are shown in blue.

pressure and temperature. For instance, Sanchez et al. found that α-Fe$_2$O$_3$ (0001) exhibits strong temperature-dependent surface reconstructions, with (1×1) patterns observed only at annealing temperatures of 1173 K or above, and the formation of a Fe$_3$O$_4$ (111) layer on the α-Fe$_2$O$_3$ (0001) surface at lower annealing temperatures. In contrast, α-Fe$_2$O$_3$ ($1\bar{1}02$) exhibits strong O$_2$ partial pressure-dependent surface reconstructions, with (1×1) patterns observed at high O$_2$ partial pressures above 10^{-6} Torr, and (2×1) patterns observed at low O$_2$ partial pressures down to 10^{-10} Torr,[76] as depicted in Figure 7.4. Unfortunately, for many years, no theoretical calculations were available to verify these surface reconstructions.[77]

Clean α-Fe$_2$O$_3$ (0001) contains three chemically distinct surface terminations, depending on location of the cleavage plane: a single iron layer (Fe–O$_3$–Fe–R), a double iron layer (Fe–Fe–O$_3$–R), and an oxygen layer (O$_3$–Fe–Fe–R).[78,79] While the highly relaxed single-layer Fe-termination (Fe–O$_3$–Fe–R) is the most stable surface configuration under ultrahigh vacuum conditions,[80] ab initio DFT calculations helped identify a ferryl (Fe double-bonded to oxygen, Fe=O) terminated surface as more favorable under higher oxygen pressure[81,82] as well as a possible pathway for creating this surface[83] that had been observed via scanning tunneling

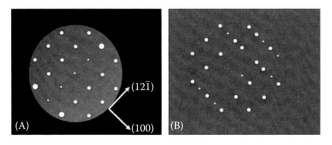

FIGURE 7.4 Schematics of the LEED patterns obtained on the α-Fe$_2$O$_3$ ($1\bar{1}02$) (1×1) (A) and (2×1) (B) surface reconstructions. (Adapted from Henderson, M.A., *Surf. Sci.*, 515, 253, 2002.)

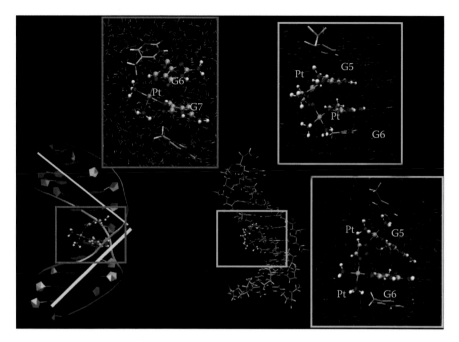

FIGURE 1.4 Cisplatin–DNA adduct and (in the orange rectangle) a close view of the cis-platin binding region; yellow lines highlight the kink induced by cisPt. Diplatin–DNA adduct and (in the blue rectangle) a close view of the N1, N2 and N1, N3 binding mode in **B** and **C** complexes.

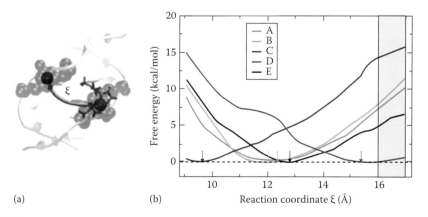

(a)   (b)

FIGURE 1.5 (a) Backbone atoms and their centers of mass (transparent and solid blue spheres, respectively) defining the reaction coordinate ξ. (b) Free-energy profiles (kcal/mol) for **A**–**E**. Small arrows indicate the free-energy minima, and the yellow bar shows the minor groove width of DNA-HMG.

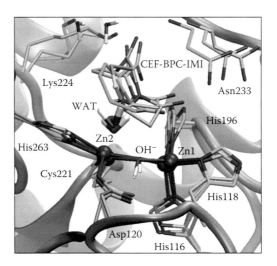

FIGURE 1.6 MD structural insights into the binding mode of di-Zn MβL CcrA in complex with benzylpenicillin, cefotaxime, and imipenem β-lactam substrates (only substrate bi-cyclic cores and residue heavy atoms are shown for sake of clarity, see also Schemes 1.1 and 1.2). (Adapted from Dal Peraro, M., Vila, A.J., Carloni, P., and Klein, M.L., *J. Am. Chem. Soc.*, 129, 2808, 2007.)

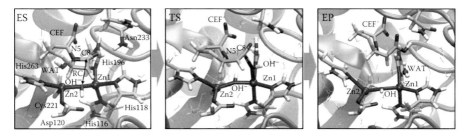

FIGURE 1.7 QM/MM structural insights into the reaction mechanism of di-Zn MβL CcrA from *B. fragilis* (atoms shown in licorice are included in the QM cell, see also Schemes 1.1 and 1.2). (Adapted from Dal Peraro, M., Vila, A.J., Carloni, P., and Klein, M.L., *J. Am. Chem. Soc.*, 129, 2808, 2007.)

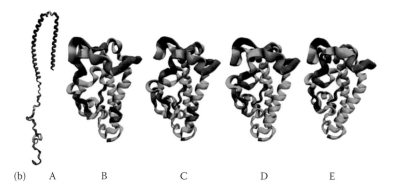

FIGURE 2.1   (b) Secondary structures of the wild-type α-synuclein protein obtained from classical MD simulations in the gas phase at various times: (A) 0 ns, (B) 5 ns, (C) 10 ns, (D) 15 ns, and (E) 20 ns. The different secondary structures are presented utilizing various colors: α-helix (purple), 3_{10} helix (orange), bridge β (blue), turn (green), coil (red).

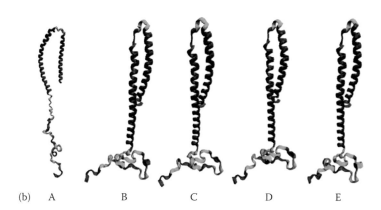

FIGURE 2.2   (b) Secondary structures of the A53T α-synuclein protein obtained from classical MD simulations in the gas phase at (A) 0 ns, (B) 5 ns, (C) 10 ns, (D) 15 ns, and (E) 20 ns. The α-helix (purple), bridge β (blue), turn (green), and coil (red) are depicted.

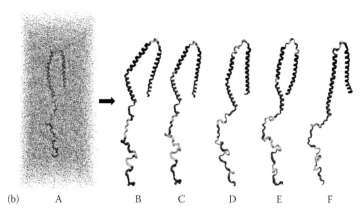

FIGURE 2.3 (b) Secondary structures of the wild-type α-synuclein in aqueous solution (A) obtained from classical MD simulations at (B) 0 ns, (C) 1 ns, (D) 2 ns, (E) 3 ns, and (F) 4 ns. The α-helix (purple), 3_{10} helix (orange), bridge β (blue), turn (green), and coil (red) are depicted.

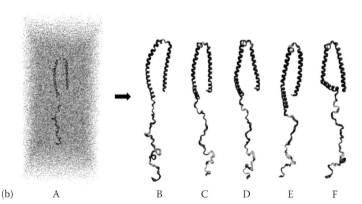

FIGURE 2.4 (b) Secondary structures of A53T α-synuclein in aqueous solution (A) obtained from classical MD simulations at (B) 0 ns, (C) 1 ns, (D) 2 ns, (E) 3 ns, and (F) 4 ns. The α-helix (purple), 3_{10} helix (orange), bridge β (blue), turn (green), and coil (red) are depicted.

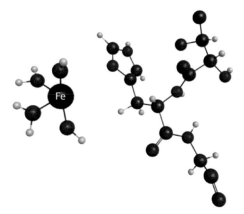

FIGURE 2.6 Pictorial representation of the Fe^{3+} ion coordination to the His50 residue of the wild-type α-synuclein protein in water at room temperature obtained from QM/MM simulations.

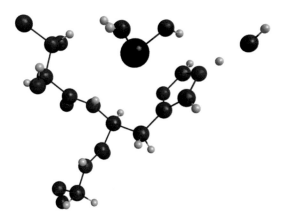

FIGURE 2.7 Pictorial representation of the Fe^{3+} ion coordination to the His50 residue of the mutant-type A53T α-synuclein protein in water at room temperature obtained from QM/MM simulations.

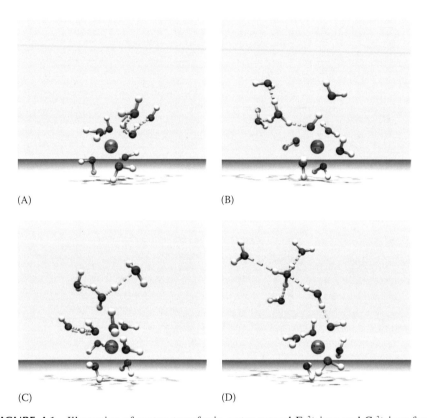

(A)

(B)

(C)

(D)

FIGURE 4.1 Illustration of proton transfer in water around Fe^{3+} ions and Cr^{3+} ions from CPMD/TPS simulations. Many water molecules are deleted to clearly show the molecular mechanism. See text for discussion of mechanism. Illustration of the mechanism of water dissociation around metal (Cr^{3+}, Fe^{3+}) ions which begins with a slightly elongated O–H bond (A). Proton migration leads to the hydroxylation of the metal ion (B). Movement of a proton into the second solvation shell leads to the formation of a Zundel complex (C). Further proton transfer leads to the formation of an Eigen complex (D). (Coskuner, O. et al., *Water Dissociation in the Presence of Metal Ions*, 7853, 2007. Copyright Wiley-VCH Verlag GmbH & Co. KGaA. Reproduced with permission.)

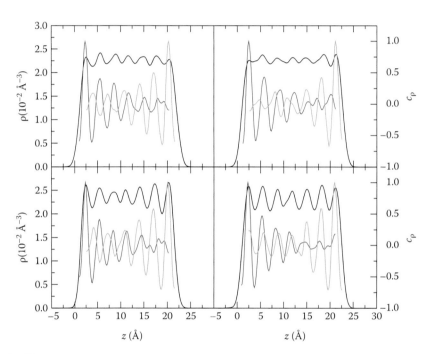

FIGURE 5.6 Density–density autocorrelation functions (c_ρ) for the Na free surface simulations. In each case, the red and green curves give the density–density autocorrelation functions referenced respectively to the left and right peaks in the atomic densities (which are shown as the black curves). The labeling of the panels is as in Figure 5.5. (Adapted with permission from Walker, B.G. et al., *J. Chem. Phys.*, 124, 174702, copyright 2006, American Institute of Physics.)

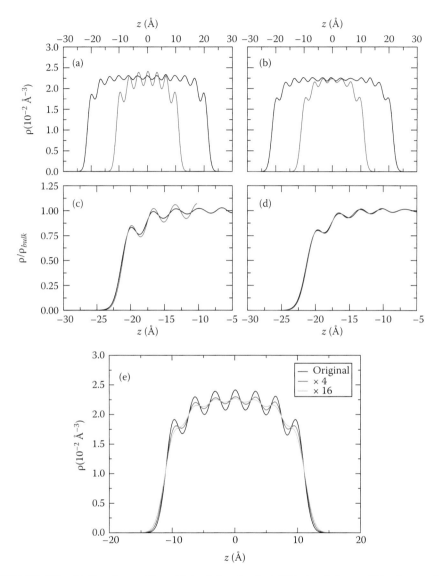

FIGURE 5.7 (a) Density profiles obtained for classical MD simulations of Na, for a cell having the same number of atoms as in the (001) ab initio simulations (red curves), and for a slab of twice the thickness (black curves), at temperature $T = 650$ K. (b) As for (a), but at temperature $T = 750$ K. (c) Surface parts of the density profiles for the slabs in (a) overlaid. (d) The surface parts of the density profiles for the slabs in (b) overlaid. (e) Density profiles for classical MD simulations of a slab having the same thickness as the (001) ab initio simulations (black curve), and for cells having 4 and 16 times the cross-sectional area (red and green curves, respectively). (Adapted with permission from Walker, B.G. et al., *J. Chem. Phys.*, 124, 174702, copyright 2006, American Institute of Physics.)

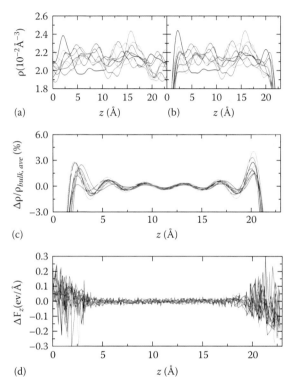

FIGURE 5.11 Friedel oscillations observed in snapshots from a bulk liquid simulation [73]. Panel (a) shows the valence charge densities for the various MD time steps taken from the bulk liquid simulation. Panel (b) shows the valence charge densities for the same MD snapshots after a vacuum region has been added to each of the unit cells. Panel (c) shows the valence charge density redistributions upon surface formation for the set of bulk MD time steps. Panel (d) shows the differences in the forces acting on atoms (i.e., the forces induced by formation of a surface). (Adapted from Walker, B.G. et al., *J. Chem. Phys.*, 124, 174702, copyright 2006; Walker, B.G. et al., *J. Phys. Condens. Matter.*, 18, L269, copyright 2006. With permission.)

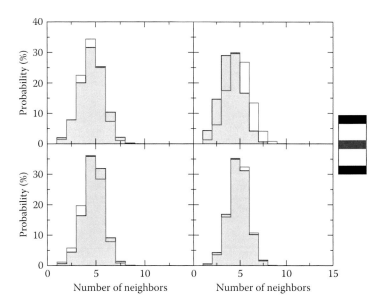

FIGURE 5.13 Nearest neighbor distributions within the different regions of the liquid surfaces simulated with ab initio molecular dynamics from Refs. [70,74]. As indicated by the bar to the right of the plots: black lines refer to the surface layers and red lines to the inner layers. The labeling of the panels is as in Figure 5.5. (Adapted with permission from Walker, B.G., *Ab Initio Molecular Dynamics Studies of Liquid Metal Surfaces*, PhD thesis, University of Cambridge, copyright 2004; Walker, B.G. et al., *J. Chem. Phys.*, 127, 134703, copyright 2007, American Institute of Physics.)

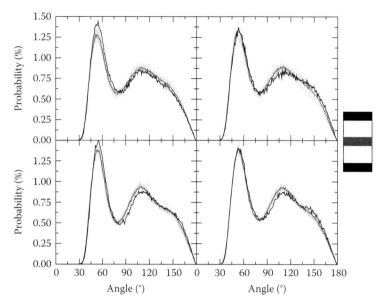

FIGURE 5.14 Distributions of bond angles determined in the different regions of the simulated liquid sodium surfaces from Refs. [70,74]. The regions referred to by the different curves are as in Figure 5.13. The labeling of the panels is as in Figure 5.5. (Adapted with permission from Walker, B.G., *Ab Initio Molecular Dynamics Studies of Liquid Metal Surfaces*, PhD thesis, University of Cambridge, copyright 2004; Walker, B.G. et al., *J. Chem. Phys.*, 127, 134703, copyright 2007, American Institute of Physics.)

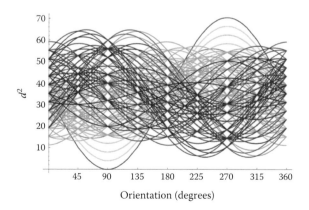

FIGURE 9.3 Distance graphs for each of the 120 possible mappings in Figure 9.2.

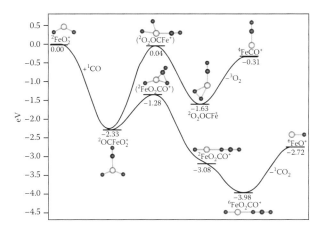

FIGURE 9.33 Reaction profile of the reaction between FeO_2^+ and CO. Relative energies are given with respect to the reactant FeO_2^+ in eV. (Reprinted from Reilly, N. et al., *J. Phys. Chem. C*, 111(51), December 1, 19088, 2007. With permission.)

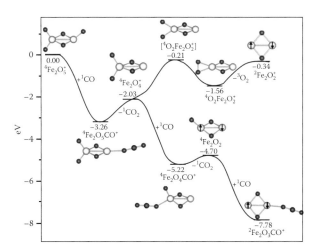

FIGURE 9.34 Reaction profile of the reaction between $Fe_2O_5^+$ and CO. Relative energies are given with respect to the reactant $Fe_2O_5^+$ in eV. The arrows in the blue Fe atoms indicate spin broken solutions. (Reprinted from Reilly, N. et al., *J. Phys. Chem. C*, 111(51), December 1, 19086, 2007. With permission.)

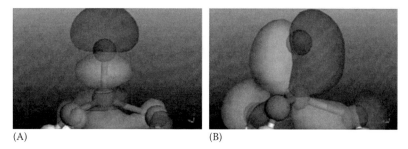

(A)   (B)

FIGURE 7.5 Ferryl (Fe=O) termination of the α-Fe$_2$O$_3$ (0001) surface with molecular orbital (MO) plots superimposed over the atomic positions. The MO in pane (A) displays bonding with a dz2-type orbital from Fe while that in (B) exhibits bonding through a dxz-type orbital. Such orbital symmetries are not available for bonding with oxygen in the case of the α-Al$_2$O$_3$ (0001) surface, which may explain the observed differences in surface terminations between Al$_2$O$_3$ and metal oxides with the same atomic structure Cr$_2$O$_3$ and Fe$_2$O$_3$.

microscopy and anticipated based on infrared reflection absorption spectroscopy.[84] This is one of the many instances where quantum chemical calculations were able to explain and inform experimental observation. Likewise, such calculations offer insight into the dissimilar behavior between Al$_2$O$_3$, Cr$_2$O$_3$, and Fe$_2$O$_3$ surface terminations. The energetically available d states with Cr and Fe metals stabilize the metal double bond to oxygen and permit that to be a favored surface termination with sufficient oxygen pressure (see Figure 7.5). However, in the case of alumina, the lack of energetically accessible d states available for bonding results in the theoretical Al=O termination being much less favorable, and thus not observed experimentally.

The experimentally observed α-Fe$_2$O$_3$ (1$\bar{1}$02) surface termination also appears to vary as a function of surface preparation, though perhaps to a lesser extent than for α-Fe$_2$O$_3$ (0001). While the stoichiometric termination is favored under UHV conditions, Lo et al. showed that an oxygen overlayer is present under high O$_2$ partial pressures approaching atmospheric pressure.[85] In the stoichiometric termination, there is a large expansion in the z direction between the layer-1 O anions and the layer-2 Fe cations, which is driven by the fivefold-coordinated layer-2 Fe cations recessing into the bulk. In addition, these two atomic layers also exhibit small, but not insignificant, in-plane (xy) motion. Both the in-plane rotation and the out-of-plane relaxation result in the formation of shorter Fe-O bonds between the under coordinated atomic layers 1 and 2, compared to the bulk α-Fe$_2$O$_3$ structure. While these atomic motions are observed using both classical energy minimizations and DFT calculations, the motion of the layer-1 O anions is more accurately modeled using a quantum mechanical approach. For instance, classical molecular dynamics simulations indicate that the layer-1 O anions recess significantly into the bulk α-Fe$_2$O$_3$, presumably so that the large negative point charges localized on the O anions may be screened by the Fe cations. In fact, density functional calculations show that the layer-1 O anions relax less significantly than predicted by classical molecular dynamics simulations. In the quantum mechanical treatment, the electrons are delocalized over the entire solid, such that it is not necessary for the surface ions to move as significantly to achieve effective charge screening.

In the oxygen-rich termination, the layer-1 O anions still recess significantly into the bulk as a result of electrostatic repulsion from the layer-i O atoms, but the layer-2 Fe cations do not relax as much into the bulk as they do in the stoichiometric termination. Since the layer-2 Fe cations are now sixfold coordinated, compared to being fivefold coordinated in the stoichiometric termination, there is less driving force for a significant relaxation of the layer-2 Fe cations into the bulk. Instead, it is the layer-i O anions that are highly unsaturated; as a result, the layer-2 Fe cations assume an increased formal oxidation state of (IV) in order to keep the slab charge neutral. While this surface termination has many structural similarities with ferryl-terminated α-Fe$_2$O$_3$ (0001), it differs in the orientation of the Fe=O bond to the surface since the topmost Fe cations are coordinated differently in α-Fe$_2$O$_3$ ($1\bar{1}02$) and α-Fe$_2$O$_3$ (0001). Both of these atomic layer relaxations are depicted in Figure 7.6.

Hematite is commonly precipitated out of standing water or mineral hot springs, or formed as a result of weathering processes in soils. Thus, the structure and reactivity of the iron (hydr)oxide–water interface are of prime importance in controlling

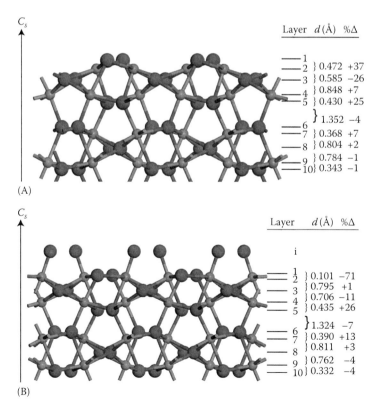

FIGURE 7.6   Stoichiometric (A) and oxygen-rich (B) terminations of the α-Fe$_2$O$_3$ ($1\bar{1}02$) surface with atomic layer relaxations in the c_s direction noted.

key environmental processes, such as the fate and transport of contaminant metals. In order to probe the 3D structure of this interface, hard x-ray scattering and spectroscopic techniques, including crystal truncation rod diffraction and x-ray reflectivity,[86–88] are used to determine the identity and ordered arrangement of atoms at the surface. In particular, the arrangement of water and other aqueous species on metal oxide surfaces can be determined with high accuracy. One drawback of these hard x-ray scattering techniques, however, is the limited sensitivity to (1) low Z elements, such as hydrogen, due to their weak scattering power, and (2) poorly ordered interfacial components, such as weakly bound/disordered sorbates and physisorbed water. Quantum mechanics-based methods, such as the DFT, and the classical mechanics-based methods, for example, molecular dynamics, are used to provide complementary and predictive models for interface structure and reactivity, which may then be used to guide experimental characterization and structural analysis.

The hydrated α-Fe_2O_3 (0001) surface possesses a unique surface termination that was recently elucidated with the help of DFT calculations and ab initio thermodynamics.[89] Trainor et al. showed that two surface terminations, $(HO)_3$–Fe–Fe–R and $(HO)_3$–Fe–H_3O_3–R are observed to coexist under H_2O partial pressures greater than 10^{-4} Torr in a roughly 2:1 ratio. These two surface terminations may coexist in a single domain, giving an overall surface stoichiometry of $(HO)_{1.2}$–$Fe_{0.4}$–$H_{1.2}O_3$–R, or exist as two separate linked domains. Figure 7.7 shows a schematic of the α-Fe_2O_3 (0001) surface, with a partial occupancy of the layer-2 Fe cations.

Ab initio thermodynamics was then used in conjunction with DFT to approximate the surface free energies of these and other surface terminations of α-Fe_2O_3 (0001).[89] These studies predicted that $(HO)_3$–Fe–H_3O_3–R is indeed the most stable surface termination under conditions of equilibrium with gas phase H_2O and O_2. However, since natural surfaces come into contact with liquid water rather than the gas phase water models commonly used in these studies, new thermodynamic models for treating interfacial water in contact with solid surfaces, including solvation and entropy effects, are ripe for future development.

The structure of the metal oxide-water interface in α-Fe_2O_3 ($1\bar{1}02$) has been the subject of much debate by experimentalists.[90–92] On one hand, the heterolytic dissociation of water into H^+ and OH^- is the natural first step allowing these species to

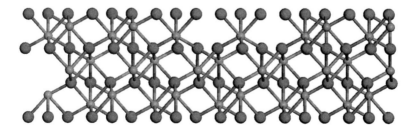

FIGURE 7.7 Possible "linked domain" model for the hydrated α-Fe_2O_3 (0001) surface (waters not shown). Only partial occupancy of the layer-2 Fe cations is observed experimentally, and the results from DFT calculations and ab initio thermodynamics are consistent with these two terminations being the most energetically stable at the oxygen partial pressures observed under environmental conditions.

bond to the undercoordinated layer-1 O anions and the layer-2 Fe cations, giving a surface termination based on $O_2–O_2–Fe_2–O_2–Fe_2–O_2–R$, where the bold atoms represent those above the stoichiometric termination. On the other hand, the partial occupancy of the layer-2 Fe cations, giving a surface termination based on $O_2–X–O_2–Fe_2–O_2–R$, has also been observed at room temperature using crystal truncation rod x-ray diffraction. Both surface terminations have been observed at room temperature, but the surface preparation procedures differed slightly. Lo et al. used DFT calculations and ab initio thermodynamics to show that while the $(H_2O)_2–X–(HO)_2–Fe_2–O_2–R$ termination is the most energetically stable at room temperature, the $(HO)_2–(HO)_2–Fe_2–O_2–Fe_2–O_2–R$ termination is more stable above 450 K.[85] These results suggest that the prevalent surface termination is highly sensitive to temperature. One plausible explanation is that molecular water desorption and surface reconstruction occur at high temperatures, thus increasing the occupancy of the layer-2 Fe cations in the $(H_2O)_2–X–(HO)_2–Fe_2–O_2–R$ surface termination to obtain $(HO)_2–Fe_2–O_2–Fe_2–O_2–R$. Upon annealing under high partial pressures of water, the surface may be hydroxylated, thus giving the $(HO)_2–(HO)_2–Fe_2–O_2–Fe_2–O_2–R$ termination. However, the results based on ab initio thermodynamics calculations on the hydroxylated surface at room temperature suggest that if the kinetic barriers to rearrangement could be overcome, then the annealed surface should reorganize back to the $(H_2O)_2–X–(HO)_2–Fe_2–O_2–R$ termination.

The electronic structure of alumina and hematite surfaces has been characterized with unprecedented accuracy at the molecular scale using DFT calculations and ab initio thermodynamics. The computed structural models, including atomic layer relaxations and surface reconstructions, are in close agreement with experimental data on the surfaces. The quantum mechanical approach for modeling metal oxide surfaces thus facilitates future explorations of chemical reactivity for diverse applications including chemical catalysis, environmental remediation, and alternative energy production. An overview of several key areas where these applications will contribute is discussed briefly in the following section.

7.3 CURRENT APPLICATIONS OF DFT TO METAL OXIDE SURFACES

Metal oxides are widely used in a diverse and fascinating array of industrial applications. For instance, metal oxides are used to passivate metal surfaces against corrosion, function as gas sensors for pollution monitoring, and control and serve as electrolytes in solid oxide fuel cells.[93–96] In this section, several examples will be presented, where quantum chemical methods have been widely used to study the structure and reactivity of metal oxide surfaces. These examples will include the chemical catalysis for hydrocarbon transformations, the environmental control of contaminant fate and transport, and materials engineering for solar energy conversion.

7.3.1 CATALYSIS

As mentioned in Section 7.1, metal oxides are often used as supports for noble metal catalyst particles, and also on their own. For instance, cerium oxide (CeO_2) is increasingly being used as a catalyst and as a catalyst support in automotive applications for

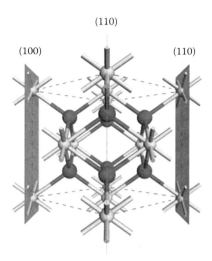

FIGURE 7.8 Bulk unit cell of ceria (CeO_2), along with one of its dominant cleavage planes (110). The oxygen positions are shown in dark and the metal ion positions are shown in the lighter atomic positions shown.

reducing NO_x emissions and converting carbon monoxide to carbon dioxide. Cerium oxide's superior catalytic properties in transforming oxygen-containing species are the result of its ability to easily become nonstoichiometric in oxygen content. As shown in Figure 7.8, cerium oxide's (110) surface is easily reduced since the oxygen atoms at the surface are all coplanar, thus facilitating a rapid diffusion of the oxygen atoms and creating oxygen vacancies.

DFT calculations have shown that the energetic stability of the cleave planes of cerium oxide follows the trend: (111)>(110)>(100). It follows then that oxygen vacancies form more easily on CeO_2 (110) than CeO_2 (111). However, the calculations also demonstrate that the initial oxygen vacancy forms in the surface atomic layer in CeO_2 (110), and in the subsurface atomic layer in CeO_2 (111).[97]

Commonly, ceria is used to support noble metal nanoparticles, such as platinum, palladium, rhodium, and gold by altering the electronic structure of these compounds. Thus, ceria is used as a cocatalyst in a number of industrial chemical applications involving hydrocarbon transformations including the water-gas shift reaction, steam reforming, and the Fischer-Tropsch reaction. Several reactants including hydrogen gas,[98] water,[99] methane,[100] methanol,[101] and carbon monoxide[102,103] adsorb very strongly and exothermically on the stoichiometric, oxidized, and reduced ceria surfaces and dissociate at oxygen vacancies. In particular, the reduced surfaces of ceria are of tremendous interest in catalysis. DFT calculations have suggested that oxygen atoms initially contained in adsorbed reactants, such as carbon monoxide, are easily incorporated into reduced surfaces. For this reason, ceria has been considered as a potential component of an integrated catalyst system for the conversion of carbon dioxide to liquid fuels.

Perhaps one of the most successful applications of DFT is in computational catalysis. In particular, the nudged elastic band[104] of Jónsson et al. is widely acknowledged to be the *de facto* method for studying reactive processes on metal oxide surface catalysts. The method seeks to find the minimum energy path to the saddle point that separates the reactant and product basins on the potential energy surface. Assuming that the reactants and products are known beforehand, a path linking the two states is created with a number of replicas of the original system initially located equidistantly along this path and connected by harmonic springs. This path is then relaxed to find the true minimum energy path by allowing the intermolecular forces to move until the transverse forces (e.g., those perpendicular to the local tangent) on the replicas vanish. The nudged elastic band method has been implemented in several plane wave codes including Dacapo,[105,106] NWChem,[107] PwSCF,[108] and VASP.[109,110]

7.3.2 Environmental Applications

The metal oxide-water interface is of tremendous interest in the study of environmental reactivity and recent quantum chemical calculations have proven to be invaluable in determining the structural and electronic factors that affect contaminant metal sorption at environmental interfaces. Of these contaminant metals, lead[51] and chromium[111] are two elements that have been recently characterized using DFT calculations.

Experimental XAS studies have shown that Pb(II) binds weakly to form outer-sphere complexes with α-Al_2O_3 (0001),[112] and strongly to form inner-sphere complexes with α-Fe_2O_3 (0001).[113] Since the two surfaces are isostructural, the differences in binding were suspected to be the result of differences in the electronic structure between transition metal-containing compounds and nontransition metal-containing compounds. Mason et al. used DFT calculations to show that in hematite, the Pb-O covalent interactions are stabilized by the partially occupied Fe d band, and the additional electronic interaction between the valence O p, Fe d, and Pb s and p orbitals.[51] Thus, the reactivity of α-Fe_2O_3 (0001) toward Pb(II) is much higher than the reactivity of α-Al_2O_3 (0001). In fact, α-Fe_2O_3 (0001) is more reactive toward Pb(II) than even α-Fe_2O_3 ($1\bar{1}02$),[113] which also forms inner-sphere complexes with Pb(II) due to structural differences and the coordination of the reactive Fe(II) sites on both surfaces.

In contrast, Cr(VI) has been shown to form only outer-sphere adsorption complexes on hydrated α-Fe_2O_3 ($1\bar{1}02$).[114] Since chromium forms chromate complexes readily, the favorable thermodynamics of hydrogen bonding and ion-dipole attraction between the complexes and the surface hydroxyls, coupled with the unfavorable energetic overlap between the valence Cr orbitals and the O p and Fe d orbitals, naturally lend themselves to the observation of outer-sphere adsorption of Cr(VI) on hydrated hematite, as observed by Yin et al.[111] Thus, quantum chemical approaches are essential for probing the effect of the electronic structure on the reactivity of metal oxide surfaces, particularly in the prediction of inner-sphere and/or outer-sphere adsorption complexes of contaminant metals on soils and sediments.

7.3.3 Photovoltaics

Another promising use of metal oxides is in thin film materials for multi-junction photovoltaic devices. The third generation of these devices aims to improve on the high processing costs of silicon and the low efficiency of non-oxide materials.[115–118] The main objection to the use of oxide materials is the large band gap of these materials, since the conversion efficiency of metal oxide semiconductors is highly dependent on photon absorption and the promotion of an electron from the filled valence band to the empty conduction band. Materials absorb all photons in the solar spectrum with energies greater than or equal to the band gap in order to create excited electron-hole pairs for initiating photochemistry. Since most metal oxide semiconductors have band gap values greater than 3.0 eV, they can only absorb in the ultraviolet portion of the spectrum. Thus, much attention has been focused on doping the metal oxide materials with both nonmetal and transition metals, and on constructing layered thin film devices with multiple junctions for capturing more of the solar spectrum.

Titanium dioxide[119] has received most of the attention from researchers using DFT to study the effect of doping on the electronic band structure and projected photovoltaic performance. For instance, both oxygen vacancies[120] and carbon[121–123] and nitrogen[124–128] dopants have been shown to increase the visible light photoresponse of TiO_2. Transition metals, such as vanadium,[129,130] iron,[130–132] and chromium[133,134] have also been shown to enhance the photoresponse of TiO_2. DFT calculations have been particularly valuable for investigating changes in the electronic band structure of TiO_2 upon nonmetal and transition metal doping. The findings of these studies and others suggest that doping introduces additional energy levels, or midgap states, into the TiO_2 band gap so that electrons may be excited from the valence band to the conduction band in a two-step process, thus lowering the energy required and shifting the absorption to the visible spectrum. Much theoretical research is still being performed on doped oxides as well as on mixed metal oxides with improved photoresponse through the narrowing of the band gap, and not just by the introduction of midgap states. Several of these compounds include the Ti−Fe−O,[132,135] Ti−Cu−O,[136] and Ti−W−O[137] systems for improved photocatalytic fuel production compared to their basic compounds.

7.4 SUMMARY AND FUTURE DIRECTIONS

As mentioned previously, the physics of the system to be studied often dictates the level of theory and model system required to address the problem of interest. Within DFT applications, challenges remain in applications where nonbonding interactions, i.e., van der Waals forces, play a significant role. Functionals and extensions of the basic density functional approaches that attempt to address these issues are active areas of research. Additionally, system sizes above several hundred atoms generally require novel techniques. Time-dependence, extreme conditions, and high coverage continue to be some of the primary computational challenges, but are developing areas of research and application.

Another problem in using quantum chemical methods to model metal oxide structure and reactivity is the separation of length and time scales. For instance, in catalysis, reactive processes occur over time scales of milliseconds and up, while molecular approaches typically max out at timescales of nanoseconds. Car–Parrinello molecular dynamics,[138,139] which is the time-integration of the equations of motion with the interatomic forces calculated "on the fly" from DFT calculations, is one approach for studying processes at finite temperatures with reactive bond-breaking and bond-forming.[140–147] Hybrid quantum mechanics/molecular mechanics methods,[148] which treat the electronically important part of the system with quantum mechanics and the rest with classical mechanics, have also been used to bridge length and time scales in modeling metal oxide surface structure and reactivity.[149]

A recent review by Hafner details some of the state of the art in DFT and post-DFT methods.[150] Pacchioni describes the challenges of addressing defective oxide materials due to the problem of self-interaction in traditional DFT calculations. He shows that hybrid functionals and DFT+U approaches are offering alternative ways to correct this issue and to approach such important problems.[151] Huang and Carter discuss recent advances in treating electron correlation by embedding configuration interaction or other calculations in DFT calculations of solids and surfaces.[152] Held provides another review of improving electron correlation through adding a dynamical mean field approach to a local density approximation DFT calculation.[153]

Multiscale modeling involves methods of bridging orders of magnitude in length and time that would otherwise be prohibitive in the straightforward application of electronic structure methods. Most traditional electronic structure applications are limited to system sizes on the order of 1000 unique atoms or less. Order N methods might increase this by a few orders of magnitude, but to address macroscopic length scales, multiscale methods are necessary. Thus, multiscale methods that include quantum mechanical effects to understand materials remain an active area of development.[154–159] Similarly, to address reactions occurring over time frames that are prohibitively long for straightforward dynamics, or to address many reaction events, techniques such as kinetic Monte Carlo may provide the answer. These are active areas of research that show promise in opening applications to much more complex systems and materials properties beyond many of those highlighted in this chapter.

ACKNOWLEDGMENTS

The authors gratefully acknowledge students Alexandria Cassady, Jennie Cook-Kollars, Nathan Fine, and Brent Sherman for research assistance related to this manuscript.

REFERENCES

1. Sholl DS. Applications of density functional theory to heterogeneous catalysis. In: Hinchliffe A, editor. *Chemical Modelling: Applications and Theory*. Volume 4. Cambridge, U.K.: The Royal Society of Chemistry; 2006. pp. 108–160.
2. Norskov JK, Bligaard T, Rossmeisl J, Christensen CH. Towards the computational design of solid catalysts. *Nature Chemistry* 2009;1(1):37–46.
3. Christensen A, Carter EA. First-principles study of the surfaces of zirconia. *Physical Review B* 1998;58(12):8050.
4. Hohenberg P, Kohn W. Inhomogeneous electron gas. *Physical Review* 1964;136(3B):B864.
5. Kohn W, Sham LJ. Self-consistent equations including exchange and correlation effects. *Physical Review* 1965;140(4A):A1133.

6. Chai J-D, Lignères VL, Ho G, Carter EA, Weeks JD. Orbital-free density functional theory: Linear scaling methods for kinetic potentials, and applications to solid Al and Si. *Chemical Physics Letters* 2009;473(4–6):263–267.

7. Aguado A, López JM. Structural and thermal behavior of compact core-shell nanoparticles: Core instabilities and dynamic contributions to surface thermal stability. *Physical Review B* 2005;72(20):205420.

8. Zhou B, Wang YA. Orbital-corrected orbital-free density functional theory. *The Journal of Chemical Physics* 2006;124(8):081107–081115.

9. Zhou B, Lignères VL, Carter EA. Improving the orbital-free density functional theory description of covalent materials. *The Journal of Chemical Physics* 2005;122(4):044103–044110.

10. Wesolowski TA. Quantum chemistry "Without Orbitals" an old idea and recent developments. *CHIMIA International Journal for Chemistry* 2004;58:311–315.

11. Zhou B, Wang YA, Carter EA. Transferable local pseudopotentials derived via inversion of the Kohn-Sham equations in a bulk environment. *Physical Review B* 2004;69(12):125109.

12. Proynov EI, Ruiz E, Vela A, Salahub DR. Determining and extending the domain of exchange and correlation functionals. *International Journal of Quantum Chemistry* 1995;56(S29):61–78.

13. Casarin M, Maccato C, Vittadini A. Chemisorption of simple inorganic/organic molecules on ZnO. *Trends in Inorganic Chemistry* 1996;4:43–77.

14. Christensen A, Jarvis EAA, Carter EA. Atomic-level properties of thermal barrier coatings: Characterization of metal-ceramic interfaces. *Chemical Dynamics in Extreme Environments, Advanced Series in Physical Chemistry*;11.

15. Woodruff DP. Should surface science exploit more quantitative experiments? *Surface Science* 2008;602(18):2963–2966.

16. Saravanan C, Markovic N, Head-Gordon M, Ross P. Multi-scale modeling of CO oxidation on Pt-based electrocatalysts. *Device and Materials Modeling in PEM Fuel Cells* 2009;113:533–549.

17. Manassidis I, De Vita A, Gillan MJ. Structure of the (0001) surface of α-Al_2O_3 from first principles calculations. *Surface Science* 1993;285(3):L517–L521.

18. Manassidis I, Gillan MJ. Structure and energetics of alumina surfaces calculated from first principles. *Journal of the American Ceramic Society* 1994;77(2):335–338.

19. Baxter R, Reinhardt P, López N, Illas F. The extent of relaxation of the α-Al_2O_3 (0001) surface and the reliability of empirical potentials. *Surface Science* 2000;445(2–3):448–460.

20. Batyrev I, Alavi A, Finnis MW. Ab initio calculations on the Al_2O_3(0001) surface. *Faraday Discussions* 1999;114(114):33–43.

21. Wang X-G, Chaka A, Scheffler M. Effect of the environment on α-Al_2O_3 (0001) surface structures. *Physical Review Letters* 2000;84(16):3650.

22. Marmier A, Parker SC. Ab initio morphology and surface thermodynamics of α-Al_2O_3. *Physical Review B* 2004;69(11):115409.

23. Jarvis EAA, Carter EA. Metallic character of the Al_2O_3(0001)-($\sqrt{31} \times \sqrt{31}$)R$\pm 9°$ surface reconstruction. *The Journal of Physical Chemistry B* 2001;105(18):4045–4052.

24. Lauritsen JV, Jensen MCR, Venkataramani K, Hinnemann B, Helveg S, Clausen BS, Besenbacher F. Atomic-scale structure and stability of the $\sqrt{31} \times \sqrt{31}R9°$ surface of Al_2O_3(0001). *Physical Review Letters* 2009;103(7):076103–076104.

25. Finnis MW. The theory of metal-ceramic interfaces. *Journal of Physics: Condensed Matter* 1996(32):5811.

26. Jaeger RM, Kuhlenbeck H, Freund HJ, Wuttig M, Hoffmann W, Franchy R, Ibach H. Formation of a well-ordered aluminium oxide overlayer by oxidation of NiAl(110). *Surface Science* 1991;259(3):235–252.

27. Becker C, Kandler J, Raaf H, Linke R, Pelster T, Drager M, Tanemura M, Wandelt K. *Oxygen Adsorption and Oxide Formation on* Ni_3Al (111). 1998; San Jose, CA. AVS 1000–1005.

28. Rosenhahn A, Schneider J, Becker C, Wandelt K. The formation of Al_2O_3-layers on $Ni_3Al(111)$. *Applied Surface Science* 1999;142(1–4):169–173.

29. Jennison DR, Verdozzi C, Schultz PA, Sears MP. Ab initio structural predictions for ultrathin aluminum oxide films on metallic substrates. *Physical Review B* 1999;59(24):R15605.

30. Bogicevic A, Jennison DR. Variations in the nature of metal adsorption on ultrathin Al_2O_3 films. *Physical Review Letters* 1999;82(20):4050.

31. Kelber JA, Niu C, Shepherd K, Jennison DR, Bogicevic A. Copper wetting of α-$Al_2O_3(0001)$: Theory and experiment. *Surface Science* 2000;446(1–2):76–88.

32. Kresse G, Schmid M, Napetschnig E, Shishkin M, Kohler L, Varga P. Structure of the ultrathin aluminum oxide film on NiAl(110). *Science* 2005;308(5727):1440–1442.

33. Jarvis EAA, Christensen A, Carter EA. Weak bonding of alumina coatings on Ni(1 1 1). *Surface Science* 2001;487(1–3):55–76.

34. Jarvis EAA, Carter EA. An atomic perspective of a doped metal-oxide interface. *The Journal of Physical Chemistry B* 2002;106(33):7995–8004.

35. Jarvis EA, Carter EA. Importance of open-shell effects in adhesion at metal-ceramic interfaces. *Physical Review B* 2002;66(10):100103.

36. Jarvis EAA, Carter EA. Exploiting covalency to enhance metal-oxide and oxide-oxide adhesion at heterogeneous interfaces. *Journal of the American Ceramic Society* 2003;86(3):373–386.

37. Casarin M, Maccato C, Vittadini A. Theoretical study of the chemisorption of CO on $Al_2O_3(0001)$. *Inorganic Chemistry* 2000;39(23):5232–5237.

38. Gamallo P, Sayos R. A density functional theory study of atomic oxygen and nitrogen adsorption over α-alumina (0001). *Physical Chemistry Chemical Physics* 2007;9(37):5112–5120.

39. Hernandez NC, Graciani J, Marquez A, Sanz JF. Cu, Ag and Au atoms deposited on the α-$Al_2O_3(0001)$ surface: A comparative density functional study. *Surface Science* 2005;575(1–2):189–196.

40. Chatterjee A, Niwa S, Mizukami F. Structure and property correlation for Ag deposition on α-Al_2O_3—A first principle study. *Journal of Molecular Graphics and Modelling* 2005;23(5):447–456.

41. Yang R, Rendell AP. First principles study of Gallium atom adsorption on the α-$Al_2O_3(0001)$ surface. *The Journal of Physical Chemistry B* 2006;110(19):9608–9618.

42. Yang R, Rendell AP. Ga cleaning of Al_2O_3 substrate: Low coverage adsorption of Ga on a hydrogen-contaminated α-$Al_2O_3(0001)$ surface. *The Journal of Physical Chemistry C* 2007;111(8):3384–3392.

43. Frank I, Marx D, Parrinello M. Structure and electronic properties of quinizarin chemisorbed on alumina. *The Journal of Chemical Physics* 1996;104(20):8143–8150.

44. Sorescu DC, Boatz JA, Thompson DL. First-principles calculations of the adsorption of Nitromethane and 1,1-diamino-2,2-dinitroethylene (FOX-7) molecules on the α-$Al_2O_3(0001)$ surface. *The Journal of Physical Chemistry B* 2005;109(4):1451–1463.

45. Bankhead M, Watson GW, Hutchings GJ, Scott J, Willock DJ. Calculation of the energy profile for the fluorination of dichloromethane over an α-alumina catalyst. *Applied Catalysis A: General* 2000;200(1–2):263–274.

46. Moskaleva LV, Nasluzov VA, Rosch N. Modeling adsorption of the Uranyl dication on the hydroxylated α-$Al_2O_3(0001)$ surface in an aqueous medium density functional study. *Langmuir* 2006;22(5):2141–2145.

47. Gomes JRB, Lodziana Z, Illas F. Adsorption of small palladium clusters on the relaxed α-$Al_2O_3(0001)$ surface. *The Journal of Physical Chemistry B* 2003;107(26):6411–6424.

48. Yang C, Li YR, Li JS, Yu WF. Density-functional study of the surface and interface of ZnO/α-Al$_2$O$_3$(0001). *Surface Review and Letters* 2004;11(6):509–513.

49. Ouyang CY, SljivanCanin Z, Baldereschi A. First-principles study of γ-Al$_2$O$_3$ (100) surface. *Physical Review B: Condensed Matter and Materials Physics* 2009;79(23):235410–235417.

50. Corral Valero M, Raybaud P, Sautet P. Influence of the hydroxylation of γ-Al$_2$O$_3$ surfaces on the stability and diffusion of single Pd atoms: A DFT study. *The Journal of Physical Chemistry B* 2006;110(4):1759–1767.

51. Mason SE, Iceman CR, Tanwar KS, Trainor TP, Chaka AM. Pb(II) Adsorption on isostructural hydrated alumina and hematite (0001) surfaces: A DFT study. *The Journal of Physical Chemistry C* 2009;113(6):2159–2170.

52. Cheng L, Ge Q. Effect of BaO morphology on NO$_x$ abatement: NO$_2$ interaction with unsupported and γ-Al$_2$O$_3$-supported BaO. *The Journal of Physical Chemistry C* 2008;112(43):16924–16931.

53. Handzlik JÇ, Sautet P. Structure of isolated Molybdenum(VI) oxide species on γ-alumina: A periodic density functional theory study. *The Journal of Physical Chemistry C* 2008;112(37):14456–14463.

54. Corral Valero M, Raybaud P, Sautet P. Interplay between molecular adsorption and metal-support interaction for small supported metal clusters: CO and C$_2$H$_4$ adsorption on Pd$_4$/γ-Al$_2$O$_3$. *Journal of Catalysis* 2007;247(2):339–355.

55. Ionescu A, Allouche A, Aycard J-P, Rajzmann M, Le Gall R. Study of γ-alumina-supported hydrotreating catalyst: I. Adsorption of bare MoS$_2$ sheets on γ-alumina surfaces. *The Journal of Physical Chemistry B* 2003;107(33):8490–8497.

56. Arrouvel C, Breysse M, Toulhoat H, Raybaud P. A density functional theory comparison of anatase (TiO$_2$)- and γ-Al$_2$O$_3$-supported MoS$_2$ catalysts. *Journal of Catalysis* 2005;232(1):161–178.

57. Bermudez VM. Energy-level alignment in the adsorption of phosphonyl reagents on γ-Al$_2$O$_3$. *Surface Science* 2008;602(11):1938–1947.

58. Bermudez VM. Computational study of environmental effects in the adsorption of DMMP, Sarin, and VX on γ-Al$_2$O$_3$: Photolysis and surface hydroxylation. *The Journal of Physical Chemistry C* 2009;113(5):1917–1930.

59. Joubert J, Delbecq F, Sautet P. Alkane metathesis by a tungsten carbyne complex grafted on gamma alumina: Is there a direct chemical role of the support? *Journal of Catalysis* 2007;251(2):507–513.

60. Raybaud P, Costa D, Corral Valero M, Arrouvel C, Digne M, Sautet P, Toulhoat H. First principles surface thermodynamics of industrial supported catalysts in working conditions. *Journal of Physics: Condensed Matter* 2008(6):064235.

61. Yourdshahyan Y, Ruberto C, Halvarsson M, Bengtsson L, Langer V, Lundqvist BI, Ruppi S, Rolander U. Theoretical structure determination of a complex aterial: κ-Al$_2$O$_3$. *Journal of the American Ceramic Society* 1999;82(6):1365–1380.

62. Fortrie R, Todorova TK, Ganduglia-Pirovano MV, Sauer J. Nonuniform temperature dependence of the reactivity of disordered VO$_x$/κ-Al$_2$O$_3$(001) surfaces: A density functional theory based Monte Carlo study. *The Journal of Chemical Physics* 2008;129(22):224710–12.

63. Carrasco J, Lopez N, Illas F. On the convergence of isolated neutral oxygen vacancy and divacancy properties in metal oxides using supercell models. *The Journal of Chemical Physics* 2005;122(22):224705–224714.

64. Carrasco J, Gomes JRB, Illas F. Theoretical study of bulk and surface oxygen and aluminum vacancies in α-Al$_2$O$_3$. *Physical Review B* 2004;69(6):064116.

65. Carrasco J, Lopez N, Sousa C, Illas F. First-principles study of the optical transitions of F centers in the bulk and on the (0001) surface of α-Al$_2$O$_3$. *Physical Review B* 2005;72(5):054109.

66. Islam MM, Diawara B, Maurice V, Marcus P. Atomistic modeling of voiding mechanisms at oxide/alloy interfaces. *The Journal of Physical Chemistry C* 2009;113(23):9978–9981.

67. Hine NDM, Frensch K, Foulkes WMC, Finnis MW. Supercell size scaling of density functional theory formation energies of charged defects. *Physical Review B: Condensed Matter and Materials Physics* 2009;79(2):024112–024113.

68. Choi M, Matsunaga K, Oba F, Tanaka I. 27Al NMR chemical shifts in oxide crystals: A first-principles study. *The Journal of Physical Chemistry C* 2009;113(9):3869–3873.

69. Pavese M, Biamino S. Mesoporous alumina obtained by combustion synthesis without template. *Journal of Porous Materials* 2009;16(1):59–64.

70. Jarvis EAA, Carter EA. A nanoscale mechanism of fatigue in ionic solids. *Nano Letters* 2006;6(3):505–509.

71. Milas I, Hinnemann B, Carter EA. Structure of and ion segregation to an alumina grain boundary: Implications for growth and creep. *Journal of Materials Research* 2008;23(5):1494–1508.

72. Milas I, Carter E. Effect of dopants on alumina grain boundary sliding: Implications for creep inhibition. *Journal of Materials Science* 2009;44(7):1741–1749.

73. Cornell RM, Schwertmann U. *The Iron Oxides: Structure, Properties, Reactions, Occurrences, and Uses*. Weinheim: Wiley-VCH; 2003. 664.

74. Rao CNR, Gopalakrishnan J. *New Directions in Solid State Chemistry: Structure, Synthesis, Properties, Reactivity and Materials Design*. Cambridge [Cambridgeshire]; New York: Cambridge University Press; 1986. x, 516 p.

75. Sanchez C, Hendewerk M, Sieber KD, Somorjai GA. Synthesis, bulk, and surface characterization of niobium-doped Fe_2O_3 single crystals. *Journal of Solid State Chemistry* 1986;61(1):47–55.

76. Henderson MA. Insights into the (1×1)-to-(2×1) phase transition of the α-Fe_2O_3(0 1 2) surface using EELS, LEED and water TPD. *Surface Science* 2002;515(1):253–262.

77. Henrich VE, Cox PA. *The Surface Science of Metal Oxides*. Cambridge: Cambridge University Press; 1994. xiv, 464 p.

78. Chambers SA, Yi SI. Fe termination for α-Fe_2O_3(0001) as grown by oxygen-plasma-assisted molecular beam epitaxy. *Surface Science* 1999;439(1–3):L785–L791.

79. Thevuthasan S, Kim YJ, Yi SI, Chambers SA, Morais J, Denecke R, Fadley CS, Liu P, Kendelewicz T, Brown GE. Surface structure of MBE-grown α-Fe_2O_3(0001) by intermediate-energy X-ray photoelectron diffraction. *Surface Science* 1999;425(2–3):276–286.

80. Wang XG, Weiss W, Shaikhutdinov SK, Ritter M, Petersen M, Wagner F, Schlögl R, Scheffler M. The hematite (alpha-Fe2O3) (0001) surface: Evidence for domains of distinct chemistry. *Physical Review Letters* 1998;81(5):1038.

81. Bergermayer W, Schweiger H, Wimmer E. Ab initio thermodynamics of oxide surfaces: O_2 on Fe_2O_3(0001). *Physical Review B* 2004;69(19):195409.

82. Rohrbach A, Hafner J, Kresse G. Ab initio study of the (0001) surfaces of hematite and chromia: Influence of strong electronic correlations. *Physical Review B* 2004;70(12):125426.

83. Jarvis EA, Chaka AM. Oxidation mechanism and ferryl domain formation on the α-Fe_2O_3 (0001) surface. *Surface Science* 2007;601(9):1909–1914.

84. Lemire C, Bertarione S, Zecchina A, Scarano D, Chaka A, Shaikhutdinov S, Freund HJ. Ferryl (Fe=O) termination of the hematite α-Fe_2O_3(0001) surface. *Physical Review Letters* 2005;94(16):166101.

85. Lo CS, Tanwar KS, Chaka AM, Trainor TP. Density functional theory study of the clean and hydrated hematite (1–102) surfaces. *Physical Review B* 2007;75(7):075425.

86. Robinson IK. Crystal truncation rods and surface roughness. *Physical Review B* 1986;33(6):3830.

87. Robinson IK, Tweet DJ. Surface X-ray diffraction. *Reports on Progress in Physics* 1992(5):599.

88. Fenter P, Sturchio NC. Mineral-water interfacial structures revealed by synchrotron X-ray scattering. *Progress in Surface Science* 2004;77(5–8):171–258.

89. Trainor TP, Chaka AM, Eng PJ, Newville M, Waychunas GA, Catalano JG, Brown GE, Jr. Structure and reactivity of the hydrated hematite (0001) surface. *Surface Science* 2004;573(2):204–224.

90. Tanwar KS, Lo CS, Eng PJ, Catalano JG, Walko DA, Brown GE, Waychunas GA, Chaka AM, Trainor TP. Surface diffraction study of the hydrated hematite (1–102) surface. *Surface Science* 2007;601(2):460–474.

91. Tanwar KS, Catalano JG, Petitto SC, Ghose SK, Eng PJ, Trainor TP. Hydrated α-Fe$_2$O$_3$ surface structure: Role of surface preparation. *Surface Science* 2007;601(12):L59–L64.

92. Catalano JG, Fenter P, Park C. Interfacial water structure on the (012) surface of hematite: Ordering and reactivity in comparison with corundum. *Geochimica et Cosmochimica Acta* 2007;71(22):5313–5324.

93. Christine F, Antoine V, Michel P, Samir M. Density functional theory calculations on microscopic aspects of oxygen diffusion in ceria-based materials. *International Journal of Quantum Chemistry* 2005;101(6):826–839.

94. Rossmeisl J, Bessler WG. Trends in catalytic activity for SOFC anode materials. *Solid State Ionics* 2008;178(31–32):1694–1700.

95. Galea NM, Kadantsev ES, Ziegler T. Studying reduction in solid oxide fuel cell activity with density functional theory—Effects of hydrogen sulfide adsorption on nickel anode surface. *The Journal of Physical Chemistry C* 2007;111(39):14457–14468.

96. Ingram DB, Linic S. First-principles analysis of the activity of transition and noble metals in the direct utilization of hydrocarbon fuels at solid oxide fuel cell operating conditions. *Journal of the Electrochemical Society* 2009;156(12):B1457–B1465.

97. Yang Z, Woo TK, Baudin M, Hermansson K. Atomic and electronic structure of unreduced and reduced CeO$_2$ surfaces: A first-principles study. *The Journal of Chemical Physics* 2004;120(16):7741–7749.

98. Chen H-T, Choi YM, Liu M, Lin MC. A theoretical study of surface reduction mechanisms of CeO$_2$ (111) and (110) by H$_2$. *ChemPhysChem* 2007;8:849–855.

99. Kumar S, Schelling PK. Density functional theory study of water adsorption at reduced and stoichiometric ceria (111) surfaces. *The Journal of Chemical Physics* 2006;125(20):204704–204708.

100. Asami K, Fujita T, Kusakabe K-i, Nishiyama Y, Ohtsuka Y. Conversion of methane with carbon dioxide into C$_2$ hydrocarbons over metal oxides. *Applied Catalysis A: General* 1995;126(2):245–255.

101. Beste A, Mullins DR, Overbury SH, Harrison RJ. Adsorption and dissociation of methanol on the fully oxidized and partially reduced (1 1 1) cerium oxide surface: Dependence on the configuration of the cerium 4f electrons. *Surface Science* 2008;602:162–175.

102. Tsuji H, Okamura-Yoshida A, Shishido T, Hattori H. Dynamic behavior of carbonate species on metal oxide surface: Oxygen scrambling between adsorbed carbon dioxide and oxide surface. *Langmuir* 2003;19(21):8793–8800.

103. Yang Z, Woo TK, Hermansson K. Strong and weak adsorption of CO on CeO$_2$ surfaces from first principles calculations. *Chemical Physics Letters* 2004;396(4–6):384–392.

104. Jónsson H, Mills G, Jacobsen KW. Nudged elastic band method for finding minimum energy paths of transitions. In: Berne BJ, Ciccotti G, Coker DF, editors. *Classical and Quantum Dynamics in Condensed Phase Simulations*. Singapore: World Scientific; 1998. pp. 385–404.

105. Hammer B, Hansen LB, Nørskov JK. Improved adsorption energetics within density-functional theory using revised Perdew-Burke-Ernzerhof functionals. *Physical Review B* 1999;59(11):7413.

106. Bahn SR, Jacobsen KW. An object-oriented scripting interface to a legacy electronic structure code. *Computing in Science & Engineering* 2002;4(3):56–66.

107. Kendall RA, Apra E, Bernholdt DE, Bylaska EJ, Dupuis M, Fann GI, Harrison RJ et al., High performance computational chemistry: An overview of NWChem a distributed parallel application. *Computer Physics Communications* 2000;128(1–2):260–283.

108. Paolo G, Stefano B, Nicola B, Matteo C, Roberto C, Carlo C, Davide C et al. QUANTUM ESPRESSO: A modular and open-source software project for quantum simulations of materials. *Journal of Physics: Condensed Matter* 2009(39):395502.

109. Kresse G, Hafner J. Ab initio molecular dynamics for liquid metals. *Physical Review B* 1993;47:558.

110. Kresse G, Furthmüller J. Efficiency of ab-initio total energy calculations for metals and semiconductors using a plane-wave basis set. *Computational Materials Science*. 1996;6:15.

111. Yin S, Ellis DE. DFT studies of Cr(VI) complex adsorption on hydroxylated hematite (1–102) surfaces. *Surface Science* 2009;603(4):736–746.

112. Bargar JR, Towle SN, Brown GE, Parks GA. Outer-sphere Pb(II) adsorbed at specific surface sites on single crystal α-alumina. *Geochimica et Cosmochimica Acta* 1996;60(18):3541–3547.

113. Bargar JR, Trainor TP, Fitts JP, Chambers SA, Brown GE. In situ grazing-incidence extended X-ray absorption fine structure study of Pb(II) chemisorption on hematite (0001) and (1–102) surfaces. *Langmuir* 2004;20(5):1667–1673.

114. Kendelewicz T, Liu P, Doyle CS, Brown GE, Jr., Nelson EJ, Chambers SA. X-ray absorption and photoemission study of the adsorption of aqueous Cr(VI) on single crystal hematite and magnetite surfaces. *Surface Science* 1999;424(2–3):219–231.

115. Gratzel M. Photoelectrochemical cells. *Nature* 2001;414(6861):338–344.

116. Green MA. *Third Generation Photovoltaics: Advanced Solar Energy Conversion*. Berlin: Springer; 2003. xi, 160 p.

117. Conibeer G. Third-generation photovoltaics. *Materials Today* 2007;10(11):42–50.

118. Kamat PV. Meeting the clean energy demand: nanostructure architectures for solar energy conversion. *The Journal of Physical Chemistry C* 2007;111(7):2834–2860.

119. Diebold U. The surface science of titanium dioxide. *Surface Science Reports* 2003;48(5–8):53–229.

120. Ihara T, Miyoshi M, Ando M, Sugihara S, Iriyama Y. Preparation of a visible-light-active TiO_2 photocatalyst by RF plasma treatment. *Journal of Materials Science* 2001;36(17):4201–4207.

121. Park JH, Kim S, Bard AJ. Novel carbon-doped TiO_2 nanotube arrays with high aspect ratios for efficient solar water splitting. *Nano Letters* 2006;6(1):24–28.

122. Ingler WB, Al-Shahr M, Kahn SUM. Efficient photochemical water splitting by a chemically modified n-TiO_2. *Science* 2002;297:2243–2245.

123. Di Valentin C, Pacchioni G, Selloni A. Theory of carbon doping of titanium dioxide. *Chemistry of Materials* 2005;17(26):6656–6665.

124. Di Valentin C, Pacchioni G, Selloni A. Origin of the different photoactivity of N-doped anatase and rutile TiO_2. *Physical Review B* 2004;70(8): 085116.

125. Asahi R, Morikawa T, Ohwaki T, Aoki K, Taga Y. Visible-light photocatalysis in nitrogen-doped titanium oxides. *Science* 2001;293(5528):269–271.

126. Di Valentin C, Finazzi E, Pacchioni G, Selloni A, Livraghi S, Paganini MC, Giamello E. N-doped TiO_2: Theory and experiment. *Chemical Physics* 2007;339(1–3):44–56.

127. Rane KS, Mhalsiker R, Yin S, Sato T, Cho K, Dunbar E, Biswas P. Visible light-sensitive yellow TiO2-xNx and Fe-N co-doped Ti1-yFeyO2-xNx anatase photocatalysts. *Journal of Solid State Chemistry* 2006;179(10):3033–3044.

128. Ghicov A, Macak JM, Tsuchiya H, Kunze J, Haeublein V, Frey L, Schmuki P. Ion implantation and annealing for an efficient N-doping of TiO_2 nanotubes. *Nano Letters* 2006;6(5):1080–1082.

129. Namiki N, Cho K, Fraundorf P, Biswas P. Tubular reactor synthesis of doped nanostructured titanium dioxide and its enhanced activation by coronas and soft X-rays. *Industrial & Engineering Chemistry Research* 2005;44(14):5213–5220.

130. Choi W, Termin A, Hoffmann MR. The role of metal ion dopants in quantum-sized TiO_2: Correlation between photoreactivity and charge carrier recombination dynamics. *The Journal of Physical Chemistry* 1994;98(51):13669–13679.

131. Mor GK, Shankar K, Paulose M, Varghese OK, Grimes CA. Use of highly-ordered TiO_2 nanotube arrays in dye-sensitized solar cells. *Nano Letters* 2006;6(2):215–218.

132. Thimsen E, Biswas S, Lo CS, Biswas P. Predicting the band structure of mixed transition metal oxides: Theory and experiment. *The Journal of Physical Chemistry C* 2009;113(5):2014–2021.

133. Ghicov A, Schmidt B, Kunze J, Schmuki P. Photoresponse in the visible range from Cr doped TiO_2 nanotubes. *Chemical Physics Letters* 2007;433(4–6):323–326.

134. Yu JC, Li GS, Wang XC, Hu XL, Leung CW, Zhang ZD. An ordered cubic Im3m mesoporous Cr-TiO_2 visible light photocatalyst. *Chemical Communications* 2006(25):2717–2719.

135. Mor GK, Prakasam HE, Varghese OK, Shankar K, Grimes CA. Vertically oriented Ti-Fe-O nanotube array films: Toward a useful material architecture for solar spectrum water photoelectrolysis. *Nano Letters* 2007;7(8):2356–2364.

136. Mor GK, Varghese OK, Wilke RHT, Sharma S, Shankar K, Latempa TJ, Choi K-S, Grimes CA. P-type Cu–Ti–O nanotube arrays and their use in self-biased heterojunction photoelectrochemical diodes for hydrogen generation. *Nano Letters* 2008;8(7):1906–1911.

137. Fernandez-Garcia M, Martinez-Arias A, Fuerte A, Conesa JC. Nanostructured Ti-W mixed-metal oxides: Structural and electronic properties. *The Journal of Physical Chemistry B* 2005;109(13):6075–6083.

138. Car R, Parrinello M. Unified approach for molecular dynamics and density-functional theory. *Physical Review Letters* 1985;55(22):2471.

139. Marx D, Hutter J. *Ab Initio Molecular Dynamics: Basic Theory and Advanced Methods*. Cambridge, U.K.: Cambridge University Press; 2009. 567 p.

140. Hass KC, Schneider WF, Curioni A, Andreoni W. First-principles molecular dynamics simulations of H_2O on α-Al_2O_3 (0001). *The Journal of Physical Chemistry B* 2000;104(23):5527–5540.

141. Tangney P, Scandolo S. How well do Car–Parrinello simulations reproduce the Born–Oppenheimer surface? Theory and examples. *The Journal of Chemical Physics* 2002;116(1):14–24.

142. Masini P, Bernasconi M. Ab initio simulations of hydroxylation and dehydroxylation reactions at surfaces: Amorphous silica and brucite. *Journal of Physics: Condensed Matter* 2002(16):4133.

143. Kresse G, Bergermayer W, Podloucky R, Lundgren E, Koller R, Schmid M, Varga P. Complex surface reconstructions solved by ab initio molecular dynamics. *Applied Physics A: Materials Science & Processing* 2003;76(5):701–710.

144. Ciacchi LC, Payne MC. "Hot-Atom" O_2 dissociation and oxide nucleation on Al(111). *Physical Review Letters* 2004;92(17):176104.

145. Tilocca A, Selloni A. O_2 and vacancy diffusion on rutile(110): Pathways and electronic properties. *Chemphyschem* 2005;6(9):1911–1916.

146. Seriani N, Pompe W, Ciacchi LC. Catalytic oxidation activity of Pt_3O_4 surfaces and thin films. *The Journal of Physical Chemistry B* 2006;110(30):14860–14869.

147. Di Valentin C, Pacchioni G, Bernasconi M. Ab initio molecular dynamics simulation of NO reactivity on the CaO(001) surface. *The Journal of Physical Chemistry B* 2006;110(16):8357–8362.

148. Gao J. Methods and applications of combined quantum mechanical and molecular mechanical potentials. In: Lipkowitz KB, Boyd DB, editor. *Reviews in Computational Chemistry*. Vol. 7. New York: VCH Publishers; 1995. pp. 119–185.

149. Ellis DE, Warschkow O. Evolution of classical/quantum methodologies: Applications to oxide surfaces and interfaces. *Coordination Chemistry Reviews* 2003;238–239:31–53.

150. Hafner J. Ab-initio simulations of materials using VASP: Density-functional theory and beyond. *Journal of Computational Chemistry* 2008;29(13):2044–2078.

151. Pacchioni G. Modeling doped and defective oxides in catalysis with density functional theory methods: Room for improvements. *The Journal of Chemical Physics* 2008;128(18):182505–182510.

152. Huang P, Carter EA. Advances in correlated electronic structure methods for solids, surfaces, and nanostructures. *Annual Review of Physical Chemistry* 2008;59(1):261–290.

153. Held K. Electronic structure calculations using dynamical mean field theory. *Advances in Physics* 2007;56(6):829–926.

154. E W, Li X. Multiscale modeling of crystalline solids. In: Yip S, editor. *Handbook of Materials Modeling*. Vol. A; 2005. pp. 1491–1506.

155. Lu H, Daphalapurkar NP, Wang B, Roy S, Komanduri R. Multiscale simulation from atomistic to continuum: Coupling molecular dynamics (MD) with the material point method (MPM). *Philosophical Magazine* 2006;86(20):2971–2994.

156. Nguyen-Manh D, Vitek V, Horsfield AP. Environmental dependence of bonding: A challenge for modelling of intermetallics and fusion materials. *Progress in Materials Science* 2007;52(2–3):255–298.

157. Gu YT, Zhang LC. A Concurrent multiscale method based on the meshfree method and molecular dynamics analysis. *Multiscale Modeling & Simulation* 2006;5(4):1128–1155.

158. Zhang X, Wang CY. Application of a hybrid quantum mechanics and empirical molecular dynamics multiscale method to carbon nanotubes. *The European Physical Journal B—Condensed Matter and Complex Systems* 2008;65(4):515–523.

159. Wang QX, Ng TY, Li H, Lam KY. Multiscale simulation of coupled length-scales via meshless method and molecular dynamics. *Mechanics of Advanced Materials and Structures* 2009;16(1):1–11.

8 Tight Binding Methods for Metallic Systems

Luis Rincón, Thomas C. Allison,
and Carlos A. González*

CONTENTS

8.1 INTRODUCTION

The atomistic description of metal systems, which is the central topic of this book, is only possible through application of quantum theory. For the great majority of the metallic systems of interest, achieving a quantum description of some phenomena requires a delicate balance between two competing requirements: the degree of realism of the physical model and the level of approximation in the quantum treatment. A third requirement is fundamental: we need a transparent interpretation of the theoretical model, because otherwise we obtain numbers without any physical significance. In the end, some compromise between these two (or three)

---

* Contribution of the National Institute of Standards and Technology.

requirements is made, and the final results depend, often strongly, on the physical and theoretical models used. The choice of physical and theoretical model is an intensely personal one, and there is frequently no best approach that should be taken. One researcher may believe that all physical problems can be reduced to systems of a few tens of atoms, which are then studied at the highest level of accuracy available. Another possibility is the simplification of the physical model in favor of accuracy. This simplification comes with the drawback that it is often very difficult to obtain a direct interpretation of the results. For many problems of interest, especially in biomolecular and nanomaterial sciences, this level of simplification is not possible. The description of a physical system in these disciplines may require the inclusion of hundreds (or even thousands) of atoms. In these cases, the expense of the required calculations at high levels of theory is prohibitive. Researchers then resort to the use of approximate theories which involve a tradeoff between accuracy and computational expense. In the field of materials science, the most commonly employed approximation for describing solids is tight binding (TB) [1–3]. In the last decade, the number of TB methodologies has increased more than any other approximate electronic structure method. A number of TB methods have been introduced for metallic systems, and it is for this reason that we give a survey of some of these methods in this chapter.

Review of the literature on applications of TB theory in metallic systems is a formidable, if not impossible, task. For this reason, we focus on two areas: (i) a survey of TB methodologies that include d-electrons rather than on TB studies of metallic systems, and (ii) a discussion of some of our own work on the structure and optical properties of gold clusters as a means to illustrate various aspects of TB theory.

To describe the TB methodology, we summarize some of the most influential TB methodologies, explore their strengths and weaknesses, and highlight the theoretical challenges in their development. We review the essential physics contained in the various approximations used to describe metal systems from the TB perspective. Throughout this chapter, TB methods are grouped into two categories: classical TB methods and the density functional-based TB.

The study of classical TB techniques begins with the seminal work of Slater and Koster that constitutes the birth of the TB method [4]. This method, and extensions to it, are very well described in many books on solid-state theory [1–3] as it is the simplest way to understand the band structure of solid metals and semiconductors. Classical TB methods are used as fitting and interpolation schemes rather than as quantitative computational tools. Their simplicity allows analytical results to be produced for a number of systems; thus it has been used extensively to provide a qualitative understanding of a wide range of electronic structure problems [3]. The most important improvement to the original Slater–Koster method came at the end of the 1980s with the advent of the calculation of TB parameters directly from first principles density functional theory (DFT) calculations [5,6]. The connection with DFT was fully exploited in a new generation of very successful TB methodologies. Prominent examples of this aproach are the Naval Research Laboratory TB method (NRL-TB) [7], the *ab initio* TB method (AITB) [8,9] and the density functional TB (DFTB) method [10,11]. These methods have become routine in the solid-state literature and the computer codes that implement them are available from their developers.

Alternatively, and despite long-recognized deficiencies, extended Huckel (EH) theory is still frequently employed in electronic structure calculations of periodic and non-periodic systems, mainly due to its simplicity and the chemical insight it provides [12,13]. In most classical EH studies, the parameters of the model are created with the (rather ambiguous) requirement of providing a reasonable description of the electronic structure of the system. In recent years a number of different EH methodologies based on DFT have been introduced and successfully applied [14,15].

In our own work, we have taken some of the lessons learned in the development of DFT TB methods and developed a new EH scheme [16,17]. Contrary to popular belief, we find that the EH methodology can provide a similar level of accuracy to competing methods in many problems, while preserving many important advantages inherent to the formalism, namely: (i) a considerable reduction in the number of fitting parameters, (ii) natural scaling laws for atomic orbitals, and (iii) a reasonable transferability of parameters that allows for their use in different chemical environments.

The most tedious part of the applications of any TB method is its parameterization, and even today, more than 50 years after the seminal work of Slater and Koster, it remains the greatest challenge to the application of TB theory. No one has developed a simple method to solve this critical problem. The conventional approach typically proceeds by identifying a series of reference data that should be reproduced by a parameter set. Such data may come from experiments, from other theoretical calculations (typically using a more accurate theoretical model), or from a combination of the two. The model must then be fit to these data using some sort of optimization technique, often at a considerable cost in human and computer time. Reducing the amount of human intervention is a highly desirable future goal. In all of the current methodologies, the parameterization process involves a trial-and-error approach requiring many iterations before suitable parameter sets are obtained. Although we do not have a definitive solution to this problem, we present some ideas on the automation of this very consuming time process in this manuscript, gleaned from the lessons learned in the course of our works in the last few years. Nevertheless, we acknowledge that much work in this area remains before a suitable method will be found.

To illustrate the use of TB methods, we compare the application of different density functional-based TB methods to the problem of unbiased global optimization of neutral atomic gold clusters [18–20]. Experiments provide only indirect information on the structure of metallic clusters. TB methods are particularly suitable for the study of metal clusters in part because the number of isomers grows rapidly with the size of the cluster and the task of identifying the lowest energy structure is a challenging computational problem. Interest in gold clusters derives from their unusual catalytic properties, in contrast with that of the bulk metal, and their selective binding to many biomolecules, including DNA. Also, gold clusters have many applications in nanoelectronics. Despite the significant effort that has been invested in this subject, there are still many open questions. One interesting question is whether the size of gold clusters changes upon transition from a planar structure to a 3D structure. In connection with this problem, recently we introduced a non-self-consistent energy correction that is based on the fluctuation of the density that can considerably improve the description of small clusters [17].

8.2   FOUNDATIONS OF THE TIGHT BINDING METHOD

In the classic paper by Slater and Koster [4], they proposed a simplification to the linear combination of atomic orbital (LCAO) method with the objective of interpolating results from first principles band structure calculations of periodic systems. It is important to understand that in the 1950s, it was impossible to perform *ab initio* computations of even the simplest solid. Along with the dramatic increase in CPU speed and memory size over the last two decades and the emergence of efficient algorithms for electronic structure calculations that have eliminated many of the original barriers to performing *ab initio* calculations, the Slater–Koster method has evolved as a powerful approach to treat molecular and solid systems, especially for metals and semiconductors, permitting studies on these systems that would be impossible at higher levels of theory.

Although the original Slater–Koster scheme was proposed for the calculation of the electronic bands of periodic solids, over time these ideas were generalized to the more robust TB method for calculation of the total energy and other electronic properties. Among the important improvements to the Slater–Koster scheme of particular historical importance are the universal TB method of Harrison that incorporates interatomic distance dependence in the classical matrix elements [21] and the inclusion of a short-range repulsive potential to the classical sum over states by Chadi [22]. These additions served to expand the range of applicability and confidence of the original TB methods.

During the 1980s, TB models were widely used in the solid state community with great success. However, the relationship of these TB models to more fundamental theories has never been clear and for this reason these models are used as fitting and interpolation schemes rather than as quantitative computational tools. In 1989, Foulkes and Haydock [5] generalized the ideas of Harris [6], clearly demonstrating that the simple TB models were approximations to the Kohn–Sham version of DFT [23]. Following the insight of Foulkes and Haydock, a new generation of DFT-based TB methods appeared in the literature.

In this section we introduce the classical TB method as found in the works of Slater and Koster [4] and of Chadi [22]. Then we discuss the DFT basis of the TB method. Knowledge of these two methods is essential in order to understand the modern realizations of the TB method which are an evolution of the theory given in the earlier works.

8.2.1   CLASSICAL TIGHT BINDING METHOD

In this section we briefly explain the classical TB method for a crystalline solid. The interested reader is referred to the book of W.A. Harrison "Elementary Electronic Structure" for a more comprehensive introduction to the fundamentals of TB theory [3].

We begin by describing the Slater–Koster method, which is derived from the original LCAO method of Bloch for electronic bands [24]. Consider a periodic unit cell containing n atoms, where the lattice vectors are denoted by R, and the ith atom is located at position r_i in the unit cell. Each atom has a set of atomic-like orbitals φ_{il}, where the index l denotes the orbital angular quantum number. Although, in general, atomic orbitals on different atoms are not orthogonal, classical TB theory assumes that they are orthogonal. Each atomic-like orbital is associated with a Bloch orbital $\Phi_{il}(k,r)$, which is a linear combination of the φ_{il} in all N cells of the crystal ($N \sim 10^{24}$) weighted by a phase factor $e^{ik \cdot R}$, where k is the Bloch wave vector,

$$\Phi_{il}(k,r) = \frac{1}{\sqrt{N}} \sum_{R}^{N} \varphi_{il}(r + r_i - R)e^{ik \cdot R}. \tag{8.1}$$

These orbitals satisfy the Bloch theorem: that is, if T_R is a translational operator along the lattice, then $T_R \Phi_{il}(k,r) = e^{ik \cdot R} \Phi_{il}(k,r)$. The crystal orbitals $\Psi_j(k,r)$ are expressed as a linear combination of Bloch orbitals as follows:

$$\Psi_j(k,r) = \sum_{il} C_{j,il}(k) \Phi_{il}(k,r), \tag{8.2}$$

where $C_{j,il}(k)$ are crystal coefficients to be determined. Clearly, $\Psi_j(k,r)$ must also satisfy the Bloch theorem. The energy of the jth crystal orbital, as a function of the wave vector, is given by

$$E_j(k) = \frac{\langle \Psi_j(k) \mid H \mid \Psi_j(k) \rangle}{\langle \Psi_j(k) \mid \Psi_j(k) \rangle} = \frac{\sum_{il,i'l'} C^*_{j,il}(k)H_{il,i'l'}(k)C_{j,i'l'}(k)}{\sum_{il,i'l'} C^*_{j,il}(k)S_{il,i'l'}(k)C_{j,i'l'}(k)}, \tag{8.3}$$

where
  H is the Hamiltonian of the solid, which is translationally invariant
  $H_{il,i'l'}(k)$ and $S_{il,i'l'}(k)$ are the matrix representation of the Hamiltonian in the
    Bloch orbital basis set and the overlap integral matrix between Bloch orbitals,
    respectively

These are defined by

$$H_{il,i'l'} = \langle \Phi_{il}(k) \mid H \mid \Phi_{i'l'}(k) \rangle, \tag{8.4}$$

$$S_{il,i'l'} = \langle \Phi_{il}(k) \mid \Phi_{i'l'}(k) \rangle. \tag{8.5}$$

The dimension of these two matrices is equal to the number of orbitals in the unit cell. In the case of an orthogonal basis of atomic orbitals, as is the case in classical TB theories, the overlap matrix is the identity matrix. The coefficients $C_{j,il}$ and crystal energies are obtained by solving the matrix eigenvalue problem,

$$H(k)C(k) = E(k)S(k)C(k). \tag{8.6}$$

Since all matrix elements depend on the wave-vector k, we solve this equation for a number of k-points in the first Brillouin zone (the unit-cell of the reciprocal lattice).

To obtain a TB expression for the Hamiltonian and overlap matrix, we write the Bloch orbitals in term of the localized atomic orbitals, obtaining

$$H_{jl,j'l'}(k) = \langle \varphi_{jl}(r) \mid H \mid \varphi_{j'l'}(r) \rangle$$

$$+ \sum_R \left[e^{ik \cdot R} \langle \varphi_{jl}(r-R) \mid H \mid \varphi_{j'l'}(r) \rangle + e^{-ik \cdot R} \langle \varphi_{jl}(r) \mid H \mid \varphi_{j'l'}(r-R) \rangle \right],$$

(8.7)

$$S_{jl,j'l'}(k) = \langle \varphi_{jl}(r) \mid \varphi_{j'l'}(r) \rangle$$

$$+ \sum_R \left[e^{ik \cdot R} \langle \varphi_{jl}(r-R) \mid \varphi_{j'l'}(r) \rangle + e^{-ik \cdot R} \langle \varphi_{jl}(r) \mid \varphi_{j'l'}(r-R) \rangle \right]. \qquad (8.8)$$

In these equations, the first term on the right-hand side corresponds to the Hamiltonian and overlap integrals between two atomic orbitals that are located in the same unit cell. This term does not depend on the wave vector. The second term, called the dispersion term, depends on the Hamiltonian and overlap integrals between two atomic orbitals located in two unit cells separated by a lattice vector R.

In order to further develop TB theory, a number of approximations, some of which date from the original work of Slater and Koster [1], are made:

1. TB methodologies are classified as orthogonal or non-orthogonal depending on the orthogonality of the atomic orbitals φ_{jl}. In early TB methods, such as the Slater–Koster method [4], the atomic orbitals are taken to be orthogonal. The assumption of orthogonality considerably simplifies the eigenvalue problem in Equation 8.6, because the overlap matrix is the identity matrix. However, it has become clear that orthogonal TB theories are less transferable, for the obvious reason that the atomic orbitals between two atoms are not, in general, orthogonal.

2. It is assumed in the TB method that only contributions from valence orbital electrons are included in the Hamiltonian and overlap matrices. Thus atoms containing valence d orbitals, as is the case for metal atoms, have a valence basis of 1s, 3p (p_x, p_y, p_z)- and 5d orbitals (d_{xy}, d_{yz}, d_{zx}, $d_{x^2-y^2}$, $d_{3z^2-r^2}$) for a total of nine orbitals. Atoms such as carbon which are represented by a sp orbital basis have 1s and 3p orbitals for a total of four orbitals. This gives rise to a significant reduction in computational expense compared to other schemes commonly employed in first principles calculations.

3. Another common approximation concerns the number of neighbor cells included in the summation of the matrix elements of Equations 8.7 and 8.8. Classical TB theories are dominated by the so-called "first neighbor" approximation, in which the summation over lattice vector is extended only over adjacent unit cells. This leads to a significant reduction in the number of non-zero matrix elements. This property can be exploited to study larger systems, to reduce the computer time required to compute a solution, or both.

4. Another assumption in TB theories is that the Hamiltonian matrix elements, $H_{il,i'l'}$, only depend on atomic centers in which orbitals (il) and $(i'l')$ are localized. This approximation is known as the "two-center approxima- tion." We note that the Hamiltonian also contains three center terms, such as the Coulomb terms, which are neglected in TB models. Slater and Koster showed that all two-center integrals depend only on the angular momentum of the orbitals and the distance between the atoms on which these orbitals are located. For an atom with a spd orbital basis (e.g., a metal), 10 indepen- dent parameters are required for the computation of their matrix elements: $ss\sigma$, $sp\sigma$, $pp\sigma$, $pp\pi$, $sd\sigma$, $pd\sigma$, $pd\pi$, $dd\sigma$, $dd\pi$, and $dd\delta$. Using these param- eters, all possible integrals within a spd basis can be computed using the Slater–Koster scheme.

All these approximations can be relaxed and improved in order to improve the quality of the TB model and, hopefully, the accuracy of the calculations. Although the original Slater–Koster scheme was developed for the calculation of the band structure of periodic solids, it was not obvious how to generalize the problem to the calculation of the total energy of the system. In 1979, Froyen and Harrison proposed an r^{-2} dependence of the matrix elements of the Hamiltonian with vary- ing interatomic distances [21]. In the same year, Chadi applied this method to the calculation of the surface energy of semiconductors [22]. In his work, Chadi proposed writing the TB total energy as a function of the atomic coordinates in the form [22]

$$E_{TB} = \sum_j \int_{BZ} \frac{dk}{(2\pi)^3} E_j(k) + \sum_{\alpha > \beta} E_{REP}(\alpha, \beta) \tag{8.9}$$

where the first term on the right-hand side is due to the TB band energies given by Equation 8.6, the summation includes all occupied electronic eigenstates and the integration is over the k points in the first Brilloiun zone (BZ). The second term is a contribution from all atom pairs and includes a short-range repulsive potential that depends on atom type. The repulsive interaction is a parametric functional derived to reproduce the cohesive energies and elastic constant of some reference crystalline materials [23].

Total energy TB calculations depend strongly on the parameterization scheme used for the band structure and the repulsive potential, and the transferability in many cases is very poor. Over the years, non-orthogonal TB has been proposed as one important means to achieving more transferable parameterizations [25,26].

In order to reduce the effort required for parameterization, more sophisticated and efficient TB schemes have been developed. Of particular importance are the linear-muffin-tin-orbital (LMTO) method of Andersen and Jepsen [27] and the DFT parameterization of Cohen, Mehl, and Papaconstantopoulos [28].

In earlier work, TB schemes were used as an interpolation method to reproduce the eigenvalue spectrum over a large number of k points. This procedure is very use- ful for the calculation of densities of states and Fermi surfaces. The determination of the parameters is usually done by a nonlinear least-squares method in which the target function to be minimized is the difference between eigenvalues from first principle calculations and the TB model.

Reference [29] is a comprehensive source of Slater–Koster TB parameters, including cases of orthogonal and non-orthogonal orbitals and two- and three-center approximations to the Hamiltonian elements for the crystal structure of 53 elements. This book also contains technical details and computer programs, as well as a short discussion of the trends in band structures along the periodic table.

The strengths of classical TB provide great incentive to find ways to overcome its shortcomings. Since these shortcomings are related to the empirical character of the TB model and the tedious process of parameterization, it is natural to seek a way to overcome these difficulties by deriving TB from a first principles theory.

Classical TB methods are computationally efficient for several reasons: the basis set is very small and thus the dimension of the eigenproblem is reduced comparing with most first principles methods (especially plane wave methods); the matrix elements are rapidly evaluated compared with the expense of evaluating large numbers of gaussian integrals in *ab initio* or DFT methods; in general, classical TB methods do not iterate the wavefunction to self-consistency which may result in considerable savings; the range of interaction is always short and thus the matrices are typically sparse which permits use of linear scaling methodologies. Any improvement to the TB method should retain these properties.

8.2.2 DENSITY FUNCTIONAL BASIS OF TIGHT BINDING METHODS

We now take a brief look at the fundamental theoretical basis underlying TB theories.

Within the pseudopotential local density functional approximation (PP-LDA), the Kohn–Sham energy of the system with density n is written as

$$E_{KS}[n] = T_S[n] + \int V_{PP-ion}(r)n(r)dr + \frac{1}{2}\int\int\frac{n(r)n(r')}{|r-r'|}drdr' + E_{xc}[n] + E_{ion-ion}, \quad (8.10)$$

where this expression includes the kinetic energy of the fictitious non-interacting system, $T_S[n]$, the interaction of the electron with the pseudopotential and ions, the Hartree electron–electron interaction, the local exchange correlation energy, $\int E_{xc}[n]n(r)dr'$ and the ion–ion repulsion, respectively. The electron density is given as a sum over the occupied orbitals, $n(r) = 2\sum_i \phi_i^*(r)\phi_i(r)$. These orbitals are eigenfunctions of the 1-electron Kohn–Sham equations,

$$\left[-\frac{1}{2}\nabla^2 + V_{PP-ion}(r) + \int\frac{n(r)}{|r-r'|}dr' + \mu_{xc}(n)\right]\phi_i(r) = \varepsilon_i\phi_i(r), \quad (8.11)$$

where $\mu_{xc}(n) = \dfrac{\delta\varepsilon_{xc}(n)}{\delta n}$. The total energy of Equation 8.10 can be rewritten using the Kohn–Sham orbital energies to yield

$$E_{KS}[n] = 2\sum_i \varepsilon_i - \frac{1}{2}\int\int \frac{n(r)n(r')}{|r-r'|}drdr' + E_{xc}[n] - \int \mu_{xc}(n)n(r)dr + E_{ion-ion}. \quad (8.12)$$

For the moment we avoid the integration over k for periodic systems of Equation 8.9. In the work of Foulkes and Haydock [5], the self-consistent density is written as

$$n(r) = n_0(r) + \delta n(r), \quad (8.13)$$

where
n_0 is a reference density
δn is the difference between the reference density and the self-consistent density, where this last density is assumed to be small in some sense

For simplicity, n_0 is taken to be a sum of atomic densities (or some other more reasonable ansatz), $n_0(r) = \sum_i^{atoms} n_i(r)$. In the following we expand the energy of Equation 8.12 at a reference density n_0 up to second order in δn,

$$E_{KS}(n) = \sum_i \varepsilon_i^0 - \frac{1}{2}\int\int \frac{n_0(r)n_0(r')}{|r-r'|}drdr' + E_{xc}[n_0] - \int \mu_{xc}[n_0]n_0dr + E_{ion-ion}$$

$$+ \frac{1}{2}\int\int \left(\frac{1}{|r-r'|} + \frac{\delta^2 E_{xc}}{\delta n(r)\delta n(r')} \right) \delta n(r)\delta n(r')drdr'. \quad (8.14)$$

In this equation, the energies ε_i^0 are the eigenvalues of the non-self-consistent Kohn–Sham equation (Equation 8.11) when $n(r) = n_0(r)$. The non-self-consistent TB approximation may be recovered from Equation 8.14 for E_{KS} by neglecting the last term involving δn and defining the repulsive TB energy of Equation 8.9 as

$$-\frac{1}{2}\int\int \frac{n_0(r)n_0(r')}{|r-r'|}drdr' + E_{xc}[n_0] - \int \mu_{xc}[n_0]n_0dr + E_{ion-ion}. \quad (8.15)$$

If the reference charge density is a superposition of atomic-like neutral charge densities, this last term can be approximated by a sum of two center contributions as long as the exchange correlation part is expanded in a cluster series. The neglect of three-center contributions can be justified by screening arguments. Since the atomic charge density corresponds to a neutral atom, the three-center electron–electron interaction mostly is canceled by the ion–ion repulsion. Due to screening, these two centers can be assumed short-range as in classical TB.

Clearly, as long as δn is negligible, all energy contributions depend only on n_0; thus this scheme is equivalent to the non-self-consistent Harris functional [6], providing a conceptual framework for the energy functional of Chadi, Equation 8.9, and a recipe for the evaluation of TB parameters from first principles calculations.

Stated in this way, non-self-consistent calculations are limited to problems in which δn is small. However, for many important problems the last term is not negligible. For example, in systems with strong charge transfer between atoms it is hard to imagine that a summation of atomic-like densities is a good approximation. Another example is the spin polarized case that frequently occurs in metallic systems in which it is crucial to consider the inclusion of changes to the density via a self-consistent procedure. To this end, a number of self-consistent TB schemes have been proposed in the literature [11,30].

8.3 MODERN TIGHT BINDING

Modern TB methodologies are based on DFT. In this section, we take a tour of three of the most cited methods: the NRL total energy TB method [7], the AITB method [8,9], and the DFTB method [10,11]. A comparison of these methods, using the case of gold clusters, is given in a subsequent section. However, it is important to note that the method which is most appropriate will depend strongly on the particular application and the particular system. We emphasize in each case the connection of the method with the Harris–Foulkes–Haydock [5,6] methodology exposed in Section 8.2.2.

8.3.1 NRL TIGHT BINDING TOTAL ENERGY METHOD

In this section we will concentrate on the TB method developed at the NRL by Cohen, Mehl, and Papaconstantopoulos [7,28,31]. Even though the NRL-TB method is not a direct descendant of the Harris–Foulkes–Haydock method, but is more "classical" in nature, the methodology is based on the Kohn–Sham DFT formulation, and for this reason it is presented in this section.

In the Kohn–Sham DFT method the total energy is given by Equation 8.10. It was shown above that this equation is closely related to the TB formulation. Note that the repulsion potential of TB methods is related to the double counting of the Hartree potential, the exchange-correlation, and the ion–ion repulsion. The value of the repulsion potential depends upon the choice of the zero of energy for the Kohn–Sham potential, which is arbitrary. In general, the zero of energy is chosen to coincide with the Fermi level energy. In the formulation of the NRL-TB methodology the Kohn–Sham potential is shifted by an amount equal to

$$V_0 = E_{REP}\frac{[n]}{N_e}, \tag{8.16}$$

where
    E_{REP} is the repulsion potential
    N_e is the number of electrons in the unit cell

Since this shift is applied to the Kohn–Sham eigenvalues, the total energy of the system becomes

$$E_{TB} = \sum_{j} \int_{BZ} \frac{dk}{(2\pi)^3} E_j(k) + N_e V_0 = \int_{BZ} \frac{dk}{(2\pi)^3} \sum_{j} \left[E_j(k) + V_0 \right]. \qquad (8.17)$$

If we now define shifted eigenvalues with the form $E'_j(k) = E_j(k) + V_0$, the total energy is just a sum over the shifted occupied eigenvalues. Cohen, Mehl, and Papaconstantopoulos [28] note that the shifted energies are in some sense "universal". That is, if two band structure calculations are sufficiently well converged, they will have the same total energy and their eigenvalues will differ by a constant. Thus, in the NRL-TB method, the total energy is calculated directly from the shifted eigenvalues without resorting to some short-range repulsive potential.

In the NRL methodology, a first principles database of band structures and total energies is constructed for several crystal structures at different unit cell volumes. Then, a nonlinear parameterization is used to generate non-orthogonal two-center TB Hamiltonian and overlap matrix elements that (approximately) reproduce the energies and eigenvalues of the database.

Of particular importance to the success of the NRL-TB method is the functional form of the Hamiltonian and overlap matrix elements. It is assumed, in contrast with other DFT based TB methods, that the diagonal terms of the Hamiltonian are sensitive to the local chemical environment. For single element systems, each atom in the crystal is associated with an embedded atom-like density

$$\rho_i = \sum_{j} \exp(-\lambda^2 r_{ij}) F(r_{ij}), \qquad (8.18)$$

where the summation extends over all neighbors of atom i within a range r_{cut}, λ is a fitting parameter, and $F(r)$ is a cut-off function

$$F(r) = \frac{\theta(r - r_{cut})}{\left[1 + \exp\left(\dfrac{r - r_{cut}}{\ell} \right) \right]} \qquad (8.19)$$

and $\theta(z)$ is a step function. Typical values for r_{cut} lie between $10.5a_0$ and $16.5a_0$ and ℓ lies between $0.25a_0$ and $0.5a_0$.

In the NRL-TB method, the diagonal on-site terms of the Hamiltonian are defined as a polynomial expansion in the embedded density by

$$H_{il,il} = a_l + b_l \rho_i^{2/3} + c_l \rho_i^{4/3} + d_l \rho_i^2. \qquad (8.20)$$

In the spirit of the two-center approximation, the non-diagonal Hamiltonian elements are expanded in an exponential polynomial form with rotations taken into account by means of the Slater–Koster formalism [4],

$$H_{il,i'l'}(r_{ii'}) = \left(e_{il,i'l'} + f_{il,i'l'} r_{ii'} + g_{il,i'l'} r_{ii',i'l'}^2 \right) \exp\left(-h_{il,i'l'}^2 r_{ii'} \right) F(r_{ii'}). \qquad (8.21)$$

A similar expression exists for the non-diagonal overlap integrals. A total of 93 (or 97) parameters are required to produce the first principles database. Parameters for the NRL-TB method are available at http://cst-www.nrl.navy.mil/bind/ and the program is available upon request from its authors.

In Section 8.4 we explore the performance of this methodology for the global optimization of gold clusters and propose an improvement.

8.3.2   *AB INITIO* TIGHT BINDING METHOD

We begin this section by describing the formalism of Sankey and Niklewski [8], following the variant due to Horsfield [9], and in Section 8.3.3 we discuss the formalism due to Frauenheim et al. [10,11]. In contrast with the NRL-TB method, these methods are derived from the Harris–Foulkes–Haydock formalism [5,6].

The method of Sankey and Niklewski, known as the *ab initio* TB method (AITB), is simply an effort to implement the non-self-consistent version of the Harris–Foulkes functional in its full form. The method is a recipe for constructing an input charge, evaluating and tabulating the Hamiltonian and overlap elements (including three center contributions), and calculating the repulsive potential from the Coulombic double counting terms without any empirical fitting (for this reason the method is termed *ab initio*). A number of technical details make this method very interesting.

Since numerous self-consistent calculations on molecules and solids have shown that the electron densities in these structures are better approximated as a superposition of compressed atomic densities, the basis set and the charge density in AITB are calculated from orbitals in which the radial function is confined to some maximum radius. This is equivalent to confining the pseudo-atoms in an infinite square well potential. Compression of the pseudo-atoms is a crucial step for the success of any DFT-based TB. The three center integrals are tabulated as a function of three variables: the bond length, the distance and the angle between the bond center and the site on which the potential is centered, and an angle.

The formalism of Horsfield [9] is an extension of the Sankey and Niklewki method [8]. The important difference is the calculation of the pseudo-atomic basis, and the way in which the exchange and correlation integrals are handled. A minimal basis set was found to be inadequate for accurate descriptions of fluorocarbons, but a double numerical basis gives better agreement with DFT calculations. Thus the orbitals were taken from the neutral atom and a positive ion in a confined potential of the form r^6, instead of an infinite square well. The integrals are evaluated in real space using the partition function [32,33], in contrast with the method of Sankey and Niklewki in which the integrals are evaluated in reciprocal space. The key point in this approach is the use of a sub-linear dependence of the functional on the charge densities, rather than on the density.

This method has been applied to crystals and clusters such as amorphous silicon, silicon clusters, silicon surfaces, carbon clusters, and a number of semiconductor systems [34]. The AITB program is implemented and available in the Fireball suite of programs by the Lewis research group at http://www.fireball-dft.org/web/fireballHome. The actual implementation includes: norm-conserving pseudopotentials (Hamann or Troullier-Martins), local and non-local functionals, double numerical basis sets, linear-scaling methods for evaluating band structure, spin-polarized functionals, a time-dependent method, and a molecular dynamics methods using various thermostats, thermodynamic integration techniques and umbrella sampling.

8.3.3 Density Functional Tight Binding Method

The DFTB method [10] and its later self-consistent charge extension (SCC-DFTB) [11] are computationally efficient approximations to the fully self-consistent Kohn–Sham DFT theory based on the work of Harris–Foulkes. This method has been successfully applied to a wide range of problems in the field of biomolecules, surfaces and interfaces, as well as point and extended defects in solid-state systems [34].

The SCC-DFTB method is based on the second-order expansion of the Kohn–Sham total energy in DFT with respect to charge density fluctuations of Equation 8.14. The zeroth-order approach is equivalent to the standard non-self-consistent TB scheme

$$E_{DFTB} = \sum_j \int_{BZ} \frac{dk}{(2\pi)^3} E_j(k) + \sum_{i>j} U(i,j),$$ (8.22)

where the first term on the right-hand side is the band energy that depends on the reference system, and the second term is the pairwise repulsive contribution that depends on the interatomic separation between two atoms. The most important difference from AITB is that the repulsive potential is fit to reproduce points on the potential energy surface of a set of reference systems. In this sense DFTB is not strictly a first principles method like AITB. However, as in AITB the Hamiltonian and overlap matrices are evaluated using the DFT method. The atomic references, which provide the non-self-consistent density, are chosen from the neutral, spin-unpolarized atoms confined by a repulsive potential. In DFTB, three center integrals are neglected, thus the method is effectively a two center TB. The absence of three center terms simplifies the parameterization considerably compared to AITB.

Besides the usual band structure and short-range repulsive terms, the final approximate Kohn–Sham energy term includes a Coulomb interaction between charge fluctuations. At large distances this accounts for long-range electrostatic forces between two point charges and approximately includes self-interaction contributions from a given atom, if the charges are located on the same atom. Due to the SCC extension, DFTB can be successfully applied to problems where deficiencies within the non-SCC standard TB approach become obvious, such as those mentioned in Section 8.3.2.

In the last few years, the DFTB method has been heavily extended to allow the calculation of optical and excited state properties. The GW formalism as well as time-dependent DFTB has been implemented [35–37]. Furthermore, DFTB has been used to calculate the Hamiltonian for electron transport using Green's function techniques [38].

The DFTB code is available from its authors at http://www.dftb-plus.info/ and is maintained by the Bremen Center for Computational Materials Science. The DFTB methodology is also included in some popular electronic structure codes such as Gaussian (http://www.gaussian.com/) and molecular modeling codes such as Amber (http://www.ambermal.org/). A number of parameters are available at http://www.dftb.org/, including parameters for bio and organic molecules, solids and surfaces, organic–inorganic hybrid systems, chalcogenide glasses, transition metals in biological systems, rare earths, and zinc oxide with organic molecules.

8.4   SELF-CONSISTENT EXTENDED HUCKEL TIGHT BINDING

The self-consistent extended Huckel tight binding (SC-EHTB) method is an approximation to the second-order expansion of the Kohn–Sham DFT total energy of Equation 8.14. Here, we consider only closed shell systems; however, extension to the open shell case is straightforward.

To derive the EHTB formalism, the Kohn–Sham orbitals are expanded into a basis of Slater–type orbitals, $\psi_i = \sum_\mu C_{\mu i}\phi_\mu$. The elements of the Kohn–Sham Hamiltonian are calculated using the EH approximations [12]: the diagonal elements are taken as the valence state ionization energies (VSIE) and the off-diagonal elements are calculated using the Wolfsberg–Helmholtz formula,

$$\hat{H}^0_{\mu\nu} = \int \phi_\mu \hat{H}^0 \phi_\nu\, dr = \frac{1}{2} K_{EH} \left(\hat{H}^0_{\mu\mu} + \hat{H}^0_{\nu\nu} \right) S_{\mu\nu}, \tag{8.23}$$

where
   $S_{\mu\nu}$ is the overlap integral over the atomic basis set and the Wolfsberg–Helmholtz constant
   K_{EH} is evaluated as the weighted distance expression of Calzaferri and Kamber [39] and Calzaferri and Rytz [40]

Turning now to the second term of Equation 8.14 within the frame of the basis expansion employed here, the density fluctuation $\delta\rho$ is given by

$$\delta\rho(r) = \sum_\mu \sum_\nu \delta P_{\mu\nu}\phi_\mu^*(r)\phi_\nu(r), \tag{8.24}$$

where

$$\delta P_{\mu\nu} = \sum_i 2(C_{\mu i}C_{\nu i} - C^0_{\mu i}C^0_{\nu i}). \tag{8.25}$$

$\delta\rho$ (Equation 8.24) is decomposed into atom-centered expansion by employing the following approximation,

$$\phi_\mu^*(r)\phi_v(r) = \begin{cases} F^\alpha(|r - R_\alpha|) & \text{if } \mu = v, \mu \in \alpha, \\ \phi_\mu^*(r)\phi_v(r) & \text{if } \mu \neq v, \mu \in \alpha, v \in \alpha, \\ \frac{1}{2}\Big[F^\alpha(|r - R_\alpha|) + F^\beta(|r - R_\beta|)\Big]S_{\mu v} & \text{if } \mu \neq v, \mu \in \alpha, v \in \beta, \end{cases}$$

(8.26)

where $F^\alpha(|r - R_\alpha|)$ is a spherical radial approximation for the density of atom α. Upon introduction of Equation 8.26, the second term of Equation 8.14 becomes,

$$\frac{1}{2}\left(\sum_{\alpha\beta}\delta q_\alpha \delta q_\beta \gamma_{\alpha\beta} + \sum_\alpha \sum_{(\mu > v)\in\alpha}\delta P_{\mu v}^2 \Gamma_{\mu v}\right),$$

(8.27)

where we define the fluctuation in the Mulliken charges as

$$\delta q_\alpha = q_\alpha - q_\alpha^0 = \frac{1}{2}\sum_{\mu\in\alpha}\sum_v (\delta P_{\mu v}S_{\mu v} + \delta P_{v\mu}S_{v\mu}),$$

(8.28)

and

$$\gamma_{\alpha\beta} = \iint F^\alpha(r)\left(\frac{1}{|r - r'|} + \frac{\delta^2 E_{xc}}{\delta\rho_0\delta\rho_0'}\right)F^\beta(r')drdr',$$

(8.29)

$$\Gamma_{\mu v} = \iint \phi_\mu^*(r)\phi_v(r)\left(\frac{1}{|r - r'|} + \frac{\delta^2 E_{xc}}{\delta\rho_0\delta\rho_0'}\right)\phi_\mu^*(r')\phi_v(r')drdr'.$$

(8.30)

Although the integrals in Equations 8.29 and 8.30 could be computed *ab initio*, in the present implementation both are given a value from semi-empirical approximations. In the case of $\gamma_{\alpha\beta}$ we use the expression [16],

$$\gamma_{\alpha\beta}(R_{\alpha\beta}) = \frac{1}{\sqrt{R_{\alpha\beta}^2 + \gamma_{\alpha\beta}^{-2}(0)}},$$

(8.31)

where $R_{\alpha\beta}$ is the distance between atomic centers α and β. The value of this integral in the limit $R_{\alpha\beta} \to 0$ is approximated as the average between the two centers $\gamma_{\alpha\beta}(0)\frac{1}{2}\big(\gamma_{\alpha\alpha}(0) + \gamma_{\beta\beta}(0)\big)$. The $\gamma_{\alpha\alpha}(0)$ are approximated as the chemical hardness of the neutral atoms. For the evaluation of $\Gamma_{\mu v}$ integrals we use the spectroscopic values of the Slater–Condon parameters as in the ZINDO semiempirical method [41,42].

The last term of Equation 8.14 is the double counting term and the ion–ion repulsion, which is approximated by a two center expansion in the form $\sum_{\alpha\beta}U_{\alpha\beta}$.

Taking in account all of these approximations, the EHTB total energy expression of Equation 8.14 is written as

$$E \sum_{i}^{occ} 2 \sum_{\mu v} C_{\mu i}^{*} C_{vi} H_{\mu v}^{0} + \frac{1}{2} \left(\sum_{\alpha\beta} \delta q_{\alpha} \delta q_{\beta} \gamma_{\alpha\beta} + \sum_{\alpha} \sum_{(\mu > v) \in \alpha} \delta P_{\mu v}^{2} \Gamma_{\mu v} \right) + \sum_{\alpha\beta} U_{\alpha\beta}. \qquad (8.32)$$

Applying the variational principle to Equation 8.32, we obtain a Kohn–Sham-like equation, $HC = ESC$, with Hamiltonian elements of the form [16]

$$H_{\mu v} = \begin{cases} -VSIE_{\mu} + \sum_{\beta} \gamma_{\alpha\beta} \delta q_{\beta} & \text{if } \mu = v, \mu \in \alpha, \\[2ex] P_{\mu v} \Gamma_{\mu v} & \text{if } \mu \neq v, \mu \in \alpha, v \in \alpha, \\[2ex] H_{\mu v}^{0} + \frac{1}{2} S_{\mu v} \sum_{\xi} (\gamma_{\alpha\xi} \gamma_{\beta\xi}) \delta q_{\xi} & \text{if } \mu \neq v, \mu \in \alpha, v \in \beta. \end{cases} \qquad (8.33)$$

Since the elements of the Hamiltonian of Equation 8.33 depend on the density matrix through the atomic charges and the off-diagonal one center elements, these equations must be solved iteratively until SCC is achieved. One important contribution of this version of the EHTB is the incorporation, through the use of spectroscopic Slater–Condon parameters, of the on-site off-diagonal elements of the Coulomb–exchange–correlation coupling matrix, in a fashion similar to that employed in the Zerner's semiempirical intermediate neglect of differential overlap (ZINDO) method [41,42].

8.5   GOLD CLUSTERS

Gold nanoclusters and gold-based complexes are currently under active investigation in various areas of nanoscience and nanotechnology due to some remarkable physical and chemical properties of gold that appear only at the nanoscale [43,44]. Gold has been shown to exhibit surprising mechanical properties at the nanoscale by forming stable nanowires with quantized electrical conductivity, which has great relevance to atomic junctions [18]. Optical absorption by gold nanoparticles depends sensitively on size, and this property can be used in labeling applications or in devising precision therapies for selective imaging and destruction of cancer cells [19]. Although totally inert as bulk solids, gold nanoparticles can be surprisingly active as catalysts in oxidation and hydrogenation reactions. As a consequence, there exist many theoretical studies on the electronic structure of gold nanoclusters [19].

Hakkine and Landman investigated neutral and anionic gold clusters, Au_{2-10}, using DFT with scalar-relativistic *ab initio* pseudopotentials and the generalized gradient approximation (GGA) [45]. Wang et al. studied low-energy conformers of gold clusters with sizes from 2 to 20 atoms [46], and Walker performed DFT calculations on neutral and cationic gold clusters with up to 9 atoms [47].

In this section, we present a systematic investigation of the applicability of TB methods to describe the physics of small gold clusters.

8.5.1 GLOBAL OPTIMIZATION OF GOLD NANOCLUSTERS

Conventional non-self-consistent TB methods only include the first and the second terms of Equation 8.14. The last term of Equation 8.14 is computed in the self-consistent charge density functional tight binding approximation (SCC-DFTB [11]) by adopting the monopolar approximation (Mulliken approximation) where the charge fluctuation, δn, is expressed as a sum of contributions per atom. In this approximation, two types of contributions are considered: on-site interactions that depend on the hardness of the atoms (on-site diagonal matrix elements) and pairwise Coulombic interactions between different centers (interatomic interactions). Note that in this model the intra-atomic interactions, due to the exchange–Coulombic interaction between different orbitals in the same center, are neglected even though it has been shown that these interactions play an important role in the transferability of the TB parameters [48,49].

Frauenheim et al. [10] have argued that for homonuclear clusters the charge density fluctuation correction can be neglected given that the total charge per atom does not depend significantly on the atomic environment. Consequently, in such cases, no significant deviation is expected from the reference electronic density. However, one may argue that in noble metal systems these on-site contributions can, in principle, lead to strong deviations from the reference electronic density due to typical orbital hybridizations, i.e., the bonding in metal systems is mainly determined by the electronic balance of the different states. In fact, it has been suggested that the tendency to planarity in gold clusters (occurring at much larger cluster sizes than in copper and silver) is due to relativistic effects that decrease the sd promotion energy and lead to hybridization of the half-filled 6s orbital with the fully occupied 5d orbitals [50]. Therefore, even though for homonuclear systems the total charge does not show a strong dependence on the atomic environment, the orbital population might [51].

In order to take into account the above considerations in our TB approximation, we incorporate the orbital intra-atomic fluctuations into the last term of Equation 8.14. The Mulliken approximation is implemented to estimate the orbital populations. This is achieved by calculating individual fluctuations instead of using total fluctuations of the electronic density, which is used in other TB approximations [10,11]. For a homonuclear system, we approximate the last term of Equation 8.14 by a sum over all nuclei

$$\frac{1}{2} \sum_{\alpha} \sum_{(\mu > \nu) \in \alpha} \delta P_{\mu\nu}^2 \gamma_{\mu\nu}, \tag{8.34}$$

where the orbital population fluctuation, $\delta P_{\mu\nu}$, is defined as

$$\delta P_{\mu\nu} = \sum_{i} 2(P_{\mu\nu} - P_{\mu\nu}^0) \tag{8.35}$$

and

$$P_{\mu\nu} = C_{\mu} C_{\nu}. \tag{8.36}$$

In Equation 8.35, the upper index 0 denotes the reference orbital population obtained from the parameterized system, and the C_μ are the TB eigenvector coefficients. The $\gamma_{\mu\nu}$ terms in Equation 8.34 correspond to the Coulomb-exchange-correlation integrals over atomic orbitals, which, for the atomic cases, are approximated using the Slater–Condon parameters of the neutral atoms. All terms in Equation 8.34 can be estimated in a non-self-consistent fashion making use of the density matrix obtained directly from the reference TB Hamiltonian, as opposed to the self-consistent procedure used in standard TB.

The model proposed in this work does not require any additional parameters as compared to conventional TB methods. Only knowledge of the reference population $P_{\mu\nu}^0$ is required. This reference orbital population varies with the chemical environment, and is usually modeled by a function that depends on the Cartesian coordinates of the system. Consequently, the reference $P_{\mu\nu}^0$ is a function of atomic environment variables given by the geometric configuration of the system. For an atom i, we define the chemical environment function in terms of an atomic local density, as proposed elsewhere [10,11].

We have tested our TB approximation on gold nanoclusters by considering the NRL-TB method as the reference TB Hamiltonian. In this method, the second term of Equation 8.14 is not explicitly considered because it is implicitly contained in the first term of this equation when the eigenvalues are shifted by a constant amount to fit the total DFT energy [31]. Furthermore, the parameterization of this TB Hamiltonian includes a wide variety of crystal structure configurations as well as different lattice constants, probing a large number of different atomic environments during the TB parameterization. This leads to a significant range of atomic local densities, making it very convenient for our TB implementation, as mentioned previously. In the case of gold, the NRL-TB parameters were obtained from a fitting procedure that considered local density approximation (LDA) calculations of the band structure of different crystal configurations (FCC, BCC, and SC) and lattice constants varied by ±20% of the equilibrium value [24,52].

We first tested the method on a gold dimer. Figure 8.1 depicts the corresponding binding energy per atom as a function of the internuclear distance obtained with the NRL-TB parameterization mentioned above, with (circles) and without (dashed curve) the perturbative correction proposed in this work. Results obtained from DFT calculations at the LDA and GGA levels combined with the LANL2DZ basis set are also depicted in Figure 8.1 (solid curve). GGA calculations were performed using the pure gradient-corrected exchange-correlation functional (PW91).

As a second test, we assess the performance of the method on the prediction of equilibrium structures of small gold clusters. The structures of small gold cluster anions and cations with up to 13 atoms have been inferred through a comparison of theoretical and experimental collision cross sections from ion-mobility measurements [53] or electron diffraction data [54]. A remarkable finding of this study was that a 2D–3D structural transition for Au_n^- occurs at the surprisingly large cluster sizes of $n = 11$ and 12. A similar result was obtained in the case of neutral gold clusters based on DFT calculations [55]. For the ionic case, the result was later confirmed

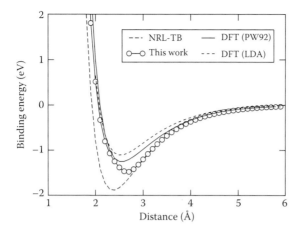

FIGURE 8.1   Binding energy per atom as a function of atomic distance for gold dimer.

through a comparison of photoelectron spectroscopy (PES) data with calculated densities of states [56].

In order to compute the geometry of the lowest energy neutral gold cluster (assumed in this work to be close to the global minimum), we make use of the basin paving optimization technique, which has been demonstrated to work quite well in the optimization of cluster structures [57]. The method is based on the combination of the optimization strategies used by basin hopping [58] and energy landscape paving [59].

The global minima of neutral gold cluster structures obtained in this work (with and without the energy correction to the NRL-TB Hamiltonian) are shown in Figure 8.2. In this figure, our optimized geometries are compared with results obtained with different DFT calculations previously reported in the literature [45,46,60,61]. In addition, the variation of the binding energy per atom as a function of the cluster size for all global minima cluster configurations found by our perturbatively corrected TB model is presented in Figure 8.3.

The results in Figure 8.1 show that the uncorrected NRL-TB Hamiltonian predicts a smaller equilibrium distance and a larger binding energy when compared to DFT calculations, even with LDA results, while the perturbatively corrected version of the same TB Hamiltonian substantially reduces these differences. This illustrates the importance of taking the on-site population fluctuation into account in the TB total energy calculation.

The most remarkable of the results depicted in Figure 8.2 is that, while the uncorrected NRL-TB Hamiltonian predicts planar global minima up to $n=20$, the perturbatively corrected NRL-TB Hamiltonian shows a transition between a 2D global minimum to the 3D structure at $n=12$, in excellent agreement with previous DFT calculations [55].

With the exception of Au_3, Au_7, and Au_{10}, the global minima for small cluster sizes ($n<12$) predicted by the uncorrected NRL-TB are similar the global minima predicted by the corrected one, but as the size increases, significant structural

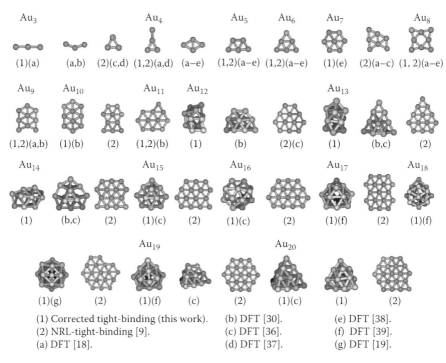

(1) Corrected tight-binding (this work).   (b) DFT [30].   (e) DFT [38].
(2) NRL-tight-binding [9].   (c) DFT [36].   (f) DFT [39].
(a) DFT [18].   (d) DFT [37].   (g) DFT [19].

FIGURE 8.2 Lowest energy neutral gold cluster structures, ranging in size from 3 to 20 atoms, obtained with the perturbatively corrected TB method. The results are compared with DFT and NRL-TB calculations.

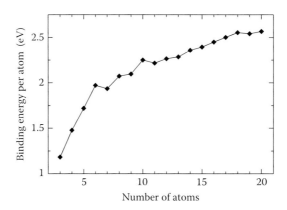

FIGURE 8.3 Binding energy per atom corresponding to the lowest energy configurations of neutral gold clusters of different sizes.

changes due to orbital rehybridization are observed. The perturbatively corrected NRL-TB geometries compare very well with those reported in recent DFT global optimization calculations on gold neutral clusters [55]. For sizes such as $n = 3$, 4, and 7, quasi-degenerate global minima are predicted by this TB, a situation that has been also observed in some DFT calculations [45,46,60,61].

The binding energy per atom ploted in Figure 8.3 exhibits oscillations between even and odd neutral clusters, and the oscillation amplitudes are larger at smaller sizes, in qualitative agreement with DFT calculations [55].

It is particularly gratifying to observe that our TB model also reproduces the ground state tetrahedral structure of Au_{20}, experimentally determined by vibrational spectroscopy [62]. The tetrahedral structure corresponds to a fragment of the FCC structure of bulk gold, which consists only of surface atoms and does not contain any inner atoms, a finding that also agrees with DFT calculations [63].

The results discussed in this work show that including the on-site orbital population fluctuation in the total energy recovers electronic structure effects that are important for the proper prediction of the energetics and structural properties of gold nanoclusters. This constitutes an improvement over other TB parameterizations and classical empirical potentials that fail to predict the correct structures and relative energetics of gold clusters, despite the explicit inclusion of the structures of some clusters in the parameterization of these methods. The perturbatively corrected TB model introduced in this work is particularly important in the case of homonuclear clusters, where charge-transfer between atoms in the system is not significant. For heteronuclear systems, charge-transfer may play a dominant role in the energy. But even in a situation of strong charge-transfer interactions, the fluctuation of the on-site atomic orbital distribution could still determine the global minimum as long as atomic orbital hybridizations are relevant.

8.5.2 Optical Spectra

Metallic nanoparticles possess unique optical, electronic, chemical, and magnetic properties that are strikingly different from those of individual atoms as well as from their bulk counterparts [44]. Colloidal solutions of noble metals like silver and gold show characteristic colors that have received considerable attention [64]. Gold nanoparticles have attracted considerable attention because of their potential technological applications. Much interest has been given to plasmon resonances of suspensions of quasi-spherical gold nanoparticles, which usually have a violet color. However, when individual gold nanoparticles come into close proximity, the color changes to blue [44]. The change in color that follows gold nanoparticle aggregation has found many applications. These systems are usually studied using classical Mie theory. This theory has traditionally been developed for calculation of the extinction spectra of single particles of highly symmetric shapes. In reality, a suspension of particles often has a much more complex structure. Also, when the interparticle distance is very small, the classical theory breaks down.

Many theoretical methodologies to perform frequency domain calculations in the context of DFT have been suggested in the literature. All of these have their merits and shortcomings [65]. If the SC-EHTB method described in the previous section is interpreted as an approximation to the Kohn–Sham energy, then the use of the TB approximation in conjunction with time-dependent DFT (TDDFT) is justifiable for the calculation of excitation energies [66]. In particular, an approach within the linear response TDDFT (LR-TDDFT) has been developed in the context of SC-EHTB [16] and SC-DFTB [35]. The TB approach to the LR-TDDFT is based on two assumptions: (i) using an ordinary, ground state TB calculation, one can obtain information similar to that obtained from Kohn–Sham DFT for the difference between the occupied and virtual orbitals, and (ii) the TDDFT Coulomb-exchange-correlation kernel can be modeled with the same kind of approximation as in the ground state TB. As in the case of the SCC-EHTB model, no exhaustive parameterization is performed; therefore, we expect our results to be qualitatively correct and provide a guide to the optical properties of the materials studied.

In our work we use the method proposed by Casidas in the context of LR-TDDFT [67]. In this formalism, the excitation energies, ϖ_I, are obtained by solving the following eigenvalue problem (for closed shells):

$$\sum_{ijkl\alpha} \left[\varpi_{ij}^2 \delta_{ik} \delta_{jl} \delta_{\sigma\tau} + 2\sqrt{\varpi_{jk}} K_{ij\sigma,kl\tau} \sqrt{\varpi_{kl}} \right] F_{ij\alpha}^I = \varpi_I^2 F_{kl\tau}^I. \tag{8.37}$$

In this equation, σ and τ are spin indices, i and k denote occupied orbitals (holes), j and l are virtual orbitals (particles), $\varpi_{ij} = \varepsilon_i - \varepsilon_j$, is the energy difference between the one-particle Kohn–Sham orbitals, and $K_{ij\sigma,kl\tau}$ are coupling-matrix elements, where

$$K_{ij\sigma,kl\tau} = \iint \psi_{i,\sigma}^*(r)\psi_{j,\sigma}(r) \left(\frac{1}{|r-r'|} + \frac{\delta^2 E_{xc}}{\delta\rho\delta\rho'} \right) \psi_{k,\tau}^*(r')\psi_{l,\tau}(r')drdr'. \tag{8.38}$$

Using the same approximation employed in the derivation of the EHTB formalism, and defining the magnetization as the difference of the spin densities $m = \rho^\alpha - \rho^\beta$, $K_{ij\sigma,kl\tau}$ can be cast in the form

$$K_{ij\sigma,kl\tau} = \sum_{\alpha\beta} q_\alpha^{ij} q_\beta^{kl} \left[\tilde{\gamma}_{\alpha\beta} + (2\delta_{\sigma\tau} - 1)m_{\alpha\beta} \right]$$

$$+ \sum_\alpha \sum_{(\mu>v)\in\alpha} \left(P_{\mu v}^{ij} P_{\mu v}^{kl} + P_{\mu v}^{ij} P_{v\mu}^{kl} + P_{v\mu}^{ij} P_{\mu v}^{kl} + P_{v\mu}^{ij} P_{v\mu}^{kl} \right) \tilde{\Gamma}_{\mu v}, \tag{8.39}$$

In Equation 8.16 we introduced the transition density matrix,

$$P_{\mu v}^{ij} = C_{\mu i} C_{vj}, \tag{8.40}$$

and the Mulliken transition charges

$$q_\alpha^{ij} = \frac{1}{2} \sum_{\mu\in\alpha} \sum_v \left(P_{\mu v}^{ij} S_{\mu v} + P_{v\mu}^{ij} S_{v\mu} \right). \tag{8.41}$$

Equation 8.16 also contains the following two electrons integrals,

$$\tilde{\gamma}_{\alpha\beta} = \iint F^\alpha(r) \left(\frac{1}{|r-r'|} + \frac{\delta^2 E_{xc}}{\delta\rho\delta\rho'} \right) F^\beta(r')drdr', \tag{8.42}$$

$$m_{\alpha\beta} = \iint F^\alpha(r) \frac{\delta^2 E_{xc}}{\delta m \delta m'} F^\beta(r') dr dr', \tag{8.43}$$

$$\tilde{\Gamma}_{\mu\nu} = \iint \phi_\mu^*(r)\phi_\nu(r) \left(\frac{1}{|r-r'|} \frac{\delta^2 E_{xc}}{\delta\rho\delta\rho'} \right) \phi_\mu^*(r')\phi_\nu(r') dr dr'. \tag{8.44}$$

The integral in Equation 8.43 is equivalent to Equation 8.31 with the final density ρ replacing the reference density ρ_0. If the charge fluctuation is sufficiently small that it can be neglected, it is possible to use the values of the ground state γ integrals as a zero-order approximation in Equation 8.43. Similarly, the integrals in Equation 8.43 involve atomic quantities that can, in principle, be obtained from DFT calculations. However, in keeping with the spirit of the semiempirical method, we compute these integrals using Slater–Condon spectroscopic parameters. Furthermore, the integral in Equation 8.44 does not involve a long-range Coulombic term, and therefore is approximated by an on-site parameter obtained from atomic DFT calculations. The excitation energies, w_I, are obtained from the eigenvalues of Equation 8.37. Despite the considerable simplification introduced in the evaluation of the TDDFT response kernel, direct diagonalization for systems with a large number of particle and hole states remains the main computational bottleneck. In our program, we make use of the DSYEVR routine implemented in LAPACK (http://www.netlib.org/lapack/) [86]. In this routine, the matrix is reduced to a tridiagonal form, and the eigenspectrum is computed using the multiple relatively robust representations method where Gram-Schmidt orthogonalization is avoided to the greatest extent possible.

Once the excitation energies are known, the corresponding oscillator strength for the Ith transition (in atomic units) can be obtained from [67]

$$f^I = \frac{2}{3} \varpi_I \left(\left| X^* S^{-1/2} F^I \right| + \left| Y^* S^{-1/2} F^I \right| + \left| Z^* S^{-1/2} F^I \right| \right). \tag{8.45}$$

where
(X,Y,Z) represents the transition dipole vector
F^I are the eigenvectors obtained by solving Equation 8.37

In Equation 8.45, S is a diagonal matrix defined as

$$S_{ij\sigma,kl\tau} = \frac{\delta_{\sigma\tau}\delta_{ik}\delta_{jl}}{\varepsilon_l - \varepsilon_k}. \tag{8.46}$$

This formalism yields a list of discrete excitation energies with their associated dipole transition intensities. A continuous spectrum can be drawn by convoluting the discrete spectrum with a Gaussian function.

Before we discuss an application of the present methodology, it is useful to check the validity of the scheme. We used our method to reproduce previous results of sodium chains obtained using TDDFT up to a chain length of 50 atoms. In these chains the interatomic distance is taken to be 2.37 Å. As more atoms are added to the chain, longitudinal and transverse modes appear. The longitudinal modes correspond to $\sigma \to \sigma^*$ transitions while the transverse modes correspond to $\sigma \to \pi$ transitions. Longitudinal transitions are of lower energy than the transverse. Due to the approximations in the calculation of the oscillator strengths in the present model, for transition between states of different symmetry, like the $\sigma \to \pi$ transverse modes, the Mulliken transition charge of Equation 8.41 vanishes exactly. For this reason we focus in the present work on longitudinal $\sigma \to \sigma^*$ transitions.

Figure 8.4 shows the length dependence of the excitation energy (Figure 8.4a) and oscillator strength (Figure 8.4b) of the main longitudinal modes. For chains shorter than 18 atoms, studied in previous TDDFT calculations [69,70], the longitudinal mode is dominated by a single absorption that is delocalized along the entire chain. The excitation energy decreases from 1.80 eV in the dimer to 0.41 eV in the chain of 18 atoms. These values are only slightly smaller than those reported previously using TDDFT [69,70]. In longer chains, two resonances show that strength and localization in the chain depend on the length. The difference in energy between these resonances is very small (0.06 eV or less for the 20 atom chain to 0.34 eV in the 50 atom chain). The lowest mode has a maximum strength at 16 atoms that decreases quickly as the chain length is increased. For chains larger than 18 atoms, a new peak appears. This absorption is mainly located in the central atoms and its strength increases with

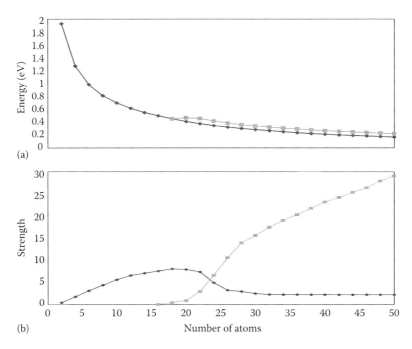

(a)

(b)

Number of atoms

FIGURE 8.4   (a) Longitudinal plasmon excitation energy and (b) dipole oscillator strength as a function of the number of sodium atoms in a linear chain.

TABLE 8.1

Five Lowest Excitation Energies (in eV) for Gold Dimer at a Distance of 2.55 Å

Expt. (a)	This Work	TDDFT/LDA (b)	TDDFT/PW91 (c)	TDDFT/B3LYP (d)
2.44	2.40	2.39	2.47	2.79
3.18	2.69	2.70	2.82	2.93
3.91	3.60	3.70	3.87	3.20
5.95	5.20	5.49	5.28	5.50
6.26	6.01	6.14	5.74	5.80

[a] Klotzbuecher, W.E. and Ozin, G.A., *Inorg. Chem.*, 19, 3767, 1980; Idrobo, J.C. et al., *Phys. Rev. B.*, 76, 205422, 2007. (b) Wang, X.J. et al., *J. Mol. Struct. (THEOCHEM)*, 9, 221, 2002. (c) Lian, K.Y. et al., *J. Chem. Phys.*, 30, 174701, 2009. (d) Datta, S., *Superlatt. Microstruct.*, 28, 253, 2000.

the atomic length. The strength of these two modes indicates that they are collective in character; however, there are no experimental measurements available for these resonances.

To assess the quality of the methodology, in Table 8.1 we present the five lowest excitation energies in the gold dimer and compare them with experimental values [71,72] and TDDFT calculations at various levels of theory: TDDFT/PW91 [73], TDDFT/B3LYP [74], and TDDFT/LDA [75]. The bond distance in these calculations was fixed at 2.55 Å. In all cases the computed values are lower than the experimental ones. The LDA results are closer to experimental values than the B3LYP and PW91 TDDFT results, with an average error of 6.3%. The differences are 6.6% for PW91 and 11% for B3LYP.

In order to cover the experimentally interesting excitation modes, we compute the excitation energies up to 7 eV for gold chains between 2 and 50 atoms. Even with the present approximations, as the chain length increases, one has to include more and more occupied and virtual Kohn–Sham states to cover the desired energy range. For 50 gold atoms, at least 500 occupied and 2000 virtual orbitals are necessary to cover an energy window of 7 eV. Of course this requires an amount of computational power that is prohibitive for large systems.

The computed excitation energies and dipole strengths for gold chains of 2–50 atoms are shown in Figure 8.5. The chains studied have a fixed internuclear separation of 2.89 Å as used in Ref. [16] for TDDFT calculations. For the gold dimer at the same separation two comparable peaks occur at 2.34 and 4.91 eV, with the second peak at twice the strength of the first. Even if the spectra become more complex as the number of atoms increases, at energies of less than 7 eV, as in the sodium chain, two main peaks are clearly seen. In all cases these peaks correspond to $\sigma(6s) \rightarrow \sigma^*$ (6p) transitions, with a small mixing with $5d(\sigma)$ orbitals. These two peaks are, in most cases, two times stronger than other 5d, or $\pi \rightarrow \pi^*$ transitions. The smaller absorption has a maximum strength at 16 gold atoms and decreases quickly with increasing chain length. As the length of the chain increases, the spectrum at

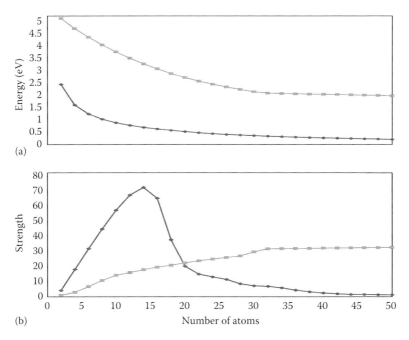

FIGURE 8.5 (a) Longitudinal plasmon excitation energy and (b) dipole oscillator strength as a function of the number of gold atoms in a linear chain.

low energy is dominated by a single absorption that is redshifted. In general, the average location of these peaks moves toward lower energies as the number of atoms increases.

There are some similarities between spectra observed in gold and sodium chains. Even though the gold spectra are more complicated due to the presence of 5d orbitals, the main features appearing in both cases are produced by longitudinal absorptions of σ symmetry. In particular, we note the appearance of two main absorptions whose location and strength depend on the size of the chain. We expect this type of transition to be dominant in most noble metal chains. It is well known that 5d orbitals are more localized, and their lower transition is of tranverse π symmetry with small oscillator strength. In both cases, the lower energy absorptions that dominate the small chains are more localized on the ends of the chain, while the large energy absorptions that dominate the large chains tend to localize on the center of the chains. In the gold chains, these three transitions are at much lower energy than in sodium and are better separated.

Previous calculations of gold and silver chains were performed up to a maximum of 20 atoms and generally possess a single longitudinal mode whose frequency depends on chain length. This dispersion is comparable to that of a propagating plasmon in a 1D electron gas model. However, as the size increases, other absorptions appear as seen in the present methodology. Also, the longitudinal excitations of s symmetry are nearly unaffected by the 5d orbital hybridization as the interband d–p transition only operates in transverse excitations of π symmetry.

8.5.3 MOLECULAR CONDUCTANCE

Computation of the electrical current in a molecule as a function of the applied voltage is of fundamental interest in studies of molecular electronics. The most rigorous approach to this problem involves use of a non-equilibrium Green's function [76] (NEGF) formalism with an *ab initio* electronic structure method (such as DFT). Of course, a TB scheme may be used for the electronic structure part of the NEGF method, and a number of implementations of this type exist [77]. The reader interested in the NEGF method and it application is encouraged to consult the literature [76].

In this section we will address a related, but considerably simpler problem. We will compute Green's functions at zero bias voltage for benzene dithiol using DFTB, EHTB, and DFT at the B3LYP/cc-pvdz level of theory. Since accurate computation of the Green's function is essential for obtaining reliable results from NEGF calculations, study of this quantity will yield some insight into the appropriateness of using a TB method with the NEGF formalism.

The Green's function of interest is calculated using [78]

$$G(E) = \sum_{\mu=1}^{N_{orb}} \sum_{i=1}^{N_L} \sum_{j=1}^{N_R} \frac{c_{i\mu}^* S_{i\mu}^* c_{j\mu} S_{j\mu}}{(E - \varepsilon_\mu - is)}, \tag{8.47}$$

where $G(E)$ is the Green's function at energy E, the sums are over the number of orbitals (N_{orb}) and the atomic centers included in the left- and right-hand fragments (N_L, N_R) of the molecule (a subset of orbitals included in the calculation), $c_{i\mu}$ is a molecular orbital coefficient, $S_{i\mu}$ is an element of the overlap matrix, ε_μ is the energy of orbital μ, and s controls the magnitude of the imaginary term. Note that this is a modification of the original method presented by González et al. [78] as we sum over all orbitals on a given atomic center and thus include the overlap in the computation. Computation of the Green's function requires knowledge of the molecular orbital coefficients and the overlap matrix, quantities that are easily extracted from many computational codes.

The Green's function was calculated in the manner described above from data extracted from the DFTB [79], EHTB, and Gaussian [80] programs. The square modulus of the Green's function versus energy is presented in Figure 8.6. We note that the TB methods show reasonable qualitative agreement for the peak at ~4.5 eV, but miss the structure above 10 eV. The EHTB results show the largest peak closer to 4.5 eV versus the DFTB results, but also show more structure around that peak compared to the B3LYP results. The differences between the TB results and the DFT results are not so surprising when we consider that the TB basis set is much smaller than the cc-pvdz basis set used for the B3LYP calculation and thus cannot show the same level of detail. In order to assess the effect of the size of the basis set, we performed a calculation at the B3LYP/STO-3G level of theory. The Green's function at this level of theory differs significantly from the B3LYP/cc-pvdz results and from the TB results as well. We conclude that the parameterization of the TB methods allows them to reproduce some of the structure of the Green's function based on B3LYP/cc-pvdz results, and that the spectra show less detail. The loss of

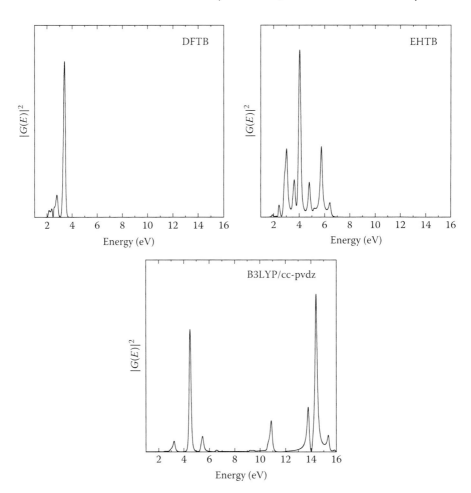

FIGURE 8.6   Plot of the square modulus of the Green's function versus energy (eV) for benzene dithiol. TB results produced by the DFTB and EHTB codes are shown along with DFT results at the B3LYP/cc-pvdz level of theory.

detail in the Green's function is a consequence of the size of the basis set. The most detail is seen for the cc-pvdz results that used 158 molecular orbitals. In contrast, the DFTB and EHTB results used 38 orbitals with a sp basis (not shown) and 48 when d functions were included.

8.6   PARALLELIZATION AND SPARSE MATRIX TECHNIQUES

TB methodologies are attractive for many reasons, one of which is that they may be evaluated much more rapidly than their *ab initio* and DFT counterparts [81] while retaining good accuracy through suitable parameterization. The efficiency with which TB methods may be applied leads naturally to applications in a number of areas including dynamics calculations and studies of large molecules.

Dynamics calculations typically require many evaluations of the potential energy and the associated nuclear gradient. Studies of large molecules similarly benefit from improvements in the efficiency of the TB method. However, studies of large molecules are additionally hampered by the amount of memory required which grows as the square of the number of orbitals. There are a number of strategies that may be employed to attack the problems listed above. Of these, we consider three: improvements in the performance of the linear algebra routines in the solution of the generalized eigenvalue problem, parallelization and sparse matrix techniques.

8.6.1 SELECTING ROUTINES FOR SOLVING THE EIGENVALUE PROBLEM

In general, the efficiency of a TB implementation depends strongly on the efficiency of the method used to solve the eigenproblem of Equation 8.6. We rewrite this equation as

$$(H - ES)C = 0, \tag{8.48}$$

where

H is the Hamiltonian matrix
E is the energy
S is the overlap matrix
C is the matrix of molecular orbital coefficients

The matrices H and S are symmetric, which makes the generalized eigenvalue problem (Equation 8.48) easier to solve. We note that if the overlap matrix, S, is the unit matrix (as in orthogonal TB methods) then the eigenvalue problem becomes a standard eigenvalue problem that is easier (i.e., faster) to solve.

The use of optimized mathematical libraries is well established in quantum chemistry. One of the most highly optimized libraries is the BLAS [82] (basic linear algebra subprograms) library. The BLAS library contains routines for vector–vector, vector–matrix, and matrix–matrix operations. Modern implementations of the BLAS are carefully optimized to exploit the multi-level memory hierarchy of modern CPUs. In doing so, optimized BLAS implementations routinely achieve high performance that is a significant fraction of the peak theoretical performance of a particular CPU. The LAPACK [83] (linear algebra package) libraries build upon the functionality offered by the BLAS to deliver a wide variety of useful algorithms including algorithms for the solution of eigenvalue problems in an efficient manner. Note that there are a number of other libraries offering functionality similar to LAPACK such as LINPACK [84] and FLAME [85]. However, we restrict our attention to the LAPACK library for the purposes of this section.

The LAPACK library offers several routines for the solution of the generalized eigenproblem: DSYGV (and the closely related DSYGVX) and DSYGVD (we restrict our attention to the double precision versions of these routines denoted by a leading "D" in the name). We assume that all eigenvalues and eigenvectors are required.

(If only eigenvalues are required then the routines offer identical performance.) The two routines mentioned above differ in the algorithm used to produce the eigenvectors. The routine DSYGVD uses a "divide and conquer" approach, significantly reducing the time required to complete the solution to the eigenvalue problem. Both of the algorithms discussed in this section scale (in number of operations) as $O(n^3)$ where n is the dimension of the matrix. In practice, the divide and conquer approach has a smaller prefactor m where performance scales as mn^3 and hence has better performance.

The procedure for solving the generalized symmetric eigenvalue problem proceeds in three major steps: (1) the generalized eigenvalue problem is reduced to a standard eigenvalue problem, (2) the matrix is reduced to tridiagonal form, and (3) the eigenvalues and eigenvectors are produced. Recently a new algorithm known as MRRR [68] (multiple relatively robust representations) for calculating eigenvalues from a tridiagonal matrix was introduced. The scaling of this algorithm is $O(n^2)$ and thus should be faster than the two algorithms mentioned above. However, because this algorithm only reduces the scaling for one of the three steps in the solution of the generalized symmetric eigenvalue problem, the overall scaling of the algorithm remains $O(n^3)$. The current release of LAPACK does not include a routine for solving the generalized symmetric eigenvalue problem that incorporates the MRRR algorithm. However, it is straightforward to modify an existing LAPACK to use the MRRR algorithm. We have done this and call the resulting subroutine DSYGVR in accordance with the LAPACK naming scheme. In practice, we have found that the performance of DSYGVD and DSYGVR are very close. Thus we plot results only for DSYGVR, which are representative of the performance characteristics of DSYGVD. This result is surprising in light of the discussion above, and illustrates the importance of testing these algorithms under realistic conditions.

A plot comparing the performance of the DSYGVX and DSYGVR routines is given in Figure 8.7. Note that the DSYGVR routine is 2.5 times faster than the DSYGVX routine at the largest problem size (8307 orbitals). In general, the DSYGVD and DSYGVR routines are appreciably faster than DSYGV(X) and their use is recommended.

In closing, we emphasize the importance of optimized BLAS libraries as the key to achieving good performance on a wide variety of scientific problems. These libraries are available from vendors [86] (e.g., ACML [86], ESSL [87], MKL [88]) and from other projects (e.g., ATLAS [89], GotoBLAS [90]). In the present work, we use the Goto BLAS library (version 2).

8.6.2 Parallelization

Again, we primarily consider solution of the eigenvalue problem as this dominates the time required for calculation of the TB energy. As before, the choice of algorithm will have a strong effect on the efficiency of this step. In the case of a parallel algorithm, efficiency will also refer to speedup on multiple processors compared to the single processor performance.

Modern parallel computers generally fall into two categories. Symmetric multiprocessor (SMP) machines are characterized by having more than one CPU (or

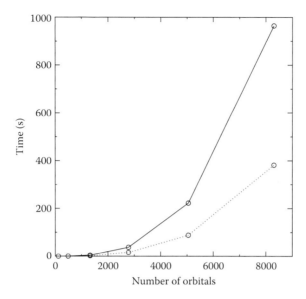

FIGURE 8.7 Performance of various LAPACK eigenvalue solvers on a single CPU for gold clusters with sizes ranging from 13 to 923 atoms (117–8307 orbitals). The solid line indicates results for the DSYGVX routine, while the dashed line indicates results for the DSYGVR routine.

CPU core) with shared memory and controlled by a single operating system. Such machines are quite common from laptops to supercomputers. Parallel machines consist of more than one node, where a node contains one or more CPUs with its own memory and operating system. (This is a rather simplified view of the diversity of computing resources currently available, but it conveniently categorizes the vast majority of the computers currently in use.) In this section, we will focus primarily on SMP resources. Many of the results will apply equally well to larger parallel machines. However, performance on these machines is highly dependent on the technology used to connect the nodes to one another for sharing data. In general, a high bandwidth, low latency interconnection is desirable. Further discussion of these architectures and their performance is outside the scope of this chapter.

We will consider two means of achieving parallel speedup on SMP architectures: multithreading and message passing. The multithreading paradigm is applicable only to SMP architectures, whereas the message passing paradigm can be applied on SMP machines or on parallel machines.

Modern, optimized implementations of the BLAS include the capability to use more than one "thread" at a time, where a thread represents an independent compute process, hence the term multithreading. Often the number of threads used is equal to the number of processors on the machine to achieve optimal performance. Optimized, multithreaded BLAS libraries typically yield very good performance with a concomitant impact on the performance of the eigenvalue solver. This is a particularly easy means to achieve parallel speedup, as usually the user is only

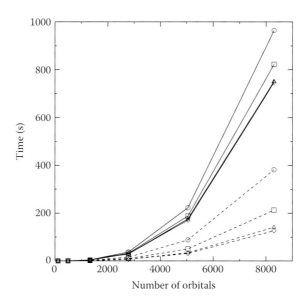

FIGURE 8.8　Performance of various LAPACK eigenvalue solvers on multiple CPUs using multithreaded BLAS for gold clusters with sizes ranging from 13 to 923 atoms (117–8307 orbitals). The solid line indicates results for the DSYGVX routine, while the dashed line indicates results for the DSYGVR routine. Symbols indicate one processor (circles), two processors (squares), four processors (triangles), and eight processors (diamonds).

required to link to a different library to take advantage of multithreading. Frequently, multithreading is the default behavior and users benefit automatically. The choice of algorithm is still important as mentioned above.

Examples of performance increases due to multithreading are shown in Figure 8.8. Again, the DSYGVR routine (and similarly DSYGVD) is much faster than the DSYGVX routine. In Figure 8.8, we may also see how the routines scale on multiple processors. The scaling for the DSYGVX algorithm is poor, achieving only a 30% speedup for the largest problem size when running on eight processors versus running on one processor. On the other hand, the DSYGVR routine (and similarly DSYGVD) shows much better scaling, achieving a speedup of 270% on four processors and 300% when running on eight processors. The parallel efficiency (speedup divided by the number of processors) of the DSYGVR routine is 90% for jobs with two processors and 68% for jobs with four processors, but drops to 38% when using eight processors. This result shows the practical limit for improving the performance of solving the eigenvalue problem on the particular computer used to produce the results. Different computer architectures will have different limits that should be explored by the user.

The above discussion focused solely on parallelism in the mathematical library. For small to medium size problems, this approach is reasonable. However, as the problem size increases, the expense of constructing the Hamiltonian and overlap matrices increases as well and will demand an increasing fraction of the total compute time. In this case, it is beneficial to parallelize construction of these matrices as well. A programming model such as OpenMP [92] is well suited for this task (though there are other methods). The OpenMP standard is implemented in a number of compilers. In our experience, application of the OpenMP parallel constructs to this case is reasonably straightforward and good speedup is achieved.

In order to parallelize TB algorithms beyond a single SMP machine, it is necessary to use some sort of message passing parallel paradigm such as the one found in the message passing interface (MPI) [92] standard. There are many implementations of this standard, both from vendors and from freely available sources (e.g., MPICH2 [93], OpenMPI [94]). Users should ensure that the MPI implementation they use is appropriate for their system for optimal performance. In the following, we will focus on parallelization of the eigenvalue solver through use of the ScaLAPACK library. We note that other libraries exist for this purpose, e.g., PETSc [95]/SLEPc [96].

One of the most important factors in achieving good parallelization when using the ScaLAPACK [97] library is data layout. The ScaLAPACK User's Guide [98] has extensive documentation on this subject that we will not repeat here. We use a 1D block cyclic data layout for the examples presented in this section with a block size of 64. A 2D block cyclic data layout is optimal for ScaLAPACK, but not recommended until the number of processors exceeds 8 [98]. Since the storage for the matrices is distributed over the MPI processes, it is natural to have each process compute its elements of the Hamiltonian and overlap matrices. This involves some bookkeeping of the mapping between global and local matrix indices, which is thoroughly described in the ScaLAPACK User's Guide. Thus, parallelization of the construction of the Hamiltonian and overlap matrices arises naturally in this algorithm.

We have implemented a parallel TB method in our EHTB code using ScaLAPACK as outlined above. In order to illustrate the potential of this method, we have computed eigenvalues and eigenvectors for a series of gold clusters from 13 to 923 atoms (with 9 orbitals per atom) using three parallel algorithms. The first algorithm is PDSYGVX from ScaLAPACK. This is the only routine in the current version of ScaLAPACK for the parallel computation of the generalized symmetric eigenvalue problem. Fortunately, several parallel solvers analogous to the solvers presented in Section 8.6.1 exist for the standard symmetric eigenvalue problem and it is straightforward to produce the corresponding parallel solver for the generalized problem. The parallel generalized symmetric eigenvalue solvers follow a simple pattern. A Cholesky factorization of the overlap matrix is formed using the ScaLAPACK routine PDPOTRF. The problem is transformed to a standard symmetric eigenvalue problem via a call to PDSYNGST. The standard symmetric eigenvalue problem is solved using PDSYEVX (or another subroutine, see below). Finally, the eigenvalues are backtransformed using PDTRSM. In order to produce a parallel solver for the generalized symmetric eigenvalue problem using the divide and conquer method, the call to PDSYEVX is replaced by a call to PDSYEVD. Similarly, in order to produce a parallel solver for the generalized symmetric eigenvalue problem using the MRRR algorithm, the call to PDSYEVX is replaced by a call to PDSYEVR [99]. At this time, PDSYEVR is not a part of the ScaLAPACK library, but it is anticipated that it will become part of ScaLAPACK in the future.

Following the results of Section 8.6.1, we anticipate that the solver based on the divide and conquer algorithm, PDSYGVD, or the MRRR algorithm, PDSYGVR, will

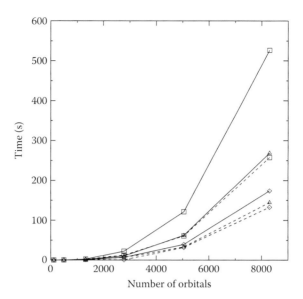

FIGURE 8.9   Peformance of various ScaLAPACK parallel eigenvalue solvers for gold clusters with sizes ranging from 13 to 923 atoms (117–8307 orbitals). The solid line indicates results for the PDSYGVX routine, while the dashed line indicates results for the PDSYGVR routine. Symbols indicate, two processors (squares), four processors (triangles), and eight processors (diamonds).

give the best performance. We compare the timings of PDSYGVX to PDSYGVR in Figure 8.9 (again, the results for PDSYGVD are similar to the results for PDSYGVR). We first note that the timings for all of the parallel routines are larger than for the corresponding multithreaded cases. We attribute this difference to overhead in the parallel implementation. This is an important result indicating that the best performance may be obtained using multithreaded BLAS when running on a single SMP node. However, we are concerned with the ScaLAPACK routines that may be run on more than one node. From Figure 8.9, the better performance of PDSYGVR versus PDSYGVX is clear. At the largest problem size, PDSYGVR is 20% more efficient than PDSYGVX. The parallel efficiency is very much different in this case, where PDSYGVX has a parallel efficiency of 69% for eight processors at the largest problem size versus 33% for PDSYGVR.

8.6.3   SPARSE MATRIX TECHNIQUES

In the previous examples of this section, we have ignored memory usage. Storage for the Hamiltonian and overlap matrices requires n^2 double precision words. Additional storage of n^2 is required for several of the algorithms for the eigenvectors. Temporary storage varies from $O(n)$ to $O(n^2)$. In Table 8.2, we show the memory requirements for storing a single matrix associated with the gold clusters we considered above as well as some larger clusters in the same family. It is clear to see that the storage requirements quickly become a problem.

TABLE 8.2

Storage Requirements (in Millions of Bytes) for a Single Matrix Associated with the TB Computation of a Gold Cluster with the Specified Number of Atoms

Number of Atoms	Matrix Dimension	Number of Non-Zeros	Density (%)	Memory for Sparse Storage	Memory for Dense Storage	Ratio (Dense/ Sparse)
13	117	8,041	58.7	0.1	0.1	1.1
55	495	74,375	30.4	0.9	1.9	2.2
147	1,323	263,011	15.0	3.0	13.4	4.4
309	2,781	638,169	8.3	7.3	59.0	8.1
561	5,049	1,264,069	5.0	14.5	194.5	13.4
923	8,307	2,201,763	3.2	25.2	526.5	20.9
1,415	12,735	3,519,215	2.2	40.3	1,237.3	30.7
2,057	18,513	5,274,949	1.5	60.4	2,614.8	43.3
2,869	25,821	7,529,793	1.1	86.3	5,086.7	59.0
3,871	34,839	10,354,207	0.9	118.6	9,260.2	78.1
5,083	45,747	13,802,299	0.7	158.1	15,966.7	101.0

In order to study systems containing thousands of atoms, sparse matrix storage techniques may be employed. The idea of sparse storage is to store only matrix elements that are non-zero. For matrices with many zero elements, this scheme can be quite efficient. However, there is an overhead associated with the pointers needed to keep track of the indices of the matrix elements. There are many schemes for sparse matrix storage. One popular matrix scheme is CSR (compressed sparse row). In this scheme, an index indicating the start of each row is kept as well as an index indicating the column of each matrix element. Thus, the storage requirement is proportional to the matrix dimension plus the number of non-zero elements in the matrix. For a nice overview of sparse matrix storage and related techniques, see Davis [100]. In Table 8.2, the sparse storage, including overhead, for a series of gold clusters is given. Note that by the time the cluster size reaches 5000 atoms, the sparse storage scheme is two orders of magnitude more efficient than the dense (n^2) storage scheme. These results are taken from our EHTB code that uses a cutoff distance (12 a_0 in this case) beyond which matrix elements are taken to be zero.

There are some drawbacks to using sparse matrix techniques. Though there are a number of libraries available that implement linear algebra operations on sparse matrices, these have not seen widespread adoption for reasons that are not clear. Another reason for the lesser popularity of sparse matrix schemes is the relatively poor performance versus optimized math libraries such as the BLAS. The primary cause of the poor performance is poor data locality and irregular data access patterns. Nevertheless, for a matrix of sufficiently large dimension (perhaps a few thousand), the performance of linear algebra operations on sparse matrices exceeds that of their dense variants. This is a consequence of the significantly smaller number of operations that must be performed. Ultimately, sparse matrix methods can scale as $O(n)$ which is highly desirable.

Sparse matrix methods hold a great deal of promise for the future of many TB applications. The DFTB+ code has recently implemented sparse storage techniques and plans to replace dense eigenvalue solvers with their sparse equivalent in the near future [79]. Zhang et al. [101] have implemented a shift-and-invert spectral transformation (SIPs) algorithm which scales as $O(n^2)$ and demonstrated its use for systems up to dimension 64,000 with a significant increase in performance over ScaLAPACK solvers. It is important to note that the SIPs algorithm is capable of computing most or all of the eigenvalues (and associated eigenvectors) that are required by the electronic structure problem. A sparse, parallel implementation of TB has been reported by Colombo and Sawyer [102]. If only a small portion of the eigenvalue spectrum is needed, the matrix dimension may go into the millions as demonstrated by Naumov et al. [103] using an algorithm implemented in NEMO-3D [104]. Computations of optical spectra, as discussed in Section 5.2, may benefit from such algorithms.

8.7   CONCLUSION

In this chapter, we have presented a summary of some of the important historical developments in the theory of TB and have presented several modern applications of this method. We have discussed some of the most important aspects of the performance of TB algorithms. Particular attention has been given to parallelization and sparse matrix methods.

REFERENCES

1. N. W. Ashcroft and N. D. Mermin, *Solid State Physics*, Harcourt, New York, 1976.
2. C. Kittel, *Introduction to Solid State Physics*, Wiley, New York, 2004.
3. W. A. Harrison, *Elementary Electronic Structure*, World Scientific, Singapore, 2004.
4. J. C. Slater and G. F. Koster, Simplified LCAO method for the periodic potential problem, *Phys. Rev.* 94, 1498 (1954).
5. W. M. C. Foulkes and R. Haydock, Tight-binding models and density-functional theory, *Phys. Rev. B* 39, 12520 (1989).
6. J. Harris, Simplified method for calculating the energy of weakly interacting fragments, *Phys. Rev. B* 31, 1770 (1985).
7. M. Mehl and D. M. Papaconstantopoulos, Applications of a tight-binding total-energy method for transition and noble metals: Elastic constants, vacancies, and surfaces of monatomic metals, *Phys. Rev. B* 54, 4519 (1996).
8. O. F. Sankey and D. J. Niklewski, Ab initio multicenter tight-binding model for molecular-dynamics simulations and other applications in covalent systems, *Phys. Rev. B* 40, 3979 (1989).
9. A. P. Horsfield, Efficient ab initio tight binding, *Phys. Rev. B* 56, 6594 (1997).
10. D. Porezag, Th. Frauenheim, Th. Kohler, G. Seifert, and R. Kaschner, Construction of tight-binding-like potentials on the basis of density-functional theory: Application to carbon, *Phys. Rev. B* 51, 12947 (1995).

11. M. Elstner, D. Porezag, G. Jungnickel, J. Elsner, M. Haugk, Th. Frauenheim, S. Suhai, and G. Seifert, Self-consistent-charge density-functional tight-binding method for simulations of complex materials properties, *Phys. Rev. B* 58, 7260 (1998).

12. R. Hoffmann, An extended Hückel theory. I. Hydrocarbons, *J. Chem. Phys.* 39, 1397 (1963).

13. T. A. Albright, J. K. Burdett, and M.-H. Whangbo, *Orbital Interaction in Chemistry*, Wiley, New York (1985).

14. J. Cerda and F. Soria, Accurate and transferable extended Hückel-type tight-binding parameters, *Phys. Rev. B* 61, 7965 (2000).

15. J. Zhao and J. P. Lu, A nonorthogonal tight-binding total energy model for molecular simulations, *Phys. Lett. A* 319, 523 (2003).

16. L. Rincon, A. Hasmy, C. A. Gonzalez, and R. Almeida, Extended Hückel tight-binding approach to electronic excitations, *J. Chem. Phys.* 129, 044107 (2008).

17. L. Rincon, A. Hasmy, T. C. Allison, and C.A. Gonzalez (unpublished work).

18. L. Rincon, A. Hasmy, M. Marquez, and C. A. Gonzalez, A perturbatively corrected tight-binding method with hybridization, *Chem. Phys. Lett.*, 2010, In Press.

19. P. Koskinen, H. Hakkinen, G. Seifert, S. Sanna, Th. Frauenheim, and M. Moseler, Density-functional based tight-binding study of small gold clusters, *New J. Phys.* 8, 9 (2006).

20. Y. Dong and M. Springborg, Global structure optimization study on Au_{2-20}, *Eur. Phys. J. D* 43, 15 (2007).

21. S. Froyen and W. A. Harrison, Elementary prediction of linear combination of atomic orbitals matrix elements, *Phys. Rev. B* 20, 2420 (1979).

22. D. J. Chadi, Atomic and electronic structures of reconstructed Si(100) surfaces, *Phys. Rev. Lett.* 43, 43 (1979).

23. W. Kohn and L. J. Sham, Self-consistent equations including exchange and correlation effects, *Phys. Rev.* 140, A1133 (1965).

24. F. Bloch, Über die Quantenmechanik der Elektronen in Kristallgittern, *Z. Phys.* 52, 555 (1928).

25. C. M. Gorinde, D. R. Bowler, and E. Hernandez, Tight-binding modelling of materials, *Rep. Prog. Phys.* 60, 1447 (1997).

26. M. Menon and K. R. Subbaswamy, Nonorthogonal tight-binding molecular-dynamics scheme for silicon with improved transferability, *Phys. Rev. B* 55, 9231 (1997).

27. O. K. Andersen and O. Jepsen, Explicit, first-principles tight-binding theory, *Phys. Rev. Lett.* 53, 2571 (1984).

28. R. E. Cohen, M. J. Mehl, and D. A. Papaconstantopoulos, Tight-binding total-energy method for transition and noble metals, *Phys. Rev. B* 50, 14694 (1994).

29. D. A. Papaconstantopoulos, *Handbook of Electronic Structure of Elemental Solids*, Plenum, New York, 1986.

30. C. Kohler, G. Seifert, and Th. Fraunheim, Density-functional based calculations for Fe(n), ($n \leq 32$), *Chem. Phys.* 309, 23 (2005).

31. D. A. Papaconstantopoulos and M. J. Mehl, The Slater–Koster tight-binding method: A computationally efficient and accurate approach, *J. Phys.: Condens. Matter* 15, R413 (2003).

32. B. Delley, An all-electron numerical method for solving the local density functional for polyatomic molecules, *J. Chem. Phys.* 92, 508 (1990).

33. A. P. Horsfield and A. M. Bratkovsky, *Ab initio* tight binding, *J. Phys.: Condens. Matter* 12, R1 (2000).

34. Th. Frauenheim, G. Seifert, M. Elstner, Z. Hajnal, G. Jungnickel, D. Porezag, S. Suhai, and R. Scholz, A self-consistent charge density-functional based tight-binding method for predictive materials simulations in physics, chemistry and biology, *Phys. Stat. Sol. (b)* 217, 41 (2000).

35. T.A. Niehaus, S. Suhai, F. Della Sala, P. Lugli, M. Elstner, G. Seifert, and T. Frauenheim, Tight-binding approach to time-dependent density-functional response theory, *Phys. Rev. B* 63, 085108 (2001).

36. T. A. Niehaus, D. Heringer, B. Torralva, and T. Frauenheim, Importance of electronic self-consistency in the TDDFT based treatment of nonadiabatic molecular dynamics, *Eur. Phys. J. D* 35, 467 (2005).

37. T. A. Niehaus, M. Rohlfing, F. Della Sala, A. Di Carlo, and T. Frauenheim, Quasiparticle energies for large molecules: A tight-binding-based Green's-function approach, *Phys. Rev. A* 71, 022508 (2005).

38. A. Pecchia and A. Di Carlo, Atomistic theory of transport in organic and inorganic nanostructures, *Rep. Prog. Phys.* 67, 1497 (2004).

39. G. Calzaferri and I. J. Kamber, Molecular geometries by the extended-Hueckel molecular orbital method: A comment, *Phys. Chem.* 93, 5366 (1989).

40. G. Calzaferri, R. J. Rytz, Electronic transition oscillator strength by the extended hueckel molecular orbital method, *Phys. Chem.* 99, 12141 (1995).

41. J. Ridley and M. C. Zerner, An intermediate neglect of differential overlap technique for spectroscopy: Pyrrole and the azines, *Theor. Chim. Acta* 32, 111 (1973).

42. J. D. Baker and M. C. Zerner, A charge-iterative Hamiltonian for molecular electronic spectra, *J. Phys. Chem.* 95, 2307 (1991).

43. P. Pyykkö, Theoretical chemistry of gold, *Angew. Chem. Int. Ed.* 43, 4412 (2004).

44. S. K. Ghosh and T. Pal, Interparticle coupling effect on the surface plasmon resonance of gold nanoparticles: From theory to applications, *Chem. Rev.* 2007, 107, 4797.

45. H. Hakkinen and U. Landman, Gold clusters (Au_N, 2<~N<~10) and their anions, *Phys. Rev. B* 62, R2287 (2000).

46. J.-L. Wang, G.-H. Wang, and J.-J. Zhao, Density-functional study of Au_n ($n = 2–20$) clusters: Lowest-energy structures and electronic properties, *Phys. Rev. B* 66, 035418 (2002).

47. A. V. Walker, Structure and energetics of small gold clusters and their positive ions, *J. Chem. Phys.* 122, 094310 (2005).

48. J. L. Mercer and M. Y. Chou, Tight-binding model with intra-atomic matrix elements, *Phys. Rev. B* 49, 8506 (1994).

40. Y. Xie and J. A. Blackman, Tight-binding model for transition metals: From cluster to solid, *Phys. Rev. B* 63, 125105 (2001).

50. E. M. Fernández, J. M. Soler, and I. L. Garzón, and L. C. Balbás, Trends in the structure and bonding of noble metal clusters, *Phys. Rev. B* 70, 165403 (2004).

51. V. Fiorentini, M. Methfessel, and M. Scheffler, Reconstruction mechanism of fcc transition metal (001) surfaces, *Phys. Rev. Lett.* 71, 1051 (1993).

52. F. Kirchhoff, M. J. Mehl, N. I. Papanicolaou, D. A. Papaconstantopoulos, and F. S. Khan, Dynamical properties of Au from tight-binding molecular-dynamics simulations, *Phys. Rev. B* 63, 195101 (2001).

53. F. Furche, R. Ahlrichs, P. Weis, C. Jacob, S. Gilb, T. Bierweiler, and M. M. Kappes, The structures of small gold cluster anions as determined by a combination of ion mobility measurements and density functional calculations, *J. Chem. Phys.* 117, 6982 (2002).

54. X. Xing, B. Yoon, U. Landman, and J. H. Parks, Structural evolution of Au nanoclusters: From planar to cage to tubular motifs, *Phys. Rev. B* 74, 165423 (2006).

55. X. B. Li, H. Y. Wang, X. D. Yang, Z. H. Zhu, and Y.-J. Tang, Size dependence of the structures and energetic and electronic properties of gold clusters, *J. Chem. Phys.* 126, 084505 (2007).

56. H. Hakkinen, B. Yoon, U. Landman, X. Li, H.-J. Zhai, and L.-S. Wang, On the electronic and atomic structures of small Au_{N^-} ($N = 4–14$) clusters: A photoelectron spectroscopy and density-functional study, *J. Phys. Chem. A* 107, 6168 (2003).

57. L. Zhan, J. Z. Y. Chen, and W.-K. Liu, Monte Carlo basin paving: An improved global optimization method, *Phys. Rev. E* 73, 015701(R) (2006).

58. D. J. Wales and J. P. K. Doye, Global optimization by basin-hopping and the lowest energy structures of Lennard-Jones clusters containing up to 110 atoms, *J. Phys. Chem. A* 101, 5111 (1997).

59. U. H. E. Hansmann and L. T. Wille, Global optimization by energy landscape paving, *Phys. Rev. Lett.* 88, 068105 (2002).

60. W. Fa, C. Luo, and J. Dong, Bulk fragment and tubelike structures of Au_N (N=2–26), *Phys. Rev. B* 72, 205428 (2005).

61. H. Gronbeck and W. Andreoni, Gold and platinum microclusters and their anions: Comparison of structural and electronic properties, *Chem. Phys.* 262, 1 (2000).

62. P. Gruene, D. M. Rayner, B. Redlich, A. F. G. van der Meer, J. T. Lyon, G. Meijer, and A. Fielicke, Structures of neutral Au_7, Au_{19}, and Au_{20} clusters in the gas phase, *Science* 321, 674 (2008).

63. E. Apra, R. Ferrando, and A. Fortunelli, Density-functional global optimization of gold nanoclusters, *Phys. Rev. B* 73, 205414 (2006).

64. S. Link and M. A. El-Sayed, Spectral properties and relaxation dynamics of surface plasmon electronic oscillations in gold and silver nanodots and nanorods, *J. Phys. Chem. B* 103, 8410 (1999).

65. M. A. L. Marques and A. Rubio, In *Time-Dependent Density Functional Theory*, Time versus frequency space techniques, M. A. L. Marques, C. A. Ulrich, F. Nogueira, A. Rubio, K. Burke, and E. K. U. Gross (Eds.) Springer, Berlin, p. 227 (2006).

66. T. A. Niehaus, Approximate time-dependent density functional theory, *J. Mol. Struct. (THEOCHEM)* 38, 914 (2009).

67. M. E. Casidas, In *Recent Advances in Density Functional Methods. Part I*, Time-dependent density functional response theory for molecules, D. Chong (Ed.), World Scientific, Singapore, p. 155 (1995).

68. (a) B. N. Parlett and I. S. Dhillon, Multiple representations to compute orthogonal eigenvectors of symmetric tridiagonal matrices, *Linear Algebra Appl.* 387 (2004). (b) I. S. Dhillon, B. N. Parlett, and C. Vömel, LAPACK working note 162: The design and implementation of the MRRR algorithm, Technical Report UCBCSD-04-1346, University of California, Berkeley, 2004.

69. J. Yan, Z. Yuan, and S. Gao, End and central plasmon resonances in linear atomic chains, *Phys. Rev. Lett.* 98, 216602 (2007).

70. J. Yan and S. Gao, Plasmon resonances in linear atomic chains: Free-electron behavior and anisotropic screening of d electrons, *Phys. Rev. B* 78, 235413 (2008).

71. G. A. Bishea and M. D. Morse, Spectroscopic studies of jet-cooled AgAu and Au_2, *J. Chem. Phys.* 95, 5646 (1991).

72. W. E. Klotzbuecher and G. A. Ozin, Optical spectra of hafnium, tungsten, rhenium and ruthenium atoms and other heavy transition-metal atoms and small clusters ($Zr_{1,2}$, $Pd_{1,2}$, $Au_{1,2,3}$) in noble-gas matrixes, *Inorg. Chem.* 19, 3767 (1980).

73. J. C. Idrobo, W. Walkosz, S. F. Yip, S. Ogut, J. L. Wang, and J. Jellinek, Static polarizabilities and optical absorption spectra of gold clusters (Au_n, n = 2–14 and 20) from first principles, *Phys. Rev. B* 76, 205422 (2007).

74. X. J. Wang, X. H. Wan, H. Zhou, S. Takami, M. Kubo, and A. J. Miyamoto, Electronic structures and spectroscopic properties of dimers Cu_2, Ag_2, and Au_2 calculated by density functional theory, *J. Mol. Struct. (THEOCHEM)* 579, 221 (2002).

75. K.-Y. Lian, P. Salek, M. Jin, and D. Ding, Density-functional studies of plasmons in small metal clusters, *J. Chem. Phys.* 130, 174701 (2009).

76. S. Datta, Nanoscale device modeling: The Green's function method, *Superlatt. Microstruct.*, 28, 253 (2000).

77. (a) F. Zahid, M. Paulsson, and S. Datta, Electrical conduction in molecules, In *Advanced Semiconductors and Organic Nano-Techniques*, H. Morkoc (Ed.), Academic Press, New York, 2003. (b) N. Neophytou, S. Ahmed, and G. Klimeck, Non-equilibrium Green's function (NEGF) simulation of metallic carbon nanotubes including vacancy defects, *J. Comput. Electron.* 6, 317 (2007). (c) D. A. Areshkin, O. A. Shenderova, J. D. Schall, S. P. Adiga, and D. W. Brenner, A self-consistent tight binding model for hydrocarbon systems: Application to quantum transport simulation, *J. Phys.: Condens. Matter* 16, 6851 (2004). (d) J. R. Reimers, G. C. Solomon, A. Gagliardi, A. Bilic, N. S. Hush, Th. Frauenheim, A. Di Carlo, and A. Pecchia, The green's function density functional tight binding (gDFTB) method for molecular electronic conduction, *J. Phys. Chem. A* 111, 5692 (2007).

78. C. Gonzalez, Y. Simon-Manso, J. Batteas, M. Marquez, M. Ratner, and V. Mujica, A quasimolecular approach to the conductance of molecule-metal junctions: Theory and application to voltage-induced conductance switching, *J. Phys. Chem. B*, 108, 18414 (2004).

79. B. Aradi, B. Hourahine, and Th. Frauenheim, DFTB+, a sparse matrix-based implementation of the DFTB method, *J. Phys. Chem. A* 111, 5678 (2007). http://www.dftb-plus.info/

80. Gaussian 03, Revision C.02, M. J. Frisch, G. W. Trucks, H. B. Schlegel, G. E. Scuseria, M. A. Robb, et al., Wallingford CT, 2004.

81. M. Elstner, The SCC-DFTB method and its application to biological systems, *Theor. Chem. Acc.* 116, 316 (2006).

82. http://www.netlib.org/blas/

83. http://www.netlib.org/lapack/

84. http://www.netlib.org/linpack/

85. http://z.cs.utexas.edu/wiki/flame.wiki/FrontPage

86. http://developer.amd.com/cpu/Libraries/acml/downloads/pages/default.aspx

87. http://www-03.ibm.com/systems/software/essl/index.html

88. http://software.intel.com/en-us/intel-mkl/

89. http://math-atlas.sourceforge.net/

90. http://www.tacc.utexas.edu/tacc-projects/

91. http://openmp.org/wp/

92. http://www.mcs.anl.gov/research/projects/mpi/

93. http://www.mcs.anl.gov/research/projects/mpich2/

94. http://www.open-mpi.org/

95. S. Balay, K. Buschelman, V. Eijkhout, W. D. Gropp, D. Kaushik, M. G. Knepley, L. Curfman McInnes, B. Smith, and H. Zhang, PETSc Users Manual—Revision 3.1, March 2010. http://www.mcs.anl.gov/petsc/petsc-as/

96. V. Hernandez, J. E. Roman, and V. Vidal, SLEPc: A scalable and flexible toolkit for the solution of eigenvalue problems, *ACM Trans. Math. Softw.* 31, 351 (2005). http://www.grycap.upv.es/slepc/

97. http://www.netlib.org/scalapack

98. L. S. Blackford, J. Choi, A. Cleary, E. D'Azevedo, J. Demmel, I. Dhillon, J. Dongarra, et al., *ScaLAPACK User's Guide*, Society for Industrial and Applied Mathematics, Philadelphia, PA, 1997.

99. C. Vömel, ScaLAPACK's MRRR Algorithm, *ACM Trans. Math. Softw.* 37, 1 (2010).

100. T. A. Davis, *Direct Methods for Sparse Linear Systems*, SIAM, Philadelphia, PA, Sept. 2006. Part of the SIAM Book Series on the Fundamentals of Algorithms.

101. H. Zhang, B. Smith, M. Sternberg, and P. Zapol, SIPs: Shift-and-invert parallel spectral transformations, *ACM Trans. Math. Softw.* 33, 1 (2007).

102. L. Colombo and W. Sawyer, A parallel implementation of tight binding molecular dynamics, SeRD-CSCS Technical Note SeRD-CSCS-TN-95–8.

103. M. Naumov, S. Lee, B. Haley, H. Bae, S. Clark, R. Rahman, H. Ryu, F. Saied, and G. Klimeck, Eigenvalue solvers for atomistic simulations of electronic structures with NEMO-3D, *J. Comput. Electron.* 7, 297 (2008).

104. https://engineering.purdue.edu/gekcogrp/software-projects/nemo3D/

9 Density Functional Calculations of Metal Clusters: Structure, Dynamics, and Reactivity

Dennis R. Salahub, Patrizia Calaminici,
Gabriel U. Gamboa, Andreas M. Köster,
and J. Manuel Vásquez

CONTENTS

9.1   INTRODUCTION

A wide variety of systems, properties, concepts, and perspectives lie within the scope of this volume on *Metallic Systems: A Quantum Chemist's Perspective.* Of course, there is no single "quantum chemist's perspective." To some, a "metallic system" implies typical bulk metallic properties, electrical conductivity, thermal conductivity, and the like. To others, the presence of a single metal atom in a molecule or complex is satisfactory for its classification as a "metallic system." In between, one finds the fascinating world of metal clusters with properties that are sometimes importantly similar to those of the bulk metal and sometimes crucially different. Both perspectives have value. Metal clusters have their own special properties, depending, for example, on cluster size, and there is a clear interest in taking advantage of them to tune the performance for a particular application, for example, catalysis. On the other hand, one can use clusters as models for either bulk systems or larger nano- and micro-clusters such as those present in working catalysts. Of course, the quality of the model will depend crucially on the size of the cluster.

    This chapter will look at metal clusters from the particular perspective of five quantum chemists who, over the years, have studied a variety of metal clusters using density functional theory (DFT) as their primary methodology (indeed a "metallic system," the homogeneous electron gas, lies at the heart of the early development of DFT). A wide variety of metals will be discussed: alkalis, free-electron metals, noble metals, and transition metals (TMs). We will also discuss a wide variety of properties: electronic structure, spectroscopies of various types, geometrical structure, vibrations, dynamics, reactivity, and reactions. The computational demands for both accuracy and speed vary widely depending on the type of metal and the property of interest.

    The performance of various DFT techniques has improved greatly over the years, and we will try to give the reader a sense of this progress through brief descriptions of and references to the methodological principle and software advances that have been made, from the Xα scattered-wave method to linear combination of Gaussian-type orbitals-Xα (LCGTO-Xα), to the use of local density functionals in early versions of the software package deMon, and, finally, to the current version deMon2k, which incorporates the latest generalized gradient approximations (GGAs) and meta-GGA functionals. While the historical perspective is, we think, both interesting and important, the chapter will focus primarily on present capabilities, recent applications, and future perspectives. This overview will be rather personal and focused on the development and applications of deMon; other chapters will provide the perspective of researchers using other software.

We have chosen to organize the chapter in a way that should put the properties front and center. Following the next section, which will define and describe succinctly the methodologies and software of interest, the following sections are each devoted to a particular property, in rough order of difficulty. For each, we will attempt to give a sense of the main hurdles that had to be overcome in previous work, examples of leading-edge research in the current time frame, and a sense of what is needed to extend the horizons even further in the future.

9.2 METHODOLOGY

9.2.1 BASIC DFT

By now, DFT is familiar to the vast majority of theoretical and computational chemists and is used by many hundreds of researchers everyday. Many good monographs and review articles may be consulted [1–7]. We will limit ourselves here to a brief overview of the foundations of DFT and to some of the options available for converting the formalism into practical methodologies and computer codes.

The cornerstone paper of DFT comes from 1964 with the proof by Hohenberg and Kohn [8] that the energy of a nondegenerate ground state is entirely fixed by the electron density. To elaborate this proof, we assume that two different external potentials $v(r)$ and $v'(r)$ can be generated from the same nondegenerate ground-state density $\rho(r)$, which determines the number of electrons in the system by simple quadrature. As a consequence, two different Hamiltonians $\hat{H}$ and $\hat{H}'$ with different eigenfunctions Ψ and Ψ' would exist, where Ψ and Ψ' would yield the same density $\rho(r)$. Taking Ψ' as a trial function for $\hat{H}$, we obtain

$$E_0 < \left\langle \Psi' \mid \hat{H} \, \Psi' \right\rangle = \left\langle \Psi' \mid \hat{H}' \, \Psi' \right\rangle + \left\langle \Psi' \mid \hat{H} - \hat{H}' \Psi' \right\rangle$$

$$= E_0' + \int \rho(r)[v(r) - v'(r)]dr. \tag{9.1}$$

On the other hand, taking Ψ as a trial for $\hat{H}'$ yields

$$E_0' < \left\langle \Psi \mid \hat{H}' \, \Psi \right\rangle = \left\langle \Psi \mid \hat{H} \, \Psi \right\rangle + \left\langle \Psi \mid \hat{H}' - \hat{H} \, \Psi \right\rangle$$

$$= E_0 + \int \rho(r)[v'(r) - v(r)]dr$$

$$= E_0 - \int \rho(r)[v(r) - v'(r)]dr. \tag{9.2}$$

Adding (9.1) and (9.2) yields the contradiction

$$E_0 + E_0' < E_0' + E_0. \tag{9.3}$$

Thus, there is enough information in $\rho(r)$ to fix the total energy through the universal functional $F[\rho(r)]$,

$$E[\rho(r)] = F[\rho(r)] + \int \rho(r) v(r) dr. \tag{9.4}$$

Here, $v(r)$ denotes the external potential of the system, which, in the absence of external fields, is given by the nuclear configuration. The unknown universal functional $F[\rho(r)]$ can further be divided into the kinetic energy $T[\rho(r)]$ and the electron–electron interaction energy $V_{ee}[\rho(r)]$. The Hohenberg-Kohn formalism is exact and includes all the difficulties and subtleties of the many-body problem. The challenge is to formulate useful approximations.

The trail to applications was blazed the following year by Kohn and Sham [9] who proposed a, still exact, reformulation of DFT that introduced the genius idea of a model reference system in which the electrons behave independently of each other. The effective one-electron Hamiltonian of such a noninteracting system is given by

$$\hat{h}_s = -\frac{1}{2} \vec{\nabla}^2 + v_s(r). \tag{9.5}$$

The introduced effective Kohn-Sham potential is chosen such that the density of the noninteracting reference system just matches the ground-state density of the real system. Because of the noninteracting nature of the reference system, it possesses a one-determinantal ground-state wave function

$$\Psi_s = \frac{1}{\sqrt{N!}} \det \left| \varphi_1, \varphi_2, \ldots, \varphi_N \right|. \tag{9.6}$$

With this wave function, the kinetic energy in the reference system can be calculated as the following expectation value

$$T_s[\rho(r)] = -\frac{1}{2} \sum_i \left\langle \varphi_i \mid \vec{\nabla}^2 \varphi_i \right\rangle. \tag{9.7}$$

Of course, $T_s[\rho(r)]$ is not the exact kinetic energy of the real system. However, it accounts for the major part. To proceed further, Kohn and Sham defined a universal functional $F_s[\rho(r)]$ for the noninteracting system very much in the same spirit as in the Hohenberg-Kohn theorem as

$$F_s[\rho(r)] = T_s[\rho(r)] + J_s[\rho(r)] + E_{XC}[\rho(r)]. \tag{9.8}$$

Here, $J[\rho(r)]$ denotes the classical Coulomb repulsion between the electrons, and the new defined exchange-correlation functional $E_{XC}[\rho(r)]$ collects all nonclassical interactions between the electrons, i.e., exchange and correlation, as well as the (small) difference between the exact and the Kohn-Sham kinetic energy. The corresponding Kohn-Sham potential is then given by

KS-DFT Choices

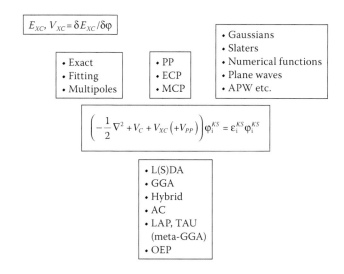

FIGURE 9.1 Kohn-Sham equations and, symbolically, some of the technical options that may be introduced.

$$v_s[\rho(r)] = v(r) + \frac{\delta J[\rho(r)]}{\delta\rho(r)} + \frac{\delta E_{XC}[\rho(r)]}{\delta\rho(r)}. \qquad (9.9)$$

The orbital approach to DFT is the workhorse of the myriad applications of DFT, from condensed matter physics to chemistry, materials science, biology, and other fields. Figure 9.1 gives a symbolic overview of different implementation options for the Kohn-Sham method. Even though it represents an independent particle approach, and the equations are still exact, the difficulty lies in finding good approximate exchange-correlation functionals and the corresponding functional derivatives, the exchange-correlation potentials.

As we will see in the following sections, great progress has been made in finding efficient and accurate algorithms for solving the KS equations and implementing them in user-friendly software suites. DFT has become a major part of mainstream quantum chemistry, and this was reflected in the awarding of the 1998 Nobel Prize in Chemistry to Walter Kohn along with John Pople, reflecting the need for both solid foundations and software that is accessible to essentially all practicing chemists (and also physicists, biologists, materials scientists, etc.).

9.2.2 Xα SCATTERED-WAVE METHOD

The earliest applications of DFT to molecular systems, including metal clusters, were performed in the 1970s, not within the formal DFT framework but, rather, within a parallel line of research that originated in solid-state physics, notably in the school of John C. Slater. As early as 1951, Slater [10] had formulated an "approximate

Hartree-Fock" theory that introduced an orbital-averaged local exchange potential that used what we now call the local density approximation (LDA), based on the homogenous electron gas model. He introduced a parameter, α, which could be adjusted, for example, to reproduce Hartree-Fock exchange energies of atoms. The resulting Xα equations are

$$\left[-\frac{1}{2}\nabla^2 + v(\boldsymbol{r}) + \int \frac{\rho(\boldsymbol{r})'}{|\boldsymbol{r} - \boldsymbol{r}'|} d\boldsymbol{r}' + v_\alpha(\boldsymbol{r}) \right] \varphi_i(\boldsymbol{r}) = \varepsilon_i \varphi_i(\boldsymbol{r}). \tag{9.10}$$

The local Xα potential $v_a(\boldsymbol{r})$ is defined as

$$v_\alpha(\boldsymbol{r}) = -\frac{3}{2}\alpha \left(\frac{3}{\pi}\rho(\boldsymbol{r}) \right)^{1/3}. \tag{9.11}$$

In order to solve these equations for finite molecules, Slater's student, Keith Johnson, adopted the so-called muffin-tin approximation from solid-state band structure methodology. The muffin-tin potential involves taking a spherical average of the potential in spheres centered at each of the nuclei and at an outer sphere that surrounds the whole molecule. In the interstitial volumes, the potential was taken to be constant. With these approximations, partial wave expansions could be used to solve the Xα equations efficiently. This is quite an ugly creature, but it turned out to be of sufficient accuracy to provide reasonably accurate orbitals, orbital energies, and other properties to allow insightful studies of quite complex systems, molecules and clusters that were beyond the reach of traditional quantum chemical methodology. Importantly, in terms of the eventual growth of DFT as a mainstream quantum chemical methodology, the Xα period allowed a first glimmer of encouragement (and much debate and push-back…) that DFT might be a part of the quantum chemistry tool box. A core group of researchers, and their students, became aware of the potential for applications of DFT, and some of them also had a background in traditional quantum chemistry. Eventually, the two worlds would come together.

Although the Xα scattered-wave method yielded remarkable results for many systems, which were beyond the reach of traditional quantum chemistry, it had important limitations, most importantly, indeed, fatally, the inability to optimize geometries and determine structures, which is certainly at the heart of chemical thinking. Overcoming this shortcoming, by bringing in concepts and techniques from traditional quantum chemistry, was the goal of the next leg of the voyage toward modern DFT methodologies and computer codes.

9.2.3 LCGTO-Xα: Bringing (Gaussian) Quantum Chemistry into DFT—Now Geometries Can Be Optimized

The problems with Xα and geometry optimizations were well recognized in the 1970s, including within the "Slater school." Two key contributions allowed enormous steps to be taken toward bringing Xα into a better competitive position with respect to traditional quantum chemistry, while keeping the eventual advantages of the density functional framework, ultimately with good approximations for the exchange-correlation terms.

The first of these got rid of the muffin-tin potential. In a key paper, Dunlap et al. [11] worked out how to bring Gaussian functions into the formalism. With the usual Linear Combinational Atomic Orbit (LCAO) approximations, the Xα equations are cast into matrix form as

$$E = \sum_{\mu,\nu} P_{\mu\nu} H_{\mu\nu} + \frac{1}{2} \sum_{\mu,\nu} \sum_{\sigma,\tau} P_{\mu\nu} P_{\sigma\tau} \langle \mu\nu \| \sigma\tau \rangle + E_{X\alpha}[\rho(r)]. \tag{9.12}$$

Here

$P_{\mu\nu}$ and $P_{\sigma\tau}$ denote elements of the density matrix

$H_{\mu\nu}$ denotes an element of the core Hamiltonian matrix

The atomic orbitals used to expand the molecular orbitals (MOs) are indicated by the Greek letters μ, ν, σ, and τ. The last term represents the Xα approximation to the exchange energy given by

$$E_{X\alpha}[\rho(r)] = -\frac{9}{8}\alpha \left(\frac{3}{11}\right)^{1/3} \int \rho(r)^{4/3} dr. \tag{9.13}$$

For the (contracted) four-center electron repulsion integral (ERI), we have introduced the following short-hand notation:

$$\langle \mu\nu \| \sigma\tau \rangle \equiv \iint \frac{\mu(r')\nu(r')}{|r-r'|} \sigma(r)\tau(r) dr' dr. \tag{9.14}$$

The calculation of these integrals represents the computational bottleneck in the above energy expression. To maintain efficiency, Dunlap et al. incorporated the variational fitting of the Coulomb potential with auxiliary functions [11–13]. It is based on the minimization of the following error:

$$\varepsilon_2 = \frac{1}{2} \iint \frac{[\rho(r)-\tilde{\rho}(r)] - [\rho(r')-\tilde{\rho}(r')]}{|r-r'|} dr' dr. \tag{9.15}$$

In the LCGTO approximation, the orbital density $\rho(r)$ is expanded in terms of atomic orbitals

$$\rho(r) = \sum_{\mu,\nu} P_{\mu\nu}\mu(r)\nu(r). \tag{9.16}$$

For the expansion of the approximated density $\tilde{\rho}(r)$, auxiliary functions are introduced. With these functions, the following linear expansion of the approximated density is obtained:

$$\tilde{\rho}(\boldsymbol{r}) = \sum_k x_k k(\boldsymbol{r}). \qquad (9.17)$$

Several suggestions for the choice of the auxiliary functions $k(\boldsymbol{r})$ exist in the literature [14–18]. In earlier deMon versions, atom-centered primitive Cartesian Gaussian auxiliary functions were used for the expansion of the approximated density. They were grouped together to auxiliary function sets that share the same exponent [15,16].

With the above LCGTO expansion for the density and the approximated density, the minimization of the ε_2 error with respect to the Coulomb fitting coefficients x_k yields

$$\frac{\partial \varepsilon_2}{\partial x_k} = -\sum_{\mu,\nu} P_{\mu\nu} \langle \mu\nu \,\|\, k \rangle + \sum_l x_l \langle l \,\|\, k \rangle \equiv 0 \ \forall k. \qquad (9.18)$$

For convenience, we cast this minimization in an inhomogeneous equation system, $\boldsymbol{G}\,\boldsymbol{x} = \boldsymbol{J}$, with the Coulomb matrix $\boldsymbol{G}$ and the Coulomb vector $\boldsymbol{J}$ containing elements of the form

$$G_{kl} = \langle k \,\|\, l \rangle; J_k = \sum_{\mu,\nu} P_{\mu\nu} \langle \mu\nu \,\|\, k \rangle. \qquad (9.19)$$

It can be shown that $\boldsymbol{G}$ is symmetric and positive definite [19]. Thus, the above equation systems can always be solved, and the corresponding fitting coefficients can be determined. This procedure is nowadays also common in wave-function approaches where it is often called the Resolution of the Identity approximation [20]. With the variational fitting of the Coulomb potential, the computational bottleneck in Xα methods shifted to the calculation of the Xα potential. For this reason, a second auxiliary function set was introduced to fit this potential, too. However, the fitting of the Xα potential (and later of the exchange-correlation potential) as suggested by Sambe and Felton [21] represents a least-squares fit on a grid. As a result, the corresponding energy is not variational. Nevertheless, approximate gradients were now available, and, therefore, structure optimizations became possible.

9.2.4 HISTORY OF DEMON

The deMon program system was developed for DFT calculations of atoms, molecules, and solids. Its first widely available version appeared in 1992. It was based on the PhD work of A. St-Amant performed under the supervision of D.R. Salahub at the Université de Montréal [22]. The name *deMon* stands for "*de*nsity of *Mon*treal." Shortly after its first appearance, the original deMon code was substantially modified for commercialization by BIOSYM Technologies. The beta-release of this version appeared in 1993. It also served as the basis for the deMon-KS series of programs developed in Montreal until 1997 [23]. The focus of this program development was mainly toward molecular property calculations with LCGTO DFT methods. Remarkable progress was achieved for the calculation of nuclear magnetic shielding tensors [24] and excited states via time-dependent density functional theory (TD-DFT) [25]. The calculation of many one-electron properties like electrostatic potentials [26], nuclear quadrupole coupling constants [27], and others was realized, too. A drawback of the first deMon-KS release, deMon-KS1, was its rather bad self-consistent field (SCF) convergence. At that time, it was not obvious if this was due only to technical problems or if the fitting of the Coulomb and exchange-correlation potential represented a fundamental problem for the SCF convergence. However, it had already been proven that the energy gradients were not variational due to the least-squares fit of the exchange-correlation potential according to Sambe and Felton [28–30].

Triggered by these concerns, the original deMon-KS version was further developed in Montpellier by A. Goursot and in Stockholm by L.G.M. Pettersson. A major achievement was the implementation of accurate numerical integration algorithms for the calculation of exchange-correlation potentials. In this way, the least-squares fit could be avoided. However, the computational demands associated with the numerical integration algorithm made this approach impractical for routine applications. Nevertheless, the variational nature of the resulting gradients resolved some of the basic problems encountered with deMon-KS1. At the beginning, the Montpellier and Stockholm efforts were independent from each other. In 1997, they merged into the deMon-KS3 program version. In cooperation with K. Hermann from the Fritz-Haber Institut in Berlin, the deMon branch from Stockholm was further modified for the calculation of core-level spectra. It is distributed under the name StoBe [31] and is particularly well suited for the calculation of core-level spectra.

Independently from the deMon development, the AllChem [32] project was started by the group of A.M. Köster in Hannover in 1995. The aim of this project was to write a well-structured DFT code for the further study of auxiliary density approaches from scratch. The first AllChem calculations appeared in 1997 in the literature. The structured programming of AllChem proved very useful for the development and testing of new auxiliary function DFT approaches and algorithms. On the other hand, AllChem was used more and more for standard computational chemistry applications. However, the primitive input structure made the program use rather cumbersome and prone to input errors.

Based upon an initiative of D.R. Salahub, A. Goursot, and A.M. Köster, the first deMon developers meeting was held in Ottawa in March 2000. At this meeting, the deMon and AllChem developers agreed to merge their codes in order to keep a Tower of Babel from rising. As a result, a new code that couples the deMon functionality and input utilities with the stable and efficient integral and SCF branch from AllChem was developed. This code, which was named deMon2k [33], was presented for the first time at the third deMon developers meeting in Geneva in 2002. Over the next 3 years, the beta version of this code was distributed and tested within the deMon developers community. The annual meetings of this community at the deMon developers workshops were used to discuss the code structure and to exchange algorithmic developments.

Since the sixth deMon developers workshop in 2005 in Dresden, deMon2k has been distributed to the public under a license agreement. The 2.2.6 release served originally as distributed version, which was later on superceded by the 2.3.1 release. Besides the standard LCGTO-DFT implementation of previous deMon versions, the 2.2.6 release contained the first stable implementation of auxiliary density functional theory (ADFT) [34] that permits calculations of large systems with modest computational resources. With the 2.3.1 release, a parallel ADFT SCF module was realized on the basis of MPI [35]. This version is now used in several research laboratories all around the world and has served as a development platform over the last 3 years. Today, ADFT is accepted as a stable and efficient alternative to conventional Kohn-Sham DFT calculations due to the increasing number of deMon2k ADFT calculations.

In February 2009, the ninth deMon developers workshop was held in Pune, India. At this workshop, the new 3.0 release of deMon2k was presented. It is a systematic further development of the 2.3.1 version. Major improvements are the implementation of an iterative solution for the fitting equations [36], of auxiliary density perturbation theory (ADPT) [37], of a hierarchical transition state finder [38], and of an extended Born-Oppenheimer molecular dynamics (BOMD) module [39,40]. Based on the demand from the deMon2k user community, we have also extended the molecular property section of the program. The new version contains internal modules for TD-DFT calculations from the group of M.E. Casida [41], nuclear magnetic resonance shielding tensor calculations employing gauge invariant atomic orbital basis sets and polarizability as well as hyperpolarizability calculations. The electronic structure analysis module has also been extended, including atoms in molecules [42] and natural bond orbital [43] analysis tools. A standardized interface to CHARMM for Quantum Mechanical/Molecular Mechanical (QM/MM) [44] calculations is available.

9.2.5 DEMON2K

An essential feature of all deMon programs is the use of auxiliary functions to enhance computational efficiency. In deMon2k, primitive Hermite Gaussian auxiliary functions are employed. Because the recurrence relations for the calculation of three-center ERIs are formulated over Hermite Gaussians, the use of Hermite Gaussian auxiliary functions considerably improves the efficiency of the ERI calculations [17,45]. The primitive Hermite Gaussian auxiliary functions are indicated by a bar. An (unnormalized) Hermite Gaussian auxiliary function centered at the atom K with exponent ζ_k has the form

$$\bar{k}(\mathbf{r}) = \left(\frac{\partial}{\partial K_x}\right)^{\bar{k}_x} \left(\frac{\partial}{\partial K_y}\right)^{\bar{k}_y} \left(\frac{\partial}{\partial K_z}\right)^{\bar{k}_z} e^{-\zeta_k(\mathbf{r}-\mathbf{K})^2} \tag{9.20}$$

Here, $\bar{k}_x$, $\bar{k}_y$, and $\bar{k}_z$ denote the angular indices of the auxiliary functions. As in earlier deMon versions, the Hermite Gaussian auxiliary functions are grouped into sets with common exponents in deMon2k. The default auxiliary function set A2 [46] possesses s and d sets. An s set contains only one s-type auxiliary function whereas a d set contains one s-, three p-, and six d-type auxiliary functions. More recently, the automatic generation of auxiliary function sets was introduced in deMon2k. In this procedure, the exponents of the auxiliary function sets are generated according to the exponent range spanned by the given basis set. These auxiliary function sets are named GEN-AX with X=2, 3, and 4. A larger X indicates that more auxiliary function sets are generated. The GEN-A2 set is for the DFT-optimized double zeta plus valence polarization (DZVP) basis set identical with the A2 auxiliary function set. Again, GEN-A2 auxiliary function sets contain only s and d sets. As an alternative, GEN-AX* sets (X=2, 3, 4) are also available in deMon2k. These sets possess the same exponents as the corresponding GEN-AX sets but also contain f and g auxiliary function sets. For a more detailed description of the GEN-AX and GEN-AX* auxiliary function sets, we refer the interested reader to the literature [47,48].

9.2.5.1 Auxiliary Density Functional Theory

As an alternative to the above-described least-squares fitting of the exchange-correlation potential, the direct use of the variationally fitted density for the calculation of the exchange-correlation energy has been investigated over the last decade [34,49]. The corresponding ADFT energy expression has the form [34]

$$E = \sum_{\mu,\nu} P_{\mu\nu} H_{\mu\nu} + \frac{1}{2}\sum_{\mu,\nu} P_{\mu\nu}\left\langle \mu\nu \,\|\, \bar{k} \right\rangle x_{\bar{k}} - \frac{1}{2}\sum_{\bar{k},\bar{l}} x_{\bar{k}} x_{\bar{l}} \left\langle \bar{k} \,\|\, \bar{l} \right\rangle + E_{XC}[\tilde{\rho}(\boldsymbol{r})]. \quad (9.21)$$

Here

μ and ν denote (contracted) atomic GTOs

$\bar{k}$ and $\bar{l}$ denote primitive Hermite Gaussian auxiliary functions

The first term represents, as usual, the one-electron core energy. The next two terms result from the variational fitting of the Coulomb potential and substitute the four-center ERI term in standard Kohn-Sham energy expressions. The last term denotes the exchange-correlation energy expression of the selected functional. It is calculated using the linear scaling auxiliary density $\tilde{\rho}(r)$. The derivatives of this energy expression with respect to the density matrix elements, keeping the fitting coefficients constant, define the corresponding ADFT Kohn-Sham matrix elements

$$K_{\mu\nu} = \frac{\partial E}{\partial P_{\mu\nu}} = H_{\mu\nu} + \sum_{\bar{k}}\left\langle \mu\nu \,\|\, \bar{k} \right\rangle x_{\bar{k}} + \frac{\partial E_{XC}[\tilde{\rho}(\boldsymbol{r})]}{\partial P_{\mu\nu}}. \quad (9.22)$$

The partial derivative of the (local) exchange-correlation energy functional with respect to the density matrix elements yields [50]

$$\frac{\partial E_{XC}[\tilde{\rho}(\boldsymbol{r})]}{\partial P_{\mu\nu}} = \int \frac{\delta E_{XC}[\tilde{\rho}(\boldsymbol{r})]}{\delta \tilde{\rho}(\boldsymbol{r})} \frac{\partial \tilde{\rho}(\boldsymbol{r})}{\partial P_{\mu\nu}} d\boldsymbol{r}. \quad (9.23)$$

As in standard Kohn-Sham DFT, the functional derivative of the exchange-correlation energy with respect to the approximated density defines the exchange-correlation potential in ADFT

$$v_{XC}[\tilde{\rho}(r)] \equiv \frac{\delta E_{XC}[\tilde{\rho}(r)]}{\delta \tilde{\rho}(r)}. \tag{9.24}$$

Due to the variational nature of the approximated density, the partial derivatives with respect to the density matrix elements can be found as

$$\frac{\partial \tilde{\rho}(r)}{\partial P_{\mu\nu}} = \sum_{\bar{k}} \frac{\partial x_{\bar{k}}}{\partial P_{\mu\nu}} \bar{k}(r). \tag{9.25}$$

With the definition of $x_{\bar{k}}$ from the variational fitting,

$$x_{\bar{k}} = \sum_{\mu,\nu} \sum_{\bar{l}} G_{\bar{k}\bar{l}}^{-1} \langle \bar{l} \, \| \, \mu\nu \rangle P_{\mu\nu}, \tag{9.26}$$

it follows for the derivative of the fitting coefficients

$$\frac{\partial x_{\bar{k}}}{\partial P_{\mu\nu}} = \sum_{\bar{l}} G_{\bar{k}\bar{l}}^{-1} \langle \bar{l} \, \| \, \mu\nu \rangle. \tag{9.27}$$

Inserting this expression into the above derivative of the approximated density yields

$$\frac{\partial \tilde{\rho}(r)}{\partial P_{\mu\nu}} = \sum_{\bar{k},\bar{l}} \langle \mu\nu \, \| \, \bar{l} \rangle G_{\bar{l}\bar{k}}^{-1} \bar{k}(r). \tag{9.28}$$

Thus, we find for the partial derivative of the (local) exchange-correlation energy functional with respect to the density matrix elements

$$\frac{\partial E_{XC}[\tilde{\rho}(r)]}{\partial P_{\mu\nu}} = \sum_{\bar{k},\bar{l}} \langle \mu\nu \, \| \, \bar{l} \rangle G_{\bar{l}\bar{k}}^{-1} \int \bar{k}(r) v_{XC}[\tilde{\rho}(r)] dr$$

$$= \sum_{\bar{k},\bar{l}} \langle \mu\nu \, \| \, \bar{l} \rangle G_{\bar{l}\bar{k}}^{-1} \langle \bar{k} \, | \, v_{XC}[\tilde{\rho}] \rangle. \tag{9.29}$$

For convenience in the notation, we now introduce a new set of fitting coefficients, which we name exchange-correlation fitting coefficients, as

$$z_{\bar{k}} \equiv \sum_{\bar{l}} G_{\bar{k}\bar{l}}^{-1} \langle \bar{l} \, | \, v_{XC}[\tilde{\rho}] \rangle. \tag{9.30}$$

The ADFT Kohn-Sham matrix elements are then given by

$$K_{\mu\nu} = \frac{\partial E}{\partial P_{\mu\nu}} = H_{\mu\nu} + \sum_{\bar{k}} \langle \mu\nu \, \| \, \bar{k} \rangle \left(x_{\bar{k}} + z_{\bar{k}} \right). \tag{9.31}$$

From this expression, we note immediately that the Kohn-Sham matrix elements in ADFT are independent from density matrix elements. As a result, only the approximated density (and the corresponding density derivatives in the case of gradient corrected functional) is numerically calculated on a grid. Because these quantities are linear scaling by construction, the necessary grid work is considerably reduced. Moreover, the sharing of auxiliary function exponents by the above discussed auxiliary function sets decreases dramatically the expensive exponential function evaluations at each grid point. Besides the efficient calculation of the three-center ERIs, the simplified grid work is the main reason for the computational efficiency of the ADFT approach. In fact, the calculation of the Kohn-Sham potential in ADFT is identical to orbital-free DFT approaches with the auxiliary function density as the basic variable. Of course, Kohn-Sham orbitals are still used in ADFT for the calculation of the kinetic energy contribution and the orbital density, which is the seed for the auxiliary density via the variational Coulomb fitting. Thus, ADFT might be seen as a combination of the basic ideas from the conventional Kohn-Sham methodology with those from orbital-free DFT. As the above derivation of the Kohn-Sham matrix as a partial derivative of the ADFT energy expression demonstrates analytic energy derivatives can be formulated straightforwardly in ADFT. This is a major difference to other approaches that use least-squares fitting techniques for the approximate calculation of the exchange-correlation energy and potential. Of course, this holds for analytic gradients too. For higher energy derivatives, the so-called coupled perturbed Kohn-Sham equations that are close analogs of the coupled perturbed Hartree-Fock equations are usually employed to solve the necessary systems of equations for the perturbed MO coefficients or density matrix elements. In order to keep the necessary memory manageable, iterative procedures are usually employed. In the next section, we describe an alternative procedure for the calculation of the perturbed density matrix elements that is based on the ADFT formalism discussed here.

9.2.5.2 Auxiliary Density Perturbation Theory

For the derivation of ADPT, we start from McWeeny's self-consistent perturbation (SCP) theory [51–56], which represents a direct approach for the calculation of the perturbed density matrix. For clarity of derivation, we restrict ourselves to perturbation-independent basis and auxiliary functions as commonly used in polarizability calculations. Under these restrictions, the perturbed closed-shell density matrix is given by the SCP formalism as [56]

$$P_{\mu\nu}^{(\lambda)} \equiv \frac{\partial P_{\mu\nu}}{\partial \lambda} = 2 \sum_i^{occ} \sum_a^{uno} \frac{K_{ia}^{(\lambda)}}{\varepsilon_i - \varepsilon_a} (c_{\mu i} c_{\nu a} + c_{\mu a} c_{\nu i}). \tag{9.32}$$

In this notation, λ denotes a perturbation parameter (e.g., an external field component in the case of polarizability calculations), and the indices i and a denote occupied (occ) and unoccupied (uno) MOs, respectively. The perturbed Kohn-Sham matrix in the MO basis is given by

$$K_{ia}^{(\lambda)} = \sum_{\mu,\nu} c_{\mu i} c_{\nu a} K_{\mu\nu}^{(\lambda)}. \tag{9.33}$$

Based on the above defined ADFT Kohn-Sham matrix, we find for the perturbed Kohn-Sham matrix

$$K_{\mu\nu}^{(\lambda)} \equiv \frac{\partial K_{\mu\nu}}{\partial \lambda} = H_{\mu\nu}^{(\lambda)} + \sum_{\bar{k}} \left\langle \mu\nu \,\|\, \bar{k} \right\rangle \left(x_{\bar{k}}^{(\lambda)} + z_{\bar{k}}^{(\lambda)} \right). \tag{9.34}$$

The perturbed exchange-correlation fitting coefficients are given by

$$z_{\bar{k}}^{(\lambda)} = \sum_{\bar{l}} G_{kl}^{-1} \left\langle \bar{l} \,\mid\, v_{XC}^{(\lambda)}[\tilde{\rho}] \right\rangle. \tag{9.35}$$

Since $\upsilon_{XC}[\tilde{\rho}]$ is a (local) functional of the (approximated) density, it follows

$$\left\langle \bar{l} \,\mid\, v_{XC}^{(\lambda)}[\tilde{\rho}] \right\rangle = \iint \bar{l}(\mathbf{r}) \frac{\delta v_{XC}[\tilde{\rho}(\mathbf{r})]}{\delta\tilde{\rho}(\mathbf{r}')} \frac{\partial\tilde{\rho}(\mathbf{r}')}{\partial\lambda} \, d\mathbf{r}\, d\mathbf{r}'$$

$$= \sum_{\bar{m}} \left\langle \bar{l} \,\middle|\, f_{XC}[\tilde{\rho}] \,\middle|\, \bar{m} \right\rangle x_{\bar{m}}^{(\lambda)}. \tag{9.36}$$

Compared to the standard LCGTO kernel integral $\langle \mu\nu | f_{XC}[\rho] | \sigma\tau \rangle$, the scaling of the ADPT kernel integral $\langle \bar{l} | f_{XC}[\tilde{\rho}] | \bar{m} \rangle$ is reduced by almost two orders of magnitude. The newly appearing exchange-correlation kernel $f_{XC}[\tilde{\rho}]$ is defined as the second functional derivative of the exchange-correlation energy

$$f_{XC}[\tilde{\rho}(\mathbf{r}),\tilde{\rho}(\mathbf{r}')] \equiv \frac{\delta^2 E_{XC}[\tilde{\rho}]}{\delta\tilde{\rho}(\mathbf{r})\delta\tilde{\rho}(\mathbf{r}')}. \tag{9.37}$$

For pure density functionals, the arguments of the approximated densities are collapsed and thus we find

$$f_{XC}[\tilde{\rho}(\mathbf{r}),\tilde{\rho}(\mathbf{r}')] \equiv \frac{\delta^2 E_{XC}[\tilde{\rho}]}{\delta\tilde{\rho}(\mathbf{r})\delta\tilde{\rho}(\mathbf{r}')}\delta(\mathbf{r}-\mathbf{r}') = \frac{\delta^2 E_{XC}[\tilde{\rho}]}{\delta\tilde{\rho}(\mathbf{r})^2} = \frac{\delta v_{XC}[\tilde{\rho}]}{\delta\tilde{\rho}(\mathbf{r})}. \tag{9.38}$$

With the explicit form for the perturbed exchange-correlation fitting coefficients, we can rewrite the perturbed Kohn-Sham matrix in terms of perturbed Coulomb fitting coefficients only

$$K_{\mu\nu}^{(\lambda)} = H_{\mu\nu}^{(\lambda)} + \sum_{\bar{k}} \left\langle \mu\nu \,\|\, \bar{k} \right\rangle \left(x_{\bar{k}}^{(\lambda)} + \sum_{\bar{k},\bar{l}} \left\langle \mu\nu \,\|\, \bar{k} \right\rangle F_{\bar{k}\bar{l}} x_{\bar{l}}^{(\lambda)} \right), \tag{9.39}$$

with

$$F_{\bar{k}\bar{l}} = \sum_{\bar{m}} G_{\bar{k}\bar{m}}^{-1} \left\langle \bar{m} \mid f_{XC}[\tilde{\rho}] \mid \bar{l} \right\rangle. \tag{9.40}$$

With the above expression for the perturbed Kohn-Sham matrix, we find for the SCP density matrix

$$P_{\mu\nu}^{(\lambda)} = 2\sum_{i}^{occ}\sum_{a}^{uno} \frac{H_{ia}^{(\lambda)}}{\varepsilon_i - \varepsilon_a} \left(c_{\mu i}c_{\nu a} + c_{\mu a}c_{\nu i} \right)$$

$$+ 2\sum_{i}^{occ}\sum_{a}^{uno}\sum_{\bar{k}} \frac{\left\langle \mu\nu \parallel \bar{k} \right\rangle x_{\bar{k}}^{(\lambda)}}{\varepsilon_i - \varepsilon_a} \left(c_{\mu i}c_{\nu a} + c_{\mu a}c_{\nu i} \right)$$

$$+ 2\sum_{i}^{occ}\sum_{a}^{uno}\sum_{\bar{k}\bar{l}} \frac{\left\langle \mu\nu \parallel \bar{k} \right\rangle F_{\bar{k}\bar{l}} x_{\bar{l}}^{(\lambda)}}{\varepsilon_i - \varepsilon_a} \left(c_{\mu i}c_{\nu a} + c_{\mu a}c_{\nu i} \right). \tag{9.41}$$

On the other hand, the derivative of the fitting equation system with respect to the perturbation parameter λ, assuming perturbation-independent basis and auxiliary function sets, yields

$$\sum_{k} G_{\bar{m}\bar{k}} x_{\bar{k}}^{(\lambda)} = \sum_{\mu,\nu} P_{\mu\nu}^{(\lambda)} \left\langle \mu\nu \parallel \bar{m} \right\rangle. \tag{9.42}$$

Inserting the above expression for $P_{\mu\nu}^{(\lambda)}$ into the perturbed fitting equation system, we obtain a system of equations for the perturbed Coulomb fitting coefficients only [37]

$$\sum_{\bar{k}} G_{\bar{m}\bar{k}} x_{\bar{k}}^{(\lambda)} = 4\sum_{i}^{occ}\sum_{a}^{uno} \frac{\left\langle \bar{m} \parallel ia \right\rangle H_{ia}^{(\lambda)}}{\varepsilon_i - \varepsilon_a}$$

$$+ 4\sum_{i}^{occ}\sum_{a}^{uno}\sum_{k} \frac{\left\langle \bar{m} \parallel ia \right\rangle \left\langle ia \parallel \bar{k} \right\rangle}{\varepsilon_i - \varepsilon_a} x_{\bar{k}}^{(\lambda)}$$

$$+ 4\sum_{i}^{occ}\sum_{a}^{uno}\sum_{\bar{k}} \frac{\left\langle \bar{m} \parallel ia \right\rangle \left\langle ia \parallel \bar{k} \right\rangle}{\varepsilon_i - \varepsilon_a} F_{\bar{k}\bar{l}} x_{\bar{l}}^{(\lambda)}. \tag{9.43}$$

In order to simplify the further notation, we now introduce the Coulomb and exchange-correlation coupling matrices A and B, as well as the perturbation vector $b^{(\lambda)}$. Their elements are defined as

$$A_{\bar{k}\bar{l}} \equiv \sum_{i}^{occ}\sum_{a}^{uno} \frac{\left\langle \bar{k} \parallel ia \right\rangle \left\langle ia \parallel \bar{l} \right\rangle}{\varepsilon_i - \varepsilon_a}, \tag{9.44}$$

$$B_{\bar{k}\bar{l}} \equiv \sum_i^{occ} \sum_a^{uno} \sum_{\bar{m},\bar{n}} \frac{\langle \bar{k} \| ia \rangle \langle ia \| \bar{m} \rangle}{\varepsilon_i - \varepsilon_a} G_{\bar{m}\bar{n}}^{-1} \langle \bar{n} | f_{XC}[\tilde{\rho}] | \bar{l} \rangle = \sum_{\bar{m}} A_{\bar{k}\bar{m}} F_{\bar{m}\bar{l}}, \quad (9.45)$$

$$b_{\bar{k}}^{(\lambda)} \equiv \sum_i^{occ} \sum_a^{uno} \frac{\langle \bar{k} \| ia \rangle H_{ia}^{(\lambda)}}{\varepsilon_i - \varepsilon_a}. \quad (9.46)$$

With these quantities, Equation 9.43 can be recast into

$$\left(G - 4A - 4b \right) x^{(\lambda)} = 4b^{(\lambda)} \Leftrightarrow x^{(\lambda)} = \left(\frac{1}{4} G - A - B \right)^{-1} b^{(\lambda)}. \quad (9.47)$$

Thus, the calculation of the perturbed fitting coefficients boils down to the solution of the above inhomogeneous equation system, which possesses the dimension of the auxiliary functions. A direct solution is feasible even for large systems. With the perturbed fitting coefficients, the perturbed Kohn-Sham matrix elements can be calculated via Equation 9.34. The perturbed density matrix elements are then obtained by the corresponding SCP equation (9.32). Therefore, the final result of an ADPT calculation is identical to the result of a coupled perturbed Kohn-Sham calculation. The major difference is the improved computational performance of ADPT calculations due to the use of the perturbed auxiliary density in the intermediate steps.

9.2.5.3 Structure Optimization and Transition State Search

Central to the theoretical description of cluster structures is the Born-Oppenheimer approximation [57], which enables the separation of electronic and nuclear motions. As a consequence, a potential energy can be assigned to each cluster structure. The resulting potential energy surface (PES) possesses various critical points. Only minima and first-order saddle points, i.e., points that possess one negative curvature and all other curvatures positive, are of interest to us. Because the PES is constructed from solutions of the electronic Schrödinger equation, it is usually not available in analytic form. This situation contributes considerably to the challenge of the theoretical study of metal clusters. Here, theoreticians face the formidable task of locating minima on the PES without any further information except the number of atoms in the metal cluster. Thus, it does not come as a surprise that (metal) cluster science presents a strong driving force for the development of local and global optimization algorithms. Despite considerable effort, global minimization algorithms [58–61] for first principles electronic structure methods are still at a very crude developmental stage. Thus, for metal clusters, local optimization algorithms based on quasi-Newton [62] restricted step methods [63] are most often employed. Here, the maximum step length is restricted to a trust region h. Within the quadratic approximation, the minimization problem then reads

$$\min_p q(p) \text{ subject to } | p | \leq h. \quad (9.48)$$

The trust region is a hypersphere defined by $|p| \leq h$, where the scalar $h > 0$ is called the trust region radius. The model function q is defined by the second-order Taylor series expansion around the local origin

$$q(p) = E + g^T p + \frac{1}{2} p^T B p, \tag{9.49}$$

where
    E is the value of the objective function, here the energy
    g is the gradient vector
    B is the corresponding Hessian matrix or an approximation to it, each of them evaluated at the expansion point

If the solution p of Equation 9.48 does not produce a sufficient decrease in the energy, the trust region is too large and needs to be reduced. After the reduction of the trust region radius, Equation 9.48 is solved again. This procedure is repeated until an acceptable decrease in the energy is achieved.

In order to solve (9.48), the following Lagrange functional is introduced

$$L(p,\lambda) = E + g^T p + \frac{1}{2} p^T B p + \frac{1}{2} \lambda \left(p^T p - h^2 \right). \tag{9.50}$$

Here, λ represents the undefined Lagrange multiplier. From the stationary condition of this Lagrange function, we find the step direction as

$$p = -(B + \lambda I)^{-1} g. \tag{9.51}$$

For the determination of the Lagrange multiplier λ, a diagonal representation is used [63–66]. In our experience, the restricted step algorithm (RSA) is very well suited for the local optimization of metal clusters. Nevertheless, it must be kept in mind that minimization is only guaranteed in an RSA with an exact Hessian calculation. However, in most cases, such an approach is not computationally efficient. For this reason, we usually apply a quasi-Newton RSA followed by a frequency analysis. In case imaginary frequencies are detected, the quasi-Newton RSA is restarted with the initial Hessian from the frequency analysis. Even in complicated cases, such an approach usually converges to minima structures within one or two restart runs [67]. For metal clusters, it is often advisable to use a level-shift [68] during the structure optimization. For this purpose, a dynamical level-shift procedure is implemented in deMon2k. It reduces the shift factor according to the SCF convergence. This approach improves considerably the SCF convergence behavior of metal systems far away from local minima without jeopardizing the description of the electronic structure.

By and large, the local optimization of stable metal cluster isomers has become a routine procedure with the above-described quasi-Newton RSA. However, the corresponding optimization of transition states still involves many challenges. For this reason, we recently developed a hierarchical transition state finder [38]. It consists of a double-ended saddle interpolation method [69], which is followed by an uphill trust region optimization [70,71]. To enhance the stability of the saddle interpolation method, we included the distance constraint between reactants and products by an undefined Lagrange multiplier in the quadratic model function rather than using modified gradients as suggested elsewhere in the literature [72]. The resulting formulas closely resemble their trust region counterparts. Thus, the developed algorithms for the solution of the trust region subproblem can also be applied to the double-ended saddle interpolation method. In the case of metal clusters, chemical intuition most often fails for the description of cluster rearrangements. For this reason, the hierarchical transition state finder represents significant progress because it needs only reactant and product structures. However, these structures must be aligned to maximum coincidence in order to guarantee the success of the hierarchical transition state finder. For this reason, an automatic alignment procedure for reactant and product structures [73] was implemented in deMon2k.

The basic idea of this automatic alignment is the selection of equivalent atoms according to the best alignment of both, reactant and product molecules. In the literature, several methods are suggested [74–78]. However, to the best of our knowledge, all of them suffer from a labeling problem. In other words, they cannot provide a unique measurement of the distance between two different structures with arbitrary atomic labeling. For this reason, we developed a new superimposing algorithm, which is capable of performing such measurements [73].

For this purpose, we represent a molecule by a set of N vectors r_A containing the spatial coordinates of N nuclei. In our tests, better results in the alignment of reactants and products were achieved using mass-weighted coordinates instead of simple coordinates, i.e., when r_A is replaced by $m_A r_A$. Defining $R_A = m_A r_A$, the two sets of vectors that represent the molecules are

$$X = \{m_1 r_1, m_2 r_2, m_3 r_3, \ldots, m_N r_N\} = \{R_1, R_2, R_3, \ldots, R_N\},$$
$$X' = \{m'_1 r'_1, m'_2 r'_2, m'_3 r'_3, \ldots, m'_N r'_N\} = \{R'_1, R'_2, R'_3, \ldots, R'_N\}. \tag{9.52}$$

The definition of the metric used to measure the separation between two structures is well known from the crystallographic literature [79]

$$d^2 = \left|X' - X\right|^2 = \sum_{A=1}^{N} \left|R'_A - R_A\right|^2. \tag{9.53}$$

At first glance, it seems easy to superimpose two molecules once the metric definition is given. However, this is not the case because the order of X and X' elements that minimizes the distance in Equation 9.53 is unknown. Fortunately, the task of determining the optimum assignment between the elements of two sets is a well-studied problem in linear optimization theory [80]. Kuhn and Munkres [81,82] developed a specialized algorithm, related to the simplex method, for the solution of the general assignment problem. This method is known as the Hungarian algorithm, which is more efficient than the simplex method and can be used without modification in our problem. The input for the algorithm is the so-called cost matrix F whose elements are given in our case by

$$F_{AB} = \left| R'_B - R_A \right|^2.$$ (9.54)

And the output is the mapping between the elements of X and X'

$$a(A) = a_A, \ a_A \in \{1, 2, 3, \ldots, N\}, \ a_A \neq a_B \quad \text{if } A \neq B,$$ (9.55)

which minimizes the total cost

$$\sum_{A=1}^{N} F_{A,a(A)} \sum_{A=1}^{N} \left| R'_{a(A)} - R_A \right|^2 = d^2.$$ (9.56)

Because only the same kind of atoms can be matched, the restriction $m'_{a(i)} = m_i$ holds here. This splits the set of coordinates into classes with the same mass. The assignment algorithm is applied separately to each of these classes. To simplify the notation, the operator $\hat{a}$ is defined by its ordering action on the X' elements as follows:

$$\hat{a}X' = \{R'_{a(1)}, R'_{a(2)}, R'_{a(3)}, \ldots, R'_{a(N)}\}.$$ (9.57)

Now, the distance associated with the mapping $\hat{a}$ is

$$d^2 = \left| \hat{a}X' - X \right|^2 = \sum_{A=1}^{N} \left| R'_{a(A)} - R_A \right|^2.$$ (9.58)

It is important to realize that the mapping $\hat{a}$, while best for the two sets of coordinates X and X', is not necessarily the overall best alignment of the molecules, because external translations and rotations can be applied to any of the two sets of coordinates. Thus, the mapping found using a particular orientation is the best only in the neighborhood of this orientation. An example of this situation is illustrated in Figure 9.2, where two two-dimensional molecules consisting of five atoms in different orientations are shown. The first molecule is identified with empty circles whereas the second molecule is identified with solid circles. As Figure 9.2 shows, in the first orientation (left side of Figure 9.2), the optimum mapping, indicated by connection lines, is different from the mapping given in the second orientation (right side of Figure 9.2), where the second molecule is rotated by 120° with respect to the first molecule that is maintained fixed. By rotating the second molecule by 90° with respect to the first one, the perfect superposition of the two molecules is reached, as can be anticipated from the left side of Figure 9.2.

With this discussion in mind, we have implemented the following alignment algorithm in deMon2k. Because the translation that minimizes the distance given by

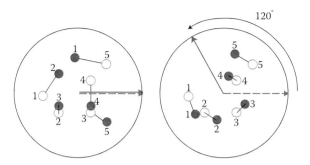

FIGURE 9.2 Two-dimensional example of mappings with different orientations of the molecule. (From Vasquez-Perez, J. et al., *J. Chem. Phys.*, 131(12), 124126-1. With permission.)

Equation 9.53 is known analytically and it is independent of the mapping used, we apply it first. Therefore, the two sets of coordinates are transformed initially as

$$X = \{R_1, R_2, R_3, ..., R_N\} - R_{CM},$$
$$X' = \{R_1', R_2', R_3', ..., R_N'\} - R_{CM},$$

(9.59)

with

$$R_{CM} = \frac{1}{M} \sum_{A=1}^{N} R_A \quad \text{and} \quad M = \sum_{A=1}^{N} m_A.$$

In these equations, R_{CM} is the center of mass. On the other hand, the optimum rotation that minimizes Equation 9.53 is mapping dependent, but can easily be calculated in a semi-analytical fashion using the quaternion representation of rotations. In this representation, the optimum rotation is given by the eigenvector $q = \{q_1, q_2, q_3, q_4\}$ associated with the lowest eigenvalue of the matrix [80]

$$
\begin{pmatrix}
\sum x_m^2 + y_m^2 + z_m^2 & \sum y_p z_m - y_m z_p & \sum x_m z_p - x_p z_m & \sum x_p y_m - x_m y_p \\
\sum y_p z_m - y_m z_p & \sum y_p^2 + z_p^2 + x_m^2 & \sum x_m y_m - x_p y_p & \sum x_m z_m - x_p z_p \\
\sum x_m z_p - x_p z_m & \sum x_m y_m - x_p y_p & \sum x_p^2 + z_p^2 + y_m^2 & \sum y_m z_m - y_p z_p \\
\sum x_p y_m - x_m y_p & \sum x_m z_m - x_p z_p & \sum y_m z_m - y_p z_p & \sum x_p^2 + y_p^2 + z_m^2
\end{pmatrix}
$$

(9.60)

Here, $(x_m, y_m, z_m) = R - R'$ and $(x_p, y_p, z_p) = R + R'$. The corresponding rotation matrix to q is found as [63,79]

$$\hat{Q} = \begin{pmatrix} q_1^2 + q_2^2 + q_3^2 + q_4^2 & 2(q_2q_3 + q_1q_4) & 2(q_2q_4 - q_1q_3) \\ 2(q_2q_3 + q_1q_4) & q_1^2 + q_3^2 + q_2^2 + q_4^2 & 2(q_3q_4 + q_1q_2) \\ 2(q_2q_4 + q_1q_3) & 2(q_3q_4 + q_1q_2) & q_1^2 + q_4^2 + q_2^2 + q_3^2 \end{pmatrix}. \tag{9.61}$$

The orientation associated with this rotation gives the least distance between the molecules, but it depends on the mapping used:

$$d_{min}^2 = \left| \hat{Q}\,\hat{a}\,X' - X \right|^2 = \sum_{A=1}^{N} \left| \hat{Q}\,R'_{a(A)} - R_A \right|^2. \tag{9.62}$$

In Figure 9.3, the distance graphs, according to Equation 9.53, for every one of the 120 permutations of the example showed in Figure 9.2 are shown. The distance, as a function of the orientation (in degrees), is measured in arbitrary length units. The lower contour of the graphs indicated by the dashed curve in Figure 9.3 is the distance given by Equation 9.58. As Figure 9.3 shows, the best orientation in this case is found at 90° where the distance between the two molecules is zero. Due to the computational demand (factorial scaling with the number of atoms), it is not feasible to test all mappings in order to find the corresponding minima on the graphs. Therefore, a probability-driven approach is suggested. In this approach, one of the molecules is rotated randomly, next the optimal mapping is computed as described, and finally the molecules are aligned with the rotation matrices $\hat{Q}_n$ according to Equation 9.62.

The random rotations are performed by random uniformly distributed rotation matrices $\hat{P}_n$ [83,84], and, therefore, we obtain

$$d_n^2 = \left| \hat{Q}_n \hat{a}_n \hat{P}_n X' - X \right|^2 = \sum_{A=1}^{N} \left| \hat{Q}_n \hat{P}_n R'_{a(A)} - R_A \right|^2. \tag{9.63}$$

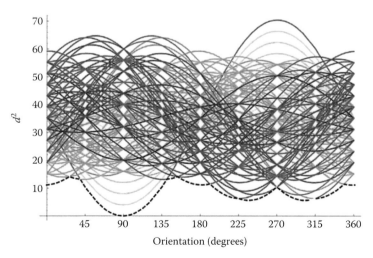

FIGURE 9.3 (See color insert) Distance graphs for each of the 120 possible mappings in Figure 9.2.

where the mapping $\hat{a}_n$ is calculated after the random rotations are applied to the coordinates. In this way, a set of distances are obtained, which are the best in the neighborhood of the orientation where they were calculated. In order to ensure the reliability of this approach, enough samples of the rotation space are needed. For this reason, the following sample algorithm is adapted. First, we make a sample $\hat{P}_1$, which yields the mapping $\hat{a}_1$ and the distance d_1. This is initially the best mapping $\hat{a}^*$ with its associated distance d^*. In a next step, we generate a new (random) sample $\hat{P}_2$ and analyze it. Then the mapping associated with the smallest distance is preserved. The process is successively repeated until the same mapping was found 50 times in a row. This has worked successfully for all our test cases.

The hierarchical transition state finder in combination with the automatic alignment procedure has proven very useful for the study of metal cluster reactions. Selected examples are presented in Section 9.6 of this chapter.

9.2.5.4   Born-Oppenheimer Molecular Dynamics

Classical molecular dynamics is a computer simulation technique where the time evolution of a set of interacting atoms is followed by integrating their classical equations of motion. Given a set of initial positions and velocities, by which the subsequent time evolution is in principle completely determined, the computer calculates a trajectory in a $6N$-dimensional phase space. In this way, a set of configurations according to an underlying statistical distribution function or statistical ensemble is obtained. Physical quantities are then represented by averages over the configuration set. Therefore, the expectation value of a physical quantity can be obtained by the arithmetic average from the corresponding instantaneous values of this quantity along the simulated trajectory. In the limit of very long simulation times, i.e., many trajectory steps, it can be expected that the phase space is fully sampled. In this limit, the averaging process yields the corresponding thermodynamic property. In practice, the simulations are always of finite length, and caution must be taken that the system is truly equilibrated.

While in a full description of the energetics and dynamics of a system, all constituents (electronic and nuclear) should be treated on equal footing quantum mechanically; physical and practical considerations motivate calculations within the Born-Oppenheimer approximation [57]. This approximation introduces a separation between the time scales of nuclear and electronic motions. Moreover, we apply the classical equations of motion for the propagation of the nuclei. Therefore, a BOMD step as discussed in this work consists of solving the static electronic structure problem, i.e., solving the time-independent Schrödinger equation, followed by propagation of the nuclei via classical molecular dynamics [85]. The resulting BOMD method is defined by

$$\hat{H}_e\Psi = E\Psi, \tag{9.64}$$

$$m_A\ddot{\mathbf{r}}_A(t) = -\nabla_A \min_{\Psi} \left\langle \Psi \left| \hat{H}_e \right| \Psi \right\rangle. \tag{9.65}$$

In the above-described BOMD step, the solution of the electronic Schrödinger equation (9.64) represents the computational bottleneck. The computational demand for this task can be considerably reduced by employing ADFT [34,86]. In an early molecular dynamics implementation in a precursor of the deMon2k program, the improved performance of the ADFT-BOMD approach has been documented [40].

The natural time evolution of a classical system of N particles in a volume V can be studied with the above-described BOMD method. In such simulations, the total energy E is a constant of motion. Although thermodynamic results can be transformed between ensembles, this is only reliable in the limit of macroscopic system sizes. Because many experiments are performed at fixed temperature instead of fixed energy, simulations in the canonical, NVT, ensemble are often desired. Thermostats are designed to ensure the molecular dynamics sampling within this ensemble, by modulating the temperature of the system in some fashion. First, we need to establish what we mean by temperature. In simulations, the "instantaneous (kinetic) temperature" is usually computed from the kinetic energy of the system using the equipartition theorem. In other words, the temperature is computed from the system's total kinetic energy. A thermostat that is coupled to the motion of the nuclei thus ensures that the average temperature of the system stays around a predefined value. By the equipartition theorem, the instantaneous temperature of the system is related to the kinetic energy as

$$T(t) = \frac{2K}{N_f k_B}, \tag{9.66}$$

where
  K is the kinetic energy of the system
  k_B is the Boltzmann constant
  N_f is the number of degrees of freedom of the system

A simple way to alter the temperature of the system is to scale the velocities of the particles (velocity scaling thermostat [87]). If the velocities are multiplied by a factor λ, then the associated temperature change can be calculated as follows:

$$\Delta T = \frac{1}{2} \sum_{A}^{N} \frac{m_A (\lambda \dot{r}_A)^2}{N_f k_B} - \frac{1}{2} \sum_{A}^{N} \frac{m_A (\dot{r}_A)^2}{N_f k_B}$$

$$= (\lambda^2 - 1) T(t). \tag{9.67}$$

Thus, it follows

$$\lambda = \sqrt{\frac{T}{T(t)}}. \tag{9.68}$$

The simplest way to control the temperature is then to multiply the velocities at each time step by the factor $\lambda = \sqrt{T/T(t)}$, where T is the desired temperature. This method is usually used to equilibrate the system. As can be seen from Figure 9.4, the convergence of the average temperature (dash line) and instantaneous temperature (gray line) to the desired temperature (300 K) is practically immediate; however, the

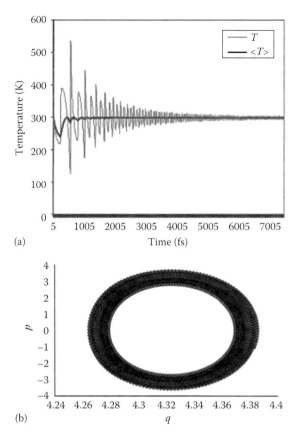

(a)

(b)

FIGURE 9.4   Temperature profile (a) and phase space distribution (b) for a copper dimer using the velocity scaling thermostat for the BOMD simulation.

points visited in phase space are not the ones expected for the canonical sampling. Therefore, the generated trajectories are not suited for the calculation of thermodynamic averages.

An alternative way to maintain the temperature is to couple the system to an external heat bath that is fixed at the desired temperature (Berendsen thermostat [88]). The bath acts as a source of thermal energy, supplying or removing heat from the system as needed.

The velocities are scaled at each step, such that the rate of change of temperature is proportional to the difference in temperature between the bath and the system

$$\frac{dT(t)}{dt} = \frac{1}{\tau}(T - T(t)). \tag{9.69}$$

Here, τ is a coupling parameter whose magnitude determines how tightly the bath and the system are coupled together. With this method, the system temperature converges exponentially toward the desired temperature. The scaling factor for the velocities is

$$\lambda = \left[1 + \frac{1}{\tau} \left(\frac{T}{T(t)} - 1 \right) \Delta t \right]^{1/2}. \tag{9.70}$$

If τ is large, the coupling will be weak. If τ is small, the coupling will be strong, and in the special case where the coupling parameter equals the time step, the algorithm becomes equivalent to the simple velocity scaling method. The advantage of this approach is that it does permit the system to fluctuate about the desired temperature. Because of its exponential convergence, this thermostat is well suited for simulated annealing. Figure 9.5 shows the behavior of the temperatures along the BOMD simulation of a copper dimer, the same system used in Figure 9.4. The exponential convergence of the average temperature to the desired one (300 K) is obvious from

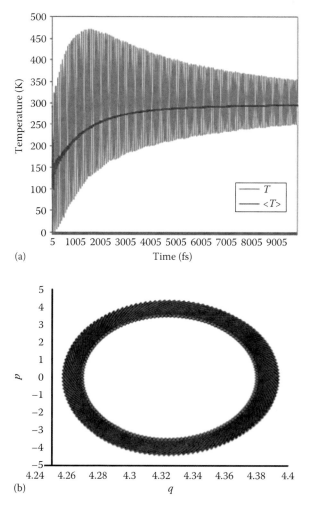

FIGURE 9.5 Temperature profile (a) and phase space distribution (b) for a copper dimer using the Berendsen thermostat for the BOMD simulation.

this figure. As Figure 9.5 shows, the phase space sampling is, however, very similar to that obtained with the previous thermostat. Thus, this thermostat is not suited for the calculation of thermodynamic averages either.

The so-called constant kinetic temperature dynamics is generated by the following equations of motion [89]:

$$\dot{r}_A = \frac{p_A}{m_A},$$

(9.71)

$$\dot{p}_A = F_A - \xi\, p_A.$$

(9.72)

The quantity ξ is a kind of "friction coefficient," which varies such that $T(t)$ is constrained to a constant value, i.e., to guarantee

$$\dot{T}(t) \propto \frac{d}{dt}\left(\sum_A p_A^2\right) = 0.$$

(9.73)

The constraint is chosen in order to perturb the classical equations of motion as little as possible. This is achieved by making ξ a Lagrange multiplier, which minimizes the difference between constrained and Newtonian trajectories. The resulting expression is

$$\xi = \frac{\sum_A p_A \cdot F_A}{\sum_A p_A^2}.$$

(9.74)

This method generates configurational properties of the canonical ensemble; however, the momentum distribution is not canonical as can be seen from the BOMD simulation of the copper dimer shown in Figure 9.6.

The last scheme for controlling the temperature discussed here is the Nosé-Hoover (chain) thermostat [90–92]. In a fictitious system that contains the real system and the thermostat, an extra degree of freedom is introduced that represents the thermal reservoir. In this way, the constant temperature simulation can be performed deterministically by a reformulation of the Lagrange equations of motion for the extended system. The additional degree of freedom is represented by the coordinate s and its conjugated momentum p_s.

The Hamiltonian describing the extended system consisting of the real system and a heat bath is given by

$$H = \frac{1}{2}\sum_A^N \frac{p_A^2}{m_A s^2} + \mathcal{V}(\{r_A\}) + \frac{p_s^2}{2Q} + \frac{N_f}{\beta}\ln s.$$

(9.75)

The first two terms are the kinetic and potential energy of the real system we are interested in. The last two terms correspond to the extra degree of freedom for the heat bath. Here, β denotes $1/(k_B T)$ with T being the temperature of the heat bath.

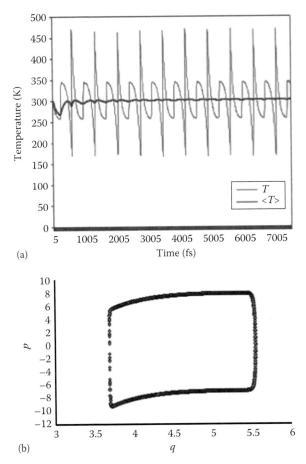

FIGURE 9.6   Temperature profile (a) and phase space distribution (b) for a copper dimer using the isokinetic thermostat for the BOMD simulation.

For some systems, the Nosé-Hoover method described above does not yield a canonical distribution in phase space, even for very long simulations [91,93]. The distribution itself has a dependence on the particles momenta, p_A, as well as the thermostat momenta, p_s. While the fluctuations of p_A are driven by a thermostat, there is nothing to drive the fluctuations of p_s. These fluctuations, which clearly occur in ergodic systems, are important in driving the system to fill phase space. This suggests to thermostat p_s and, by analogy, the thermostat of p_s plus its thermostat, etc., thus forming a chain of thermostats.

The Hamiltonian for the new extended system is then given by

$$H = \frac{1}{2}\sum_A^N \frac{p_A^2}{m_A s^2} + \mathcal{V}(\{r_A\}) + \sum_{i=1}^n \frac{p_{si}^2}{2Q} + \frac{N_f}{\beta}\ln s_1 + \sum_{i=2}^n \frac{\ln s_i}{\beta},\qquad(9.76)$$

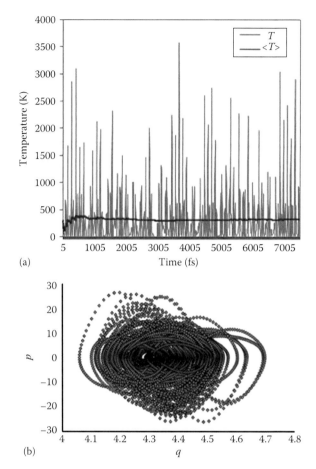

(a)

(b)

FIGURE 9.7 Temperature profile (a) and phase space distribution (b) for a copper dimer using the Nosé-Hoover chain thermostat for the BOMD simulation.

where only the first thermostat interacts directly with the real system. This method increases the ergodic behavior of the system and, therefore, leads to the correct canonical distribution in phase space as the BOMD copper dimer simulation in Figure 9.7 shows. Furthermore, the instantaneous temperature fluctuations are occurring during the full simulation.

These four schemes are implemented in the most recent version of deMon2k. For the calculation of temperature-dependent molecular properties, we have made good experiences with the Nosé-Hoover chain thermostat [39].

9.3 GEOMETRY OPTIMIZATION

9.3.1 Free Metal Clusters (V_3, Cu, Ni)

Over the last years, the knowledge about free transition-metal (TM) clusters has grown considerably, due to many reasons such as the development of novel experimental techniques, new theoretical methods, and high-speed computers at reasonable cost (see, e.g., Refs. [94–98] and references therein). However, determining the geometrical structure of TM clusters remains still a difficult task, and, with only a few exceptions, the structures are still unsolved. Only with detailed structural information can models for reactive processes of these clusters be developed. Given the lack of experimentally available data for these systems, theoretical methods have to be used to determine the structure. A major drawback for the theoretical investigations of TM clusters is the required accuracy in the treatment of electron correlation that significantly increases the computational effort in Hartree-Fock-based methods. The development of density functional methods and related computer codes has changed this situation considerably, and it is now possible to perform reliable calculations of small clusters at reasonable computational cost [30,99–107].

The possibility of combining theoretical calculations with available experimental data from pulsed-field-ionization-zero-electron-kinetic-energy (PFI-ZEKE) photoelectron spectroscopy to resolve the structure of small TM clusters has attracted, in the past few years, more and more attention. A combination of DFT and harmonic Franck-Condon factor calculations has been used to determine the structures of some substituted TM clusters [108–111]. The PFI-ZEKE technique was also used to determine accurate adiabatic ionization potentials of bare V_3 and V_4 clusters [112]. The observation of a PFI-ZEKE spectrum of V_3 with well-resolved vibrational bands has opened the possibility for a structure determination via the above-described technique.

Additional relevant literature [113–142] about vanadium clusters may be summarized as follows. Divanadium has been experimentally well determined [113,119]. The uncertainty concerning the nature of its chemical bonding has been brilliantly cleared up with detailed gas-phase electronic spectroscopy experiments by Langridge-Smith et al. [114]. Under high resolution, the rotational structure of the observed bands could be assigned, and the ground state of V_2 was determined to be $^3\Sigma_g^-$, with a bond length of 1.77 Å. Later on, Spain et al. extended this work into the infrared [115]. Moreover, Spain and Morse have used two-color resonant two-photon ionization (R2PI) spectroscopy to measure the dissociation energy of V_2, and a value of $D_0 = 2.75$ eV was obtained [116]. The matrix-isolated resonance Raman spectra of divanadium have been extensively investigated by Moskovits et al. [117], and a frequency of 537 cm^{-1} was assigned to the V_2 ground state. Later on, James et al. recorded photoionization efficiency (PIE) spectra to determine the adiabatic ionization potential of V_2 as 6.36 eV [118]. They were also able to assign the ground state of V_2^+ as $^4\Sigma_g^-$ [118]. With the available information of the intermediate states and ionization potentials, Hackett et al. measured the PFI-ZEKE spectra of the V_2^+ ground state [118]. Simard et al. carried out a detailed analysis of the rotational spectra and determined the bond length [98] of the $^4\Sigma_g^-$ ground state to be 1.73 Å. For larger vanadium clusters, up to 22 atoms, ionization threshold energies have been reported [120]. Vanadium clusters have also been included in several reactivity studies [121–128]. Stern-Gerlach measurements of vanadium clusters have been performed by Douglass et al., and small moments have been determined

[129]. The kinetic energy dependence of the collision-induced dissociation (CID) of V_n^+ ($n = 2$–20) with xenon using a guided ion beam mass spectrometer has been measured [130]. The evolution of the electronic structure of V_n clusters has been probed by photoelectron spectroscopy [131]. The electronic structure of vanadium cluster anions was also studied by photoelectron spectroscopy [132].

Different theoretical works [133–141] exist on vanadium clusters, too. Most of them are model calculations, and only a few studies have been performed with DFT methods. As far as calculations are concerned, the first theoretical investigation on vanadium clusters was carried out by Salahub and Messmer with a spin-polarized SCF-Xα-SW method [133]. Liu et al. used SCF molecular orbital theory and DFT to show that vanadium could become magnetic if its size and dimensions were constrained [134]. Walch and Bauschlicher carried out detailed calculations for several TM trimers and used the results to predict the character of the vanadium trimer [135]. They expected a strong 3d orbital bonding and guessed that V_3 should be formed by the combination of three excited $4s^1 3d^4$ vanadium atoms. Dreyssé et al. investigated the antiferromagnetic coupling between nearest neighbor atoms in vanadium clusters and slabs [136]. With the local spin density (LSD) approximation and a Gaussian orbital basis set, Lee and Callaway studied the electronic structures and the magnetism of V_9 and V_{15} clusters with bcc symmetry [137,138]. Zhao et al. studied the structural and magnetic properties of small vanadium clusters in the framework of tight-binding theory [139]. Grönbeck and Rosén [140] performed a density functional study of small vanadium clusters in the range from V_2 to V_8 using a gradient-corrected exchange-correlation functional and a triple zeta basis with polarization functions. They predicted an acute triangle ground-state structure for the neutral vanadium trimer.

Another density functional study of small neutral and cationic vanadium clusters in the range from V_2 to V_9 using a 6-311G basis was performed by Wu and Ray [141]. Contrary to the work of Grönbeck and Rosén, they found a quartet ground state with an apex angle of $66°$ for the neutral vanadium trimer and a close-lying equilateral triangle doublet excited state [142]. From their vanadium dimer calculations, Wu and Ray also conclude that the results are quite sensitive to the choices of the basis and the exchange-correlation functional.

This brief summary shows the uncertainties in the determination of the ground-state structures of small TM clusters based only on calculations. Over the last few years, we have performed several combined theoretical and experimental studies on the determination of ground-state structures of different small free TM clusters such as the V_3 cluster [143], small copper clusters [105,144,145], and small nickel [146] clusters. The main results we obtained are now briefly reviewed.

The neutral and cationic vanadium trimers were optimized with a DFT method [143]. The GGA of Perdew and Wang (PW86) [147] and Perdew (P86) [148,149] in conjunction with the local exchange-correlation proposed by Vosko, Wilk, and Nusair (VWN) [150] has been used. A newly developed DZVP basis set optimized for gradient-corrected functionals [47,48] was used for the vanadium atom. We have optimized this kind of basis set for all 3d TM elements. Compared to the DZVP basis set optimized for local functionals, the new basis set improves the accuracy of the low-lying atomic energy levels of the 3d TM atoms [48]. We have named these new basis sets as DZVP-GGA [47]. They are available in the basis set file of the deMon2k code and can be downloaded directly from the deMon2k web page at http://www.demon-software.com/public_html/download.html#basissets

The experimental PFI-ZEKE spectrum was simulated using multidimensional Franck-Condon factors calculated from the geometries and harmonic frequencies of the obtained structures of V_3 and V_3^+. The comparison between the experimental and theoretical PFI-ZEKE spectra establishes unequivocally the ground-state structure of the V_3 cluster as an equilateral doublet. This represented the first work in which the structure of a pure TM metal cluster has been determined from the combination of DFT calculations, Franck-Condon simulations, and ZEKE photoelectron spectroscopy [143]. Figure 9.8 summarizes our restricted open shell Kohn-Sham (ROKS) optimized V_3 and V_3^+ structures. For neutral V_3, we find an equilateral triangle $^2A_1'$ ground state. The optimized bond distance is 2.17 Å. Our D_{3h} vanadium trimer ground state differs from the C_{2v} structures reported in previous DFT studies [140,141]. This is due to the different orbital ordering at the Fermi level. In fact, with the VWN-optimized DZVP basis set [46], we obtain an acute triangle doublet ground state for the vanadium trimer. Only 0.03 eV above the $^2A_1'$ ground state, we find an acute triangle $^4A_2$ state (Figure 9.8). The $^4A_2$ quartet is formed by the excitation of one of the e'' electrons into an unoccupied e' orbital. As a result, a Jahn-Teller distortion into C_{2v} symmetry occurs. Other structures with higher spin multiplicities such as the sextet depicted in Figure 9.8 are energetically well separated from these two structures.

Similar to the neutral trimer, we find an equilateral triangle ground state for V_3^+. This structure can be assigned to a $^3A_2'$ state. Wu and Ray also found in their calculations [141] the same V_3^+ ground state. The optimized bond distances of 2.18 Å in the cationic ground state are very close to those of the neutral trimer (see Figure 9.8).

The calculated adiabatic ionization potential is 5.61 eV in good agreement with the experimental value of 5.49 eV [112,120]. Above the $^3A_2'$ state, we find a $^1A_1'$ singlet. This singlet is, at 0.18 eV, reasonably well separated from the $^3A_2'$ ground state. Higher spin multiplicities, like the quintet (0.52 eV above the $^3A_2'$ ground state) depicted in Figure 9.8 are considerably higher in energy. Therefore, the assignment of the V_3^+ ground state to the $^3A_2'$ state depends less critically on the choice of the exchange-correlation functional and basis set. The MO analysis shows that the assignment of the V_3 ground state depends on the ordering of the d a_1' and e' orbitals

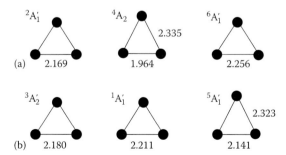

FIGURE 9.8 Ground-state structures of V_3 (a) and V_3^+ (b) clusters. Bond lengths are in Å. (Reprinted from Calaminici, P. et al., *J. Chem. Phys.*, 114(9), March 1, 4036, 2001. With permission.)

[143], which in turn depends on the relative stability of the atomic d orbitals. The well-known trend in DFT calculations to overestimate the stability of the atomic d functions in TM elements results in V_3 in an extra (artificial) stabilization of the a_1' orbital. As a result, a Jahn-Teller distortion of the V_3 ground state may occur if the ordering of the a_1' and e' orbitals has exchanged. Whether this distortion is real or artificial cannot be concluded from the calculations alone. Therefore, the vanadium trimer ground-state structure cannot be assigned based on these calculations. Fortunately, Hackett and coworkers recorded a well-resolved PFI-ZEKE spectrum of V_3 [112]. The simulation of this spectrum from our DFT results allowed us to assign the V_3 ground-state structure. For this purpose, we have calculated the harmonic frequencies of the optimized neutral and cationic vanadium trimers [143]. From these frequencies and the corresponding normal mode vectors, we calculated harmonic Franck-Condon factors and simulated the PFI-ZEKE spectrum of V_3. The simulated spectrum will be shown and discussed in Section 9.4.

Another goal in our studies of TM clusters is to explain the experimentally observed pattern in stability and other molecular properties on the basis of electronic structure of these systems. Here we present an example for small copper clusters with up to 10 atoms [144,145]. Structure optimizations were performed in the LDA with the VWN exchange-correlation functional, and all-electron basis sets were employed. The same functional was used for the frequency analysis in order to distinguish between stable structures and transition structures. LDA studies on small copper clusters with up to five atoms [105] have already shown that the DZVP basis together with the A2 auxiliary function set [46] supplies very reliable geometries for structure optimizations. Since such calculations are insufficient for energetic considerations, single point energies were finally calculated using the same basis and auxiliary function set with the exchange-correlation functional PW86-P86 [146–148]. This approach has already proven reliable for small copper clusters with up to five atoms [105]. The corresponding cluster structures are depicted in Figure 9.9 showing all bonds with bond lengths up to 2.6 Å.

The neutral hexamer is the largest copper cluster ground state with a two-dimensional structure. At the GGA level, the Cu_6 C_{5v} isomer is only slightly higher in energy. In contrast to the hexamer, the most stable neutral heptamers have three-dimensional structures. We find in our calculations a pentagonal bipyramid of D_{5h} symmetry as the global minimum for Cu_7 both at the LDA and GGA levels. Compared with the C_{5v} structure of Cu_6, the ring distances of the Cu_7 cluster are slightly shortened from 2.41 to 2.39 Å and the cap-ring distance is substantially lengthened from 2.32 to 2.39 Å. Still, the cap-cap distance is only 2.50 Å, indicating a bond between the capping atoms. Another relatively stable Cu_7 isomer is the depicted C_{3v} structure in Figure 9.9. This structure was obtained from the relaxation of a crystal fragment. During this relaxation, the average bond length decreased from 2.56 to 2.39 Å. Planar structures of the heptamer were found considerably higher in energy and were excluded as possible ground states from the study.

To arrive at new cluster structures, the capping of stable structures seems to be a reasonable procedure. Starting from the Cu_7 bipyramid, a capping of one of the triangles has been proposed and calculated on the LDA level [151,152]. In our study, we found that such a structure is not stable and relaxes. The global minimum of Cu_8 has

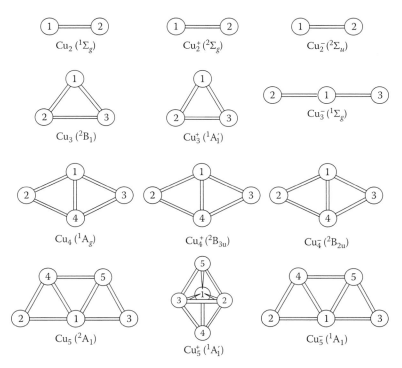

FIGURE 9.9 Ground-state structures of neutral, cationic, and anionic copper clusters with up to five atoms. (Reprinted from Jug, K. et al., *J. Chem. Phys.*, 116(11), March 15, 4497, 2002. With permission.)

C_{2v} symmetry (Figure 9.9). This global minimum can be obtained from the relaxation of a capped triangle or a capped ring bond structure of the heptamer ground state. For Cu_8, a T_d structure is found to be only 0.5 kcal/mol higher than the C_{2v} structure. This structure has a cuboidal cage form compared with the more spherical cage form of the C_{2v} structure. The search for stable structures of larger clusters is increasingly more difficult because of the increasing number of possible arrangements for isomers. We have therefore used the strategy of capping smaller clusters or solid state fragments. In the case of Cu_9, a starting structure for optimization can be obtained by single capping of Cu_8 structures or double capping of Cu_7 structures. The first procedure was accompanied by extremely strong relaxations, whereas the second procedure resulted in all cases in Cu_9 minima. The global minimum has C_s symmetry. It is derived from the Cu_7 bipyramid by double capping of two adjacent upper triangles. The five-membered ring and the bond between the capping atoms are preserved (Figure 9.10). A C_{2v} structure with a slightly higher energy of 1.1 kcal/mol is obtained from double capping of adjacent upper and lower triangles. The C_{3v} isomer of Cu_9 depicted in Figure 9.10 was obtained from the relaxation of a crystal fragment. The average bond distance of this isomer is similar to the C_{3v} isomer of Cu_7 and is still considerably shorter than the bulk distance of 2.56 Å.

The strategy of capping leads to stable structures also for Cu_{10} clusters. The global minimum has D_{4d} symmetry. Like the global minimum of Cu_8, this structure is a

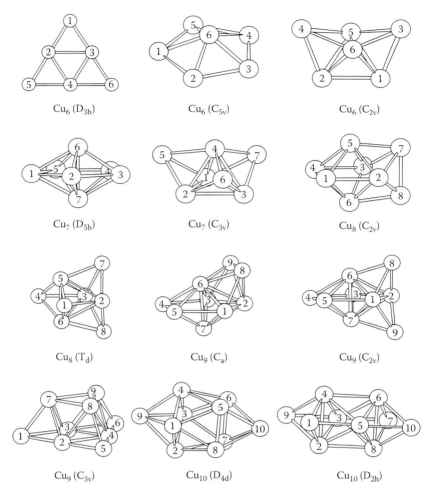

FIGURE 9.10 Structures of the most stable neutral copper clusters with 6–10 atoms. (Reprinted from Jug, K. et al., *J. Chem. Phys.*, 116(11), March 15, 4497, 2002. With permission.)

cage. The same structure was obtained in [152]. It can be obtained by bicapping the T_d structure of Cu_8 over the two squares. A D_{2h} structure lies 8 kcal/mol higher in energy. The optimized structures of the global minima of the larger cationic Cu_n^+ and Cu_n^- clusters with $n = 6$–10 are depicted in Figure 9.11.

Turning the discussion to the positively charged copper clusters, from Figures 9.9 through 9.11, we see that Cu_5, Cu_6, and Cu_9 undergo the most substantial structural changes upon ionization. The first two clusters acquire a three-dimensional structure compared to the planar structure of the neutral molecules. In the case of Cu_9, the second lowest-lying neutral structure is found to be more favorable for the cation.

Compared to the neutral species, the most significant structural changes for the anionic systems are observed for Cu_3^-, Cu_6^-, and Cu_9^- (Figure 9.11). The first system

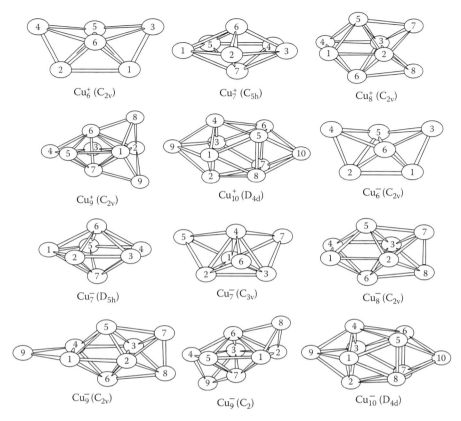

FIGURE 9.11 Structures of the most stable charged copper clusters with 6–10 atoms. (Reprinted from Jug, K. et al., *J. Chem. Phys.*, 116(11), March 15, 4497, 2002. With permission.)

becomes linear, Cu_6^- acquires a three-dimensional structure like the corresponding cation, and the Cu_9^- ground state represents a bond ring capping of Cu_8, which is not stable for the neutral nonamer.

For the optimized structures of the neutral and ionic clusters, binding energies per atom were calculated at the GGA level. It was pointed out in the literature that basis set superposition errors (BSSEs) may be important for a quantitative calculation of cluster binding energies [153]. Therefore, we have calculated the BSSE per atom using the approximate correction of Jansen and Ross [154]. It was shown [153] that the accuracy of this approximation is within 0.2 kcal/mol per atom for the basis set used. We have also calculated the zero point energies (ZPE) of the clusters. The values for the neutral clusters are listed in Table 4 of Ref. [144]. The BSSE per atom increases from 1.4 kcal/mol for Cu_2 to 5.9 kcal/mol for Cu_{10}. The ZPE per atom is much less size dependent and considerably smaller. For Cu_2 and Cu_3, direct measurements of the atomization energies by fluorescence [155] and Knudsen cell mass spectrometry [156] are available. The corrected binding energies (CBE) per atom are in good agreement with these direct measurements (see Table 4 of Ref. [143]). The binding energies of the larger neutral clusters were derived from CID experiments of

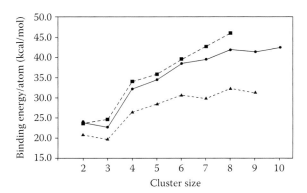

FIGURE 9.12 Binding energies per atom (kcal/mol) for neutral copper clusters. Circles, calculated; squares, experiment with anionic clusters; triangles, experiment with cationic clusters.

anionic [157] and cationic [158] copper clusters using electron affinities or threshold detachment energies [159] and ionization potentials [160–163] from the literature. In Figure 9.12, the derived binding energies from the CID of anionic (squares) and cationic (triangles) copper clusters are compared with our calculated values (circles).

As Figure 9.12 shows, our calculated values are in better agreement with the experimental values of the anionic copper cluster [157] than with those of the cationic systems [158].

Among the homonuclear 3d TM clusters, Ni systems have received considerable attention both experimentally and theoretically. It is not our purpose here to review the corresponding literature. We point the interested reader to references [146] and [164–181]. In particular, the localized behavior of the unfilled 3d electrons in nickel aggregates results in a great complexity of the electronic structure, which has motivated many groups to study this effect and its influence on the system properties. As a result, nickel clusters have constituted one of the most studied clusters since they are excellent systems for exploring new theoretical approaches [146,167–181].

However, despite the availability of a large number of articles on small neutral nickel clusters, much less attention has been paid to the study of structures and properties of their corresponding cationic and anionic systems. Most DFT calculations on small nickel clusters have been performed using basis sets optimized for the local level of theory. In Ref. [146], a particular effort was made to answer some of the questions related to the lack of experimental information about the structure and electronic properties of small cationic and anionic nickel clusters up to the pentamer. For this purpose, newly GGA-optimized basis sets for the nickel atom were employed and the most extensive theoretical investigation performed on the ground and low-lying states of the neutral, cationic, and anionic Ni_n clusters up to the heptamer was reported. For this study, different topologies and spin multiplicities were considered in order to scan as much of each PES as possible and to find the lowest energetic structures. The stability of the structures with respect to geometrical distortions was confirmed by a frequency analysis. Structural parameters, harmonic frequencies, binding energy, ionization potential, and electron affinity were reported

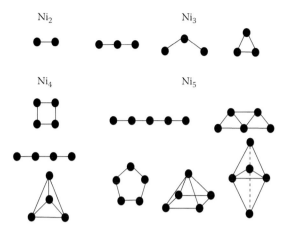

FIGURE 9.13 Different topologies considered as initial structures for the geometry optimization of Ni_n, Ni_n^+, and Ni_n^- ($n=2$–5) clusters. (Reprinted from Lopez Arvizu, G. et al., *J. Chem. Phys.*, 126(19), May 21, 194102-1, 2007. With permission.)

as well [146]. The calculated values were compared with available data from the literature. The GGA basis sets employed turn out to be very important for the correct ground-state structure determination of nickel clusters. Full geometry optimizations were performed considering as initial structures the different topologies illustrated in Figure 9.13.

In Figures 9.14 and 9.15, the results of the LDA and GGA geometry optimizations, structures, bond lengths in Å, and multiplicities of the ground states of neutral, cationic, and anionic nickel clusters are presented.

The calculations of the clusters reported in Figure 9.14 have been performed with the original DZVP-LDA basis set as well as with the newly developed DZVP-GGA and TZVP-GGA basis sets in combination with the VWN functional. The DZVP-LDA bond lengths are given without parentheses while the bond lengths obtained with the DZVP-GGA and TZVP-GGA basis sets are given in round and square brackets, respectively. The results displayed in Figure 9.15 have been obtained with the newly developed DZVP-GGA and TZVP-GGA basis sets in combination with the PW86 functional.

A comparison of Figures 9.14 and 9.15 shows that depending on the methodology used, structural changes in the clusters may occur as in the case of the nickel trimer and of the nickel pentamer. In fact, for Ni_3^-, the VWN functional in combination with all employed basis sets predicts a distorted triangular structure (Figure 9.14), whereas with the newly developed DZVP-GGA and TZVP-GGA basis sets in combination with the PW86 functional, an almost linear ground-state structure is found (Figure 9.15). This was the first time that such a structure was predicted for the anionic nickel trimer [146]. The angle formed by the three Ni atoms is around 165°.

In order to gain more insight into the nickel trimer ground states, the orbital filling pattern with increasing charge of the cluster was investigated. In Figures 9.16 and 9.17, the schematic correlation diagrams of the Kohn-Sham MOs between cationic and neutral Ni_3 clusters and between neutral and anionic trimers are presented,

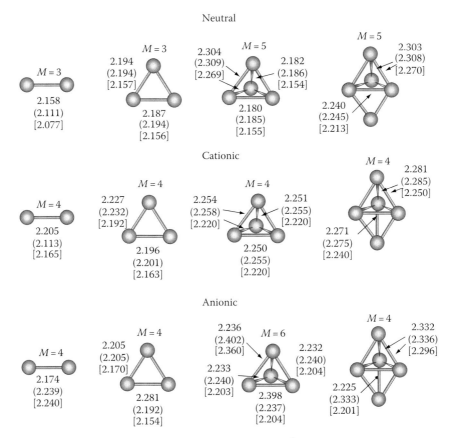

FIGURE 9.14 Ground-state structures, bond lengths (in Å), and multiplicities of Ni_n, Ni_n^+, and Ni_n^- ($n = 2–5$) clusters. The original DZVP-LDA basis set as well as the newly developed DZVP-GGA, parentheses, and TZVP-GGA, square brackets, basis sets in combination with the VWN functional were employed. (Reprinted from Lopez Arvizu, G. et al., *J. Chem. Phys.*, 126(19), May 21, 194102-1, 2007. With permission.)

respectively. The ground state of the neutral trimer is formed by adding an extra electron (indicated by a dotted arrow) to the singly occupied antibonding 42 MO of the cationic trimer (left side of Figure 9.16). This orbital results in the doubly occupied 40 MO of the neutral system (right side of Figure 9.16). The crossing of different MOs (indicated by dashed arrows) occurs, and the two unpaired electrons of the neutral Ni_3 ground state occupy the 42 and 43 MOs (Figure 9.16).

On the other hand, the anionic trimer ground state is formed by adding an extra electron to the singly occupied 42 MO of the neutral system (Figure 9.17). The crossing of several MOs occurs in this case, too (Figure 9.17). The three unpaired electrons occupy the MOs 42, 43, and 44 in the anionic trimer (Figure 9.17).

The VWN calculations predict the bipyramidal structure on the quintet PES as the ground-state structure for the neutral nickel pentamer. This structure is followed by a pyramidal structure on the septet PES. With the DZVP-GGA/PW86 and

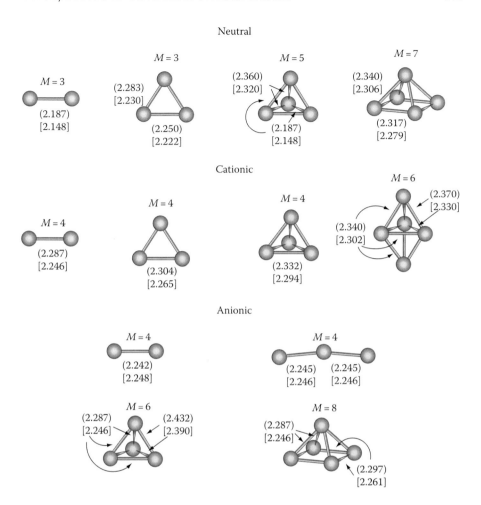

FIGURE 9.15 Ground-state structures, bond lengths (in Å), and multiplicities of Ni_n, Ni_n^+, and Ni_n^- ($n = 2$–5) clusters. The newly developed DZVP-GGA, parentheses, and TZVP-GGA, square brackets, basis sets in combination with the PW86 functional were employed. (Reprinted from Lopez Arvizu, G. et al., *J. Chem. Phys.*, 126(19), May 21, 194102-1, 2007. With permission.)

TZVP-GGA/PW86 methods, this result is inverted and the pyramidal structure on the septet PES is the favored one (Figure 9.15). The energy difference between these two structures is 0.16 eV with the DZVP-GGA/PW86 method and 0.22 eV with the TZVP-GGA/PW86 calculation, respectively.

For these nickel clusters, adiabatic ionization potentials and adiabatic electron affinities were also presented (Tables II and IV of Ref. [146]). The results were compared with the experimental data available in the literature. The trends observed in the energetic properties showed agreement with the ones determined experimentally, demonstrating that the structural assignment of the clusters can be performed with good resolution over the ionization potential [146]. For unequivocal assignment of

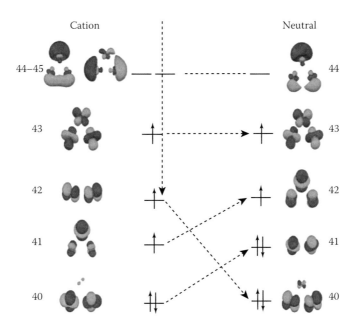

FIGURE 9.16 Schematic molecular orbital correlation diagram from the cationic to the neutral ground-state trimers. The adding of the extra electron is indicated by a dotted arrow. (Reprinted from Lopez Arvizu, G. et al., *J. Chem. Phys.*, 126(19), May 21, 194102-1, 2007. With permission.)

the ground-state structure of small anionic nickel clusters, experiments based on negative-ion vibrationally resolved spectroscopy would be highly desirable.

9.3.2 Metal Cluster Oxides (V_3O, Nb_3O)

The gas-phase spectroscopic study of TM oxides is of experimental and theoretical interest. Experimentally, properties related to the metal–oxygen bond are crucial for understanding chemisorption and catalytic activities of metal oxides. In particular, vanadium oxides are important in catalysis, and the investigation of vanadium–oxygen chemical bonding is of great interest. Theoretically, molecules containing TMs are rather challenging due to the open d shells. Simple TM oxide molecules provide ideal systems for the investigation of the reliability of theoretical methods. The study of TM oxide species in the gas phase by photoelectron spectroscopy offers a unique opportunity to probe the electronic structure of the isolated molecules. Dyke et al. reported the first photoelectron spectroscopy experiment of neutral VO [182]. Negative ion photoelectron spectroscopy was applied to the group 5 metal trimer monoxides V_3O, Nb_3O, and Ta_3O, yielding insight into the bonding of these early TM cluster oxides [128]. The experimental data reported in this work provide a benchmark for computational studies of partially ligated early TM clusters. In particular, the vibrationally resolved 488 nm negative ion photoelectron spectrum of V_3O provides measurements of its electron affinity and vibrational frequencies, which can be summarized as follows. The electron affinity of V_3O is 1.218 ± 0.008 eV

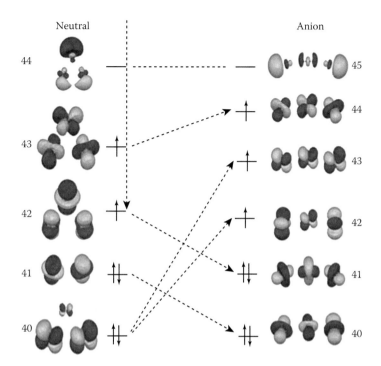

FIGURE 9.17 Schematic molecular orbital correlation diagram from the neutral to the anionic ground-state trimers. The adding of the extra electron is indicated by a dotted arrow. (Reprinted from Lopez Arvizu, G. et al., *J. Chem. Phys.*, 126(19), May 21, 194102-1, 2007. With permission.)

[128]. The metal–oxygen stretching frequency is $750 \pm 20\,cm^{-1}$ for the neutral V_3O and $770 \pm 20\,cm^{-1}$ for V_3^-. Lower frequencies of $415 \pm 15\,cm^{-1}$ and $340 \pm 15\,cm^{-1}$ were also found to be active in V_3O. Moreover, the V_3O spectrum indicates that the extra electron in the anionic system occupies essentially a nonbonding orbital and that the neutral and anionic clusters have planar structures.

Motivated by this experimental work, we have performed an all-electron LCGTO-KS-DFT study of structural and spectroscopic properties of both neutral and anionic vanadium trimer monoxide and neutral, cationic, and anionic niobium trimer monoxide [183–185]. The calculated values of the adiabatic electron affinity and the vibrational frequencies were compared with those recently measured [128]. The theoretical study was extended to the low-lying states of both neutral and anionic vanadium trimer monoxides, and a combined theoretical and experimental approach for the determination of the V_3O and V_3O^- ground-state structures was presented.

The same approach used earlier that allowed us to simulate the ZEKE spectrum and to determine the ground-state structure of the V_3 cluster [143] was used in order to simulate the experimental negative ion photoelectron spectrum using multidimensional Franck-Condon factors calculated from the geometries and harmonic frequencies of the structures obtained for V_3O and V_3O^-. The comparison between the experimental and theoretical negative ion photoelectron spectrum establishes

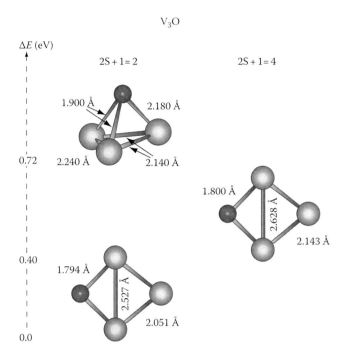

FIGURE 9.18   Ground-state and low-lying structures of V_3O clusters. Bond lengths are in Å and energy differences are in eV. (Reprinted from Calaminici, P. et al., *J. Chem. Phys.*, 118(11), March 15, 4913, 2003. With permission.)

unequivocally the ground-state structure of V_3O and V_3O^- as a planar C_{2v} structure on the doublet and singlet PES, respectively [183]. Detailed information about how the computations were performed is given in Ref. [184]. In Figures 9.18 and 9.19, the results of the optimized structures of V_3O and V_3O^- are, respectively, summarized. The calculated bond distances are reported in Å.

Both V_3O and V_3O^- ground states have planar C_{2v} structures with the oxygen atom bridging two vanadium atoms (Figures 9.18 and 9.19). This result is in perfect agreement with the hypothesis that the neutral and anionic V_3O clusters have planar structures with doubly bridging oxygen atoms [128]. The ground-state structure of neutral V_3O could be assigned from our calculation to a $^2B_2$ state [128,183]. The calculated vanadium–oxygen bonds are 1.794 Å, the two equal vanadium–vanadium bonds are 2.051 Å, and the other metal–metal bond is longer by about 0.5 Å (Figure 9.18). A planar C_{2v} structure in the quartet PES is found at 0.4 eV above the ground-state structure. For this structure, the vanadium–oxygen bonds are very similar to those of the ground-state structure while the metal–metal bonds are slightly different (0.1 Å longer). A three-dimensional C_s structure in the doublet PES lies 0.72 eV above the ground-state cluster. In this structure, the oxygen atom bridges two vanadium atoms giving two equal V–O bonds of 1.90 Å and one V–O bond of about 0.3 Å longer. Two V–V bonds are 2.14 Å and the V–V bond on which the oxygen atom is bridged in 2.24 Å (Figure 9.18). We assigned a $^1A_1$ state as the

V_3O^-

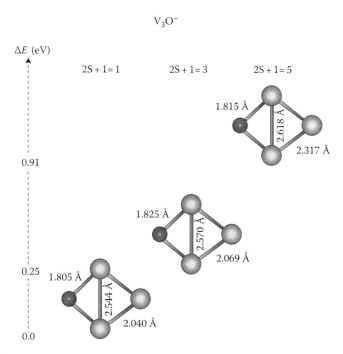

ΔE (eV)

$2S + 1 = 1$ $2S + 1 = 3$ $2S + 1 = 5$

1.815 Å   2.618 Å   2.317 Å

0.91

1.825 Å   2.570 Å   2.069 Å

0.25   1.805 Å   2.544 Å   2.040 Å

0.0

FIGURE 9.19 Ground-state and low-lying structures of V_3O^- clusters. Bond lengths are in Å and energy differences are in eV. (Reprinted from Calaminici, P. et al., *J. Chem. Phys.*, 118(11), March 15, 4913, 2003. With permission.)

ground-state structure of the anionic system. The equilibrium geometries of both ground-state structures are very similar (see Figures 9.18 and 9.19). A triplet $^3B_2$ state is found 0.25 eV above the anionic ground-state structure. Depending on the coordinate system orientation, this structure could be also assigned as a $^3B_1$ state, which has been proposed as the anionic ground-state structure in the literature [128]. A planar C_{2v} structure in the quintet PES is found at 0.91 eV above the anionic ground state (Figure 9.19).

In order to further characterize the optimized ground-state structures of V_3O and V_3O^-, the harmonic vibrational frequencies have been calculated. The results of the frequencies calculated for the ground-state structures, the assignment of each normal mode, and its symmetry are collected in Table I of Ref. [183]. The calculated atomization energy of V_3O is -13.45 eV [183]. The binding energy of the oxygen atom on the vanadium trimer is -8.77 eV [183]. The calculated adiabatic electron affinity is equal to 1.30 eV [183]. This value is in very good agreement with the experimental value of 1.21 ± 0.008 eV from negative ion photoelectron spectroscopy [128]. From the harmonic vibrational frequencies of the ground-state structures and the corresponding normal mode vectors, we calculated harmonic Franck-Condon factors and simulated the experimental negative ion photoelectron spectra of V_3O. The reliability of the ground-state assignment was therefore proven by the simulation of the experimental negative ion photoelectron spectrum [184].

DFT calculations of neutral and anionic niobium tri-mer monoxides have been performed [184] employing scalar quasi-relativistic effective core potentials (ECPs). A valence space of 6 and 13 electrons for the oxygen and niobium atoms was employed, respectively. The topology and geometrical parameters of the obtained ground-state structures of Nb_3O (top), Nb_3O^+ (middle), and Nb_3O^- (bottom) are reported in Figure 9.20. In this figure, the calculated bond distances are reported in Å. As Figure 9.20 shows, the neutral, cationic, and anionic ground states have a C_{2v} structure. This result is in per-fect agreement with the experimental hypothesis that the neutral and anionic Nb_3O clusters are planar with doubly bridging oxygen atoms [128]. Moreover, we also note that the equilibrium geometries of the neutral and anionic ground-state structures are very similar (Figure 9.20). The ground-state structure of the neutral Nb_3O could be assigned from our calculation to a $^2B_2$ state. We assigned a $^1A_1$ state as the ground-state structure of the anionic system. Similar results were found for the ground-state structure of the neutral and anionic V_3O cluster [128,183]. Both, neutral and anionic ground-state structures, are followed by planar structures in the quar-tet ($^4A_1$) and triplet ($^3A_1$) PES, respectively. The quartet state lies 1.08 eV above the neutral ground state while the energy difference between the singlet and the triplet anionic states is 0.49 eV.

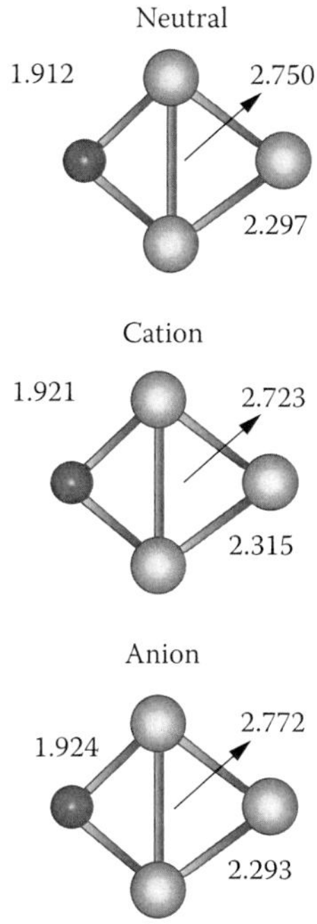

FIGURE 9.20 Ground-state structures of Nb_3O (top), Nb_3O^+ (middle), and Nb_3O^- (bottom) clusters. All bond lengths are in Å. (Reprinted from Calaminici, P. et al., *J. Chem. Phys.*, 121(8), August 22, 3558, 2004. With permission.)

9.4 VIBRATIONAL ANALYSIS

9.4.1 V_3

In order to assign the ground-state structure of the V_3 cluster, we simulated its PFI-ZEKE spectrum. For this purpose, we have calculated the harmonic frequencies of the optimized neutral and cationic ground-state vanadium trimer. For the neutral $^2A_1'$ state, we obtain a degenerate E' mode of 255 cm^{-1} and an A_1' mode of 421 cm^{-1}. The calculated frequencies for the $^4A_2$ quartet state are 144, 259, and 428 cm^{-1}. For the cationic $^3A_2'$ ground state, we find an E' mode of 253 cm^{-1} and an A_1' mode of 434 cm^{-1}. The obtained frequencies for the excited cationic $^1A_1'$ state are 266 and 422 cm^{-1} [143].

Figure 9.21 shows the experimental PFI-ZEKE spectrum recorded at room tem-perature (a) and the simulated one for the transition from the neutral $^2A_1'$ to the cat-ionic $^3A_2'$ state. The simulation was performed at room temperature (b) and at 700 K (c). The theoretical frequencies are reported relative to the experimental adiabatic ionization potential, which correspond to the origin of the PFI-ZEKE spectrum at 44,342 cm^{-1}.

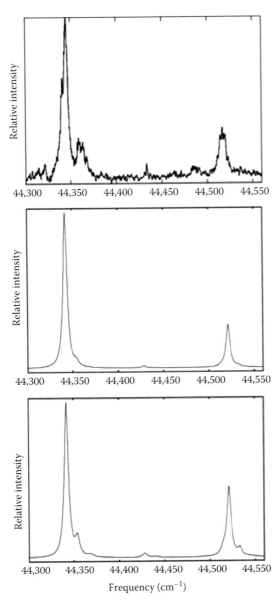

FIGURE 9.21 Experimental (a) and simulated PFI-ZEKE spectrum for the $^2A'_1 \rightarrow {}^3A'_2$ transition at room temperature (b) and 700 K (c). Both simulated spectra were convoluted with a 5 cm^{-1} FWHM Lorentzian line shape to simulate the (rotational) width of the experimental bands. (Reprinted from Calaminici, P. et al., *J. Chem. Phys.*, 114(9), March 1, 4036, 2001. With permission.)

As Figure 9.21 shows, the agreement between the experimental and theoretical PFI-ZEKE spectrum is very satisfying. Therefore, the reliability of the ground-state assignment of vanadium trimer is proven by the simulation of the experimental PFI-ZEKE spectrum.

9.4.2 Nb$_3$O

The vibrationally resolved 488 nm negative ion photoelectron spectrum of Nb$_3$O provides measurements of its electron affinity and vibrational frequencies, which can be summarized as follows. The electron affinity of Nb$_3$O is 1.393 ± 0.006 eV [128]. The metal–oxygen stretching frequency is 710 ± 15 cm^{-1} for the neutral Nb$_3$O cluster. Lower symmetric modes are also active, with frequencies of 320 ± 15 cm^{-1} for the neutral Nb$_3$O and 300 ± 20 cm^{-1} for its anion. The Nb$_3$O spectrum indicates that the extra electron in the anionic system occupies essentially a nonbonding orbital and that the neutral and anionic clusters have very similar structures. We have presented computational results for structural and spectroscopic properties of both neutral and anionic niobium trimer monoxides [185]. We used DFT calculations in combination with scalar quasirelativistic ECPs. In order to test the accuracy of the used ECPs in the framework of the Kohn-Sham method, the PFI-ZEKE spectrum of Nb$_3$O [108] was simulated from the harmonic vibrational frequencies of the neutral and cationic ground-state structures and the corresponding normal mode vectors. Figure 9.22 shows the comparison between the experimental PFI-ZEKE spectrum (a) recorded at 300 K (top) and at 100 K (bottom) and the corresponding spectra obtained by our simulation (b).

As Figure 9.22 shows, the agreement between the experimental and theoretical PFI-ZEKE is very satisfying. Moreover, we notice that our simulated spectra are in better agreement with the experimental ones than the simulated spectra reported in

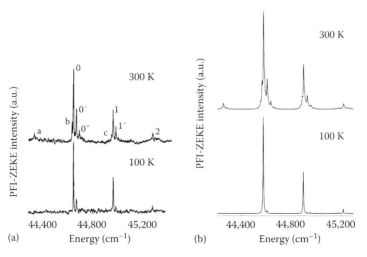

FIGURE 9.22 Experimental (a) and simulated (b) PFI-ZEKE spectra of Nb$_3$O at 300 and 100 K. The position of the 0–0 band has been shifted to the experimental value of 44,578 cm^{-1}. (Reprinted from Calaminici, P. et al., *J. Chem. Phys.*, 121(8), August 22, 3558, 2004. With permission.)

TABLE 9.1

Peak Positions, Assignment, and Relative Intensity, Given as $v_{ion}^{neutral}$, of the Simulated $Nb_3O^- \rightarrow Nb_3O +$ e^- Photoelectron Spectrum at 300 K

Our Notation (Neutral)	Assignment	Spacing from Origin (cm⁻¹)	Relative Intensity
3	3_1^0	−310 (−295)	0.017 (0.036)
	0_0^0	0	1
3	3_0^1	316 (320)	0.135 (0.164)
3	3_0^2	631 (635)	0.001 (0.001)
6	1_0^1	738 (710)	0.037 (0.033)
	$1_0^1 3_0^1$	1054 (1025)	0.077 (0.012)

Source: Reprinted from Calaminici, P. et al., *J. Chem. Phys.*, 121(8), August 22, 3558, 2004. With permission.

Note: Experimental data from Ref. [128] are given in parentheses.

Ref. [108]. This is probably due to the difference in bond lengths calculated for the neutral Nb_3O system (see Figure 9.20 of this work and Figure 9.4 of Ref. [108]). We account this to the improved accuracy of the numerical integration of the exchange-correlation potential in deMon2k.

In Ref. [185], we have compared the calculated spacing from the origin and the relative intensities for the most relevant peaks with experiment. We list in Table 9.1 the peak positions, the assignment, and relative intensity of the simulated photoelectron spectrum at 300 K together with the experimental data. For easy comparison, we have used the same labeling as in Ref. [128]. The experimental values are given in parentheses. As Table 9.1 shows, the calculated spacing of all peaks is in good agreement with the observed data. Also the intensities of the hot band and the fundamentals show fair agreement with experiment. The intensity of the non-Franck-Condon transition $1_0^1 3_0^1$, however, is much too large in our simulation. We have observed a similar behavior also in the case of the V_3O study [184]. Our studies indicate that these disagreements can be attributed to small errors in the force field of the anion. The calculated adiabatic electron affinity is 1.39 eV. This value is in excellent agreement with the experimental reported value of 1.393 ± 0.006 eV from negative ion photoelectron spectroscopy [128].

9.5 POLARIZABILITIES (LI, NA, CU, FE, NI)

9.5.1 SIMPLE METAL SYSTEMS (LI, NA, CU)

The static polarizabilities of atoms and free clusters have been extensively studied both theoretically and experimentally (we refer the reader to Refs. [186,187] and references therein). The static polarizability α describes the induced dipole moments by static external electric field in a molecular cluster. It represents one of the most important observables for the study of the electronic structure of clusters, since it is very sensitive to the delocalization of valence electrons, as well as the structure and shape. Despite numerous investigations on metal clusters, over many years, static polarizabilitity measurements were only available for alkali-metal clusters such as sodium, lithium, and potassium [188–191]. Because of their particular configuration, homo-nuclear alkali-metal clusters are often considered to be the simplest metal clusters, so that they have become the prototype systems for understanding size effects in metal clusters. The experimental work of Knight et al. [188] by electric deflection techniques has shown that the optical properties of alkali metal aggregates, such as sodium and potassium clusters, follow a general trend toward the bulk value. Knight et al. have also shown the existence of a pronounced size dependency in the polarizability of small clusters. In fact, for sodium aggregates, from the atom to the pentamer, the polarizability per atom has an oscillating behavior. After the pentamer, it decreases to a minimum for the octamer and it increases then again for the nonamer. Potassium-cluster polarizabilities closely follow the same pattern [188]. Later on, Benichou et al. [189] have measured static electric polarizabilities of lithium clusters up to 22 atoms by deflecting a well-collimated beam through a static inhomogeneous transverse electric field. In order to avoid any systematic error in their experiment, Benichou et al. also carried out measurements of sodium cluster polarizabilities. The values they obtained were in close agreement with those previously measured by Knight et al. [188]. The work of Benichou et al. shows that the trend of the polarizability per atom of small lithium clusters differs from those of small sodium and potassium clusters. In fact, a sharp decrease by about a factor of two from the monomer to the trimer is observed. For larger sizes, $n \geq 4$, the polarizability per atom decreases slowly. For $n \geq 15$, a marked oscillation is observed. The sudden transition of the polarizability from atom to trimer and the following slow variation were interpreted by Benichou et al. [189] as a sign that the electronic delocalization already appears in Li_4 or Li_5 clusters. The electronic configuration of the noble metals Cu, Ag, and Au is characterized by a closed d shell and a single valence electron and therefore closely related to that of the alkali metals. In view of this prominent characteristic, clusters of noble metals are expected to exhibit certain similarities to alkali-metal clusters.

Around 10 years ago, we presented the first theoretical study of an all-electron gradient-corrected density functional study of static polarizability and polarizability anisotropy for copper clusters up to the nonamer [192]. The calculations of polarizabilities and polarizability anisotropies were performed at the ground-state structures found on each PES with the expression

$$\bar{\alpha} = \frac{1}{3}[\bar{\alpha}_{xx} + \bar{\alpha}_{yy} + \bar{\alpha}_{zz}]. \tag{9.77}$$

The obtained results of the static polarizabilities were compared with experimental data for sodium and lithium clusters. All details concerning the computations are given in Ref. [192]. The static mean polarizabilities and the polarizability anisotropies of copper clusters up to the nonamer were calculated with the PB86 functional at the optimized geometries obtained by using the VWN functional [192]. In Figure 9.23, the mean polarizability per atom of copper clusters is plotted.

As this figure shows, going from the atom up to the pentamer, the mean polarizability per atom has an oscillating behavior. From the pentamer, the polarizability per atom decreases with a minimum value for the octamer. It increases then again, going from the octamer to the nonamer.

In Figures 9.24 and 9.25, the experimental polarizability per atom of sodium and lithium clusters up to the nonamer is plotted, respectively.

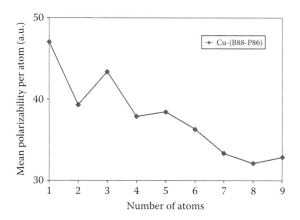

FIGURE 9.23 Calculated mean polarizability per atom of Cu_n ($n=2-9$). The calculations were performed using the B88-P86 functional. (Reprinted from Calaminici, P. et al., *J. Chem. Phys.*, 113(6), August 8, 2199, 2000. With permission.)

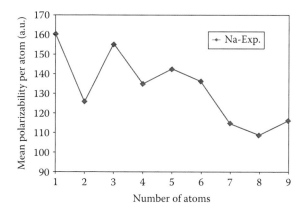

FIGURE 9.24 Experimental mean polarizability per atom of Na_n ($n=2-9$). Values are from Ref. [188]. (Reprinted from Calaminici, P. et al., *J. Chem. Phys.*, 113(6), August 8, 2199, 2000. With permission.)

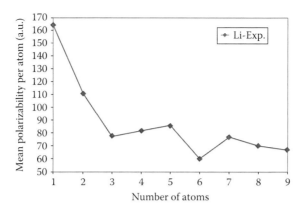

FIGURE 9.25 Experimental mean polarizability per atom of Li_n ($n = 2$–9). Values are from Ref. [189]. (Reprinted from Calaminici, P. et al., *J. Chem. Phys.*, 113(6), August 8, 2199, 2000. With permission.)

The values of the experimental polarizability per atom of sodium clusters are those reported by Knight et al. [188], while the experimental polarizability data of lithium clusters are those reported by Benichou et al. [189]. The comparison of Figures 9.23 through 9.25 shows that the calculated polarizability per atom of copper clusters, going from the atom to the nonamer, presents the same trend as experimentally observed by Knight et al. [188] for sodium clusters. However, the mean polarizability per atom of copper clusters is about three times smaller. This result indicates that in copper clusters, the electrons are more strongly attracted by the nuclei than in the sodium clusters. Therefore, their electronic structure is more compact. We are confident that the reliability of the absolute values for the calculated polarizabilities of the copper clusters is in the same range as for the previously studied sodium clusters [193]. Thus, deviations of less than 10% for the absolute values can be expected. The theoretical prediction of the trend of the mean polarizabilities per copper atom is believed to be reliable. In fact, it was confirmed by several other theoretical studies [194–196] too. However, the experimental work of Knickelbein [197] on copper cluster polarizabilty has shown large discrepancies between the experimental and theoretical polarizability value of Cu_9. Further experimental and theoretical studies are needed to resolve this mismatch.

9.5.2 TRANSITION METAL SYSTEMS (FE, NI)

In the previous section, we showed that the electronic structure of small copper clusters is more closely related to that of sodium rather than to that of lithium clusters. However, generally, the TMs are characterized by an open *d* shell, and, therefore, the prediction of their electronic properties is even more complicated. Moreover, these calculations also provide a means of testing computational methods and are, therefore, very interesting for theoreticians. From a theoretical point of view, the calculation of properties of iron- and nickel-containing systems presents a challenging goal for any quantum chemical methodology. The investigation of electronic and geometric structure of these systems was the subject of several DFT investigations in the last 10 years or so, and it would not be appropriate to review all of them here.

Among all the available works on iron systems, those published by Salahub and Chrétien [198] and Gutsev and Bauschlicher [199] are particularly interesting. Both works present results of extensive studies performed on neutral, cationic, and anionic small iron clusters, concluding that the determination of the ground-state structure of small iron clusters depends on the choice of the correlation functional.

Despite the fact that many works on these systems are available, additional experimental information is required in order to unequivocally assign the ground-state structure of small iron clusters.

Since the static polarizability is a property directly related to the electronic properties of clusters, polarizability measurements would be highly valuable in definitively answering questions about the electronic structure of these systems. However, no experimental studies on the electric properties of these small-sized systems are available. Therefore, we carried out theoretical studies on the static polarizability of small iron and nickel clusters in order to gain more insight into their electronic structure and to guide future experimental investigations [200,201]. For both, iron and nickel atoms, newly developed basis sets were employed for the polarizabilitiy calculations. For these calculations, the finite field approach of Kurtz et al. [202] implemented in the deMon2k code was used. The reliability of this approach was extensively tested for various systems [192,193,203–207], including sodium and copper clusters as well as iron-containing molecules. These studies demonstrated that DFT methods can be used in order to predict fairly accurate polarizability values for metallic and TM systems. In Figure 9.26, the mean polarizability per atom (in a.u.) of the studied iron clusters up to the tetramer is plotted.

As Figure 9.26 shows, going from the atom up to the tetramer, the mean polarizability per atom has an oscillating behavior. The comparison of this plot with those of Figures 9.23 through 9.25 shows that the calculated polarizability per atom of small iron clusters, going from the atom to the tetramer, presents the same trend as that experimentally observed by Knight et al. [188] for sodium clusters. However,

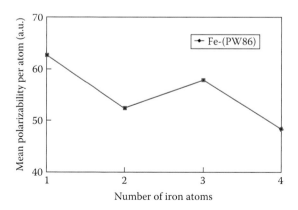

FIGURE 9.26   Mean polarizability per atom of Fe$_n$ ($n = 1$–4).

the mean polarizability per atom of iron clusters is about two times smaller. This result shows that in iron clusters, the electrons are more strongly attracted by the nuclei than in the sodium clusters as expected. Therefore, their electronic structure is more compact. We have already observed this same behavior in the study of polarizabilities of copper clusters [192]. The calculated polarizability anisotropies for Fe_n ($n = 2$–4) are reported in Table I of Ref. [200]. The polarizability anisotropy increases from the dimer to the trimer by about more than two times and increases slightly, going from the trimer to the tetramer.

A few years ago, the electric dipole polarizabilities of nickel clusters in the range from 12 to 58 atoms were measured via a molecular beam deflection experiment [208]. This work showed that nickel aggregates adopting closely packed or quasi-spherical structures correspond to local minima cluster polarizabilities, while icosahedral structures with missing atoms correlate to local maxima. For other clusters such as Ni_{21} and Ni_{22}, no correlation between the structure and the polarizability was observed [208]. So far, polarizability values of smaller nickel systems are available only for the Ni atom [209,210]. In order to gain more insight into the structure–polarizability relationships of these clusters and to guide future experimental investigations, a study of the static polarizability focusing on cluster sizes smaller than 10 atoms is needed. Recently, a very extensive investigation on neutral, cationic, and anionic small nickel clusters up to the pentamer was performed [146], employing newly developed nickel all-electron basis sets optimized for the GGA [48]. As previously discussed, it was demonstrated that by combining this GGA basis set for the Ni atom with GGA functionals, reliable structures for small nickel clusters can be obtained. The electric properties of small nickel clusters were computed at the ground-state structures. In Figure 9.27, the mean polarizability per atom of the studied nickel clusters is displayed.

As Figure 9.27 shows, a sharp decrease from the monomer to the trimer can be observed. From the trimer to the tetramer, the static polarizability per atom increases substantially; and from the tetramer to the pentamer, it decreases to a value very similar to the one calculated for the trimer. As we can see, the behavior of the static

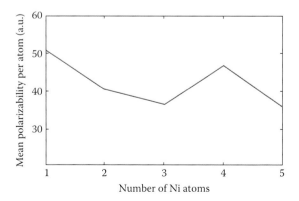

FIGURE 9.27   Mean polarizability per atom of Ni_n ($n = 1$–5). (Reprinted from Calaminici, P., *J. Chem. Phys.*, 128(16), April 28, 164317-1, 2008. With permission.)

polarizability per atom of nickel aggregates is different from the one predicted in other TM clusters of similar size as iron and copper clusters. Interestingly, we notice that this trend looks very similar to the one experimentally observed for small lithium clusters [189]. In fact, in their experimental work on lithium clusters, Benichou et al. observed a sharp decrease by about a factor 2 from the lithium monomer to the lithium trimer. The sudden transition of the polarizability from atom to trimer was interpreted by the authors as a sign that the electronic delocalization already appears in Li_4 and Li_5 clusters. However, the mean polarizability per atom of nickel clusters is about three times smaller.

9.5.3 TEMPERATURE DEPENDENCE OF SODIUM CLUSTER POLARIZABILITIES

Over the last two decades, different error sources have been investigated in order to resolve the long-standing discrepancy between theoretical and experimental sodium cluster polarizabilities, but the situation remains unresolved. More recently, it has been speculated that the mismatch between calculated and measured sodium cluster polarizabilities is due to finite temperature effects [211–213]. In fact, this idea was already mentioned in the original experimental work [188]. However, a systematic study of the temperature dependence of sodium cluster polarizabilities at a reliable first-principles all-electron level of theory has never been performed. In this work, we address this question. In order to study the dynamics of small sodium clusters at finite temperatures, BOMD calculations are performed at the PBE/DZVP/ A2 level of theory. For each cluster, from the dimer to the nonamer, 18 trajectories are recorded in a temperature range from 50 to 900 K with intervals of 50 K. Each trajectory has a length of 220 ps and was recorded with a time step of 2 fs. Similar statistics have already been successfully applied to determine the melting temperatures of sodium clusters with LDA pseudo-potential DFT molecular dynamics. The temperature in the canonical BOMD simulation was controlled by a Nosé-Hoover chain thermostat. In order to study the temperature dependence of the sodium cluster polarizabilities, the polarizability tensor was calculated along the recorded trajectories. For this purpose, the first 20 ps of each trajectory were discarded, and α was then calculated in 100 fs time steps along the remaining 200 ps. Due to the computational demand of the analytic polarizability calculation, we employed the LDA exchange-correlation kernel, which represents the second functional derivative of the exchange-correlation energy with respect to the density. Thus, the computational level for the calculation of the temperature-dependent part of the cluster polarizabilities is VWN/TZVP-FIP/GEN-A2*. The temperature-dependent mean sodium cluster polarizability is then calculated as

$$\bar{\alpha}(T) = \bar{\alpha}^{PBE}(0) + \delta\bar{\alpha}^{VWN}(T), \tag{9.78}$$

with

$$\bar{\alpha}(T) = \frac{1}{3}[\bar{\alpha}_{xx}(T) + \bar{\alpha}_{yy}(T) + \bar{\alpha}_{zz}(T)]. \tag{9.79}$$

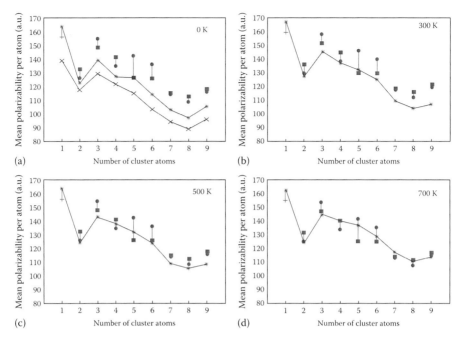

FIGURE 9.28 Experimental (atom: Molof et al. [215]; circle: Knight et al. [214]; square: Rayane et al. [191]) and theoretical mean polarizabilities per atom [a.u.] of Na_n ($n=2$–9) clusters. The theoretical values are calculated at the VWN/TZVP-FIP/GEN-A2* (crosses) and PBE/TZVP-FIP/GEN-A2* (stars) level of theory. These values are connected in order to guide the eye. The individual graphs display calculated polarizabilities at 0 (a), 300 (b), 500 (c), and 700 K (d). The cluster temperature in the experiment is estimated to be between 400 and 800 K. (Reprinted from *J. Phys. Chem. A*, 112(47), November 1, 11969, 2008. With permission.)

This approximation assumes that the temperature dependency of $\bar{\alpha}(T)$, namely $\delta\bar{\alpha}(T)$, is the same at the PBE and VWN level of theory. It should be remembered that the geometries are derived from PBE BOMD calculations. In Figure 9.28, the calculated cluster polarizabilities at 0 K (a), 300 K (b), 500 K (c), and 700 K (d) are depicted. As this figure shows, the individual cluster polarizabilities increase with temperature. Somewhere between 500 and 700 K, the calculated $\bar{\alpha}(T)$ per atom match the experimental data sets. In particular, the comparison of the calculated and experimental $\bar{\alpha}(T)$ per atom at 700 K for Na_7, Na_8, and Na_9, for which excellent agreement between the different experimental data sets exists, is very satisfying. It is important to mention that the experimental cluster temperature is estimated to be between 400 and 800 K. Moreover, it is interesting to note that the oscillating behavior of the polarizability per atom for the smaller clusters, which was observed in the original measurement [214] and also appears, less pronounced, in the $T=0$ K PBE calculations (Figure 9.28a, stars), disappears at higher temperatures. Instead, the per polarizabilities atom decreases monotonically from Na_3 to Na_8. Therefore, the finite temperature polarizabilities show no characteristic oscillations for open and closed shell systems. Instead, they only reflect the shell closing at the dimer and octamer consistent with the jellium model [214]. Our BOMD calculations also show that cluster fragmentations are not important for the cluster polarizabilities. Such fragmentations occur in our simulations above 800 K. The change in the trend of the polarizability per atom for the small sodium clusters with increasing temperature is due to the different temperature dependencies of the individual cluster polarizabilities.

In Figure 9.29a, the temperature dependencies of the individual cluster polarizabilities are presented. In this graph, the behavior of Na_4 is particularly interested. Over a large temperature range up to 600 K, the $\delta\bar{\alpha}(T)$ value for this cluster changes almost linearly with the temperature. A closer analysis reveals that in this temperature range, the D_{2h} rhombic structure of Na_4 rearranges only in the molecular plane. At 600 K and above, three-dimensional rearrangements occur. In this case, the temperature dependence of the polarizability directly reflects the dynamics of the cluster rearrangement. In Figure 9.29b, typical snapshots of the highest occupied Na_4 MOs at 200 K (left) and 750 K (right) during the cluster rearrangements are shown. At low temperature, the s-type Na_4 orbital (nodeless) remains delocalized over the full system (lower orbital on the left side of Figure 9.29b), whereas at higher temperature, this orbital correlates with the HOMO forming two Na_2 s-type orbitals. As a consequence, the two Na_2 fragments can rearrange over three-dimensional transition states. Figure 9.29a also shows that $\delta\bar{\alpha}(T)$ increases considerably faster with temperature for Na_4 than for Na_5. As a result, the bump in the static $T=0$ K PBE polarizabilities at the pentamer (Figure 9.28a) disappears for the finite temperature polarizabilities.

In conclusion, we have shown that the calculated $\bar{\alpha}(T)$ per atom match well with the available experimental data sets at around 700 K. Thus, the long-standing discrepancy between theory and experiment is resolved by inclusion of finite temperature effects in the electronic structure calculation. The calculated finite temperature sodium cluster polarizabilities show characteristic minima at the dimer and octamer as expected from the jellium model. However, individual molecular structures besides these two are not resolved in the calculated finite temperature sodium cluster polarizabilities.

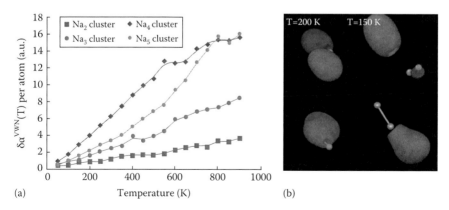

(a)

(b)

FIGURE 9.29 (a) Change of the mean polarizability per atom with temperature for Na_n clusters with $n=2$–5. (b) The two highest occupied orbitals of Na_4 at 200 K (left) and 750 K (right), respectively. (Reprinted from *J. Phys. Chem. A*, 112(47), November 1, 11969, 2008. With permission.)

9.6   REACTIONS

In this final section, we will review briefly two studies in, arguably, the most difficult area of all, namely, the direct calculation of chemical reactions involving TM species. The first example, aimed at understanding the gas-phase synthesis of benzene from acetylene on small iron clusters, was performed some 7 or 8 years ago, and it highlights both the promise and the difficulty of DFT calculations on even small TM clusters. Modern techniques for locating transition states were not available in deMon at that time. The second example involves iron oxide clusters reacting with carbon monoxide, and it benefits from the improvements that are now implemented in deMon2k.

9.6.1   FORMATION OF BENZENE FROM ACETYLENE ON SMALL ($N = 1-4$) FE CLUSTERS

We will review briefly three papers by Chrétien and Salahub that together provide considerable insight into the complex systems and reactions involved in making even a "simple" molecule like benzene using iron clusters as the catalyst.

The study was motivated by early experimental work by Irion and collaborators [216] who found in beam experiments that benzene could be produced from Fe^+, Fe_4^+, and Fe_5^+, but not from Fe_2^+, Fe_3^+, or larger clusters (up to Fe_{13}^+). The small size of the active clusters seemed very attractive for a theoretical study, even though we were well aware of the potential structural and electronic complexity of a small cluster such as Fe_4^+.

The first paper [198] treated the isolated iron clusters and involved an extensive series of calculations on Fe_n^+, Fe_n, and Fe_n^- clusters. Careful attention was paid to all of the computational parameters (exchange-correlation potentials, basis sets, integration grids, number of unpaired spins, starting geometries, SCF convergence, etc.), and we tried to establish reasonable (conservative) error bars for the calculations. In the end, comparing available calculations and experimental values, we proposed that an error bar on the relative energy of the various states should be set at around 50 kJ/mol (12 kcal/mol), and we have kept all states within this error bar of the computed ground state in the subsequent studies of the reactions.

Table IV of Ref. [198] reproduced here as Table 9.2 will give the reader a sense of the extent of the calculations for Fe_4, Fe_4^+, and Fe_4^-. Table 9.3 summarizes the available experimental results for D_0, IP, and EA of all of the clusters up to the tetramer. While the agreement is "reasonable" in particular concerning trends with cluster size, it is clear that quantitative agreement is still beyond the reach of GGA DFT. While our approach is not perfect (see Ref. [198] for a discussion on the limiting factors), there is enough encouragement in the performance of the methodology to allow the study of reactions on the clusters, provided that the quantitative limitations are kept in mind.

The second paper from Chrétien and Salahub [217] looks at the adsorption of acetylene and vinylidene on the clusters and has, as the end result, the elimination of vinylidene as a reactant, thus greatly simplifying the computational task when three

TABLE 9.2

Relative Energies (in kJ/mol), Optimized Structures (Distances in Å and Angles in Degrees), and Vibrational Frequencies (in cm^{-1}) of Low-Lying States of Fe$_4$, Fe$_4^+$, and Fe$_4^-$ Calculated with PP86

Method	IS (Symmetry)[a]	State	ΔE	d_{12}	d_{13}	d_{34}	θ_{314}	Θ_{132}	Θ_{3124}	ω_i
								Fe$_4$		
B	TET(Cs)	$^{15}A''$	0.0	2.402	2.395	2.195	54.5	60.2	64.0	97(a1); 107(a1); 232(a2); 233(b1); 234(b2); 332(a1)[b]
A+B	TET(C2v)	$^{15}A_2$	0.8	2.632	2.273	2.637	70.9	70.9	90.7	
B	TET(C2v)	$^{15}B_2$	3.9	2.367	2.341	2.438	62.8	60.8	74.3	
A+B	TET(C2v)	$^{13}A_1$	8.0	2.245	2.352	2.245	57.0	57.0	65.8	167(a2); 181(a1); 183(b1); 186(b2); 276(a1); 361(a1)[b]
B	TET(Cs)	$^{13}A''$	11.5	2.283	2.294	2.423	63.8	59.7	75.0	
B	TET(C2v)	$^{15}A_1$	14.0	2.276	2.389	2.325	58.2	56.9	67.2	
A+B	RHO(D2h)	$^{13}B_{3g}$	19.9	2.437	2.234	3.745	113.9	66.1	180.0	95(b1u); 163(b1g); 202(ag); 238(b2u); 290(b3u); 325(ag)[c]
A+B	RHO(D2h)	$^{15}B_{2g}$	24.2	2.598	2.263	3.706	109.9	70.1	180.0	74(b1u); 175(ag); 194(b1g); 203(b2u); 278(ag); 283(b3u)[c]
B	TET(C2v)	$^{17}B_2$	24.3	2.437	2.389	2.478	62.5	61.3	74.2	
A+B	TET(C2v)	$^{17}A_1$	26.3	2.355	2.438	2.356	57.8	57.8	67.0	142(a2); 177(b2); 180(b1); 203(a1); 250(a1); 316(a1)[b]
B	RHO(D2h)	$^{15}A_u$	28.5	2.595	2.289	3.773	111.0	69.0	180.0	
A+B	RHO(C2v)	$^{13}A_1$	29.4	2.292	2.269	3.316	93.9	60.7	115.6	67(a1); 143(a2); 186(a1); 232(b2); 270(b1); 349(a1)[b]
A+B	RHO(C2v)	$^{15}A_1$	29.6	2.431	2.300	3.805	111.7	63.8	154.1	70(a1); 159(a2); 162(b2); 177(a1); 296(b1); 303(a1)[b]
B	SQR(D2h)	$^{15}B_{1g}$	34.3	2.306	2.306	2.306	45.0	45.0	0.0	
B	RHO(D2h)	$^{15}A_g$	37.3	2.517	2.295	3.839	113.5	66.5	180.0	
B	RHO(C2v)	$^{13}A_2$	48.0	2.547	2.252	3.578	105.2	68.9	148.9	
B	REC(D2h)	$^{15}B_{2g}$	57.4	2.361	2.232	2.361	43.4	43.4	0.0	

(continued)

TABLE 9.2 (continued)

Relative Energies (in kJ/mol), Optimized Structures (Distances in Å and Angles in Degrees), and Vibrational Frequencies (in cm⁻¹) of Low-Lying States of Fe_4, Fe_4^+, and Fe_4^- Calculated with PP86

IS Method	(Symmetry)[a]	State	ΔE	d_{12}	d_{13}	d_{34}	θ_{314}	Θ_{132}	Θ_{3124}	ω_i
B	LIN(D2h)	$^{15}\Delta_u^d$	201.1	4.607	2.093	2.577	0.0	180.0	0.0	179(e); 255(t1); 368(a1)[c]
								Fe_4^+		
A+B	TET(C2v)	$^{12}A_1$	0.0	2.301	2.301	2.301	60.0	60.0	70.5	
B	TET(C2v)	$^{14}A_2$	30.3	2.483	2.353	2.246	57.0	57.0	70.3	92(a1); 151(a2); 178(a1); 231(b1); 252(b2); 345(a1)[b]
A+B	RHO(C2v)	$^{14}B_2$	31.6	2.353	2.291	2.830	76.3	76.3	145.4	
B	TET(C2v)	$^{14}B_2$	33.7	2.299	2.410	2.222	54.9	54.9	73.6	
B	TET(C2v)	$^{14}B_2$	43.0	2.407	2.347	2.327	59.4	61.7	70.1	
B	TET(Cs)	$^{10}A'$	91.9	2.037	2.322	2.219	57.1	52.0	64.1	
A	TET(C1)	^{10}A	92.1	2.049	2.347	2.223	57.2	51.8	64.6	110(a); 129(a); 165(a); 192(a); 304(a); 404(a)[f]
B	REC(D2h)	$^{14}B_{2u}$	104.1	2.247	2.347	2.247	43.8	43.8	0.0	
B	SQR(D2h)	$^{14}B_{1g}$	111.2	2.248	2.248	2.248	45.0	45.0	0.0	
B	LIN(D2h)	$^{16}\Delta_g^d$	254.0	4.772	2.188	2.584	0.0	180.0	0.0	
							2	Fe_4^-		
B	TET(Cs)	$^{16}A''$	0.2	2.363	2.373	2.422	61.4	59.9	72.5	
B	TET(C2v)	$^{16}B_1$	0.3	2.388	2.389	2.312	57.9	60.0	67.9	
B	TET(Cs)	$^{16}A''$	4.3	2.411	2.323	2.449	63.6	61.8	74.9	
A+B	TET(C2v)	$^{16}A_2$	5.5	2.476	2.329	2.463	63.9	64.2	77.3	141(a1); 159(a1); 206(a2); 233(b1); 233(b2); 322(a1)[b]
A+B	RHO(D2h)	$^{14}A_u$	28.0	2.514	2.243	3.716	111.8	68.2	180.0	99(b1u); 175(b1g); 200(ag); 218(b2u); 296(b3u);303(ag)[c]
A+B	TET(C2v)	$^{14}A_2$	35.6	2.409	2.303	2.407	63.0	63.1	75.6	160(a1); 180(a2); 197(b2); 198(b1); 214(a1); 341(a1)[b]

B	TET(Cs)	$^{14}A'$	39.8	2.360	2.357	2.235	56.6	59.9	66.5	
A+B	RHO($C2v$)	$^{14}B_1$	48.3	2.433	2.267	3.367	95.9	64.9	123.2	64($a1$); 165($a1$); 174($a2$); 205($b1$); 269($b2$); 321($a1$)[b]
B	RHO($C2v$)	$^{14}B_2$	53.6	2.648	2.265	3.545	103.0	71.5	149.4	
B	REC($D2h$)	$^{14}B_{3u}$	80.7	2.352	2.215	2.352	46.7	46.7	0.0	
B	SQR($D2h$)	$^{14}B_{1g}$	88.3	2.267	2.267	2.267	45.0	45.0	0.0	
A+B	RHO($C2v$)	$^{12}A_2$	94.4	2.478	2.223	3.683	111.9	67.8	173.2	64($a1$); 102($a2$); 203($a1$); 227($b1$); 258($b2$); 319($a1$)[b]
A	RHO($C1$)	$^{10}A_2$	143.0	2.873	2.167	2.874	86.0	85.9	137.2	68(a); 130(a); 244(a); 248(a); 342(a); 369(a)[f]
B	RHO($C2v$)	$^{10}B_1$	144.2	2.728	2.109	3.026	91.7	80.7	140.5	
B	TET($C2v$)	$^{18}B_2$	194.1	2.458	2.409	2.515	62.9	61.3	74.7	
B	LIN($D2h$)	$^{14}\Sigma_o^{+d}$	200.5	6.719	2.063	2.593	0.0	180.0	0.0	

Source: Salahub, D. et al., *Phys. Rev. B*, 66, 155425, 2002. With permission.

[a] Initial structure (IS) defined in Figure 1 of Ref. [198]. (TET=tetrahedral; RHO=rhombus SQR=square; REC=rectangular LIN=linear) and symmetry group imposed during the optimization with method B between parentheses.

[b] Attribution based on $C2v$.

[c] Attribution based on $D2h$.

[d] Attribution based on $D'h$.

[e] Attribution based on Td.

[f] Attribution based on $C1$.

TABLE 9.3

Binding Energies (D_0), Adiabatic Ionization Potentials (IP), and Adiabatic Electronic Affinities (EA) (in kJ/mol) for Small Iron Clusters Calculated with PP86

	Calculated	Experimental
		D_0	
Fe-Fe	189	114
Fe_2-Fe	192	184
Fe_3-Fe	248	211
Fe^+-Fe	299	268
Fe_2^+-Fe	275	169
Fe_3^+-Fe	274	216
Fe^--Fe	183	183 ± 9
Fe_2^--Fe	253	235
Fe_3^--Fe	297	245
		IP	
Fe	814	762 ± 1
Fe_2	689	608 ± 1
Fe_3	606	622 ± 5
Fe_4	598	618 ± 10
		EA	
Fe	115	14.6 ± 0.3
Fe_2	113	87.0 ± 0.8
Fe_3	173	138 ± 6
Fe_4	203	172 ± 6

Source: Salahub, D. et al., *Phys. Rev. B*, 66, 155425, 2002. With permission.

acetylene molecules must come together to form benzene, the subject of the third paper [218].

A very large number of calculations had to be performed in order to explore the binding of acetylene or vinylidene in the various local minima that result, for all of the low-lying (<50 kJ/mol) states of the clusters. Figure 9.30 shows the starting geometries.

Figure 9.31 shows the various low-lying structures found for the 4-atom clusters along with their relative energies and the number of unpaired spins.

Note that for the positive clusters, the lowest vinylidene structure is about 25 kJ/mol above the ground state. Moreover, we have calculated the transition states for isomerization of acetylene to vinylidene in the gas phase and on the clusters (using Gaussian94 to locate the transition states and fix their geometries, followed by

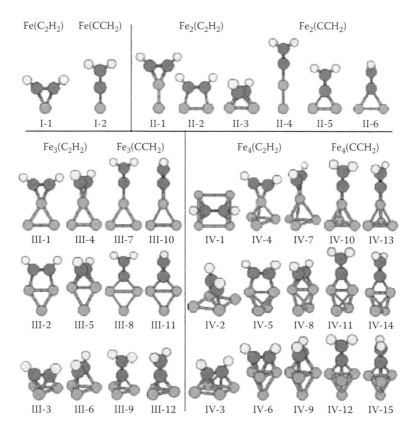

FIGURE 9.30 Initial structures used to study the adsorption of an acetylene or a vinylidene molecule on iron clusters Fe_n/Fe_n^+ ($n = 1–4$). (Reprinted from Chretien, S. et al., *J. Chem. Phys.*, 119(23), December 5, 12279, 2003. With permission.)

single-point calculations using deMonKS3P4). For the gas-phase reaction, activation energies of 182.7 and 199.3 kJ/mol are obtained with the PP86 and PLAP3 functionals, respectively. These two DFT results bracket the CCSD(T) value of 187.4 kJ/mol. The values of the isomerization energy and the activation energy are shown in Table VI of Ref. [217], shown here as Figure 9.32. The missing values are due to our inability to converge some of the transition states with Gaussian94. The activation energies are all very high and greater than the energy that would be available from the adsorption of a single acetylene molecule on the clusters. Consequently, we believe that vinylidene will not be involved in the reaction mechanism.

In the third paper, Chrétien and Salahub go as far as the then current technology would take them toward mapping out the mechanism for the formation of benzene from acetylene on the small clusters. A schematic representation of the mechanisms considered is shown in Figure 9.33.

All of these steps were considered in a truly extensive (but of course not exhaustive) series of calculations that involved some 381 starting structures and more than 2000 unconstrained geometry optimizations for the various spin states. Some transition

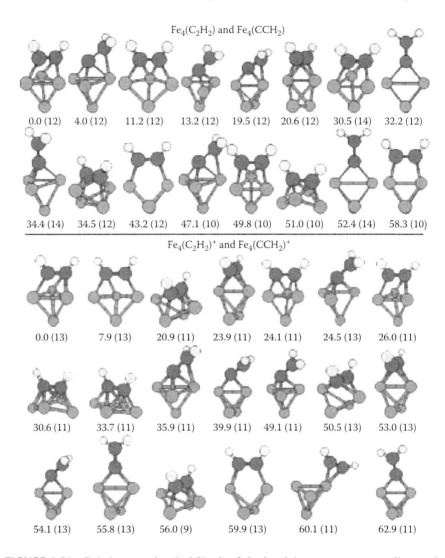

Fe$_4$(C$_2$H$_2$) and Fe$_4$(CCH$_2$)

0.0 (12)   4.0 (12)   11.2 (12)   13.2 (12)   19.5 (12)   20.6 (12)   30.5 (14)   32.2 (12)

34.4 (14)   34.5 (12)   43.2 (12)   47.1 (10)   49.8 (10)   51.0 (10)   52.4 (14)   58.3 (10)

Fe$_4$(C$_2$H$_2$)$^+$ and Fe$_4$(CCH$_2$)$^+$

0.0 (13)   7.9 (13)   20.9 (11)   23.9 (11)   24.1 (11)   24.5 (13)   26.0 (11)

30.6 (11)   33.7 (11)   35.9 (11)   39.9 (11)   49.1 (11)   50.5 (13)   53.0 (13)

54.1 (13)   55.8 (13)   56.0 (9)   59.9 (13)   60.1 (11)   62.9 (11)

FIGURE 9.31 Relative energies (in kJ/mol) of the low-lying states corresponding to the adsorption of an acetylene or a vinylidene molecule on Fe$_3$ and Fe$_3^+$ calculated with the PP86 functional (numbers between parentheses indicate the number of unpaired electrons in the complex). (Reprinted from Chretien, S. et al., *J. Chem. Phys.*, 119(23), December 5, 12279, 2003. With permission.)

states were located (again using Gaussian94) for the first part of the reaction mechanism and for the case of a single Fe atom, with only two acetylene molecules. For the final addition of the third acetylene and for the clusters, only a thermodynamic analysis could be performed. At the time these calculations were performed, there was no parallel version of deMon, and the extensive calculations that would have been required to locate all of the transition states were beyond our computational resources. The results for Fe$_4^+$ are shown in Table V of Ref. [217] (Figure 9.34).

TABLE 9.4 (Table VI of the Reference)
Isomerization (ΔE_{iso}) and Activation
($\Delta E_{act}^{\ddagger}$) Energies (in kJ/mol) for the
Transformation of Acetylene to
Vinylidene on Iron Clusters, Fe_n^{q+} ($n = 1 - 4$
and $q = 0 - 1$) Calculated with the PP86
and the PLAP3 Functionals

		ΔE_{iso}		$\Delta E_{act}^{\ddagger}$	
q	n	PP86	PLAP3	PP86	PLAP3
0	1	62.7	53.4	233.9	249.2
0	2	−12.5		225.9	258.7
0	3	13.0	−22.9	252.8	288.1
0	4	4.0	1.0[a]		
1	1	77.9		224.2	236.1
1	2	46.9	24.1[b]	277.8	308.3
1	3	25.7	3.4[c]		
1	4	24.5	18.9[d]		

Source: Reprinted from Chretien, S. et al., *J. Chem. Phys.*, 119(23), December 5, 12279, 2003. With permission.

[a] The lower state of $Fe_4(C_2H_2)$ is different from the one with the PP86.
[b] The lower state of $Fe_2(C_2H_2)^+$ is different from the one with the PP86.
[c] The lower state of $Fe_3(CCH_2)^+$ is different from the one with the PP86.
[d] The lower state of $Fe_4(C_2H_2)^+$ is different from the one with the PP86.

Analysis of these results indicates that with respect to the proposed mechanism (Figure 9.33), the formation of a cyclobutadiene complex is unlikely, the cyclobutadiene complexes being much less stable than the complexes containing two acetylene molecules or an n-C_4H_4 ligand. The calculations also show that the formation of benzene is thermodynamically possible.

The above studies have shed light on several aspects of the reaction mechanism, but they have not allowed us to understand the reason that benzene is formed on Fe^+ and Fe_4^+ but not on Fe_2^+ or Fe_3^+. The answer does not appear to lie in the thermodynamics of the systems, so we must await a future study that will look deeply at the kinetics.

9.6.2 OXIDATION OF CO WITH CATIONIC IRON OXIDE CLUSTERS

Over the last years, several combined experimental and theoretical studies on the oxidation of CO to CO_2 with TM oxide clusters were realized with deMon2k,

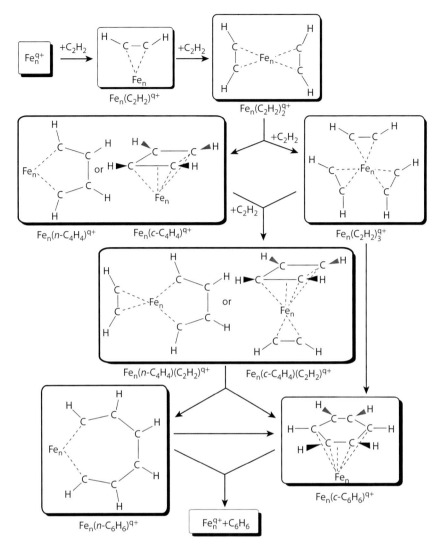

FIGURE 9.32 Schematic representation of the mechanism studied for the formation of benzene from acetylene on iron clusters, Fe_n^{q+} ($n = 1$–4 and $q = 0$–1). (Reprinted from Chretien, S. et al., *J. Chem. Phys.*, 119(23), December 5, 12291, 2003. With permission.)

employing the hierarchical transition state finder described in 2.5.3 [219–222]. Here we will review in more detail the reactions of FeO_2^+ and $Fe_2O_5^+$ with CO. The calculations were performed with the gradient corrected PBE functional [223]. For C and O, the LDA-optimized DZVP basis set [46] in combination with the GEN-A2 auxiliary function set [48] was used. For the Fe atom, the Wachters basis set [224,225], including f functions, in combination with a GEN-A2* auxiliary function set was employed. In a first step, the various reactants, products, and intermediates that are

TABLE 9.5 (Table V of the Reference)
Relative Energies (in kJ/mol) Corrected for ZPE and BSSE for the
Formation of Benzene from Acetylene on Fe_4^+ Calculated with
the PP86 Functional

	N_s				
	7	9	11	13	15
$Fe_4(C_2H_2)^+ + 2C_2H_2$		−128.5	−167.4	−191.5	−130.0
$Fe_4(C_2H_2)_2^+ + 1\ C_2H_2$		−340.3	−372.9	−364.1	
$Fe_4(n\text{-}C_4H_4)^+ + 1\ C_2H_2$		−365.7	−404.0	−407.2	−360.6
$Fe_4(c\text{-}C_4H_4)^+ + 1\ C_2H_2$		−304.5	−309.4	−336.2	−199.7
$Fe_4(C_2H_2)_3^+$	−594.3	−611.7	−599.3	−545.2	
$Fe_4(C_2H_2)(n\text{-}C_4H_4)^+$	−617.0	−592.8	−620.9	−609.4	
$Fe_4(C_2H_2)(c\text{-}C_4H_4)^+$		−520.3	−553.2	−481.9	
$Fe_4(n\text{-}C_6H_6)^+$		−623.9	−646.6	−598.6	
$Fe_4(c\text{-}C_6H_6)^+$		−761.9	−788.8	−709.4	

Source: Reprinted from Chretien, S. et al., *J. Chem. Phys.*, 119(23), December 5, 2003. With permission.

observed experimentally by gas-phase mass spectrometry were optimized. In order to find the global minima for these clusters, the configuration space was sampled by local optimizations starting from various initial geometries with different spin multiplicity. All optimized minima as well as transition states are characterized by frequency analysis.

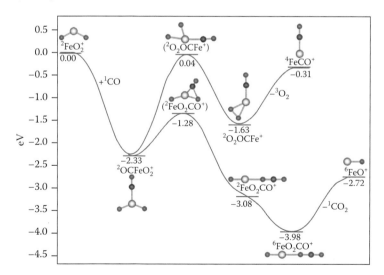

FIGURE 9.33 (See color insert.) Reaction profile of the reaction between FeO_2^+ and CO. Relative energies are given with respect to the reactant FeO_2^+ in eV. (Reprinted from Reilly, N. et al., *J. Phys. Chem. C*, 111(51), December 1, 19088, 2007. With permission.)

Figure 9.35 displays the calculated reaction profile for the reaction between FeO_2^+ and CO. Oxygen atoms are given in red, carbon atoms in black, and iron atoms in blue. Transition states are indicated by structural formulas in brackets. The leading superscript on the structural formulas indicates the multiplicity of the corresponding PES. The reaction profile in Figure 9.35 explains the experimentally observed product formation of FeO^+ and $FeCO^+$. The reaction is initialized by the attachment of CO to FeO_2^+, forming the intermediate $OCFeO_2^+$. This process occurs without barrier, i.e., spontaneously. As is well known from metal carbonyls, the CO binds over the carbon to the iron atom. The $OCFeO_2^+$ intermediate can further react over two different reaction channels. Along the energetically higher reaction channel, a second intermediate with a preformed O_2 unit is obtained. The corresponding transition state lies 0.04 eV above the FeO_2^+ reference energy. This high-lying transition state is critical for the formation of $FeCO^+$, which is obtained by oxygen release from the previous intermediate. The oxygen release also results in a change of the multiplicity of the PES of the iron system, from the doublet to the quartet. As Figure 9.35 shows, the formation of $FeCO^+$ from FeO_2^+ and CO under oxygen release is overall exothermic. However, the high-lying transition state limits the $FeCO^+$ production. This is consistent with experiment where the $FeCO^+$ production is observed after FeO^+.

In fact, the second reaction path from the $OCFeO_2^+$ intermediate leads to FeO^+ over a low energy transition state. The product on the doublet energy surface is a linear molecule with a preformed CO_2 unit. As the low-energy transition state in Figure 9.35 shows, the preformed CO_2 unit is produced by the insertion of an oxygen atom into the iron carbonyl bond. The resulting product can be further stabilized by spin flips from the doublet to the sextet surface. The CO_2 is finally released in order to form the experimentally observed FeO^+ product. Thus, the oxidation of CO to CO_2 is performed by the formal reaction

$$FeO_2^+ + CO \rightarrow FeO^+ + CO_2.$$

In Figure 9.36, the calculated reaction profile for $Fe_2O_5^+$ with CO is depicted. As in the previous case, the initial attachment of CO to the cationic iron oxide cluster is without barrier. However, in this case, the CO attaches not directly to the iron but rather to an external oxygen atom. As a result, the formation of a CO_2 subunit is immediate. The $^4Fe_2O_5CO^+$ intermediate releases CO_2. Beyond the $^4Fe_2O_4^+$ product of this elimination, the reaction profile splits. The higher energy reaction path leads, by oxygen elimination, to $^4Fe_2O_2^+$. The formation of the O_2 subunit inside the $Fe_2O_4^+$ system proceeds over the depicted transition state $O_2Fe_2O_2^+$ on the quartet energy surface. The spin flip to the doublet surface occurs with the release of oxygen and the formation of the final product of this reaction branch, $^2Fe_2O_2^+$. The low energy reaction path is initialized by a second CO attachment, now to the $^4Fe_2O_4^+$ intermediate. Again, CO attaches to an external oxygen atom. The corresponding CO_2 release then forms the $^4Fe_2O_3^+$ intermediate that by itself serves again as a CO adsorbate forming $^2Fe_2O_3CO^+$. This species releases CO_2 and, thus, forms $^2Fe_2O_2^+$ through a second reaction channel. As Figure 9.36 shows, no oxygen insertion

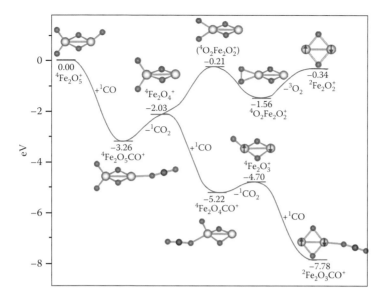

FIGURE 9.34 (See color insert.) Reaction profile of the reaction between $Fe_2O_5^+$ and CO. Relative energies are given with respect to the reactant $Fe_2O_5^+$ in eV. The arrows in the blue Fe atoms indicate spin broken solutions. (Reprinted from Reilly, N. et al., *J. Phys. Chem. C*, 111(51), December 1, 19086, 2007. With permission.)

occurs in the reaction of CO with $Fe_2O_5^+$. Similar results are obtained for higher iron oxides, too [220].

9.7 CONCLUDING REMARKS AND PERSPECTIVES

This chapter has taken a rather personal view of metallic systems, focusing on recent technical improvements to the deMon2k software and giving some examples from our own work in the area of metal clusters. The examples are, for the most part, small to very small clusters. Indeed, for TMs, already at three or four atoms, the challenges are substantial. Alkali and noble metals are considerably more forgiving; but, even for these, the task of locating all of the possible local minima, and calculating relevant properties, remains in the nonroutine category.

Progress over the last decade has been substantial. Almost all of the examples treated here could not have been done only a few years ago, at least not with present day efficiency and ease. Structures can now be optimized and transition states located, thanks to the efficient implementation of rigorous state-of-the-art algorithms. Vibrational analysis is now, for the most part, routine, notwithstanding the usual difficulties with imaginary frequency modes that can arise in metal clusters when there are a number of competing structures with close-lying excited states. In such cases, spin contamination must be considered, too. The cases of V_3, V_3O, and Nb_3O provided compelling examples of the power of combining experimental data with computations, yielding definitive ground-state assignments. The accurate calculation of polarizabilities presents its own challenges to DFT and ADFT methods. Here the need to consider finite temperature effects (through BOMD for the case of sodium clusters) contains a lesson that, we believe, has considerable generality. Finally, the use of DFT in the arena of TM cluster catalysis has been explored. The catalytic clusters are still very small in these studies, and the general reliability of the methodology still requires further validation. Nonetheless, the results most likely provide valid insight into the catalytic mechanisms.

Although the field has progressed greatly since the Xα-scattered wave days, much remains to be done. The currently available functionals are still not entirely adequate for the case of many-electron, multi-determinantal systems such as the TM clusters. Even the meta-GGA and hybrid functionals only represent some sort of mean-field approximation. Perhaps the future will see a better combination of DFT and wave function methodology; work in this direction is certainly needed. Although the systems reviewed in this chapter have considerable complexity, they are all in the "small" category when one considers applications in the main forefront areas of experimental science: materials science, nanoscience, and systems biology. Although we have not reviewed work in these disciplines in this chapter, many of the deMon developers are highly interested in the next level of complexity, and considerable progress has been made toward finding the proper role for DFT and deMon within a broader context of multi-method multiscale modeling. The tight-binding DFT method, which has been implemented in deMon2k, allows systems with up to a few hundred atoms to be treated (so far not including TMs) in full dynamics mode. Various coarse-graining approaches are being explored as is QM/MM methodology (with CHARMM and deMon being linked, in the first instance) as well as QM/QM′ approaches, which combine real DFT with TBDFT. The results of such coarse-grained approaches will, in the fullness of time, be linked to even more "macroscopic" approaches, such as the kinetic Monte Carlo method, to finally yield predictions on rate processes that have DFT results at the heart of the methodology, but also have the other techniques that are needed for a full description of the complex systems.

We hope that this personal overview will give the reader a sense of what has been accomplished in the area of DFT calculations for metal clusters over the last two decades or so. More importantly, we hope that some readers will rise to the challenge of taking the theory to the next level, either along some of the lines mentioned in the previous paragraph or along other directions that have not been listed or have not been thought of yet. We think the next decade will be a most exciting one!

REFERENCES

1. R. G. Parr, W. Yang, *Density-Functional Theory of Atoms and Molecules,* Oxford University Press, New York (1989).
2. R. M. Dreizler, E. K. U. Gross, *Density Functional Theory*, Springer-Verlag, Berlin (1990).
3. M. Levy, J. P. Perdew, *Density Functional Methods in Physics*, Plenum, New York (1985).

4. D. R. Salahub, A. Goursot, J. Weber, A. M. Köster, A. Vela, *Theory and Applications of Theoretical Chemistry: The First Forty Years*, Elsevier, Amsterdam, the Netherlands (2005).

5. D. R. Salahub, R. Fournier, P. Mlynarski, I. Papai, A. St-Amant, J. Ushio, *Density Functional Methods in Chemistry*, Springer, New York (1991).

6. A. M. Köster, P. Calaminici, Z. Gómez, U. Reveles, *Reviews of Modern Quantum Chemistry, A Celebration of the Contribution of Robert G Parr*, World Scientific Publishing Co., Singapore (2002).

7. A. M. Köster, A. Goursot, D. R. Salahub, *Comprehensive Coordination Chemistry II: From Biology to Nanotechnology*, Vol. 1, Elsevier, Amsterdam, the Netherlands (2003).

8. P. Hohenberg, W. Kohn, *Phys. Rev.* **136**, B864 (1964).

9. W. Kohn, J. Sham, *Phys. Rev.* **137**, A1697 (1965).

10. J. C. Slater, *Phys. Rev.* **81**, 385 (1951).

11. B. I. Dunlap, J. W. D. Connolly, J. R. Sabin, *J. Chem. Phys.* **71**, 4993 (1982).

12. J. W. Mintmire, B. I. Dunlap, *Phys Rev. A* **25**, 88 (1982).

13. J. W. Mintmire, J. R. Sabin, S. B. Trickey, *Phys. Rev. B* **26**, 1743 (1982).

14. E. J. Baerends, D. E. Ellis, P. Ros, *Chem. Phys.* **2**, 41 (1973).

15. J. Andzelm, E. Radzio, D. R. Salahub, *J. Comp. Chem.* **6**, 520 (1985).

16. J. Andzelm, N. Russo, D. R. Salahub, *J. Chem. Phys.* **87**, 6562 (1987).

17. A. M. Köster, *J. Chem. Phys.* **118**, 9943 (2003).

18. S. B. Trickey, J. A. Alford, J. C. Boettger, *Computational Materials Science, Theoretical and Computational Chemistry*, Vol. 15, Elsevier, Amsterdam, the Netherlands (2004).

19. A. M. Köster, Habilitation thesis, Universität Hannover, Hannover, Germany (1998).

20. O. Vahtras, J. Almlöf, M. W. Feyereisen, *Chem. Phys. Lett.* **213**, 514 (1993).

21. H. Sambe, R. H. Felton, *J. Chem. Phys.* **62**, 1122 (1975).

22. A. St-Amant, D. R. Salahub, *Chem. Phys. Lett.* **169**, 387 (1990).

23. M. E. Casida, C. Daul, A. Goursot, A. M. Köster, L. G. M. Pettersson, E. Proynov, A. St.-Amant et al., deMon-KS Version 3.4, deMon Software, Montréal, Quebec, Canada (1996).

24. V. G. Malkin, O. L. Malkina, M. E. Casida, D. R. Salahub, *J. Am. Chem. Soc.* **116**, 5898 (1994).

25. C. Jamorski, M. E. Casida, D. R. Salahub, *J. Chem. Phys.* **104**, 5134 (1996).

26. A. M. Köster, M. Leboeuf, D. R. Salahub, *Molecular Electrostatic Potentials, Concepts and Applications*, Elsevier, Amsterdam, the Netherlands (1996); pp. 105–142.

27. A. M. Köster, P. Calaminici, N. Russo, *Phys. Rev. A* **53**, 3865 (1996).

28. L. Versluis, T. Ziegler, *J. Chem. Phys.* **88**, 322 (1988).

29. J. Andzelm, E. Wimmer, *Physica B* **172**, 307 (1991).

30. J. Guan, M. E. Casida, A. M. Köster, D. R. Salahub, *Phys. Rev. B* **52**, 2184 (1995).

31. K. Hermann, L.G.M. Pettersson, M. E. Casida, C. Daul, A. Goursot, A. M. Köster, E. Proynov, A. St-Amant, D. R. Salahub. Contributing authors: V. Carravetta, H. Duarte, C. Friedrich, N. Godbout, J. Guan, C. Jamorski, M. Leboeuf, M. Leetmaa, M. Nyberg, S. Patchkovskii, L. Pedocchi, F. Sim, L. Triguero, A. Vela, StoBe-deMon version 2.2 (2006).

32. A. M. Köster, M. Krack, M. Leboeuf, B. Zimmermann, AllChem, Universität Hannover, Hannover, Germany (1998).

33. A. M. Köster, P. Calaminici, M. E. Casida, R. Flores-Moreno, G. Geudtner, A. Goursot, T. Heine et al., The deMon developers, CINVESTAV, Mexico-City (2006). http://www.demon-software.com.

34. A. M. Köster, J. U. Reveles, J. M. del Campo, *J. Chem. Phys.* **121**, 3417 (2004).

35. G. Geudtner, F. Janetzko, A. M. Köster, A. Vela, P. Calaminici, *J. Comput. Chem.* **27**, 483 (2006).

36. J. L. Morales, J. Nocedal, *SIAM J. Optim.* **10**, 1079 (2000).

37. R. Flores-Moreno, A. M. Köster, *J. Chem. Phys.* **128**, 134105 (2008).
38. J. M. del Campo, A. M. Köster, *J. Chem. Phys.* **129**, 024107 (2008).
39. G. U. Gamboa, P. Calaminici, G. Geudtner, A. M. Köster, *J. Phys. Chem. A* **112**, 11969 (2008).
40. S. Krishnamurty, T. Heine, A. Goursot, *J. Phys. Chem. B* **107**, 5728 (2003).
41. A. Ipatov, A. Fouqueau, C. Perez del Valle, F. Cordova, M. E. Casida, A. M. Köster, A. Vela, *J. Mol. Struct. Theochem.* **762**, 179 (2006).
42. R. F. W. Bader, *Atoms in Molecules: A Quantum Theory*, Oxford University Press, New York (1990).
43. F. Weinhold, C. R. Landis, *Valency and Bonding: A Natural Bonding Orbital Donor-Acceptor Perspective*, Cambridge University Press, Cambridge (2003).
44. B. Lev, R. Zhang, A. de la Lande, D. R. Salahub, S. Y. Noskov, *J. Comput. Chem.*, **31**, 1015 (2010).
45. J. C. Boettger, S. B. Trickey, *Phys. Rev. B* **53**, 3007 (1996).
46. N. Godbout, D. R. Salahub, J. Andzelm, E. Wimmer, *Can. J. Chem.* **70**, 560 (1992).
47. P. Calaminici, R. Flores-Moreno, A. M. Köster, *Comp. Lett. (COLE)* **1**, 164 (2005).
48. P. Calaminici, F. Janetzko, A. M. Köster, R. Mejia-Olvera, B. Zuniga-Gutierrez, *J. Chem. Phys.* **126**, 044108 (2007).
49. D. N. Laikov, *Chem. Phys. Lett.* **281**, 151 (1997).
50. I. M. Gel'fand, S. V. Fomin, *Calculus of Variations*, Prentice Hall, Englewood Cliffs, NJ (1963).
51. R. McWeeny, *Phys. Rev.* **126**, 1028 (1962).
52. G. Dierksen, R. McWeeny, *J. Chem. Phys.* **44**, 3554 (1966).
53. R. McWeeny, G. Dierksen, *J. Chem. Phys.* **49**, 4852 (1968).
54. W. T. Raynes, J. L. Dodds, R. McWeeny, J. P. Riley, *Mol. Phys.* **33**, 611 (1977).
55. R. McWeeny, J. L. Dodds, A. J. Sadlej, *Mol. Phys.* **34**, 1779 (1977).
56. R. McWeeny, *Methods of Molecular Quantum Mechanics*, 2nd reprinting, Academic Press, London (2001).
57. M. Born, J. R. Oppenheimer, *Ann. Phys.* **84**, 457 (1927).
58. R. L. Johnston, *Dalton Trans.* 4193 (2003).
59. D. J. Wales, H. A. Scheraga, *Science* **285**, 1368 (1999).
60. J. P. K. Doye, D. J. Wales, M. A. Miller, *J. Chem. Phys.* **109**, 8143 (1998).
61. I. L. Garzón, K. Michaelian, M. R. Beltrán, A. Posada-Amarillas, P. Ordejón, E. Artacho, D. Sánchez-Portal, J. M. Soler, *Phys. Rev. Lett.* **81**, 1600 (1998).
62. W. C. Davidon, *SIAM, J. Optim.* **1**, 1 (1991).
63. R. Fletcher, *Practical Methods of Optimizations*, Wiley & Sons, New York (1980).
64. K. Levenderg, *Q. Appl. Math.* **2**, 164 (1944).
65. D. W. Marquardt, *SIAM J.* **11**, 431 (1963).
66. J. Nocedal, S. J. Wright, *Numerical Optimization*, Springer-Verlag, New York (1999).
67. J. M. del Campo, A. M. Köster, *Croatica Chem. Acta* **82**, A029 (2009).
68. V. R. Saunders, I. H. Hillier, *Int. J. Quantum Chem.* **8**, 699 (1973).
69. M. J. S. Dewar, E. F. Healy, J. J. P. Stewart, *J. Chem. Soc., Faraday Trans.* 2 **80**, 227 (1984).
70. P. Culot, G. Dive, V. H. Nguyen, J. M. Ghuysen, *Theor. Chim. Acta* **82**, 189 (1992).
71. J. M. Boffil, *J. Comp. Chem.* **15**, 1 (1994).
72. J. Y. Abashkin, N. Russo, *J. Chem. Phys.* **100**, 4477 (1994).
73. J. M. Vásquez-Pérez, G. U. Gamboa Martínez, A. M. Köster, P. Calaminici, *J. Chem. Phys.*, **131**, 124126 (2009).
74. J. W. M. Nissink, M. L. Verdonk, J. Kroon, T. Mietzner, G. Klebe, *J. Comput. Chem.* **18**, 638 (1997).
75. X. Gironés, D. Robert, R. Carbsó-Dorca, *J. Comput. Chem.* **22**, 255 (2001).
76. J. Mestres, D.C. Roher, J.M. Maggiora, *J. Comput. Chem.* **18**, 934 (1997).

77. M. T. Barakat, P.M. Dean, *J. Comput. Aided Mol. Design* **4**, 295 (1990).
78. D. M. Bayada, R. W. Simpson, A. P. Johnson, C. Laurenco, *J. Chem. Inf. Comput. Sci.* **32**, 680 (1992).
79. S. K. Kearsley, *Acta Cryst. A* **45**, 208 (1989).
80. G. Carpento, S. Martello, P. Toth, *Ann. Operations Res.* **13**, 193 (1988).
81. H. W. Kuhn, *Naval Res. Logistics Q.* **2**, 83 (1955).
82. J. Munkres, *J. Soc. Indust. Appl. Math.* **5**, 32 (1957).
83. J. Arvo, *Graphics Gems III.4*, Academic Press, New York (1992); pp. 117–120.
84. K. Shoemake, *Graphics Gems III.6*, Academic Press, New York (1992); pp. 124–132.
85. R. N. Barnett, U. Landman, A. Nitzan, G. Rajagopal, *J. Chem. Phys.* **94**, 608 (1991).
86. F. Janetzko, A. Goursot, T. Mineva, P. Calaminici, R. Flores-Moreno, A. M. Köster, D. R. Salahub, *Structure Determination of Clusters: Bridging Experiment and Theory*, in Nanoclusters: A bridge across disciplines, Eds. P. Jena, A. Castleman Jr. Elsevier, Amsterdam (2010).
87. L. V. Woodcock, *Chem. Phys. Lett.* **10**, 257 (1971).
88. H. J. C. Berendsen, J. P. M. Postma, W. F. van Gunsteren, J. Hermans, *J. Chem. Phys.* **81**, 3684 (1984).
89. D. J. Evans, W. G. Hoover, B. H. Failor, B. Moran, A. J. C. Ladd, *Phys. Rev. B* **28**, 1016 (1983).
90. S. Nosé, *J. Chem. Phys.* **81**, 511 (1984).
91. W. G. Hoover, *Phys. Rev. A* **31**, 1695 (1985).
92. G. J. Martyna, M. L. Klein, M. Tuckerman, *J. Chem. Phys.* **97**, 2635 (1992).
93. S. Toxvaerd, O. H. Olsen, *Ber. Bunsenges. Phys. Chem.* **94**, 274 (1990).
94. M. D. Morse, *Chem. Rev.* **86**, 1049 (1986).
95. J. Koutecký, P. Fantucci, *Chem. Phys. Rev.* **86**, 539 (1986).
96. D. R. Salahub, in *Ab Initio Methods in Quantum Chemistry-II*, edited by K.P. Lawley Wiley, New York (1987); p. 447.
97. J. A. Alonso, *Chem. Rev.* **100**, 637 (2000).
98. B. Simard, S. A. Mitchell, D. M. Rayner, D. S. Yang, in *Metal-Ligand Interactions in Chemistry, Physics and Biology*, NATO Science Series **546**, Kluwer Academic Publishers, Dordrecht, the Netherlands (2000); p. 239.
99. M. Castro, D. R. Salahub, *Phys. Rev. B* **47**, 10955 (1993).
100. M. Castro, D. R. Salahub, *Phys. Rev. B* **49**, 11842 (1994).
101. L. Goodwin, D. R. Salahub, *Phys. Rev. A* **47**, 774 (1993).
102. A. M. James, P. Kowalczyk, R. Fournier, B. Simard, *J. Chem. Phys.* **99**, 8504 (1993).
103. A. Martínez, A. Vela, D. R. Salahub, P. Calaminici, N. Russo, *J. Chem. Phys.* **101**, 10677 (1994).
104. P. Calaminici, N. Russo, M. Toscano, *Z. Phys. D* **33**, 281 (1995).
105. P. Calaminici, A. M. Köster, N. Russo, D. R. Salahub, *J. Chem. Phys.* **105**, 9546 (1996).
106. H. Kietzmann, J. Morenzin, P. S. Bechthold, G. Ganteför, W. Eberhardt, D. S. Yang, P. A. Hackett, R. Fournier, T. Pang, C. Chen, *Phys. Rev. Lett.* **77**, 4528 (1996).
107. M. Castro, C. Jamorski, D. R. Salahub, *Chem. Phys. Lett.* **271**, 133 (1997).
108. D. S. Yang, M. Z. Zgierski, D. M. Rayner, P. A. Hackett, A. Martínez, D. R. Salahub, P. N. Roy, T. Carrington Jr., *J. Chem. Phys.* **103**, 5335 (1995).
109. D. S. Yang, M. Z. Zgierski, A. Berces, P. A. Hackett, P. N. Roy, A. Martínez, T. Carrington Jr., D. R. Salahub, R. Fournier, C. Chen, T. Pang, *J. Chem. Phys.* **104**, 10663 (1996).
110. D. S. Yang, M. Z. Zgierski, A. Berces, P. A. Hackett, A. Martínez, D. R. Salahub, *Chem. Phys. Lett.* **227**, 71 (1997).
111. D. S. Yang, M. Z. Zgierski, P. A. Hackett, *J. Chem. Phys.* **108**, 3591 (1998).
112. D. S. Yang, A. M. James, D. M. Rayner, P. A. Hackett, *Chem. Phys. Lett.* **231**, 177 (1994).

113. A. M. James, P. Kowalczyk, E. Langlois, M. D. Campbell, A. Ogawa, B. Simard, *J. Chem. Phys.* **101**, 4485 (1994).

114. P. R. R. Langridge-Smith, M. D. Morse, G. P. Hansen, R. E. Smalley, *J. Chem. Phys.* **80**, 593 (1984).

115. E. M. Spain, J. M. Behm, M. D. Morse, *J. Chem. Phys.* **96**, 2512 (1992).

116. E. M. Spain, M. D. Morse, *J. Chem. Phys.* **96**, 2479 (1992).

117. C. Cosse, M. Fouassier, T. Mejean, M. Tranquille, D. P. DiLella, M. Moskovits, *J. Chem. Phys.* **73**, 6076 (1980).

118. D. S. Yang, A. M. James, D. M. Rayner, P. A. Hackett, *J. Chem. Phys.* **102**, 3129 (1995).

119. L. Russon, S. Heidecke, M. Morse, *J. Chem. Phys.* **91**, 1061 (1989).

120. D. M. Cox, R. L. Whetten, M. R. Zakin, D. J. Trevor, K. C. Reichmann, A. Kaldor, *Optical Science and Engineering Ser. 6, Advances in Laser Science*, AIP Conf. Proc. **146**, New York (1986); Vol. I, p. 527.

121. Y. M. Hamrick, M. D. Morse, *J. Phys. Chem.* **93**, 6494 (1989).

122. D. M. Cox, K. C. Reichmann, D. J. Trevor, A. Kaldor, *J. Chem. Phys.* **88**, 111 (1988).

123. M. R. Zakin, D. M. Cox, R. O Brickman, A. Kaldor, *J. Phys. Chem.* **93**, 6823 (1989).

124. N. Shinji, S. Yasutomo, K. Noriyoshi, F. Kiyokazu, K. Koji, *Chem. Phys. Lett.* **158**, 152 (1989).

125. G. Dietrich, K. Kuznetsov, S. Lützenkirchen, L. Schweikhard, J. Ziegler, *Int. J. Mass Spectrom. Ion Process.* **157/158**, 319 (1996).

126. G. Dietrich, K. Dasgupta, S. Lützenkirchen, L. Schweikhard, J. Ziegler, *Chem. Phys. Lett.* **252**, 141 (1996).

127. L. Holmgren, A. Rosén, *J. Chem. Phys.* **110**, 2629 (1999).

128. S. M. E. Green, S. Alex, N. L. Fleischer, E. L. Millam, T. P. Marcy, D. G. Leopold, *J. Chem. Phys.* **114**, 2653 (2001).

129. D. C. Douglass, J. P. Bucher, L. A. Bloomfield, *Phys. Rev. B* **45**, 6341 (1992).

130. C. X. Su, D. A. Hales, P. B. Armentrout, *J. Chem. Phys.* **99**, 6613 (1993).

131. W. Hongbin, S. R. Desai, L. S. Wang, *Phys. Rev. Lett.* **77**, 2436 (1996).

132. M. Iseda, T. Nishio, S. Y. Han, H. Yoshida, A. Terasaki, T. Kondow, *J. Chem. Phys.* **106**, 2182 (1997).

133. D. R. Salahub, R. P. Messmer, *Surf. Sci.* **106**, 415 (1981).

134. F. Liu, S. N. Khanna, P. Jena, *Phys. Rev. B* **43**, 8179 (1991).

135. S. P. Walch, C. W. Bauschlicher Jr., *J. Chem. Phys.* **83**, 5735 (1985).

136. H. Dreyssé, J. Dorantes-Davila, A. Vega, L.C. Balbas, S. Bouarab, H. Nait-Laziz, C. Demangeat, *J. Appl. Phys.* **73**, 6207 (1993).

137. K. Lee, J. Callaway, *Phys. Rev. B* **48**, 15358 (1993).

138. K. Lee, J. Callaway, *Phys. Rev. B* **49**, 13906 (1994).

139. J. Zhao, X. Chen, Q. Sun, F. Liu, G. Wang, K. D. Lain, *Physica B* **215**, 177 (1995).

140. H. Grönbeck, A. Rosén, *J. Chem. Phys.* **107**, 10620 (1997).

141. X. Wu, A. K. Ray, *J. Chem. Phys.* **110**, 2437 (1999).

142. A. K. Ray, Private communication.

143. P. Calaminici, A. M. Köster, N. Russo, P. N. Roy, T. Carrington Jr., D. R. Salahub, *J. Chem. Phys.* **114**, 4036 (2001).

144. K. Jug, B. Zimmermann, P. Calaminici, A. M. Köster, *J. Chem. Phys.* **116**, 4497 (2002).

145. K. Jug, B. Zimmermann, A. M. Köster, *Int. J. Quantum Chem.* **90**, 594 (2002).

146. G. López-Arvizu, P. Calaminici, *J. Chem. Phys.* **126**, 194102 (2007).

147. J. P. Perdew, Y. Wang, *Phys. Rev. B* **33**, 8800 (1986).

148. J. P. Perdew, *Phys. Rev. B* **33**, 8822 (1986).

149. J. P. Perdew, *Phys. Rev. B* **34**, 7406 (1986).

150. S. H. Vosko, L. Wilk, M. Nusair, *Can. J. Phys.* **58**, 1200 (1980).

151. C. Massobrio, A. Pasquarello, A. dal Corso, *J. Chem. Phys.* **109**, 6626 (1998).

152. C. Massobrio, A. Pasquarello, R. Car, *Chem. Phys. Lett.* **238**, 215 (1995).

153. P. Fuentealba, Y. Simón-Manso, *Chem. Phys. Lett.* **314**, 108 (1999).
154. H. B. Jansen, P. Ross, *Chem. Phys. Lett.* **3**, 140 (1969).
155. E. A. Rohlfing, J. J. Valentini, *J. Chem. Phys.* **84**, 6560 (1986).
156. K. Hilpert, K.A. Gingerich, Ber. Bunsenges. *Phys. Chem.* **84**, 379 (1980).
157. V. A.Spasov, T. H. Lee, K. M. Ervin, *J. Chem. Phys.* **112**, 1713 (2000).
158. O. Ingólfsson, U. Busolt, K. I. Sugawara, *J. Chem. Phys.* **112**, 4613 (2000).
159. J. Ho, K. M. Ervin, W. C. Lineberger, *J. Chem. Phys.* **93**, 6987 (1990).
160. C. E. Moore, *Natl. Bur. Stand.* **2**, 111 (1971).
161. A. M. James, G. W. Lemire, P. R. R. Langridge-Smith, *Chem. Phys. Lett.* **277**, 503 (1994).
162. M. B. Knickelbein, *Chem. Phys. Lett.* **192**, 129 (1992).
163. D. E. Powers, G. S. Hansen, M. E. Geusic, D. L. Michalopulos, R. E. Smalley, *J. Chem. Phys.* **78**, 2866 (1983).
164. E. K. Parks, K. P. Kerns, S. J. Riley, *J. Chem. Phys.* **114**, 2228 (2001).
165. M. B. Knickelbein, *J. Chem. Phys.* **116**, 9703 (2002).
166. F. Liu, R. Liyanage, P. B. Armentrout, *J. Chem. Phys.* **117**, 132 (2002).
167. F. A. Reuse, S. N. Khanna, *Eur. Phys. J. D.- At., Mol. Opt. Phys.* **6**, 77 (1999).
168. T. L. Wetzel, DePristo, *J. Phys. Chem.* **105**, 572 (1996).
169. F. Ruette, C. Gonzalez, *Chem. Phys. Lett.* **359**, 428 (2002).
170. S. N. Khanna, P. Jena, *Chem. Phys. Lett.* 336, 467 (2001).
171. S. N. Khanna, M. R. Beltran, P. Jena, *Phys. Rev. B* **64**, 235419 (2001).
172. G. M. Pastor, R. Hirsch, B. Muhlschlegel, *Phys. Rev. B* **53**, 10382 (1996).
173. G. A. Cisneros, M. Castro, D. R. Salahub, *Int. J. Quantum Chem.* **75**, 847 (1999).
174. G. L. Estiu, M. C. Zerner, *J. Phys. Chem.* **100**, 16874 (1996).
175. N. Desmarais, C. Jamorski, F. A. Reuse, S. N. Khanna, *Chem. Phys. Lett.* **294**, 480 (1998).
176. Z. Xie, Q. M. Ma, Y. Liu, Y.-C. Li, *Phys. Lett. A* **342**, 459 (2005).
177. M. C. Michelini, R. P. Diez, A. H. Jubert, *Int. J. Quantum Chem.* **85**, 22 (2001).
178. P. Calaminici, M. R. Beltrán, *Comp. Lett. (COLE)* **1**, 172 (2005).
179. K. Michaelian, N. Rendon, I. L. Garzón, *Phys. Rev.* **60**, 2000 (1999).
180. M. Boyukata, Z. B. Guvenc, S. Ozcelik, J. Jellinek, *Int. J. Mod. Phys. C* **16**, 295 (2005).
181. S. Ozcelik, Z. B. Guvenc, P. Durmus, J. Jellinek, Surf. Sci. **566**, 377 (2004).
182. J. M. Dyke, B.W.J. Gravenor, M.P. Hastings, A. Morris, *J. Phys. Chem.* **89**, 4613 (1985).
183. P. Calaminici, A. M. Köster, *Int. J. Quantum Chem.* **91**, 317 (2003).
184. P. Calaminici, A. M. Köster, D. R. Salahub, *J. Chem. Phys.* **118**, 4913 (2003).
185. P. Calaminici, R. Flores-Moreno, A. M. Köster, *J. Chem. Phys.* **121**, 3558 (2004).
186. U. Kreibig, M. Vollmer, *Optical Properties of Metal Clusters*, Springer, Berlin (1995).
187. K. D. Bonin, V. V. Kresin, *Electric-Dipole Polarizabilities of Atoms, Molecules and Clusters*, World Scientific, Singapore (1997).
188. W. D. Knight, K. Clemenger, W. A. de Heer, M. Y. Saunder, *Phys. Rev. B* **31**, 2539 (1985).
189. E. Benichou, R. Antoine, D. Rayane, B. Vezin, F. W. Dalby, Ph. Dugourd, M. Broyer, C. Ristori, F. Chandezon, B. A. Huber, J. C. Rocco, S. A. Blundell, C. Guet, *Phys. Rev. A* **59**, R1 (1999).
190. R. Antoine, D. Rayane, A. R. Allouche, M. Aubert-Frécon, E. Benichou, F.W. Dalby, Ph. Dugourd, M. Broyer, C. Guet, *J. Chem. Phys.* **110**, 5568 (1999).
191. D. Rayane, A. R. Allouche, E. Benichou, R. Antoine, M. Aubert-Frecon, P. Dugourd, M. Broyer, C. Ristori, C. Chandezon, B. A. Hubert, C. Guet, *Eur. Phys. J. D* **9**, 243 (1999).
192. P. Calaminici, A. M. Köster, A. Vela, K. Jug, *J. Chem. Phys.* **113**, 2199 (2000).
193. P. Calaminici, K. Jug, A.M. Köster, *J. Chem. Phys.* **111**, 4613 (1999).
194. P. Jaque, A. Toro-Labbe, *J. Chem. Phys.* **117**, 3208 (2002).

195. Z. Cao, Y. Wang, J. Zhu, W. Wu, Q. Zhang, *J. Phys. Chem. B* 106, 9649 (2002).
196. M. Yang, K. Jackson, *J. Chem. Phys.* 122, 184317 (2005).
197. M. Knickelbein, *J. Chem. Phys.* **120**, 10450 (2004).
198. S. Chrétien, D.R. Salahub, *Phys. Rev. B* **66** 155425 (2002).
199. G. L. Gutsev, C. W. Bauschlicher Jr., *J. Phys. Chem. A* **107** 7013 (2003).
200. P. Calaminici, *Chem. Phys. Lett.* **387**, 253 (2004).
201. P. Calaminici, *J. Chem. Phys.* **128**, 164317 (2008).
202. H. A. Kurtz, J. J. P. Stewart, K. M. Dieter, *J. Comp. Chem.* **11**, 82 (1990).
203. P. Calaminici, K. Jug, A. M. Köster, *J. Chem. Phys.* **109**, 7756 (1998).
204. P. Calaminici, K. Jug, A.M. Köster, V. E. Ingamells, M. G. Papadopoulos, *J. Chem. Phys.* **112**, 6301 (2000).
205. P. Calaminici, A. M. Köster, K. Jug, C. Arbez-Gindre, C. G. Screttas, *J. Comp. Chem.* **23**, 291 (2002).
206. K. Jug, S. Chiodo, P. Calaminici, A. Avramopoulos, M. G. Papadopoulos, *J. Phys. Chem.* **107**, 4172 (2003).
207. P. Calaminici, *Chem. Phys. Lett.* **374**, 650 (2003).
208. M. B. Knickelbein, *J. Chem. Phys.* **115**, 5957 (2001).
209. D. R. Lide (ed.), *Handbook of Chemistry and Physics*, 74th ed., CRC, Boca Raton, FL (1994).
210. R. Pou-Amérigo, M. Merchán, I. Nebot-Gil, P. O. Widmark, B. O. Roos, *Theor. Chim. Acta* **92**, 149 (1995).
211. K. R. S. Chandrakumar, T.K. Ghanty, S.K. Ghosh, *J. Chem. Phys.* **120**, 6487 (2004).
212. S. Kümel, J. Akola, M. Manninen, *Phys. Rev. Lett.* **84**, 3827 (2000).
213. S. A. Blundell, C. Guet, R. R. Zope, *Phys. Rev. Lett.* **84**, 4826 (2000).
214. W. D. Knight, K. Clemenger, W. A. de Heer, M. Y. Saunder, W. A. Chou, M. L. Cohen, *Phys. Rev. Lett.* **52**, 2141 (1984).
215. R. W. Molof, T. M. Miller, H. L. Schwartz, B. Benderson, J. T. Park, *J. Chem. Phys.* **61**, 1816 (1974).
216. P. Schnabel, M. P. Irion, K. G. Weil, *J. Phys. Chem.* **95**, 9688 (1996).
217. S. Chrétien, D.R. Salahub, *J. Chem. Phys.* **119**, 12279 (2003).
218. S. Chrétien, D.R. Salahub, *J. Chem. Phys.* **119**, 12291 (2003).
219. N. M. Reilly, J. U. Reveles, G. E. Johnson, S. N. Khanna, A. W. Castleman Jr., *Chem. Phys. Lett.* **435**, 295 (2007).
220. N. M. Reilly, J. U. Reveles, G. E. Johnson, S. N. Khanna, A. W. Castleman Jr., J. *Phys. Chem. A* **111**, 4158 (2007).
221. N. M. Reilly, J. U. Reveles, G. E. Johnson, J. M. del Campo, S. N. Khanna, A. M. Köster, A. W. Castleman Jr., *J. Phys. Chem. C* **111**, 19086 (2007).
222. G. E. Johnson, J. U. Reveles, N. M. Reilly, E. C. Tyo, S. N. Khanna, A. W. Castleman Jr., *J. Phys. Chem. A* **112**, 11330 (2008).
223. J. P. Perdew, K. Burke, M. Ernzerhof, *Phys. Rev. Lett.* **77**, 3865 (1996).
224. A. J. H. Wachters, *J. Chem. Phys.* **52**, 1033 (1970).
225. C. W. Bauschlicher Jr., S. R. Langhoff, L. A. Barnes, *J. Chem. Phys.* **91**, 2399 (1989).

10 Density Functional Theory Calculations on Cobalt and Platinum Transition Metal Clusters

Ali Sebetci

CONTENTS

10.1 INTRODUCTION

10.1.1 TRANSITION METAL CLUSTERS AND THEIR IMPORTANCE

Nanoparticles, which are aggregates of identical or different atoms or molecules ranging from a few to many millions, are called atomic or molecular clusters [1–6]. They constitute a separate class of materials in between the bulk on one hand and the molecular state on the other. Their chemical, optical, electrical, physical, and magnetic properties can be significantly different from the respective bulk materials. For instance, while bulk gold appears yellow in color, nanosized gold particles appear in red, or while bulk gold does not react with oxygen, gold clusters do. Thus, materials that are commonly chemically inert in bulk can become catalytically active [7], nonmagnetic bulk materials can show magnetic behavior when scaled down to nanostructures [8], and metallic materials can become semiconductors [9]. The main reasons for such property changes are as follows: in the regime of clusters, gravitational forces become negligible and electromagnetic forces begin to dominate, quantum mechanical effects dominate the motion and energy instead of the classical mechanics, the surface-to-volume ratio becomes much greater, and random molecular motion becomes more important. In contrast to molecules, nanoclusters do not have a fixed size or composition. For example, the vitamin C molecule (see Figure 10.1a) contains six oxygen, six carbon, and eight hydrogen atoms, which are placed at certain positions. On the other hand, platinum clusters (see Figure 10.1b and c) may contain any number of constituent particles and present a variety of morphologies for a fixed size.

Clusters can be classified in many different ways: they are either homogeneous (composed of only one type of atom or molecule) or heterogeneous (composed of different types of atoms or molecules). They may be neutral or charged. They can be held together by ionic bonds as in NaCl clusters, covalent bonds as in B clusters, van der Waals forces as in He or C_{60} clusters, or metallic bonds as in Au and Pt clusters. Since the electrons in small metal clusters are not as free as electrons of bulk metals, the metallic bonds holding small metal clusters together are more likely to be covalent bonds. Another useful categorization of clusters is defined with respect

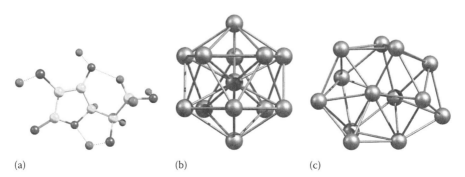

(a)   (b)   (c)

FIGURE 10.1 Geometric structures of vitamin-C molecule and two isomers of Pt_{13} clusters. (a) Vitamin-C, (b) Pt_{13} (1st isomer), and (c) Pt_{13} (second isomer).

to their size-dependent property changes [10]. Clusters with thousands of atoms in the range of 10 nm diameter and more are said to be in the "scalable regime" where various properties change smoothly with the cluster radius, and tend to the bulk limit as the size (diameter) increases. On the other hand, in the so-called non-scalable regime, where clusters have no more than a few hundred particles and 1–3 nm diameter, changes in chemical and physical properties are not predictable and are strongly size dependent. In this non-scalable size regime, the removal or addition of a single atom to the cluster can result in significant changes to properties such as geometric and electronic structure, binding energy (BE), melting temperature, or catalytic activity [11].

Size-dependent properties make clusters a separate class of material. Currently, understanding the mechanism that determines properties of small particles, predicting the changes of such properties with respect to the bulk or to the molecular limit, and controlling and manipulating objects of a few nanometer in size are the key issues of nanoscience and nanotechnology.

Metal clusters are "finite groups of metal atoms participating in direct metal-metal bonds" [12]. Small transition metal clusters (TMC) attract a lot of attention because of their potential use as catalysts, components of nanodevices, magnetic data storage materials, and in novel cluster-based materials. For example, platinum is one of the most important elements in the heterogeneous catalysis of hydrogenation as well as in the catalysis of the CO, NO, and hydrocarbons. It is currently the preferred oxidation and reduction catalyst for low-temperature polymer electrolyte membrane (PEM) fuel cells [13].

10.1.2 EXPERIMENTAL METHODS FOR TMC

Metallic clusters can be generated in a variety of ways: in the gas phase, in solution, in a matrix, or supported on a substrate [5,14]. In a recent review by Ferrando et al. [15], methods for generating mono- and bimetallic nanoclusters were listed as (a) molecular beams [16] where interaction-free clusters are produced by the processes of the vaporization of atoms or molecules in the gas phase, nucleation of atoms or molecules to form a cluster nucleus, and addition of more atoms or molecules to the initially formed nucleus; (b) chemical reductions [17] where metal salts are dissolved in an appropriate solvent in the presence of a surfactant or polymeric ligands, which passivate the cluster surface; (c) ion implantation (metal nanoclusters embedded in insulating matrices) [18]; (d) electrochemical synthesis [19] for generating mono- and bimetallic nanoparticles in solution; (e) radiolysis [20] where metal ions are reduced by γ-ray irradiation of water; (f) sonochemical synthesis [21] where nanoparticles are induced by "sonication" of a solution of metal ions by high-intensity ultrasound; and (g) biosynthesis [22] by the self-assembly of biomolecules.

The most commonly applied techniques for the characterization of nanoclusters are mass spectrometry, electron diffraction microscopy, and x-ray spectroscopy as well as other forms of spectroscopy. Clusters in a molecular beam can be ionized by electron impact or laser and then be deflected by an electric field. Since the deflections occur with respect to the mass of the clusters, it is possible to study the mass abundance of clusters in the cluster beam. The peaks that occur in the mass abundance

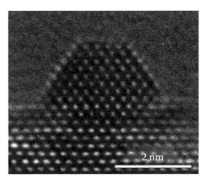

FIGURE 10.2 HREM image of a truncated octahedral Au cluster deposited on MgO. (Reprinted with permission from Pauwels, B., Van Tendeloo, G., Bouwen, W., Theil Kuhn, L., Lievens, P., Lei, H., Hou, M. *Phys. Rev. B*, 62, 10383, 2000. Copyright 2000 by the American Physical Society.)

spectrum occur at "magic numbers" due to extra thermodynamic or kinetic stability at these cluster sizes, which may be explained in terms of the electronic or atomic packing effects [23]. As x-ray diffraction is generally used to study the size, structure, and chemical composition of particles supported on surfaces [24], electron diffraction is widely used on free clusters to get information on the geometry, average size, and temperature [25]. Electron and scanning probe microscopes are powerful techniques for investigating nanoparticles. Transmission electron microscopy (TEM) is useful for studying metal nanoclusters dispersed onto organic substrates. It is possible to get resolution down to the Angstrom level by high-resolution TEMs (HRTEM, HREM), as shown in Figure 10.2 [26]. In scanning transmission electron microscopy (STEM), an electron beam is scanned across the sample [27]. Scanning electron microscopy (SEM) is similar to STEM, but the image of the sample surface is produced by the secondary electrons emitted by the surface following excitation by the primary electron beam [28]. In atomic force microscopy (AFM), a fine tip is brought into close contact with the sample and senses the small repulsive force between the probe tip and the surface [29], while in the scanning tunneling microscopy (STM) the fine tip is again brought extremely close to the surface and a voltage is applied between the tip and the sample so that a tunneling current flows [30]. Since the binding energies of metal atom core electrons distinguishably change with the atomic number, high-energy x-ray radiation is particularly useful for studying metallic nanoclusters [31]. X-ray absorption spectroscopy (XAS), extended x-ray absorption fine structure (EXAFS), near-edge x-ray absorption fine structure (NEX-AFS), x-ray absorption near-edge structure (XANES), x-ray photoelectron spectroscopy (XPS), and Auger electron spectroscopy (AES) are useful x-ray spectroscopic techniques for studying nanoclusters. Other spectroscopic techniques such as ultraviolet-visible (UV-vis) [32], infrared (IR), photoelectron, surface-enhanced Raman, Mossbauer, nuclear magnetic resonance (NMR), and electrochemical NMR (EC-NMR) spectroscopies are also used for the characterization of metallic nanoparticles. In addition, magnetic properties of free TMC can be studied by Stern–Gerlach deflection experiments [33], whereas those of larger, surface-supported clusters can be measured using superconducting quantum interface device (SQUID) magnetometer [34] or magnetic force microscopy [35].

10.1.3 COMPUTATIONAL METHODS FOR TMC

Computational studies of nanoparticles have developed into an important part of cluster science, which contributes information that often complements experimental studies. These studies can be separated into two groups: those investigating equilibrium (static or very low temperature) properties and those studying dynamical properties. Finding the most stable structure for a given cluster size and composition [36–38] and describing its electronic and magnetic properties [39] are the basic goals of equilibrium calculations. Transitions between structural motifs, phase changes [40–42], diffusion, collision, segregation, and growth kinetics are related to the dynamical properties of interest. Both types of calculations are extremely interesting and challenging.

A critical consideration in any theoretical investigation is the means by which intra- and interatomic interactions are described. The cluster configuration energy as a function of atomic coordinates, for example, the potential energy surface (PES) is determined by describing intra- and interatomic forces. The methods for computation of the PES range from simple atom–atom model potentials (such as the Morse potential) and many-body semiempirical methods (such as embedded atom method [EAM] or the Sutton-Chen potential) to parameter free ab initio calculations. First principles methods like density functional theory (DFT) [43] can provide sufficiently high accuracy to be applied to a wide variety of systems with sizes up to a few hundred atoms for symmetric structures [44], whereas semiempirical methods are less computationally intensive but exhibit accuracy, which is strongly system dependent [3,4,6,15]. The accuracy of the first principles methods highly depends on the choice of basis set and the method used for the description of the electronic interactions.

A complete exploration of the PES by means of first principles methods is presently not feasible even for clusters containing a few tens of atoms, while simply guessing the most stable structures of TMCs is very difficult and nontrivial [15]. Thus, large databases of possible cluster structures constructed by a much more complete sampling of the PES with semiempirical potentials are used as the starting point for further ab initio calculations.

Since TMCs have many valence electrons, close-lying states and mostly open d-shells, a treatment of electron correlation is required. Additionally, relativistic effects become important for the heavier transition metals. DFT is one of the most effective and widely used methods in the study of TMCs. While highly accurate multi-determinant quantum chemical methods [45] can treat about a dozen atoms, modern DFT methods [44,46] can describe clusters with up to a few hundred atoms. In a recent review, Alonso [47] has reported electronic and atomic structures and magnetism of bare TMCs noble metal, nickel, iron, niobium, titanium, vanadium, and chromium clusters.

In this chapter, by focusing on size-selected clusters consisting of up to a few tens of atoms, that is, in the size regime where properties change drastically by addition or subtraction of a single atom or even a single electron, we discuss the current state of understanding of geometric, electronic, and magnetic properties of neutral and anionic small cobalt clusters, and the effect of spin-orbit (SO) coupling on chemical and physical properties of small platinum clusters.

10.2　STRUCTURAL, ELECTRONIC, AND MAGNETIC PROPERTIES OF NEUTRAL AND ANIONIC SMALL COBALT CLUSTERS

Co clusters are one of the most interesting examples of TMCs due to their magnetic properties [48] and importance in magnetic storage devices. Although it was reported by a photoionization experiment [49] that Ni_n and Co_n clusters consisting of 50–800 atoms have icosahedral structures, the structures of smaller Co_n ($n \leq 50$) clusters were not well identified. Similarly, chemical reactions of Co_n with water and ammonia [50] indicate icosahedral structures for both bare and ammoniated cluster in the size range of $n = 50$–100. However, it was also mentioned that small, bare clusters may adopt a variety of structures [51]. Yoshida et al. [52] obtained the geometry and electronic structures of Co_n^- anions ($n = 3$, 4, and 6) by comparing the measured and calculated photoelectron spectra. The magnetic moments and adiabatic magnetization of bare Co_n clusters ($20 \leq n \leq 300$) were investigated via Stern–Gerlach experiments by different groups [48,53]. They found that the magnetic moment per atom for small Co_n clusters are significantly larger than the reported bulk value ($1.7 \mu_B$ [54]).

Theoretical investigations of cobalt clusters include an evolutive algorithm based on the semiempirical Gupta potential [55], tight-binding molecular dynamics [56], and pseudo-atomic [57–61] and all-electron DFT calculations [62–65]. Results in the literature are not in good agreement with each other. The discrepancies among DFT calculations may be attributed to differing basis sets and exchange-correlation (xc) functionals and the incomplete searches for the global minimum. Due to the scarcity of experimental study, these inconsistencies still remain unresolved. We are not aware of any higher level non-DFT work for these clusters.

10.2.1　COMPUTATIONAL DETAILS

We performed [66] DFT calculations for Co_n clusters ($n \leq 6$) within the generalized gradient approximation (GGA) to investigate the putative ground state by relaxing geometric structures starting from a large number of initial candidate geometries including those reported in the literature [57–65] for a large number of different spin multiplicities. The NWChem 5.0 program package [67] was used to perform geometry optimizations and total energy calculations. The CRENBL [68] basis set and effective core potential (ECP) for Co were employed whereby the outermost 17 electrons of the free Co atom ($3s^2 3p^6 3d^7 4s^2$) were treated as valence electrons. The GGA exchange functional of Becke [69] and the correlation [70] functional of Lee, Yang, and Parr (BLYP) were used to compute the PES. Geometries were optimized without imposing any symmetry constraints. Spin-polarized calculations were performed for the first $3n + 1$ spin multiplicities, where n is the number of atoms in the clusters.

10.2.2 Results and Discussion

10.2.2.1 Co_2 and Co_2^-

Existing experimental predictions for the BE of the Co_2 dimer are contradictory: while an earlier mass spectroscopic experiment [71] estimated a BE of 1.69 ± 0.26 eV, a more recent collision-induced dissociation experiment [72] yielded an upper bound of 1.32 eV, and finally a photodissociation spectra [73] yielded a value in the range of 0.7–1.4 eV. The mass spectroscopy study [71] predicted a dimer bond length of 2.31 Å calculated from Pauling radius and a total magnetic moment of $4 \mu_B$. Estimation via Badger's rule gives a bond length of 2.02 Å [74]. Hales et al. [72] measured the ionization potential (IP) of the dimer as 6.42 eV, which was previously estimated as 6.26 ± 0.16 eV [75]. The vibrational frequency of Co_2 was reported by different groups as 290 cm^{-1} [76], 280 ± 20 cm^{-1} [77], and 296.8 ± 5.4 cm^{-1} [78]. The results for the properties of Co_2 from previous theoretical calculations are reported in Table 10.1 along with experimental estimations.

In agreement with previous first principles calculations [57–65], we find the quintet spin multiplicity as the ground state, which was observed in experiments [71]. Triplet and septet states belong to the first two exited states that lie 0.74 and 0.90 eV

TABLE 10.1

Comparisons of the Present Results for Co_2 with Those of the Previous Works Reported in the Literature

	Reference	Binding Energy (eV)	Bond Length (Å)	Vibrational Frequency (cm^{-1})	Ionization Potential (eV)
Experimental	[71]	1.69 ± 0.26	2.31		
	[72]	≤ 1.32			6.42
	[73]	0.7–1.4			
	[75]				6.26 ± 0.16
	[76]			290	
	[77]			280 ± 20	
	[78]			296.8 ± 5.4	
Theoretical	[66]	1.71	2.13	329	6.84
	[57]	2.85	2.04		
	[58]	2.35	1.99	373	7.49
	[59]	1.50	2.41	230	
	[60]	0.87	1.96		5.97
	[61]	2.90	1.96		
	[62]	2.26	2.01	342	7.48
	[63]	2.26	1.95	421	
	[64]	5.08	2.14		
	[65]	5.54	2.13		

Source: Reprinted from *Chem. Phys.*, 354, Sebetci, A., Cobalt clusters (Co_n, $n <= 6$) and their anions, 196–201, Copyright (2008). With permission from Elsevier.

above the ground state, respectively. The experimental bond length of the neighboring dimers Fe_2 and Ni_2 are 1.87 [79] and 2.155 Å [80], respectively. Our calculated value of 2.13 Å for Co_2 is slightly less than the Ni_2 dimer bond length and in between the values obtained from Pauling radius [71] and Badger's rule [74]. Previous various theoretical results for Co_2 bond length present values ranging from 1.95 to 2.41 Å. We calculated the IP of the dimer as 6.84 eV in better (or similar) agreement with experiment [72] than previous theoretical studies (see Table 10.1). The ground spin multiplicity of the Co_2^+ ion was calculated as the sextet state, which had previously been determined by ESR spectroscopy [81]. The calculated bond length of the dimer anion (Co_2^-) is 2.19 Å (sextet state), which exceeds that of the neutral dimer by 0.06 Å. This increase in bond length has been estimated experimentally as 0.08 ± 0.02 Å [77]. The calculated BE for the Co_2 dimer is 1.71 eV with respect to the atomic ground state while other theoretical estimations span in the larger range of 0.87–5.54 eV. Our calculated vibrational frequency of 329 cm⁻¹ is also closer to the experimental values than previous ab initio studies (see Table 10.1).

Mulliken population analysis gives a charge distribution of $3d^{7.82}4s^{1.18}$ for each Co atom in the neutral dimer, which indicates that 3d orbitals participate in bonding. The symmetry and energy levels of the dimer molecular orbitals are depicted in Figure 10.3. The point group symmetry of the dimer was determined as D_{4h}. The electron density isosurfaces and the types of the bonds of each molecular orbital are given in Figure 10.4. It is evident from Figure 10.4 that the hybridization of on-site 4s and $3d_{z^2}$ orbitals constructs two sd_{z^2} hybrid atomic orbitals (the z-axis is the bond axis). (These atomic orbitals on each of the Co atoms contribute to two bonding (A_{1g}) and two antibonding (A_{2u}) molecular orbitals (see Figure 10.4) The d_{xz} and d_{yz} orbitals form π bonds. Since the π bond formed by d_{xz} orbitals is symmetric to the one formed by d_{yz} orbitals, the dimer has the degenerate bonding E_u and antibonding E_g π molecular orbitals. Finally, d_{xy} and $d_{x^2-y^2}$ atomic orbitals form the δ-bonds: the

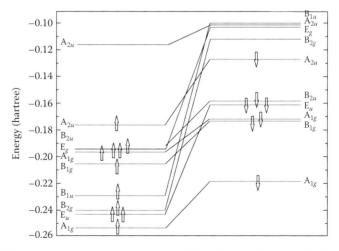

FIGURE 10.3 Energy levels and symmetries of Co_2 dimer molecular orbitals. (Reprinted from *Chem. Phys.*, 354, Sebetci, A., Cobalt clusters (Co_n, $n \leq 6$) and their anions, 196–201, Copyright (2008). With permission from Elsevier.)

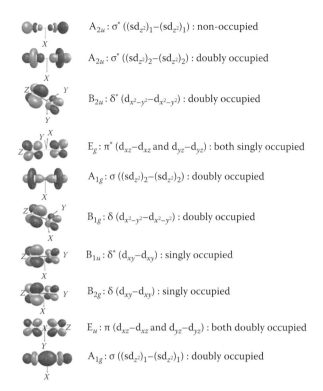

A_{2u} : σ^* $((sd_{z^2})_1-(sd_{z^2})_1)$: non-occupied

A_{2u} : σ^* $((sd_{z^2})_2-(sd_{z^2})_2)$: doubly occupied

B_{2u} : δ^* $(d_{x^2-y^2}-d_{x^2-y^2})$: doubly occupied

E_g : π^* $(d_{xz}-d_{xz}$ and $d_{yz}-d_{yz})$: both singly occupied

A_{1g} : σ $((sd_{z^2})_2-(sd_{z^2})_2)$: doubly occupied

B_{1g} : δ $(d_{x^2-y^2}-d_{x^2-y^2})$: doubly occupied

B_{1u} : δ^* $(d_{xy}-d_{xy})$: singly occupied

B_{2g} : δ $(d_{xy}-d_{xy})$: singly occupied

E_u : π $(d_{xz}-d_{xz}$ and $d_{yz}-d_{yz})$: both doubly occupied

A_{1g} : σ $((sd_{z^2})_1-(sd_{z^2})_1)$: doubly occupied

FIGURE 10.4 Co_2 dimer molecular orbitals. (Reprinted from *Chem. Phys.*, 354, Sebetci, A., Cobalt clusters (Co_n, $n \leq 6$) and their anions, 196–201, Copyright (2008). With permission from Elsevier.)

bonding B_{2u} and antibonding B_{1u} δ-bonds are formed by the d_{xy} orbital and the bonding B_{1g} and antibonding B_{2u} δ-bonds are formed by $d_{x^2-y^2}$ atomic orbitals. Since the B_{2g}: δ $(d_{xy}-d_{xy})$, B_{1u}: δ^* $(d_{xy}-d_{xy})$, E_g: π^* $(d_{xz}-d_{xz})$, and E_g: π^* $(d_{yz}-d_{yz})$, molecular orbitals have higher energy than orbitals in the minority spin channel (the high lying A_{2u} is not occupied in majority spin state) they are singly occupied. Thus, the ground state was found with the following electronic assignment in terms of the infinite symmetry group: $1\sigma_g^2 1\pi_u^4 1\delta_g^1 1\delta_u^1 2\delta_g^2 2\sigma_g^2 1\pi_g^2 2\delta_u^2 1\sigma_u^2$.

10.2.2.2 Co_3 and Co_3^-

We found that the linear structure of Co_3 trimer is more stable than the triangular ones (see Figure 10.5) in contrast to the findings of Jamorski et al. [62], Castro et al. [63], Fan et al. [57], Pereiro et al. [58], and Datta et al. [61]. To the best of our knowledge, only a single study performed by Ma et al. in the literature is in agreement with our results [64]. Fan et al. [57] and Pereiro et al. [58] found the lowest energy geometry of Co_3 to be an equilateral triangle, while Jamorski et al. [62], Castro et al. [63], and Datta et al. [61] suggested an isosceles triangle. Previous calculations for the trimer are summarized in Table 10.2. In Ref. [81] it is reported that the spin resonance spectra of Co_3 in an Ar/Kr matrix indicates a triangular structure with a total magnetic moment of 5 or $7\mu_B$. We estimate that the lowest energy structure of the trimer is the

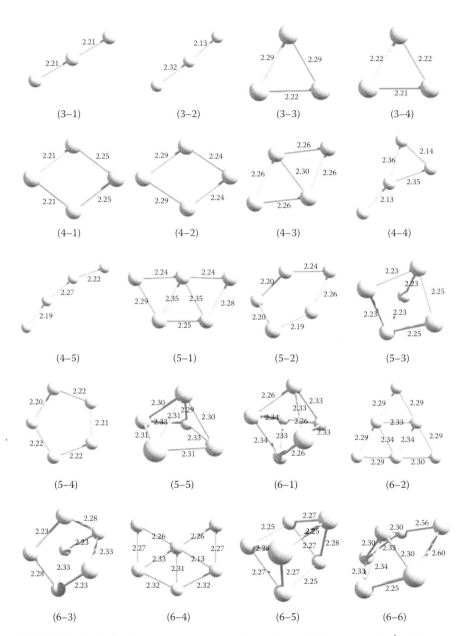

FIGURE 10.5   Relaxed geometries of neutral Co_n clusters (distances are in Å). (Reprinted from *Chem. Phys.*, 354, Sebetci, A., Cobalt clusters (Co_n, $n \leq 6$) and their anions, 196–201, Copyright (2008). With permission from Elsevier.)

TABLE 10.2

Comparisons of the Obtained Results for Co_3 Along with Results from the Literature (Those of the Previous Works)

Reference	Symmetry	Spin Moment (μ_B)	Binding Energy (eV)	Bond Length (Å)
[66]	$D_{\infty h}$	7	3.60	2.21
[57]	D_{3h}	7	2.55	2.28
[58]	C_{3v}	7	1.68	2.04
[61]	C_{2v}	5	5.34	2.10, 2.19, 2.19
[62,63]	C_{2v}	5	4.60	2.12, 2.12, 2.24
[64]	$D_{\infty h}$	7	8.68	2.27

Source: Reprinted from *Chem. Phys.*, 354, Sebetci, A., Cobalt clusters (Co_n, $n <= 6$) and their anions, 196–201, Copyright (2008). With permission from Elsevier.

linear ($D_{\infty h}$) configuration with a bond length of 2.21 Å in the octet state. The vibrational frequencies were calculated as 80, 190, and 237 cm^{-1}. Another linear structure with C_s symmetry in the quartet state was found as the second lowest energy structure where the bond lengths are 2.13 and 2.32 Å. The energy separation between these two structures is 0.16 eV. The two isosceles triangles (C_{2v}) have similar energies, which is 0.68 eV higher than the ground-state energy. The bond lengths are 2.22, 2.29, and 2.29 Å, and 2.22, 2.22, and 2.21 Å, which resemble an equilateral triangle.

Structural assignments, point group symmetries, total spin moments, BEs, highest occupied and lowest unoccupied molecular orbital (HOMO-LUMO) gaps, and the lowest and the highest vibrational frequencies of Co_3 and all other minima of Co_n clusters ($n \leq 6$) are depicted in Table 10.3. The charge distribution of a single Co atom in the lowest energy structure of the trimer ($3d^{8.07}4s^{0.92}$) is slightly different from that of the dimer. The HOMO-LUMO gap of the ground-state geometry is particularly high, 1.04 eV. This is not unusual for a DFT code since many common xc-functionals can mismatch HOMO-LUMO gaps of the order of electron volt. Yoshida et al. [52] reported that Co_3^- has a linear structure with a bond length between 2.25 and 2.50 Å based on photoelectron spectroscopy measurements. We found a linear structure with a bond distance of 2.29 Å as the lowest energy structure of the anionic trimer, which is consistent with the experimental measurements [52]. The ground-state linear structure of the anionic trimer is a septet state. The isosceles triangular configuration of Co_3^- is found to be in the nonet state, which has an energy of 0.46 eV higher than the linear configuration where the bond lengths are 2.14, 2.33, and 2.33 Å.

10.2.2.3 Co_4 and Co_4^-

Five different initial geometries were considered for the tetramer: tetrahedron, square, rhombus, planar Y-like, and linear. The lowest energy structure is found to be an out-of-plane rhombus (C_s) with a spin moment of $10 \mu_B$. It has a BE of 1.37 eV/atom and HOMO-LUMO gap of 0.94 eV. One pair of the bond lengths in this structure is

TABLE 10.3

Isomeric Structure Properties of Co$_n$ ($n = 2 - 6$) Clusters

	Structure	Symmetry	Spin Moment (μ_B)	Total BE (eV/atom)	HOMO-LUMO Gap (eV)[a]	$w_l - w_h$ (cm^{-1})[b]
(2-1)	Linear	$D_{\infty h}$	4	0.85	0.41	329
(3-1)	Linear	$D_{\infty h}$	7	1.20	1.04	80, 237
(3-2)	Linear	C_s	3	1.09	0.65	34, 282
(3-3)	Isosceles triangle	C_{2v}	7	0.98	0.57	151, 305
(3-4)	Isosceles triangle	C_{2v}	5	0.97	0.51	112, 318
(4-1)	Out-of-plane rhombus	C_s	10	1.37	0.94	16, 260
(4-2)	Rhombus	C_{2v}	10	1.35	0.62	80, 260
(4-3)	Rhombus	D_{2h}	8	1.35	0.73	97, 299
(4-4)	Y-like	C_s	8	1.31	0.66	19, 320
(4-5)	Arc-like	C_s	10	1.22	0.71	47, 245
(5-1)	W-like	C_s	11	1.54	0.81	98, 297
(5-2)	Capped rhombus	C_1	11	1.51	0.77	74, 285
(5-3)	Bipyramid	D_{3h}	13	1.50	0.71	114, 301
(5-4)	Pentagon	C_1	11	1.49	0.68	91, 244
(5-5)	Pyramid	C_{4v}	11	1.49	0.54	77, 314
(6-1)	Distorted octahedron	D_3	14	1.71	0.58	94, 296
(6-2)	Triangle	D_{3h}	12	1.67	0.85	89, 272
(6-3)	Face-capped bipyramid	C_2	14	1.67	0.54	82, 284
(6-4)	W-like	C_s	12	1.64	0.58	69, 334
(6-5)	Trigonal prism	D_{3h}	14	1.63	0.54	85, 286
(6-6)	Pentagonal pyramid	C_1	8	1.57	0.55	66, 294

Source: Reprinted from *Chem. Phys.*, 354, Sebetci, A., Cobalt clusters (Co$_n$, $n <= 6$) and their anions, 196–201, Copyright (2008). With permission from Elsevier.

[a] β spin.

[b] Lowest and highest vibrational frequencies.

2.21 Å and the other pair is 2.25 Å. The next two lowest energy configurations are planar rhombus structures with C_{2v} and D_{2h} symmetries, which have spin moments of 10 and 8 μ_B, respectively. They are nearly degenerate having a BE of 1.35 eV/atom. The fourth isomer has a Y-like structure in the nonet state with an energy 0.06 eV/atom higher than the ground state. The fifth locally stable isomer of the tetramer is a linear structure, as shown in Figure 10.5. An initially tetrahedral structure becomes an out-of-plane rhombus after optimization, which is consistent with the calculations of Fan et al. [57] and Ma et al. [64]; however, it is in contradiction to the findings of Datta et al. [61]. Yoshida and coworkers [52] predicted a tetrahedral structure as the ground state of Co_4^- anion by starting from a T_d tetrahedron and a D_{4h} square without considering a rhombus. We found a planar rhombus with a bond length of 2.28 Å in the decet state as the ground state of the tetramer anion. One of the reasons of why the various results disagree with one another is the incompleteness of the global minimum search. Most of the DFT codes perform only local geometry optimizations. When an author does not consider the morphology of the lowest energy structure as the starting configuration of the local geometry optimization, that specific structure may be missed. The lowest and the highest vibrational frequencies of the anion are 113 and 241 cm^{-1}.

10.2.2.4 Co_5 and Co_5^-

For Co_5, seven different initial geometries were investigated: pyramid, bipyramid, capped tetrahedron, X-like, W-like, V-like, and pentagon. A planar W-like structure was found as the lowest energy geometry with a spin magnetic moment of 11 μ_B. The next low-lying minima are out-of-plane bridge-side capped rhombus, bipyramid, pentagon, and pyramid (see Figure 10.5), respectively. Except the bipyramidal structure, which has 13 μ_B spin moment, all the minima have the same magnetic moment of 11 μ_B. The ground-state BE of the neutral pentamer was calculated as 1.54 eV/atom. This structure has the highest HOMO-LUMO gap (0.81 eV) as well. The HOMO-LUMO energy separation is considered as a parameter for chemical stability since it is favorable to add electrons to a low-lying LUMO and to receive electrons from a high-lying HOMO. Although HOMO-LUMO gaps calculated via DFT are not so reliable in magnitude, they can provide reasonable estimations for the relative size of the gaps. The energy separations between the next low-lying minima and the ground state are 0.03, 0.04, 0.05, and 0.05 eV/atom, respectively. Fan et al. [57] and Ma et al. [64] identified the pyramidal structure, and Pereiro et al. [58], Datta et al. [61], and Castro et al. [63] found the bipyramidal structure as the ground-state geometry. Thus, the W-like structure found in this study is reported as the ground state for the first time. Although the literature [57,60,61] reports ground-state spin moment as 13 μ_B, Ma et al. [64] calculated it as 11 μ_B, in agreement with the present result. The bond lengths for each of the minima can be found in Figure 10.5. The lowest energy structure of the anionic pentamer has the same morphology with the neutral pentamer (the W-like structure) with the spin moment of 10 μ_B. The addition of an electron reduces the bond lengths from 2.25, 2.35, and 2.35 Å (see the structure (5-1) in Figure 10.5) to 2.18, 2.34, and 2.34 Å and stretches those of 2.24, 2.24, 2.29, and 2.28 Å to 2.25, 2.26, 2.39, and 2.37 Å, respectively. The lowest and the highest vibrational frequencies of the anionic pentamer are 105 and 296 cm^{-1}.

10.2.2.5 Co_6 and Co_6^-

At the end of several optimizations initially starting with 13 different structural motifs of Co_6, we ended up with six local minima structures in the energetic order of distorted octahedron, planar triangle, face capped bipyramid, planar W-like structure, trigonal prism, and pentagonal pyramid. The lowest energy structure is a distorted octahedron with a D_3 symmetry and a spin moment of 14 μ_B. Its BE is 1.71 eV/atom and the HOMO-LUMO gap is 0.58 eV. The second low-lying minimum structure is a planar triangular with a BE of 1.67 eV/atom which has the highest gap (0.85 eV) among all the hexamer isomers. The planar stable triangular and W-like structures identified in the present study are not considered in the two most recent investigations [61,64]. The spin moments of hexamer isomers are either 14 or 12 μ_B except for the pentagonal pyramid for which it is 8 μ_B where the magnetic moment of one of the Co atoms couples antiferromagnetically with that of the others. A pentagonal pyramidal structure with a spin moment of 12 μ_B was identified as the first-order transition state in our calculations. Ma and coworkers [64] reported an octahedron with D_{4h} symmetry as the lowest energy structure where the bond lengths ranged from 2.38 to 2.40 Å. They [64] have calculated the BE of the Co dimer as 5.08 eV, which is nearly twice that of the experimental data. Our calculated bond lengths are slightly shorter (from 2.26 to 2.34 Å). Datta et al. [61] found an octahedral structure with a total spin moment of 14 μ_B for the ground state. Their calculated bond length, 2.27 Å, is similar to our prediction. However, the photoelectron spectroscopy study of Yoshida et al. [52] predicted a pentagonal pyramid with a bond distance of 2.75 ± 0.1 Å to be the most probable structure for the Co_6^- anion. However, they [52] considered only two structures; a pentagonal pyramid with C_{5v} symmetry and an octahedron with O_h symmetry in their spectral calculations. We obtained a distorted octahedron with D_3 symmetry as the lowest energy structure for the hexamer anion. As indicated in Yoshida et al.'s [52] study, a little change in the structure brings a great difference in the spectral feature. Thus, the distorted octahedral structure should be tested experimentally. The spin magnetic moment of the lowest energy structure is 13 μ_B in the ground state. The bond distances are between 2.25 and 2.64 Å and its lowest and highest vibrational frequencies are 70 and 294 cm^{-1}, respectively. The next isomer of the anion is a nearly equilateral triangle as in the case of the neutral cluster. The pentagonal pyramidal anion structure is about 0.11 eV/atom higher in energy than the lowest energy structure.

In summary, for neutral Co_n (3 ≤ n ≤ 6) clusters, a linear structure, an out-of-plane rhombus, a planar W-like structure, and a distorted octahedron were identified as the global minima geometries, respectively. The addition of an electron to the neutral cluster does not change the geometric motifs significantly. Thus, anionic clusters have the same morphology with the neutral ones with slightly stretched or compressed bond lengths. As the number of atoms in the Co clusters increases, the total spin magnetic moment rises except the dimer. The ground-state total spin moment (M) of the neutral Co clusters in the present work obeys the rule that $M = 2n + a$, where $a = 1$ if n is odd and $a = 2$ if n is even. For the anionic clusters, the total spin moment is equal to $M - 1$, again with the exception that anionic dimer has a spin moment of 5 μ_B.

10.3   THE EFFECT OF SPIN-ORBIT COUPLING ON THE STRUCTURAL, ELECTRONIC, AND MAGNETIC PROPERTIES OF SMALL PLATINUM CLUSTERS

As the size of the Pt particles decreases, their catalytic activities tend to increase because of the increased surface area of the smaller particles [82]. Lineberger and coworkers [83,84] measured the electronic spectra of small platinum and palladium dimers and trimers by negative ion photodetachment spectroscopy and Eberhardt and coworkers [85] obtained the valence- and core-level photoemission spectra of mass-selected monodisperse Pt_n ($n=1-6$) clusters. In a recent theoretical calculation, we studied bare and hydrogenated Pt_nH_m ($n=1-5$, $m=0-2$) clusters within the scalar-relativistic DFT formalism [39]. Relevant literature can be found in the literature [39] for previous experimental and theoretical investigations. In addition, Saenz et al. [86] worked on the interaction of Pt clusters with molecular oxygen. Futschek et al. [87] presented ab initio density functional studies of structural and magnetic isomers of Ni_n and Pt_n clusters with up to 13 atoms. Seivane and Ferrer [88] analyzed the impact of the magnetic anisotropy on the geometric structure and magnetic ordering of small atomic clusters of palladium, iridium, platinum, and gold from two to five, six, or seven atoms depending on the element. Bhattacharyya and Majumder [89] reported the growth pattern and bonding trends in Pt_n ($n=2-13$) clusters and concluded in their first principles study that small Pt_n clusters have planar geometries and that a structural transition to nonplanar geometries occurs at $n=10$. Similarly, Huda et al. [90] predicted that SO coupling leads to planar structures of small Pt clusters.

10.3.1 COMPUTATIONAL DETAILS

We used [91] the NWChem 5.0 program package [67] to perform geometry optimizations and total energy calculations using DFT. The CRENBL [92] basis set, ECPs, and the spin orbit (SO) operator for Pt were employed where the outermost 18 electrons of the free Pt atom ($5s^25p^65d^96s^1$) are treated as valence electrons. The generalized gradient approximation exchange functional of Becke [93] and the Lee–Yang–Parr correlation functional [70] (B3LYP) was chosen as the hybrid xc functional. In our calculations with SO operator, the chosen hybrid functional produced better results for small Pt clusters than the GGA functional. The default convergence criteria $1\,\mu E_h$ for energy and $0.5\,mE_h$/Bohr for geometry optimization were used. Geometries were optimized without imposing any symmetry constraints. Spin-polarized calculations were performed for the first five spin multiplicities (from singlet to nonet). The SO coupling DFT calculations were performed in a Kramers-unrestricted approach, which means that spinors do not have to have the same spinor energy.

10.3.2 RESULTS AND DISCUSSION

10.3.2.1 Pt and Pt_2

First, we discuss the properties of Pt and Pt_2 to assess the accuracy of the chosen B3LYP/CRENBL method. The ground-state Pt atom was found in the triplet state ($5d^96s^1$) for the non-SO coupling case in agreement with the experimental results [94]. The excitation energy of the singlet ($5d^{10}6s^0$) state was calculated as $0.510\,eV$ (neglecting SO coupling), which can be compared with a spin-averaged experimental value of $0.478\,eV$ [94]. By including SO effects, the excitation energy of the closed-shell configuration is $0.881\,eV$ compared with an experimental value of $0.761\,eV$ [94].

Both of these calculated excitation energies are much better than the ones obtained by Fortunelli in 1999 [95], where BPW91 GGA functional was employed with a double-zeta-plus-polarization basis set. The SO splitting of the 5d orbital of the Pt atom in its singlet state was found to be 1.131 eV, in reasonable agreement with the value of 1.256 eV calculated using the all-electron fully relativistic DFT code FPLO [96] via employing Perdew–Wang [97] local xc functional. The IP of Pt was calculated as 9.319 eV (SO included) in contrast with an experimental value of 8.958 eV [98]. Huda et al. [90] calculated the IP of Pt as 9.381 eV in their DFT study with the projected augmented wave (PAW) method. The calculated electron affinity (EA) is 1.806 eV, which is comparable to the experimental value of 2.123 ± 0.001 eV [99]. The discrepancy between the calculated and experimental values of the EA may be reduced when the experimental value is corrected for SO effects, which can be estimated to decrease it by about 0.2 eV [100].

The ground-state spin multiplicity of the Pt_2 dimer is found to be a triplet (not including SO coupling). The bond lengths of the Pt dimer were calculated as 2.373 and 2.406 Å for non-SO and SO calculations, respectively, while the experimental value [101] is 2.333 Å. The corresponding binding energies are 1.354 (non-SO) and 1.238 eV/atom (SO) with respect to the atomic ground state. The experimental BE is 1.57 ± 0.01 eV/atom [102]. Thus, the present method with SO effects slightly overestimates the Pt dimer bond length and therefore underestimates the BE. However, the primary goal of the present work is to investigate the effect of the SO term on the relative stability of the isomers rather than providing accurate quantitative results. Both of the calculated vibrational frequencies of 237 (non-SO) and 219 cm^{-1} (SO) agree very well with the measured value of 223 cm^{-1} [103].

10.3.2.2 Pt_4

We report the relative stability of Pt_4 isomers with and without SO coupling for different spin multiplicities in Table 10.4. The identical results in Tables 10.4 through 10.6 with the same ΔE and spin moment for SO cases correspond to identical local minima structures. Each SO calculation was performed after a non-SO one starting from the charge and spin densities obtained in the non-SO calculation. Thus, the identical SO results for a particular size and isomer mean that the different spin multiplicities of the non-SO case converge to the same SO electronic state in the relaxation process. When the SO coupling effect is included in the calculations, the spin (S) and orbital angular momentum (L) are not conserved separately, instead the total angular momentum $J = L + S$ is conserved. If relativistic effects are considered as a perturbation, then one can consider $L^2 = L(L+1)$ and $S^2 = S(S+1)$ as being conserved [104]. We use the term spin moment in Tables 10.4 through 10.7 for SO cases in this sense. The geometric structures and bond lengths of each isomer of Pt_4 for their most stable spin multiplicity are given in Figure 10.6.

Similar to the results of our previous study on small bare and hydrogenated Pt clusters [39] where we did not employ SO coupling effects and similar to results presented in Refs. [87,105], the calculated lowest energy structure of the Pt tetramer is a tetrahedron regardless of whether SO coupling is included. When SO effects are not considered, the ground-state spin multiplicity of the tetrahedron is 3 and it has the point group symmetry C_{3v}. On the other hand, when SO effects are taken into

TABLE 10.4

Relative Stability of Pt₄ Isomers Predicted Utilizing Non-SO and SO DFT Calculations

Isomer	Structure	Non-SO Spin Moment (μ_B)	ΔE (eV)	SO Spin Moment (μ_B)	ΔE (eV)
(4-1)	Tetrahedron	2	0.000	1.84	0.026
		4	0.109[a]	3.64	0.000
		6	1.110	3.64	
		8	2.239	3.64	
(4-2)	Rhombus	0	0.576	2.08	0.133
		2	0.471	2.11	
		4	0.206	2.10	
		6	0.945	2.09	
(4-3)	Square	0	0.232	0.00	0.357
		2	0.321	1.77	0.493
(4-4)	Y-like	0	1.410	0.54	0.410
		2	0.892	1.50	0.458
		4	0.551	3.51	0.354
		6	1.134	3.47	0.337

Source: Reprinted from *Phys. Chem. Chem. Phys.*, 354, Sebetci, A., Does spin orbit coupling effects favour planar structures for small platinum clusters, 921–925, Copyright (2009). With permission from The Royal Chemical Society.

[a] Saddle point.

account, tetramers with quintet, septet, and nonet initial spin multiplicity converge to the same spin moment of 3.64 μ_B and this state is 26 meV lower in energy than the state with a 1.84 μ_B spin moment. The point group symmetry of the optimized structure in this case is D_{2d}. Our prediction of a tetrahedron as the ground-state geometry differs from those reported in the literature [89,90,106]. A common feature of these studies is that they employ plane wave codes (Refs. [89,90] employ the PAW method of the code VASP and Refs. [106,107] employ the code CPMD [108]). These studies predict that the rhombus isomer is the lowest energy structure with a spin multiplicity of 5. However, Futschek et al. [87] obtained a distorted tetrahedron similar to our results using VASP so the source of the discrepancy is not clear. According to the present calculations on the non-SO case, the quintet state is the ground magnetic state of the rhombus, but its total energy is 0.206 eV higher than that of the tetrahedron. For the SO case, the rhombus whose singlet, triplet, quintet, and septet initial spin multiplicities converge to a spin moment between 2.08 and 2.11 μ_B has an energy of 0.133 eV higher than the tetrahedron. The optimized structures of the rhombus in both non-SO and SO cases have C_{2v} symmetry. As in the case of the dimer, SO effects slightly stretch the bond lengths. Both of these structures are out of plane where the angles between the triangular planes are 113° (non-SO) and 105° (SO).

TABLE 10.5

Relative Stability of Pt_5 Isomers Predicted Utilizing Non-SO and SO DFT Calculations

Isomer	Structure	Non-SO		SO	
		Spin Moment (μ_B)	ΔE (eV)	Spin Moment (μ_B)	ΔE (eV)
(5-1)	Pyramid	2	0.460	3.46	0.530[a]
		4	0.149	3.79	0.144
		6	0.011	5.58	0.000
		8	0.816	7.28	0.757[a]
(5-2)	Capped tetrahedron	0	0.920	0.00	0.855
		2	0.428	2.38	0.500
		4	0.000	3.67	0.167
		6	0.336	5.18	0.278
(5-3)	W-like	0	0.661	0.00	0.609
		4	0.169	3.58	0.173
		6	0.536	2.03	0.193
(5-4)	Bipyramid	0	1.011	0.02	0.807
		2	0.319	2.05	0.339[a]
		4	0.186	3.56	0.247
		6	0.609	3.57	
		8	0.954	3.97	0.388
(5-5)	Capped square	2	0.674	2.43	0.363
(5-6)	X-like	0	1.319	0.00	0.713
		2	0.930	1.83	0.524
		8	2.859	3.39	0.490

Source: Reprinted from *Phys. Chem. Chem. Phys.*, 354, Sebetci, A., Does spin orbit coupling effects favour planar structures for small platinum clusters, 921–925, Copyright (2009). With permission from The Royal Chemical Society.

[a] Saddle point.

Thus, inclusion of the SO coupling strengthens the non-planarity of the rhombus in contrast to the findings of Huda [90]. The third and fourth lowest energy isomers of the Pt tetramer are the square and Y-like (see Figure 10.6) planar structures, respectively. The total energy of the singlet square (non-SO) is 0.232 eV higher than that of the global minimum. SO effect increases this energy difference to a value of 0.357 eV, which is dissimilar to the general so-called trend that SO effects decrease the energy differences between the isomers. For the Y-like isomer, the energy separations from the lowest energy structure are 0.551 (non-SO quintet) and 0.337 eV (SO 3.47 μ_B).

10.3.2.3 Pt_5

We have identified six different stable isomers of Pt_5 clusters and reported their relative stabilities in Table 10.5 and their structures and bond lengths in Figure 10.7. Although a bridge-side-capped tetrahedron with a spin multiplicity of 5 was obtained

TABLE 10.6

Relative Stability of Pt$_6$ Isomers Predicted Utilizing Non-SO and SO DFT Calculations

Isomer	Structure	Non-SO		SO	
		Spin Moment (μ_B)	ΔE (eV)	Spin Moment (μ_B)	ΔE (eV)
(6-1)	Trigonal prism	4	0.246	3.68	0.285
		6	0.00	5.43	0.000
		8	0.168	5.41	
(6-2)	Triangle	2	0.923	1.63	0.219
		4	0.957[a]	2.69	0.209
		6	0.390	3.44	0.114
		8	1.831	3.43	
(6-3)	Face-capped pyramid	2	0.404	2.50	0.333
		4	0.221	4.75	0.155
		6	0.150	4.75	
(6-4)	Double square	6	0.684	5.38	0.756
		8	2.031	0.74	0.334
(6-5)	Bridge-capped pyramid	0	0.865	0.00	0.781
		6	0.644	5.42	0.338
		8	0.899	5.13	0.467
(6-6)	Face-capped bipyramid	0	1.418	0.00	1.044
		2	0.598	1.74[b]	0.686[b]
		8	0.369	7.01[b]	0.355[b]
(6-7)	Octahedron	0	1.326	0.00	1.620
		2	1.132	4.83	0.509
		6	1.041	2.44	0.711
(6-8)	W-like	0	0.728	0.00	0.763[a]
		2	1.301	4.02	0.520
		6	0.530	4.06	
(6-9)	Double-capped tetrahedron	4	0.398	3.66	0.649
		6	0.782	5.08	0.678
(6-10)	Bridge-capped bipyramid	0	0.744	0.00	0.663
		4	0.349	3.54[b]	0.801[b]
(6-11)	Double-capped square	0	0.941	1.31	0.807
		2	0.669	1.83	0.729
		4	0.728	3.22	
		6	0.880	3.22	0.696
		8	1.483	3.21	

Source: Reprinted from *Phys. Chem. Chem. Phys.*, 354, Sebetci, A., Does spin orbit coupling effects favour planar structures for small platinum clusters, 921–925, Copyright (2009). With permission from The Royal Chemical Society.

[a] Saddle point.
[b] Convergence cannot be achieved.

TABLE 10.7

Isomeric Structure Properties of Pt$_n$ ($n=4$–6) Clusters with and without SO Coupling

	Symmetry		Spin Moment (μ_B)		Total BE (eV/atom)		SO Energy (eV/atom)	HOMO-LUMO Gap (eV)		w_l–w_h (cm^{-1})[a]	
	Non-SO	SO	Non-SO	SO	Non-SO	SO		Non-SO[b]	SO	Non-SO	SO
(4-1)	C_{3v}	D_{2d}	2	3.64	2.12	1.91	0.21	1.60	1.16	80, 215	99, 215
(4-2)	C_{2v}	C_{2v}	4	2.10	2.07	1.88	0.19	1.89	1.43	50, 213	44, 206
(4-3)	D_{4h}	D_{4h}	0	0.00	2.06	1.83	0.24	1.58	1.49	9, 198	41, 187
(4-4)	C_s	C_1	4	3.49	1.99	1.83	0.16	2.32	1.90	42, 234	48, 241
(5-1)	C_{2v}	C_{2v}	6	5.58	2.32	2.13	0.19	1.58	1.63	47, 209	51, 197
(5-2)	C_{2v}	C_1	4	3.67	2.32	2.09	0.23	1.66	1.40	39, 212	46, 198
(5-3)	C_1	C_1	4	3.58	2.29	2.09	0.20	1.78	1.56	41, 229	46, 230
(5-4)	D_3	C_1	4	3.56	2.28	2.08	0.21	1.81	1.24	71, 223	69, 210
(5-5)	C_1	C_1	2	2.43	2.19	2.05	0.13	1.95	1.30	50, 229	50, 228
(5-6)	C_1	C_1	2	3.39	2.14	2.03	0.11	2.06	1.75	30, 269	34, 246
(6-1)	C_{2v}	C_1	6	5.42	2.48	2.29	0.20	1.00	1.17	19, 197	11, 184
(6-2)	C_1	C_1	6	3.44	2.42	2.27	0.15	3.08	1.39	37, 239	26, 242
(6-3)	C_1	C_1	6	4.75	2.46	2.26	0.20	1.34	1.32	39, 206	38, 201
(6-4)	C_{2v}	C_2	6	0.74	2.37	2.23	0.14	1.96	1.34	31, 230	34, 245
(6-5)	C_1	C_1	6	5.42	2.38	2.23	0.15	1.66	1.59	32, 216	22, 204
(6-6)	C_1	C_1	8	7.01	2.42	2.23	0.20	2.58	1.33	35, 200	c
(6-7)	C_1	C_1	6	4.83	2.31	2.20	0.11	0.87	1.30	72, 209	49, 194
(6-8)	C_1	C_1	6	4.04	2.40	2.20	0.20	2.44	1.38	36, 238	47, 226
(6-9)	D_{2d}	D_{2d}	4	3.66	2.42	2.18	0.24	1.82	1.27	28, 192	27, 196
(6-10)	C_1	C_1	4	0.00	2.43	2.18	0.25	1.63	1.33	41, 217	48, 217
(6-11)	C_2	C_1	2	3.22	2.37	2.17	0.20	1.51	1.28	30, 216	30, 229

Source:　Reprinted from *Phys. Chem. Chem. Phys.*, 354, Sebetci, A., Does spin orbit coupling effects favour planar structures for small platinum clusters, 921–925, Copyright (2009). With permission from The Royal Chemical Society.

[a] Lowest and highest vibrational frequencies.

[b] α spin.

[c] Convergence cannot be achieved.

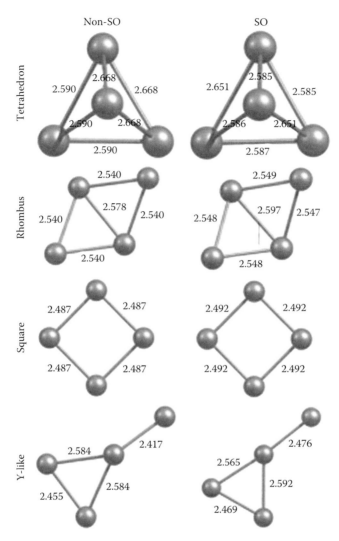

FIGURE 10.6 Relaxed geometries of Pt_4 isomers for the most stable spin multiplicity of each isomer with and without SO coupling effects (distances are in Å). (PCCP Owner Societies, Sebetci, A., Does spin orbit coupling effect favour planar structures for small platinum clusters, *Phys. Chem. Chem. Phys.*, 11, 921, 2009. Reproduced by permission of The Royal Chemical Society.)

as the lowest energy structure for the non-SO case in agreement with our previous calculations [39], SO coupling effects favor a pyramid with $5.58\,\mu_B$ spin moment. The energy separation between the capped tetrahedron and the pyramid in the former case is only 11 meV, which may be considered within the accuracy of the calculations (remember that the energy convergence criteria is chosen as about 0.03 meV). On the other hand, SO coupling effects favor the pyramidal structure over the tetrahedral structure by as much as 167 meV. While the lowest energy structure obtained for the

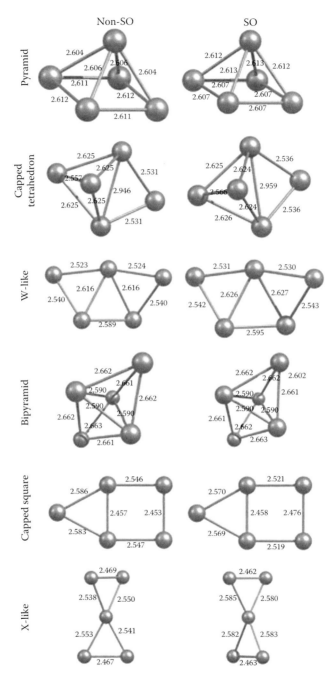

FIGURE 10.7 Relaxed geometries of Pt$_5$ isomers for the most stable spin multiplicity of each isomer with and without SO coupling effects (distances are in Å). (PCCP Owner Societies, Sebetci, A., Does spin orbit coupling effect favour planar structures for small platinum clusters, *Phys. Chem. Chem. Phys.*, 11, 921, 2009. Reproduced by permission of The Royal Chemical Society.).

pyramidal isomer contradicts the findings of Bhattacharyya and Majumder [89], Huda et al. [90], Grönbeck and Andreoni [106], Xiao and Wang [109], Saenz et al. [86], and Futschek et al. [87], it agrees with the predictions of Majumder et al. [105], Yang et al. [110], and Seivane and Ferrer [88]. Bhattacharyya and Majumder [89] and Huda et al. [90] predicted a planar bridge-side-capped structure, Grönbeck and Andreoni [106] and Saenz et al. [86] predicted a planar W-like structure, Xiao and Wang [109] and Futschek et al. [87] predicted a trigonal bipyramid as the lowest energy configuration. Even if it is assumed that the search for the lowest energy structure is performed satisfactorily in all of these studies, since the exact xc-functional of DFT is not known, such contradictions in the results obtained by different groups seem to be solved only by the help of accurate experimental investigations.

The third low-lying isomer is an out-of-plane, W-like structure (see Figure 10.7) with a spin moment of $3.58\,\mu_B$ with an energy of $173\,meV$ higher than the global minimum. Previously, we identified a trapezoidal structure as the fifth lowest lying isomer [39]. However, in that study we considered only its singlet and triplet states. Thus, the discrepancy can be attributed to the limitation of the previous calculations for the first two spin multiplicities. For the discerning between the contradictory results, the first thing to be assessed is the completeness of the study. The fourth energetically favorable structure is a bipyramid with a spin moment of $3.57\,\mu_B$. The planar, bridge-side-capped square structure, which had been predicted as the global minimum of Pt_5 clusters [89,90] by a plane wave code, was obtained as the fifth lowest energy isomer in our calculations. Its energies of 0.674 (non-SO) and 0.373 eV (SO) are higher than the ones for the lowest energy structure. Finally, a nonplanar X-like geometry was identified as a stable isomer with the highest total energy. Unlike most of the other isomers where the magnetic moment of the cluster for the most stable magnetic state in SO case is similar to that of non-SO case, the most stable magnetic state of the X-like structure in SO case ($3.39\,\mu_B$) could be obtained from an initial nonet state, which gives very high relative energy (2.859 eV) in non-SO calculations.

The effect of SO coupling on the bond lengths of Pt_5 clusters is not constant. For the pyramidal structure, while the interatomic distances between the apex atom and the atoms on the square plane are stretched, the distances between the atoms on the square plane are diminished. For the capped tetrahedron, as SO effects make the triangle constructed by the capping atom and the two atoms bonded to the capping atom a bit larger, all other bond lengths are kept nearly the same. All bond lengths of W-like structure and most of them in X-like structures are elongated by SO coupling; on the contrary, SO coupling does not have a significant effect on the bond length of the bipyramid and even causes contractions in the bond length of the capped square isomer.

10.3.2.4 Pt_6

The relative stabilities of 11 different isomers of Pt_6 can be found in Table 10.6. Their bond lengths and structures have been given in Figure 10.8. The most stable structure of Pt_6 has been obtained as a trigonal prism in the septet state for non-SO case and a state with a spin moment of $5.43\,\mu_B$ for the SO case. Xiao and Wang [109] predicted a planar double-capped square, Bhattacharyya and Majumder [89] identified a planar triangular structure, Futschek et al. [87] obtained a face-capped pyramid as the lowest energy isomer of Pt_6. In our calculations, the triangle is the

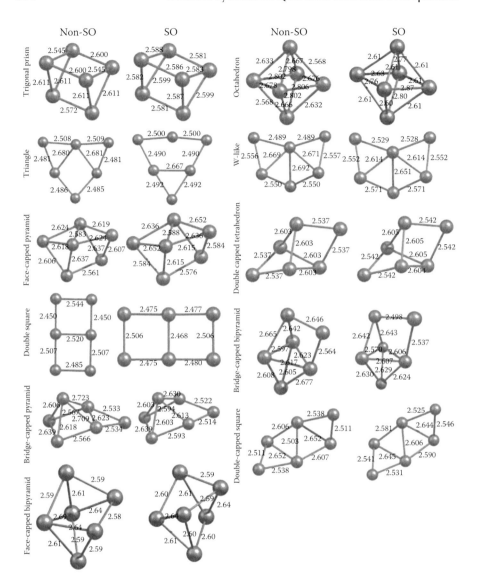

FIGURE 10.8 Relaxed geometries of Pt_6 isomers for the most stable spin multiplicity of each isomer with and without SO coupling effects (distances are in Å). (PCCP Owner Societies, Sebetci, A., Does spin orbit coupling effect favour planar structures for small platinum clusters, *Phys. Chem. Chem. Phys.*, 11, 921, 2009. Reproduced by permission of The Royal Chemical Society.)

second lowest energy structure with an energy of 0.113 eV higher (SO case) than the prism. The face-capped pyramid is the third lowest energy isomer with an energy of 0.154 eV above the ground state, while an out-of-plane double-capped square was obtained with an energy of 0.696 eV higher than the lowest energy state. Unlike Pt_4 and Pt_5 clusters, the SO coupling effect changes the energetic ordering of most of the

Pt_6 isomers. When SO effects are not included in the calculations, the second and the third isomers are predicted as the face-capped pyramid and the bridge-capped bipyramid, respectively. Similarly, the double-capped tetrahedron is lower in energy for the non-SO case than for the SO case. On the other hand, the SO effect significantly decreases the relative energies of planar or quasi-planar structures. For instance, the energy (0.390 eV) of the septet triangle becomes 0.113 eV, the energy (2.031 eV) of the nonet double square becomes 0.334 eV, and the energy (0.880 eV) of double-capped square becomes 0.696 eV due to SO coupling. The large difference between the relative energies of double square in the nonet state (2.031 eV for non-SO case and 0.334 eV for SO case) can be explained by the fact that SO coupling significantly changes the initial spin multiplicity, which converges to a spin moment of $0.74 \mu_B$. Thus, we agree with Huda et al. [90] on the considerable effect of SO coupling on these Pt clusters via increasing the stability of planar structures. However, we do not agree with either Huda et al. [90] or Bhattacharyya and Majumder [89] that the planar structures are the most stable isomers of small Pt_n ($n = 4-6$) clusters. In our calculations, the lowest energy structures of Pt_n ($n = 4-6$) clusters are all three dimensional. This conclusion is supported not only by the results of Futschek et al. [87] but also by the findings of Tian et al. [111], who calculated the structural properties of the Pt_7 cluster using DFT with both Gaussian and plane-wave basis sets and obtained a three-dimensional coupled tetragonal pyramid as the global minimum, which can be constructed by adding an atom to the center of one of the rectangular faces of the triangular prism. As in the case of Pt_5, while the SO coupling increases some of the bond lengths of Pt_6 clusters, it decreases some other bond lengths.

10.3.2.5 Most Stable Isomers of Pt_4–Pt_6

Point group symmetries, spin magnetic moments, total binding energies per atom, SO coupling energies per atom (difference in the BE between non-SO and SO cases), HOMO-LUMO gaps and the lowest and highest vibrational frequencies for the most stable spin states of each isomer of all Pt clusters studied in this work can be found in Table 10.7. SO coupling effects always reduce BE since they make a larger contribution to the atomic energy than to the cluster energy. As the cluster size increases from 4 to 6, the BE per atom increases as well (from 2.122 to 2.483 eV for the non-SO case and from 1.914 to 2.286 eV for the SO case). In contrast to the results of Huda et al. [90], the SO coupling energy per atom decreases from 0.208 to 0.193 eV when the size of the clusters changes from 4 to 5. For the lowest energy structure of Pt_6, the SO coupling energy per atom is 0.196 eV. Except the first isomers of Pt_5 and Pt_6 and the seventh isomer of Pt_6, the HOMO-LUMO gaps are smaller due to SO coupling. For the non-SO case, planar or quasi-planar (4-4), (5-5), (5-6), (6-2), (6-4), (6-8) structures have significantly large HOMO-LUMO gaps when one considers that SO coupling reduces these large gaps. In general, the SO coupling effect does not change the vibrational frequencies significantly. The distorted structures may arise merely due to the single-determinant wavefunction instead of a wavefunction, which is "state-averaged" over the components in degenerate cases.

In summary, we studied the effect of SO coupling on small Pt clusters, Pt_n ($n = 4-6$). Four isomers of Pt_4, six isomers of Pt_5, and eleven isomers of Pt_6 were calculated with and without SO effects. We found that SO coupling effects have a considerable impact on these clusters, which can change the energetic ordering of the isomers. Although the stability of planar structures increases with the inclusion of SO coupling, it cannot make these planar structures the most stable isomers. The lowest energy structures of Pt_4, Pt_5, and Pt_6 clusters are predicted to be the tetrahedron, pyramid, and trigonal prism structures, respectively. In general, SO coupling reduces both total binding energies and HOMO-LUMO gaps.

10.4 CONCLUSIONS

In this chapter, we first give a brief and introductory report on the TMC and their importance, in general, and then list the relevant, currently available experimental and computational methods. By focusing on size-selected small Co and Pt clusters consisting of up to a few atoms, which is the size regime where properties can change drastically by addition or subtraction of a single atom or even a single electron, we discuss the current state of understanding of geometric, electronic, and magnetic properties of neutral and anionic small cobalt clusters, and the effect of SO coupling on the properties of small platinum clusters. The present study and references herein should be considered as just a few examples of the very active, fascinating, and promising research area of TMC. We believe that our calculations on small Co and Pt clusters contribute to the search of global minima, electronic, and magnetic states of these clusters.

REFERENCES

1. *Clusters of Atoms and Molecules*; Haberland, H., Ed.; Springer: Berlin, Germany, 1994; Vols. I and II.
2. *Large Clusters of Atoms and Molecules*; Martin, T.P., Ed.; Kluwer: Dordrecht, the Netherlands, 1996.
3. *Theory of Atomic and Molecular Clusters*; Jellinek, J., Ed.; Springer: Berlin, Germany, 1999.
4. *Atomic and Molecular Clusters*; Johnston, R.L., Ed.; Taylor & Francis: London, U.K., 2002.
5. *Nanoparticles: From Theory to Application*; Schmid, G., Ed.; Wiley-VCH: Weinheim, Germany, 2003.
6. Baletto, F.; Ferrando, R. *Rev. Mod. Phys.* 2005, *77*, 371.
7. Valden, M.; Lai, X.; Goodman, D.W. *Science.* 1998, *281*, 1647.
8. Magyar, R.J.; Mujica, V.; Marquez, M.; Gonzalez, C. *Phys. Rev. B* 2007, *75*, 144421.
9. Yan, L.; Seminario, J.M. *Int. J. Quantum Chem.* 2007, *107*, 440.
10. Arenz, M.; Landman, U.; Heiz, U. *Chemphyschem.* 2006, *7(9)*, 1871.
11. Boyen, H.-G.; Kastle, G.; Weigl, F.; Koslowski, B.; Dietrich, C.; Ziemann, P.; Spatz, J.P. et al., *Science.* 2002, *297*, 1533.
12. Cotton, F.A. *Q. Rev. Chem. Soc.* 1966, 466.
13. *Handbook of Fuel Cells: Fundamentals, Technology, Applications*; Vielstich, W.; Lamm, A.; Gasteiger, H.A.; Ed.; Wiley: West Sussex, 2003.
14. Burda, C.; Chen, X.-B.; Narayanan, R.; El-Sayed, M.A. *Chem. Rev.* 2005, *105*, 1025.
15. Ferrando, R.; Jellinek, J.; Johnston, R.L. *Chem. Rev.* 2008, *108*, 845.
16. *Cluster Beam Synthesis of Nanostructured Materials*; Milani, P.; Iannotta, I.; Ed.; Springer: Berlin, Germany, 1999.
17. Bonnemann, H.; Richards, R.M. *Eur. J. Inorg. Chem.* 2001, *10*, 2455.

18. Mattei, G.; Maurizio, C.; Mazzoldi, P.; D'Acapito, F.; Battaglin, G.; Cattaruzza, E.; de Julian Fernandez, C.; Sada, C. *Phys. Rev. B* 2005, *71*, 195418.
19. Reetz, M.T.; Helbig, W.; Quaiser, S.A. In *Active Metals*; Furstner, A., Ed.; VCH: Weinheim, Germany, 1996; p. 279.
20. Doudna, C.M.; Bertino, M.F.; Blum, F.D.; Tokuhiro, A.T.; Lahiri-Dey, D.; Chattopadhay, S.; Terry, *J. J. Phys. Chem. B* 2003, *107*, 2966.
21. Mizukoshi, Y.; Fujimoto, T.; Nagata, Y.; Oshima, R.; Maeda, Y. *J. Phys. Chem. B* 2000, *104*, 6028.
22. Brayner, R.; Coradin, T.; Fievet-Vincent, F.; Livage, J.; Fievet, F. *New J. Chem.* 2005, *29*, 681.
23. Martin, T.P. *Phys. Rep.* 1996, *273*, 199.
24. Renouprez, A.J.; Lebas, K.; Bergeret, G.; Rousset, J.L.; Delichere, P. *Stud. Surf. Sci. Catal.* 1996, *101*, 1105.
25. Hall, B.D.; Hyslop, M.; Wurl, A.; Brown, S.A. In *Fundamentals of Gas-phase Nanotechnology*; Kish, L., Granqvist, C.G., Marlow, W., Siegel, R.W., Ed.; Kluwer: Dordrecht, the Netherlands, 2001; p. 1.
26. Pauwels, B.; Van Tendeloo, G.; Bouwen, W.; Theil Kuhn, L.; Lievens, P.; Lei, H.; Hou, M. *Phys. Rev. B* 2000, *62*, 10383.
27. Voyles, P.M.; Muller, D.A.; Grazul, J.L.; Citrin, P.H.; Gossmann, H.-J.L. *Nature* 2002, *416*, 826.
28. Goia, D.V.; Matijevic, E. *New J. Chem.* 1998, *22*, 1203.
29. Fain, S.C. Jr.; Polwarth, C.A.; Tait, S.L.; Campbell, C.T.; French, R.H. *Nanotechnology* 2006, *17*, S121.
30. Samori, P. *J. Mater. Chem.* 2004, *14*, 1353.
31. Russell, A.E.; Rose, A. *Chem. Rev.* 2004, *104*, 4613.
32. Toshima, N.; Yonezawa, T. *New J. Chem.* 1998, *22*, 1179.
33. Bansmann, J.; Baker, S.H.; Binns, C.; Blackman, J.A.; Bucher, J.-P; Dorantes-Davila J.; Dupuis, V. et al. *Surf. Sci. Rep.* 2005, *56*, 189.
34. Nguyen, H.L.; Howard, L.E.M.; Giblin, S.R.; Tanner, B.K.; Terry, I.; Hughes, A.K.; Ross, I.M.; Serres, A.; Bürckstümmer, H.; Evans, J.S.O. *J. Mater. Chem.* 2005, *15*, 5136.
35. Sul, Y.C.; Liu, W.; Yue, L.P.; Li, X.Z.; Zhou, J.; Skomski, R.; Sellmyer, D.J. *J. Appl. Phys.* 2005 *97*, 10J304.
36. Sebetci, A.; Guvenc, Z.B. *Surf. Sci.* 2003, *525*, 66.
37. Sebetci, A.; Guvenc, Z.B. *Eur. Phys. J. D* 2004, *30*, 71.
38. Sebetci, A.; Guvenc, Z.B. *Modelling Simul. Mater. Sci. Eng.* 2005, *13*, 683.
39. Sebetci, A. *Chem. Phys.* 2006, *331*, 9.
40. Sebetci, A.; Guvenc, Z.B.; Kokten, H. *Int. J. Mod. Phys. C* 2004, *15*, 981.
41. Sebetci, A.; Guvenc, Z.B. *Modelling Simul. Mater. Sci. Eng.* 2004, *12*, 1131.
42. Sebetci, A.; Guvenc, Z.B. Kokten. H. *Comput. Mater. Sci.* 2006, *35*, 192.
43. Hohenberg, P.; Kohn, W. *Phys. Rev.* 1964, *136*, B864.
44. Kumar, V.; Kawazoe, Y. *Phys. Rev. B* 2008, *77*, 205418.
45. Dai, D.; Balasubramanian, K. *J. Chem. Phys.* 1995, *103*, 648.
46. Kruger, S.; Vent, S.; Nortemann, F.; Staufer, M.; Rosch, N. *J. Chem. Phys.*, 2001, *115*, 2082.
47. Alonso, J.A. *Chem. Rev.*, 2000, *100*, 637.
48. Xu, X.; Yin, S.; Moro, R.; de Heer, W.A. *Phys. Rev. Lett.* 2005, *95*, 237209.
49. Pellarin, M.; Baguenard, B.; Vialle, J.L.; Lerme, J.; Broyer, M.; Miller, J.; Perez, A. *Chem. Phys. Lett.* 1994, *217*, 349.
50. Parks, E.K.; Klots, T.D.; Winter, B.J.; Riley, S.J. *J. Chem. Phys.* 1993, *99*, 5831.
51. Parks, E.K.; Winter, B.J.; Klots, T.D.; Riley, S.J. *J. Chem. Phys.* 1992, *96*, 8267.
52. Yoshida, H.; Terasaki, A.; Kobayashi, K.; Tsukada, M.; Kondow, T. *J. Chem. Phys.* 1995, *102*, 5960.

53. Douglass, D.C.; Cox, A.J.; Bucher, J.P.; Bloomfield, L.A. *Phys. Rev. B* 1993, *47*, 12874.
54. *Introduction to Solid State Physics*, Kittel, J., Ed.; Wiley, New York, 1996, 7th ed.
55. Rodriguez-Lopez, J.L.; Aguilera-Granja, F.; Michaelian, K.; Vega, A. *Phys. Rev. B* 2003, *67*, 174413.
56. Andriotis, A.N.; Menon, M. *Phys. Rev. B* 1998, *57*, 10069.
57. Fan, H.-J.; Liu, C.-W.; Liao, M.-S. *Chem. Phys. Lett.* 1997, *273*, 353.
58. Pereiro, M.; Baldomir, D.; Iglesias, M.; Rosales, C.; Castro, M. *Int. J. Quantum Chem.* 2001, *81*, 422.
59. Barden, C.J.; Rienstra-Kiracofe, J.C.; Schaefer III, H.F. *J. Chem. Phys.* 2000, *113*, 690.
60. Pereiro, M.; Mankovsky, S.; Baldomir, D.; Iglesias, M.; Mlynarski, P.; Valladares, M.; Suarez, D.; Castro, M.; Arias, J.E. *Comput. Mater. Sci.* 2001, *22*, 118.
61. Datta, S.; Kabir, M.; Ganguly, S.; Sanyal, B.; Saha-Dasgupta, T.; Mookerjee, A. *Phys. Rev. B* 2007, *76*, 014429.
62. Jamorski, C.; Martinez, A.; Castro, M.; Salahub, D.R. *Phys. Rev. B* 1997, *55*, 10905.
63. Castro, M.; Jamorski, C.; Salahub, D.R. *Chem. Phys. Lett.* 1997, *271*, 133.
64. Ma, Q.-M.; Xie, Z.; Wang, J.; Liu, Y.; Li, Y.-C. *Phys. Lett. A* 2006, *358*, 289.
65. Feng, R.-J.; Xu, X.-H.; Wu, H.-S. *J. Magn. Magn. Mater.* 2007, *308*, 131.
66. Sebetci, A. *Chem. Phys.* 2008, *354*, 196.
67. Bylaska, E.J.; de Jong, W.A.; Kowalski, K.; Straatsma, T.P.; Valiev, M.; Wang, D.; Apra, E.; Windus, T.L.; Hirata S. et al., "*NWChem, A Computational Chemistry Package for Parallel Computers, Version 5.0*" (2006), Pacific Northwest National Laboratory, Richland, WA, USA.
68. Hurley, M.M.; Pacios, L.F.; Christiansen, P.A.; Ross, R.B.; Ermler, W.C.; *J. Chem. Phys.* 1986, *84*, 6840.
69. Becke, A.D. *Phys. Rev. A* 1988, *38*, 3098.
70. Lee, C.; Yang, W.; Parr, R.G. *Phys. Rev. B* 1988, *37*, 785.
71. Kant, A.; Strauss, B. *J. Chem. Phys.* 1964, *41*, 3806.
72. Hales, D.A.; Su, C.-X.; Lian, L.; Armentrout, P.B. *J. Chem. Phys.* 1994, *100*, 1049.
73. Russon, L.M.; Heidecke, S.A.; Birke, M.K.; Conceicao, J.; Morse, M.D.; Armentrout, P.B. *J. Chem. Phys.* 1994, *100*, 4747.
74. Weisshaar, J.C. *J. Chem. Phys.* 1988, *90*, 1429.
75. Parks, E.K.; Klots, T.D.; Riley, S.J. *J. Chem. Phys.* 1990, *92*, 3813.
76. DiLella, D.P.; Limm, W.; Lipson, R.H.; Moskovits, M.; Taylor, K.V. *J. Chem. Phys.* 1982, *77*, 5263.
77. Leopold, D.G.; Lineberger, W.C. *J. Chem. Phys.* 1986, *85*, 51.
78. Dong, J.G.; Hu, Z.; Craig, R.; Lombardi, J.R.; Lindsay, D.M. *J. Chem. Phys.* 1994, *101*, 9280.
79. Montano, P.A.; Shenoy, G.K. *Solid. State. Commun.* 1980, *35*, 53 (1980).
80. Morse, M.D.; Hansen, G.P.; Langridgesmith, P.R.R.; Zheng, L.S.; Geusic, M.E.; Michalopoulos, D.L.; Smalley, R.E. *Phys. Rev. B* 1984, *80*, 5400.
81. Van Zee, R.J.; Hamrick, Y.M.; Li, S.; Weltner, W. Jr. *Chem. Phys. Lett.* 1992, *195*, 214.
82. Xu, Y.; Shelton, W.A.; Schneider, W.F. *J. Phys. Chem. A* 2006, *110*, 5839.
83. Ho, J.; Polak, M.L.; Ervin, K.M.; Lineberger, W.C. *J. Chem. Phys.* 1993, *99*, 8542.
84. Ervin, K.M.; Ho, J.; Lineberger, W.C. *J. Chem. Phys.* 1988, *89*, 4514.
85. Eberhardt, W.; Fayet, P.; Cox, D.M.; Fu, Z.; Kaldor, A.; Sherwood, R.; Sondericker, D. *Phys. Rev. Lett.* 1990, *64*, 780.
86. Saenz, L.R.; Balbuena, P.B.; Seminario, J.M. *J. Phys. Chem. A* 2006, *110*, 11968.
87. Futschek, T.; Hafner, J.; Marsman, M. *J. Phys.: Condens. Matter* 2006, *18*, 9703.
88. Seivane, L.F.; Ferrer, *J. Phys. Rev. Lett.* 2007, *99*, 183401.
89. Bhattacharyya, K.; Majumder, C. *Chem. Phys. Lett.* 2007, *446*, 374.
90. Huda, M.N.; Niranjan, M.K.; Sahu, B.R.; Kleinman, L. *Phys. Rev. A* 2006, *73*, 053201.
91. Sebetci, A. *Phys. Chem. Chem. Phys.* 2009, *11*, 921.

92. Ross, R.B.; Powers, J.M.; Atashroo, T.; Ermler, W.C.; LaJohn, L.A.; Christiansen, P.A. *J. Chem. Phys.* 1990, *93*, 6654.

93. Becke, A.D. *J. Chem. Phys.* 1993, *98*, 5648.

94. Moore, C.E. *Natl Bur. Stand. (US) Circ.* 1971, 35, 467.

95. Fortunelli, A. *J. Mol. Struct. (THEOCHEM)* 1999, *493*, 233.

96. Koepernik, K.; Eschrig, H. *Phys. Rev. B* 1999, *59*, 1743.

97. Perdew, J.P.; Wang, Y. *Phys. Rev. B* 1992, *45*, 13244.

98. Marijnissen, A.; ter Meulen, T.T.; Hackett, P.A.; Simard, B. *Phys. Rev. A* 1995, *52*, 2606.

99. Gibson, N.D.; Davies, B.J.; Larson, D.J. *J. Chem. Phys.* 1993, *98*, 5104.

100. Hotop, H.; Lineberger, W.C. *J. Phys. Chem. Ref. Data* 1985, *14*, 731.

101. Airola, M.B.; Morse, M.D. *J. Chem. Phys.* 2002, *116*, 1313.

102. Taylor, S.; Lemire, G.W.; Hamrick, Y.M.; Fu, Z.; Morse, M.D. *J. Chem. Phys.* 1988, *89*, 5517.

103. Fabbi, J.C.; Langenberg, J.D.; Costello, Q.D.; Morse, M.D.; Karlsson, L. *J. Chem. Phys.* 2001, *115*, 7543.

104. *Magnetism in Condensed Matter*, Blundell, S., Ed.; Oxford University Press, New York, 2001.

105. Majumdar, D.; Dai, D.; Balasubramanian, K.; *J. Chem. Phys.* 2000, *113*, 7919; *Chem. Phys.* 2000, *113*, 7928.

106. Grönbeck, H.; Andreoni, W. *Chem. Phys.* 2000, *262*, 1.

107. Kresse, G.; Furthmuller, *J. Comput. Mater. Sci.* 1996, *6*, 15.

108. Car, R.; Parrinello, M. *Phys. Rev. Lett.* 1985, *55*, 2471.

109. Xiao, L.; Wang, L. *J. Phys. Chem. A* 2004, *108*, 8605.

110. Yang, S.H.; Drabold, D.A.; Adams, J.B.; Ordejon, P.; Glassford, K. *J. Phys.: Condens. Matter* 1997, *9*, L39.

111. Tian, W.Q.; Ge, M.; Sahu, B.R.; Wang, D.; Yamada, T.; Mashiko, S. *J. Phys. Chem. A* 2004, *108*, 3806.

11 Exploring the Borderland between Physics and Chemistry: Theoretical Methods in the Study of Atomic Clusters

Yamil Simón-Manso,* Carlos A. González,* and Patricio Fuentealba

CONTENTS

Experimental and theoretical studies of atomic clusters have not only proven to be a challenging research theme, but have also proven to be of great practical importance in developing new technologies [1, and references therein]. The physical and chemical properties of atomic clusters lie between the properties of molecular systems and solids and are often unique. This chapter describes a number of theoretical methodologies currently applied to study small atomic clusters. Several current methods for optimizing cluster geometries, such as genetic algorithm, simulated annealing, and

---

* Contribution of the National Institute of Standards and Technology.

the big bang method are described and illustrated. We discuss the theoretical tools used in cluster studies to describe the chemical bonding and to predict chemical reactivity. In particular, we emphasize the use of the electron localization function (ELF) and the Fukui function for these purposes. We also include a brief discussion of recent reports of unusual physical properties of metallic clusters and the application of density functional theory (DFT) methods to shed light on the physics of these phenomena. Of special interest is magnetism observed in clusters and films of gold, a typically diamagnetic metal.

11.1 INTRODUCTION

In recent times, there have been many investigations that lie at the border between physics and chemistry. It is not new to these sciences; it has happened in the past with many other disciplines, such as the kinetic theory of gases, low temperature studies, theory of solutions, the study of physical properties of compounds, molecular structure studies [2, and references therein]. There is invariably a great deal of work that must be done before disciplines at the frontier of two sciences become an integral part of one another. The experimental and theoretical study of atomic clusters is a modern discipline that lies between physics and chemistry. It is a discipline in the making; simply finding an unambiguous definition of atomic clusters is a difficult task.

Most researchers "define" a cluster similar to the definition given by Connerade et al. [3], "a group of atoms (structural subunits) bound together by interatomic forces is called a cluster. There is no qualitative distinction between small clusters and molecules, except perhaps that the binding forces must be such as to permit the molecule (system) to grow much larger by stacking more atoms or molecules (subunits) of the same type if the system is to be called a cluster." It is worth mentioning that from the chemist's perspective, any arrangement of atoms without a covalent bond should not be considered a molecule. A typical covalent bond implies binding energies greater than 15 kcal/mol and bond distances smaller than 2.0 Å. In fact, the nature of chemical bonding in clusters can vary strongly. For instance, magnesium dimer is a van der Waals system, as are Mg_3 and Mg_4. Magnesium clusters with more than 10 atoms are more covalent in nature and the largest clusters at some point should experience a transition to metallic bonding characteristic of a solid. In reality, the binding forces in atomic clusters may be metallic, covalent, ionic, hydrogen-bonded, or van der Waals in character.

The structural and compositional diversity of clusters is huge. All the units (components) of a cluster are not necessarily atoms or monatomic ions. For example, a cluster may be an arrangement of a metal core with one or several atoms surrounded by molecules, so-called ligands, or may be exclusively formed by molecules. It should be noted that ligands play a stabilizing role in chemically synthesized metal clusters, and consequently, ligands substantially modify the physical and chemical properties of the clusters. Throughout this chapter we shall refer to bare clusters (atomic clusters) with a single component, unless otherwise stated. The properties of the atomic clusters can vary from atomic to bulk properties, but are also often unique depending on their size and elemental composition. Many clusters with sizes in the range of nanoparticles, between 1 and 100 nm, and thin films have shown unexpected properties and have found use in areas such as catalysis and magnetic storage [1,4,5].

The presence of dangling bonds (unsatisfied valences) on the surface makes clusters chemically very active. In general, clusters may split or combine with other particles more rapidly than normal molecules. This means that they are much more reactive, and therefore, more difficult to stabilize for study in the laboratory with conventional methods. Hence, most experimental measurements are carried out in the gas phase at low pressures and require sophisticated instrumentation. Experimental difficulties are one of the main reasons that theoretical work on clusters is so valuable. As an added benefit, research in this field has united chemists and physicists, both experimental and theoretical, to study a huge variety of phenomena in a very detailed and synergistic manner. One of the first successful attempts to rationalize an experimental observation of cluster phenomena was the explanation of the existence of magic numbers in the abundance spectrum of alkali metal clusters [6], which was concurrently explained by the very simple theoretical jellium model [7,8]. Magic numbers arise as a consequence of filling molecular orbitals to form a closed-shell species like the atomic noble gases, giving them extra stabilization energy with a large HOMO–LUMO gap. However, it was also realized early on that to explain the variation of most of the properties of a given cluster with respect to the number of atoms, it is necessary to use theoretical models that take into account the quantum nature of the electronic structure [9]. Fortunately, around the same time DFT emerged as a robust and efficient methodology to perform calculations of the electronic structure of atoms, molecules, and clusters of moderate size [10,11].

We briefly discuss some key steps that most researchers perform in conducting theoretical studies of atomic clusters. This chapter does not provide a systematic study (or review) of this topic but a short description of our own and related research. It is set out in five sections, beginning with this introduction. In Section 11.2, we use DFT calculations to describe clusters as entities that lie between atoms and solids with unique properties depending on their size, chemical nature, and symmetry. Section 11.3 describes several methods of finding different stable isomers of a given cluster. There are various stochastic methodologies and evolutionary algorithms currently used for this task; we shall discuss the genetic algorithm, simulated annealing, and the so-called "big bang" methods. The chemical bonding and reactivity of clusters are discussed in Section 11.4. We present the ELF as a very robust tool to understand chemical bonding in clusters and the Fukui function as an effective means of predicting their chemical reactivity. Finally, in Section 11.5 we discuss some interesting properties of clusters such as the magnetic moment observed in clusters and surfaces of gold, a typically diamagnetic metal, from a theoretical point of view.

11.2 CLUSTERS: BETWEEN ATOMS AND BULK

Recently, the study—both experimental and theoretical—of the transition from atomic and molecular clusters to bulk solids has become accessible for almost any type of material. Atomic clusters represent an intermediate state between molecules and bulk solids and are very useful as models of nanoparticles, surfaces, or solids to study phenomena such as physical adsorption, chemisorption, plasmon excitations, and magnetism. However, this modeling might be incomplete or even incorrect due to the fact that the electronic properties of clusters change dramatically with size and frequently experience many transitions such as metal-insulator, color absorption, and collective excitations [3, and references therein]. The fact that the chemical and physical properties of clusters do not change monotonically with their sizes complicates the exploration of trends and applications in this area. Most cluster properties show great fluctuations and very irregular dependence on size. Even so, the mean nearest-neighbor coordination numbers vary with the cluster sizes; thus, the cluster properties shift gradually from surface dominated to volume dominated.

Estimating the number of atoms that are needed in a cluster to mimic a nanoparticle, an infinite surface or the bulk solid is an important part of modeling atomic clusters. However, other factors need to be considered for proper modeling. Frequently it is necessary to impose additional constraints on the models to avoid unphysical situations. For example, in theoretical calculations where the cluster is intended to represent the bulk or a bulk region, it should be subjected to crystal symmetry constraints [12]. The crystal symmetry determines the symmetry of the band structure due to the commutation between symmetry operations and the crystal Hamiltonian [13]. Fully optimized cluster geometries frequently deviate from the bulk symmetry, yielding substantial differences in most of the calculated cluster properties. Also, many small clusters experience spontaneous symmetry-breaking events, such as Pierls' distortions (Jahn–Teller effect), that must be considered [14]. In magnetic systems, certain restrictions on the spin symmetry may be necessary [15]. Although the complete theoretical description of a quantum many-body problem, even of moderate size (less than 100 atoms), is very difficult, DFT is able to deliver reliable results in many cluster problems. DFT is possibly the only ab initio correlated method that can be implemented and used with reasonable computational cost for describing the electronic structure of molecular systems and solids.

In Figure 11.1 we show the binding energy per atom (BE) dependence on the cluster size (N) of a representative metal (lithium) and a transition metal (Cu). Binding energies are calculated using

$$\text{BE} = \frac{E[\text{cluster}] - NE[\text{atom}]}{N} \qquad (11.1)$$

where BE, E, and N stand for binding energy, total energy, and number of atoms, respectively. The all-electron calculations were done at the BP86/6-31G(d) level of theory, which implies use of a DFT electronic structure method with Becke's 1988 exchange functional [16] and Perdew's gradient-corrected correlation functional method (BP86) [17] in conjunction with the Pople basis set 6-31G with an additional set of polarization d-functions added to all nonhydrogen atoms [18]. The clusters used in this calculation are portions of the face-centered cubic structure,

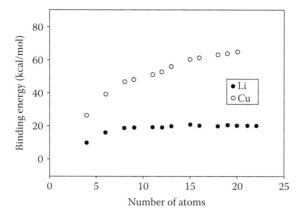

FIGURE 11.1 Binding energy per atom (kcal/mol) versus number of cluster atoms for lithium (Li) and copper (Cu) atomic clusters.

beginning with a tetrahedron and successively adding atoms to form new tetrahedrons. For example, 12 atoms surrounding the reference atom complete the first-neighbor shell. The bond distances are fixed to the values of the bulk metals, 2.56 Å for copper and 3.11 Å for lithium. The calculated binding energies change smoothly with cluster size when the models are constrained to the crystal symmetry (or a subgroup) and reach saturation values that depend on the nature of the metal. As shown in Figure 11.1, BE in the case of lithium clusters levels off more quickly than in the case of copper because alkali metals have energy bands very similar to those of free electrons [13]. The saturation values of the binding energies found by polynomial extrapolation differ by 8–10 kcal/mol from the bulk values of approximately 29 kcal/mol for lithium and 80 kcal/mol for copper [19], but most of these differences are attributable to basis set superposition error and a small part is due to an embedded effect that can be corrected with an external potential with the crystal symmetry [12,20]. The fluctuations of the second finite differences of the BE versus N dependence ($\Delta^2 E/\Delta N^2$) are very small, and only the smallest clusters (four and six atoms) show significant deviations, according to their lower symmetry.

We also have explored some properties (not shown in the figure), such as ionization potentials, electron affinities, and HOMO–LUMO gaps with similar results [12]. It is worth mentioning that the number of atoms needed to reach the saturation values depends on the property itself and not only on the nature of the metal.

In summary, a general cluster model should contain at least the minimum number of atoms to represent the set of properties to be studied. According to our calculations, a lower bound to the critical cluster size necessary to represent bulk properties and transitions depends on the nature of the metal and crystal symmetry and should always be tested against reliable experimental values. At the same time, these calculations reveal that the fluctuations observed in the size-dependent cluster properties are due to the fact that actual cluster does not preserve the full symmetry of the crystal. Most clusters, regardless of the experimental technique used to create them, suffer dynamic rearrangements leading to more stable structures. Thus, an important part of all cluster studies is to find the optimum structures in the configuration space spanned by the nuclear coordinates. In the next section we discuss some of the most common methods used to optimize the geometry of clusters.

11.3 OPTIMIZATION METHODS APPLIED TO CLUSTER STUDIES: FINDING THE MOST STABLE ISOMERS

In general terms, the problem of finding stable isomer structures is equivalent to the problem of finding minima on a multidimensional hypersurface. Every local minimum represents the geometry of an isomer and the global minimum is the most stable one (i.e., the geometry with the lowest total energy). The total energy is a function of the nuclear coordinates. There are $3N-6$ variables ($3N-5$ in the case of a linear molecule), where N is the number of atoms. In general, geometry optimization is a nontrivial mathematical problem that does not lend itself to an analytic solution. Therefore, it is necessary to use numerical techniques. There are two main types of optimization techniques, those that follow the gradient (of the total energy) to reach a minimum energy and stochastic techniques. (Recently, some very interesting techniques based on quantum molecular dynamics have emerged, but these will not be addressed here [21].) The most traditional methods used in quantum chemistry are gradient-following techniques and are implemented in almost all computational codes. The most successful gradient-following techniques are approximations to Newton's method and are termed quasi-Newton methods. These need a very limited number of energy and gradient evaluations compared to other approaches and are therefore very efficient. However, they are dependent on the initial geometry and locate only one stationary point at a time. They cannot jump from one minimum to another. Hence, if one has a good initial guess of the geometry of the most stable isomer, a quasi-Newton optimization is likely the most effective method. Nevertheless, one cannot be sure to have reached the global minimum. In the case of optimization of "normal" molecules, where the chemical bonding rules reliably predict the geometry of the most stable isomer, quasi-Newton methods are very useful. However, for a cluster with a moderate number of atoms, for instance Si_9, the chemical rules of bonding are not as helpful and the number of isomers with energies close to the global minimum energy is high. For Si_9 there are at least 14 such isomers [22].

On the other hand, stochastic methods are based on a random search on the potential energy hypersurface (PEH). As a consequence, they are able to jump from one minimum to another, allowing the location of various isomers in one run. They are also, in principle, independent of the initial guess. However, this should be verified by performing a series of similar optimizations starting at different locations on the PEH of the system under study. The drawback of stochastic methods is that they generally need many more evaluations of the energy function than quasi-Newton methods, which makes it difficult to apply them at a high level of electronic structure theory. Usually, one starts with a low level, in many cases classical molecular dynamics, and then proceeds to perform high-level calculations of the minima identified at the lower level. There are a variety of stochastic methodologies, and each has several variants. We will briefly describe three of them: genetic algorithm, simulated annealing, and the big bang method.

11.3.1 GENETIC ALGORITHM

Genetic algorithm techniques are based on the ideas of Darwinian biological evolution and have recently been widely used in the optimization of atomic clusters [23,24]. One starts by defining a *genome* or string that represents a candidate solution to the problem. In the case of atomic clusters, the choice is simple: the genome is a set of coordinates for each atom of the cluster. One starts with an initial population of genomes selected randomly. This population will be propagated to produce more "fit" species by applying "natural selection" rules; in the case of clusters this means combining clusters with low energies to (hopefully) produce new cluster geometries, which lead to still lower cluster energies. In order to increase sampling of the search space, one defines operators simulating crossover and mutation and applies these to the developing population. The lowest energy clusters are produced based on the principle of "survival of the fittest." At each step, population members with energy above a given threshold are deleted from the population, and species with low energies are allowed to reproduce. In this way, after a given number of generations, one should obtain structures of lower energy. In the case of clusters, this means low-lying isomers.

The initial population of individuals, represented as a set of atomic coordinates, is generated randomly. In practice, different constraints can be imposed to avoid searching very unphysical regions of the hypersurface. The coordinates should be constrained to lie inside a box of the expected dimensions of the cluster, and any pair of atoms can neither be closer than a given distance nor separated by more than a given distance. The specific criteria for desired fitness, selection rules, and crossover and mutation probabilities are specific to every genetic algorithm implementation. In this part of the work, we followed the algorithm presented in Ref. [25]. The number of times one needs to evaluate the energy can become significantly large. Therefore, it is a common practice to produce all generations in the genetic algorithm cycles using an empirical molecular orbital method, such as MSINDO [26]. It is also important to note that each individual is an optimized geometry within the semiempirical method. When the genetic algorithm optimization is complete, the resulting isomers are re-optimized using a high-level electronic structure theory, usually Kohn–Sham DFT including an exchange-correlation functional via a conventional gradient-following quasi-Newton optimization technique.

As an example, in Ref. [22] the genetic algorithm was used to find 14 isomers of the Si_9 cluster. The eight most important isomer structures are presented in Figure 11.2. All of the structures were optimized using the B3PW91 functional with the Stuttgart pseudopotential [27]. Since the corresponding basis set does not contain diffuse and polarization functions, the basis was augmented with diffuse s- and p-functions and one set of d-polarization function from the Sadlej basis set [28]. To ensure that the optimized structures are stationary points on the molecular potential energy surface,

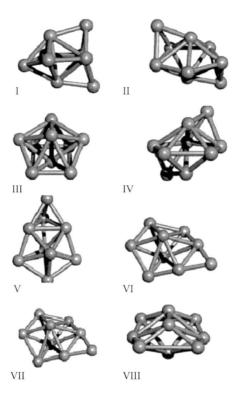

FIGURE 11.2 Geometries of the silicon cluster (Si_9) isomers generated using a genetic algorithm technique.

TABLE 11.1

Energy of the HOMO, Binding Energy per Atom, Dipole Polarizability, and Energy Gap for the Low-Lying Isomers of Si_9

Cluster	Symmetry	ε_H (a.u.)	E_B/atom (eV/atom)	α (A/atom)	Gap (eV)
I	C_s	−0.223	−3.055	4.73	2.77
II	C_2	−0.212	−2.994	4.91	2.07
III	C_{2v}	−0.219	−2.988	4.80	2.77
V	C_2	−0.210	−2.945	4.79	1.99
VI	C_s	−0.207	−2.931	4.90	2.37
VII	C_{2v}	−0.205	−2.828	5.27	2.37
VIII	C_{2v}	−0.195	−2.788	5.31	1.99

vibrational frequencies were calculated and found to be positive indicating a minimum geometry. The structures are given in Figure 11.2 in the order of increasing energy. Some of the properties of these clusters are displayed in Table 11.1. One can see that there is less than a 0.23 eV difference in the BE between the first and the last isomer. BE correlates reasonably well with the energy of the HOMO. However,

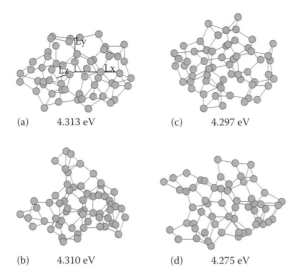

(a)   4.313 eV   (c)   4.297 eV

(b)   4.310 eV   (d)   4.275 eV

FIGURE 11.3 Geometries of the four most stable isomers of Si_{60} and their respective BE in eV.

there does not seem to be any correlation with the dipole polarizability or with the HOMO–LUMO gap [22].

Using the same methodology as in Ref. [29], the following larger silicon clusters were studied: Si_{40}, Si_{46}, and Si_{60} (the last one was included in this study due to its possible similarity to C_{60}). Figure 11.3 shows the four most stable isomers of Si_{60}. Structure (b) corresponds to the structure reported in Ref. [30]. However, the calculations of Ref. [29] show that at this level of theory, the structure has five imaginary frequencies. All of the structures in Figure 11.3 are very close in binding energy making it difficult to predict the most stable isomer. However, it is clear that for the Si_{60} cluster there is no cage structure similar to the fullerene structure observed for carbon atom clusters. In addition, it is observed that the silicon cluster structures are more prolate in shape than their carbon counterparts.

As discussed in Ref. [29], one significant weakness of the methodology consists of the use of a semiempirical method, MSINDO, to evaluate the atomic clusters. For example, silicon clusters of the endohedral type, where some silicon atoms are hypervalent, are not properly described given that the semiempirical methodology is not parametrized to treat hypervalent silicon atoms.

11.3.2 SIMULATED ANNEALING

In this section, we discuss simulated annealing techniques [31,32] as implemented by Perez et al. [33]. In this methodology the system is allowed to evolve inside a box of a given length. The method is initialized with randomly selected cluster geometries, and the energy of each atomic cluster is calculated using some quantum chemical method. In this case, the energy is evaluated at each step at the HF/Lanl2z level of theory. Every generated structure is then subject to one or more acceptance

TABLE 11.2

Typical Simulated Annealing Parameters

Parameter	Li_5	Li_6	Li_7
Generated structures	49	111	33
With $\Delta E < 0$	14	10	8
With $\Phi(\Delta E) < P(\Delta E)$	35	101	25
Finally located minima	2	3	2

tests. If the energy is lowered, $\Delta E \leq 0$, the new structure is accepted; otherwise, when $\Delta E > 0$, the structure is accepted if $\Phi(\Delta E) < P(\Delta E)$ where $P(\Delta E) = \exp(-\Delta E/k_B T)$ is the Boltzmann probability distribution function and $\Phi(\Delta E) = |\Delta E/E_j|$, where E_j is the energy of the structure being evaluated. (Most simulated annealing algorithms are based on the Metropolis sampling method that simply compares $P(\Delta E)$ to a random number between 0 and 1.) If neither of the tests is satisfied, the new structure is not accepted, the parent structure is subjected to another random modification, and the acceptance procedure is repeated. At every temperature there are a maximum number of generated structures that satisfy neither of the two acceptance criteria. This number is reduced as the simulated annealing temperature is decreased. The method for reducing the temperature is called the quenching schedule; the simplest method is one that decreases the temperature by a constant amount per simulated annealing iteration. A successful run generates a number of possible structures of different energies. Structures with energies no more than 0.02 hartrees above the lowest energy of the set are selected for a final optimization. As an example, Table 11.2 shows typical parameters for the calculation of Li_q $(q = 5\text{–}7)$ clusters.

Notice that the modification of the acceptance criteria in principle allows for a relatively exhaustive sampling of potential energy search that could lead to the generation of the most physically relevant structures. Figure 11.4 shows the results of a typical simulated annealing optimization of the Li_7^+ cluster. Most of the observed structures are sampled early on in the simulated annealing procedure as a result of producing Markov chains generated at high temperatures, which allows the algorithm to "jump" over potential energy barriers with relative ease. As the annealing temperature is reduced, the number of accepted structures tends toward the minimum energy set of structures, reducing the scope of the sampling.

In Ref. [33] the procedure described above was applied to $Li_q^{-,0,+}$ $(q = 5\text{–}7)$ clusters and some new structures were reported. Most of the isomers agree very well with previous work [34–36]. The method was also applied to the more complex binary Li_5Na cluster where six isomers were found [33].

11.3.3 BIG BANG METHOD

The "big bang" method [37,38] is the simplest of the three methodologies discussed in this chapter, and the only one specifically designed for geometry optimization of

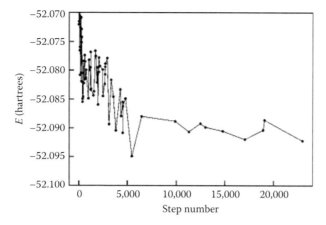

FIGURE 11.4   The quenching route in the optimization of the Li_7^+ cluster.

clusters. It starts with a relatively large population of geometrical structures that is chosen randomly and enclosed in a very small volume. Then, using standard geometry optimization procedures, one proceeds to relax the structures by following the force (negative gradient). Since at the beginning all the atoms are very close to each other, the repulsive forces among them are very large and consequently they tend to "explode," landing in distant parts of the hypersurface. By starting with a large population, one can be certain to explore a large portion of the hypersurface. By optimizing individual structures, the procedure produces different local minima including, hopefully, the global minima. The method is very simple to implement, but curiously has not been adopted by the chemistry community. In the implementation developed in Ref. [39], $1000N$ random structures were generated, where N is the number of atoms in the cluster. These were confined to a volume proportional to the covalent radius of the atom and the structures were optimized using the semiempirical MSINDO method [26]. The optimized structures where then optimized at a higher level of theory.

In this way, more than 70 different isomers of Li_n ($n = 3$–20) have been found [39]. In Table 11.3, the symmetry of the reported structures is compared with other work [40,41]. More than 50 new structures are reported. The big bang methodology has been also applied to sodium and potassium clusters [39].

The three methodologies discussed in this section are all based on the use of stochastic techniques. In each of the methods there is no guarantee of reaching the global minimum, and one should perform the minimization various times with different starting structures to be sure that a consistent set of minimum-energy isomers is obtained. In general, the choice of which method to use is more a matter of convenience, especially with respect to the computational facilities, than a formal decision based on theoretical grounds. In the cases discussed here, a detailed comparison of the small members of the series indicates that the three methodologies discussed above are able to locate the lowest energy isomers and the differences lie more in the efficiency of the implementation rather than in some formal advantage of the method.

TABLE 11.3
Isomers of Lithium Clusters

Li_n	This Work (Big Bang)	Jones[40]	Gardet[41]
3	$C_{2v}{}^P$, $D_{\infty h}{}^L$	$C_{2v}{}^P$, $D_{\infty h}{}^L$	$C_{2v}{}^P$
4	$D_{2h}{}^P$	$D_{2h}{}^P$	$D_{2h}{}^P$
5	C_{2v}, $C_{2v}{}^P$	C_{2v}, $C_{2v}{}^P$	C_{2v}, $C_{2v}{}^P$
6	D_{4h}, C_{5v}, D_{3h}, (D_{2d})	D_{4h}, C_{5v}, D_{3h}	D_{4h}, C_{2v}, D_{3h}
7	D_{5h}, C_{3v}, (C_s)	D_{5h}, C_{3v}	D_{5h}
8	T_d, (C_{3v}), (C_s)	D_{5h}, T_d, C_{2v}	T_d, C_{2v}, C_s
9	C_{4v}, C_{2v}, C_s	C_s, C_{4v}	C_{2v}, C_s
10	(D_{2d}), (C_s), (C_{4v}), (T_d)	C_1, C_{2V}	C_1, C_{2V}
11	C_2, (C_{2v}), (C_2), (C_s)		C_2
12	C_s, (C_s), (C_{2v})		C_s
13	(C_s), (C_s), (C_2), (C_1), (C_1), (C_{3v})		C_s, C_1
14	(C_2), (C_s), (C_{3v}), (C_s), (C_1), (C_1)		
15	(C_{4v}), (C_1), (C_2), (C_s), (C_1), (C_s), (C_s)		
16	(C_s), (C_1), (C_1), (C_1), (C_1)		
17	(C_1), (C_s), (C_1)		
18	(C_s), (C_1), C_{5v}, (C_1), (C_1), (C_1), (C_s), (C_1), (C_{4v})		C_{5v}, D_{5h}, D_{3h}, C_1
19	(C_{2v}), (C_s), (C_1), (C_1)		D_{5h}
20	C_s, (C_1), C_{2v}, (C_1), (C_1), (C_1)		C_{2v}, C_s

11.4 ANALYSIS OF THE CHEMICAL BONDING AND REACTIVITY OF ATOMIC CLUSTERS

Most atomic clusters possess very labile bonds that do not follow the ordinary chemical rules describing the bonding and geometry of ordinary molecules such as hydrocarbons. For instance, they can present a lower or higher coordination than the one dictated by the number of valence electrons: the simple octet rule is almost never followed. Therefore, the number of possible isomers increases dramatically (usually exponentially) with the number of atoms, and chemical intuition does not help in predicting the most stable isomer. This presents two issues to the theoretical calculations. First, how can one find the most stable isomers (global minima) when most optimization techniques do not guarantee that a global minimum will be found? Second, how can one understand the bonding between the atoms in a cluster? Once the geometric parameters of the lowest energy isomers of a given cluster are known, the task is to try and understand the way the atoms are bonded to one another. This knowledge permits the prediction of new geometries and the prediction of the class of reactivity the cluster can undergo. Unfortunately, this is not an easy task, and until recently there has been no simple model for understanding bonding in atomic clusters. Generally, the Lewis electron pair model does not work and one has atomic clusters that are either electron deficient or hypervalent. Often it is difficult to predict the multiplicity of the ground state. In addition, the widely popular Mulliken population analysis tends to predict the incorrect electron charge distribution in atom pairs, and even the more robust natural bond order method can fail in cases of extremely delocalized bonds that are common in atomic clusters. One of the most useful theoretical tools for analyzing bonding in atomic clusters is the ELF.

The ELF was originally proposed by Becke and Edgecombe [42] and is defined as

$$\text{ELF}(\vec{r}) = \left[1 + \left(\frac{T[\rho(\vec{r})]}{T_0[\rho(\vec{r})]} \right)^2 \right]^{-1} \tag{11.2}$$

where $T[\rho(\vec{r})] = T_s[\rho(\vec{r})] - T_w[\rho(\vec{r})]$ is the kinetic energy density difference between the kinetic energy density of the noninteracting system $T_s[\rho(\vec{r})]$ with density $\rho(\vec{r})$, and the von Weizsacker kinetic energy density, $T_w[\rho(\vec{r})]$. In addition, $T_0[\rho(\vec{r})]$ indicates the kinetic energy density of the noninteracting electron gas. The important quantity in Equation 11.2 is the term $T[\rho(\vec{r})]$ that can be interpreted [43] as the excess kinetic energy density due to the Pauli exclusion principle. A simple analysis of Equation 11.2 indicates that the function $\text{ELF}(\vec{r})$ varies between 0 and 1. The interpretation of the function is that the region of the space where the ELF has a value close to 1 corresponds to the regions where it is most probable to find a localized electron pair; regions with a low value of the ELF (≈ 0.5) correspond to regions where the electrons are delocalized. Hence, the interpretation of ELF isosurfaces allows us to understand bonding in clusters. For a more detailed description of the ELF, see Ref. [42].

As an example of the utility of the ELF, we show in panel (a) of Figure 11.5 the position of the atoms of one of the isomers of the cluster of Li_6. The lines in this

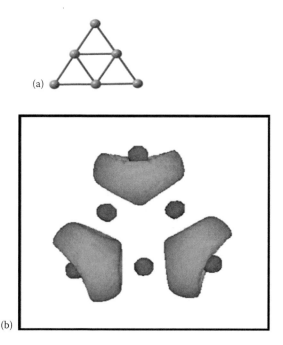

(a)

(b)

FIGURE 11.5   (a) Geometry and (b) ELF isosurface for the lithium cluster with six atoms.

figure serve to guide the eyes and they do not represent electrons in the sense of the Lewis structures. The cluster has only six valence electrons and the question is: how are the six atoms bonded to one another with only six electrons? The ELF isosurface depicted in panel (b) gives a clear answer. There are three equivalent regions where the electrons are most probably localized forming *two-electron three-centered* bonds.

The applications of the ELF have been very successful in all fields of chemistry and physics [43,44]. Recently, the ELF has been used to understand and quantify the concept of aromaticity in metallic clusters [45,46]. Unfortunately, its use for clusters formed by heavy atoms is not free of complications and the issue remains open to interpretation.

Once the geometry of a given cluster and the types of bonds it forms are understood, the next step is to predict how the cluster will react in the presence of a given atom or molecule. There have been various studies of the reactivity of clusters. For example, the bonding of a hydrogen atom to a lithium cluster has been analyzed using the ELF [47]. The reactivity of a particular silicon cluster, Si_4, with a Ga atom has been studied using reactivity indices defined within DFT [48,49]. The topology of the frontier orbitals was used in Refs. [48,49] to propose simple rules to predict the binding sites in Au and Ag clusters. The bonding and reactivity between H and the Al_{13} cluster has been studied using the Fukui function [52]. It is very important to understand how a cluster forms and how it will react in presence of other species. Once the most stable isomers of an atomic cluster are found, it is necessary to calculate as accurately as possible properties that may be compared to experimental measurements. Only in cases where the theoretical calculations agree with the results of the experimental measurements, can one have confidence in the prediction of the most stable isomer. From the considerations above, it is clear that the theoretical study of atomic clusters is a rather cumbersome task involving a variety of theoretical methodologies.

In Ref. [53], a recently proposed version of the local Fukui function has been applied to the study of the reactivity of silicon clusters in presence of a hydrogen atom. The Fukui function has been defined by Parr and Yang [54] as

$$f(\vec{r}) = \left(\frac{\partial \mu}{\partial v(\vec{r})} \right)_N \tag{11.3}$$

where
  μ is the chemical potential
  $v(\vec{r})$ is the external potential

and the derivative is taken at a constant number of electrons N. Using the frozen orbital approximation, the derivative in Equation 11.3 can be evaluated as the square of the frontier molecular orbital—the square of the HOMO if the derivative is approximated from the left and the square of the LUMO if the derivative is approximated from the right. The distinction is due to the known discontinuity of the density as N passes through an integer value [48]. Hence, the Fukui function can be approximated as

$$f^{\pm}(\vec{r}) = \left|\phi(\vec{r})\right|^2 \tag{11.4}$$

where $\phi(\vec{r})$ is the frontier molecular orbital, and the sign $\pm$ means the HOMO or LUMO, respectively. A condensed form of this function using the integration of the proper Fukui function over its own basins has been proposed as a better alternative than using population analysis [55]. In Ref. [53] this condensed form has been used to study the reaction of hydrogen atom with silicon clusters. The theoretical predictions of the sites of most probable attack have been confirmed by computational simulations where the genetic algorithm has been used to find the most stable clusters of Si_nH. Given that the chemical potential, defined as the negative of the absolute electronegativity, has a value of $-7.25\,eV$ for the hydrogen atom and is in the range of -4.5 to $-5.0\,eV$ for silicon clusters, it is expected that the silicon clusters will donate charge to the hydrogen atom and therefore the Fukui function should be calculated with the square of the HOMO.

On the left column in Figure 11.6 the geometrical structures of the clusters of Si_3, Si_4, Si_5, and Si_7 have been depicted, while in the right column the respective Fukui

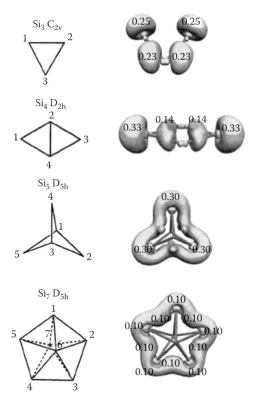

FIGURE 11.6 Geometries of the clusters of Si_3, Si_4, Si_5, and Si_7 and the corresponding Fukui function isosurface (on the right).

function isosurfaces and their condensed values over the basins are also shown. They are calculated as the integral of the Fukui function over the respective volume. Larger numbers indicate more reactive regions. Therefore, one expects that a hydrogen atom would bind the cluster in the zone where the condensed Fukui function has the largest value. In fact, in Ref. [53] it was found that the predicted cluster was always one of the most stable ones. Exceptions occur when the hydrogenated cluster drastically alters the Si_n skeleton of the Si_nH cluster with respect to the initial silicon cluster. A similar methodology has also been used to explain and predict the reactivity of copper clusters with molecular oxygen [56].

11.5 UNIQUE PHYSICAL PROPERTIES OF CLUSTERS AND THEIR RELATION TO THE CORRESPONDING SURFACE AND BULK PROPERTIES

As mentioned above, atomic clusters show many distinct properties from those of isolated atoms or bulk solids that can be very useful in practical applications. At the same time, there is the possibility to analyze known phenomena from a new perspective. For example, the discovery of giant plasmon resonances in metal clusters [57,58] and in fullerenes [59,60] allowed scientists to study the transition from the classical Mie picture of plasmon oscillations to the quantum limit or to detect cluster deformations by the value of the splitting of plasmon resonance frequencies [3, and references therein]. In this section, we briefly discuss another very interesting phenomenon: the onset of magnetism in bare and ligand-coated clusters of gold, a typical diamagnetic metal. We emphasize the usefulness of clusters as "transition structures" that facilitate a better understanding of magnetism in molecular and extended systems (surfaces or solids).

One of the first observations of magnetism in gold was reported by Zhang and Sham [61] in their x-ray spectroscopic study of alkane-thiolated nanoclusters. Later, Crespo et al. [62] used x-ray absorption near-edge structure measurements to show the appearance of ferromagnetism accompanied by room-temperature hysteresis in thiol-capped Au nanoparticles with a diameter of 1.4 nm. This rather surprising finding led to the proposal that the strong chemisorption of ligands into relatively small nanoparticles of bulk diamagnetic materials could induce permanent magnetism as a result of electronic structure effects involving interactions between the sp-orbitals of the adsorbate and the 5d-orbitals of the Au atoms lying on the surface of the nanoparticle. This mechanism was later questioned by the results reported by Hori et al. [63] based on superconducting quantum interference device magnetometer (SQUID) measurements of a series of Au nanoparticles protected by strongly interacting thiols such as dodecane thiol (DT) and weakly coupled ligands such as polyacrylonitrile (PAN), polyallyl amine hydrochloride (PAAHC), and polyvinyl pyrrolidone (PVP). In this study, the authors found that the strong chemisorptive interaction between the sulfur atom in DT and the Au atoms on the surface of the nanoparticle induced a spin singlet state that significantly decreased the magnetization of the system (although it did not quench it completely) when compared to the corresponding magnetization measured in Au nanoparticles capped with weakly coupled ligands such as PAN,

PAAHC, and PVP. In the same study, Hori et al. report that the measured magnetization is strongly size dependent, increasing with particle diameter at the smaller nanoparticle sizes, peaking at approximately 3 nm for Au-thiol nanoparticles, and subsequently decreasing with increasing nanoparticle size, keeping with the fact that as the Au nanoparticle size increases, its configuration approaches that of the bulk lattice [63]. Additionally, a closely related phenomenon has been observed in gold surfaces: the occurrence of magnetism when organic molecules are self-assembled as monolayers on a surface [64]. The observed magnetism has been attributed to charge transfer between the organic layer and the metal substrate [64].

These effects are somehow intriguing given that gold is a diamagnetic metal, but perhaps not totally unexpected. A close examination of Stoner's model [65] suggests that the emergence of "unexpected" magnetic ordering in transition metal clusters in confined spaces at the nanoscale is possible. In this model, the paramagnetic susceptibility χ is determined by the density of d-states at the Fermi level $N(E_F)$ and the exchange function J:

$$\chi = \frac{\mu_0 \mu_B^2 N(E_F)}{(1 - JN(E_F))} \tag{11.5}$$

where
  μ_0 is the permeability of free space
  μ_B is the Bohr magneton (9.27×10^{-24} J/T)

The spatial confinement produces a narrower d-band and eventually the center of the band is shifted closer to the Fermi level. For $JN(E_F) > 1$, the term $1 - JN(E_F)$ in Equation 11.5 becomes negative, consequently generating ferromagnetic instability and thus a magnetic order.

Several previous experimental studies have shown that the onset of magnetism at the nanoscale occurs in a complicated way with clusters frequently having unusual magnetic properties not clearly related to the strength and orientation of the metal bulk magnetization. Stern–Gerlach molecular-beam deflection experiments [66–71] with bare clusters of transition metals show either high-field deflection indicative of superparamagnetism or symmetric broadening indicative of locked moment behavior. These clusters exhibit susceptibilities significantly larger in magnitude than those expected based on the extrapolation of the susceptibility of the bulk solids. It is worth mentioning also that molecules comprising a large number of coupled paramagnetic centers capped by ligands (e.g., Mn_{12} and Fe_8 families) are known to show unambiguous evidence of quantum size effects in magnets [72].

Magnetism in bare (uncapped) gold nanoclusters has been explored in our group [73] from a spin-dependent DFT perspective with scalar relativistic effects included via the use of pseudopotentials. In this study, three different DFT exchange-correlation energy functionals were used: the Perdew, Burke, and Ernzerhof [74] generalized gradient approximation (PBE) and the hybrid functionals B3LYP [75] and PBE1PBE [76]. In addition, two different basis sets were chosen: a single valence plus polarization with the Stuttgart [77] effective core potential (SVP/STUTT) basis (27 basis functions comprised of 55 primitive gaussians) and the LANL2DZ [76] basis set developed by the Los Alamos group (24 basis functions comprised of 44 primitive gaussians) with its respective effective core potential. The calculated electronic structures of gold nanoclusters of different sizes (Au$_n$, with $n=2$, 14, 28, 38, 56, and 68) reveal that they exhibit (with the exception of Au$_2$) a core-shell geometric arrangement of Au atoms and that permanent size-dependent spin polarization appears without geometry relaxation for bare clusters even though bulk gold is diamagnetic [73]. The spin-polarized ground states for clusters are favorable due to the hybridization of the s and d orbitals, and bare octahedral clusters are expected to be magnetic for cluster sizes of approximately 38 atoms and larger. Much larger clusters will be diamagnetic when the surface-to-volume ratio is small and the core diamagnetism prevails.

Making use of these findings, we developed a spin–spin Ising interaction model [79] that explains the origin of the size dependency of magnetization in Au clusters. This model combines the bulk diamagnetic response of the core with surface (shell) ferromagnetism behavior (as suggested by the results in Ref. [73]). In this model, the maximum entropy formalism is used in order to obtain an average temperature-dependent magnetization of bare Au nanoparticles within a mean-field theory. Accordingly, the total Hamiltonian H_T can be partitioned into a core H_c and a surface H_s contribution:

$$H_T = H_c + H_s \tag{11.6}$$

with

$$H_c = -\sum_{i<j}^{N_c}\sum_j^{N_c} J_{ij}^c \cdot \vec{S}_i \cdot \vec{S}_j - \frac{1}{2}\sum_i^{N_s}\sum_j^{N_c} J_{ij}^{sc} \cdot \vec{S}_i \cdot \vec{S}_j - \sum_i^{N_c} k_c \cdot \vec{S}_{z_i}^2 \tag{11.7}$$

and

$$H_s = -\sum_{i<j}^{N_s}\sum_j^{N_s} J_{ij}^s \cdot \vec{S}_i \cdot \vec{S}_j - \frac{1}{2}\sum_i^{N_s}\sum_j^{N_c} J_{ij}^{sc} \cdot \vec{S}_i \cdot \vec{S}_j - \sum_i^{N_s} k_s \cdot \left|\hat{n} \cdot \vec{S}_i\right|^2 \tag{11.8}$$

In Equations 11.7 and 11.8, N_c and N_s indicate the number of core and surface Au atoms, respectively, such that the total number of atoms $N=N_c+N_s$. The interaction terms are partitioned into $N_s(N_s-1)$ surface atom interactions, N_sN_c core atom interactions with the surface atoms and $N_c(N_c-1)$ core atoms interactions with coupling functions J^s, J^{cs}, J^{sc}, and J^c, respectively (with $J_{ij}^x > 0$ for ferromagnetic interactions). In addition, the last terms in Equations 11.7 and 11.8 describe the anisotropy term spin with the Hamiltonian assumed to be aligned along the radial direction $\hat{n}_r$, with k_c and k_s being the anisotropy constants for the core and surface spins, respectively. After some algebraic manipulations making use of the maximum spin entropy approach (see Ref. [79] for details), the following expression for the total magnetic moment per atom μ_t is obtained

$$\mu_t = \frac{N_s}{N_t}\left(1 - v \cdot J^s \cdot N_s(N_t - N_s)\right)^{\frac{1}{3}} \tag{11.9}$$

where the surface magnetic moment μ_s can be obtained self-consistently from the following expressions:

$$\mu_s = \tanh\left(\frac{1}{k_B T}\left[\frac{1}{g}J^s \cdot N_s \cdot \mu_s + g \cdot h_z\right]\right) \tag{11.10}$$

$$\mu_s = \tanh(\xi)$$

$$\xi = \frac{1}{k_B T}\left(\frac{1}{g}J^s \cdot N_s \cdot \mu_s + g \cdot h_z\right) \tag{11.11}$$

In expressions (11.9) through (11.11), J^s indicates the average surface exchange, g is related to the gyromagnetic constant, h_z is the effective field in the preferred z direction, and the constant v is given by

$$v = \frac{3^{1/3}e^2}{12\pi^{4/3}m_e c^2}\tilde{\lambda} \tag{11.12}$$

with
   e and m_e indicating the charge and mass of an electron
   c the speed of light
   $\tilde{\lambda}$ is a phenomenological constant

Figure 11.7 depicts a plot of the magnetic moment per atom as a function of the particle size for a typical gold nanoparticle. The results show how this simple Ising model reproduces qualitatively the size dependence behavior observed experimentally by Hori et al. [63].

We have also performed studies aimed at understanding the origins of magnetic behavior in gold upon chemisorption as well as the effect of different ligands on the magnetic moment. Using a simple quantum chemical model based on finite perturbation theory (FPT) combined with DFT calculations on a Au cluster with a two-layer slab of 13 atoms (9 in the first layer and 4 in the second layer), Gonzalez et al. [80] were able to suggest ideas regarding the electronic structure origin of the observed magnetism in Au cluster–ligand systems with different chemical linkers and provided a theoretical basis to rationalize the experimental observation indicating that magnetism can be induced in gold as a result of the chemisorption of

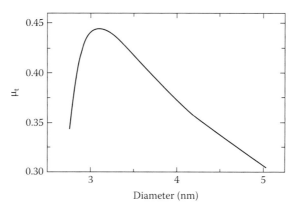

FIGURE 11.7   Total magnetic moment per atom as a function of nanoparticle diameter (nm) for Au_n.

S-linked ligands, whereas weakly coupled linkers such as N do not affect the dia-magnetic behavior of Au surfaces. In particular, it was shown that for some link-ers such as sulfur, a spin symmetry breaking occurs that lowers the energy and leads to preferential spin density localization on the gold atoms neighboring the chemisorption site. These preliminary results seem to be in agreement with the conjecture proposed by Crespo et al. [62] that the interaction between S and Au orbitals is responsible for the onset of magnetism in thiol-capped gold nanoclus-ters. However, in order to confirm this proposal, similar calculations using the FPT model should be performed on larger clusters with core-shell structures typical of Au nanoparticles.

Our group has also performed electronic structure calculations on larger clusters (Au_{23} and Au_{55}) in order to reproduce and study the magnetic behavior of high elec-tric dipole moment thiopolypeptide R-helix linkers (consisting of eight L-glycine units), chemisorbed on the (111) Au surfaces [81]. The wave function broken sym-metry method (BS-UDFT) was used for this purpose [82]. The results of this study indicate a strong correlation between the magnetic behavior of the adsorbate–cluster system and the orientation of the electric dipole of the R-helix and charge trans-fer at the molecule–metal interface. Upon chemisorption, dipole moments may be quenched or enhanced with respect to the gas phase value. The results of this study indicate that the strongest reduction in dipole moment accompanied with net charge transfer from the Au surface leads to a very stable magnetic state. Additionally, it was found that the magnetic properties of these systems are strongly dependent on the size and geometrical structure of the Au cluster under consideration.

In addition to the cluster calculations, we have also studied the effect of benze-nethiol chemisorption on the magnetic properties of gold films [83] via DFT calcu-lations on extended systems. Gold films have been modeled by constructing slabs from the Au (111) lattice, containing from one to four layers, with 3×3 atoms in each layer. Spin-polarized DFT calculations have been performed using the SIESTA package [84]. This code allows electronic structure calculations of extended systems using pseudopotentials with periodic boundary conditions, where numerical atomic orbitals are used as basis sets to solve the single-particle Kohn–Sham equations. Our calculations were performed within the PBE-generalized gradient approximation of DFT, using norm-conserving Troullier–Martins pseudopotentials [85] as implemented in the SIESTA software. We use a double-zeta basis set with polarization orbitals on all the atoms [86]. The real-space mesh used for the calculation of the charge density and evaluation of real-space integrals is defined by a "mesh cut-off" parameter, defining the maximum energy of plane waves that can be represented on the mesh without aliasing. This parameter was set to 350 Ry for all calculations. The supercell dimensions were kept fixed for all calculations ($a = b = 8.65 \text{ Å}$, $c = 30 \text{ Å}$).

Figure 11.8 depicts a plot of the variation of the magnetic moment with the number of layers in the gold films. As observed in this plot, the total magnetic moment decays exponentially as the number of layers increases, approaching a value of zero for four layers. This magnetic behavior has recently been observed experimentally in gold films [4]. In the case of the four-layer supercell capped with benzenethiol, our calculations show an onset of a finite magnetic moment of approximately $0.025 \, \mu_B$ per absorbed molecule. The fact that this value is smaller than the experimental one (on the order of $10 \, \mu_B$; see Ref. [64] for details) is not surprising given the limitation of the size of the unit cell used in our calculations and also due to the fact that our calculations do not explicitly include important spin–orbit coupling effects. Finally, population analysis of this system suggests that the majority of "spin up" electrons are predominantly localized on the surface and shared equally among the atoms. Small "diamagnetic" contributions on the bulk atoms are also found.

Overall, our theoretical calculations show that chemisorbed thiolates on gold surfaces induce magnetism, basically due to a local Pauli repulsion between the sulfur and the gold atoms in the neighborhood of the chemisorption site. However, the results also show that bare Au clusters have an intrinsic tendency to exhibit magnetic behavior, suggesting that it is quite possible that the origin of chemisorption-induced magnetism on Au surfaces might indeed be different from a simple spin symmetry breaking on the surface atoms in Au clusters with core-shell structures. It should be noted that despite the interesting results of this study, other aspects of this problem

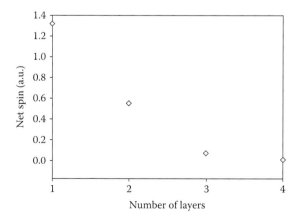

FIGURE 11.8 Net magnetic moment versus the number of layers in the supercell.

remain unclear, including the origin of the very small hysteresis and the high anisotropy of this magnetism. In order to answer these questions, more elaborate and theoretical models should be developed.

Although not an exhaustive review, this chapter briefly illustrates the potential benefits of applying theory and quantum chemistry calculations in understanding diverse intricate physical and chemical phenomena related to metallic clusters. More than a predictive tool replacing experimental measurements, these techniques should be considered as an important complement to measurement sciences and metrology aimed at helping with the interpretation of experimental results and the proposal of new measurement campaigns.

ACKNOWLEDGMENT

Patricio Fuentealba thanks Fondecyt, grant 1080184, Chile, and the NIST foreign visitor program for financial support.

REFERENCES

1. G. P. Wiederrecht (ed.), *Handbook of Nanofabrication*, Elsevier, Boston, MA, 2010.
2. F. Papanelopoulou, Between physics and chemistry: Early-low-temperature research, 1877–1908, in: *Sixth International Conference on the History of Chemistry*, Leuven, Belgium, 2007.
3. J. P. Connerade, A. V. Solov'yov, and W. Greiner, *Europhysicsnews* 33 (6), 200–202, 2002.
4. S. Reich, G. Leitus, and Y. Feldman, *Appl. Phys. Lett.*, 88, 222502, 2006.
5. D. J. Sellmyer, M. Yu, R. A. Thomas, Y. Liu, and R. D. Kirby, Nanoscale design of films for extremely high density magnetic recording, *Phys. Low-Dimensional Struct.*, 1/2, 1998. *Proceedings of the Ninth Joint MMM/Intermag Conference*, Anaheim, CA, *J. Appl. Phys.*, 95 (pt. 2), 2004.
6. W. Knight, K. Clemenger, W. A. de Heer, W. Saunders, M. Chou, and M. Cohen, *Phys. Rev. Lett.*, 52, 2141, 1984.
7. W. Ekardt, *Phys. Rev. Lett.*, 52, 1925, 1984.
8. M. Brack, *Rev. Mod. Phys.*, 65, 677, 1993.
9. W. de Heer, *Rev. Mod. Phys.*, 65, 612, 1993.
10. P. Hohenberg and W. Kohn, *Phys. Rev.*, 136, B864, 1964.
11. W. Kohn and L. Sham, *Phys. Rev.*, 140, A1133, 1965.
12. Y. Simón-Manso, Doctoral thesis, Pontifical Catholic University of Chile, Santiago, Chile, 1998.
13. W.A. Harrison, *Electronic Structure and the Properties of Solids: The Physics of the Chemical Bond*, Dover Publications, Inc., New York, 1989.
14. Z. Yu and J. Almlof, *J. Phys. Chem.*, 95, 9167–9169, 1991.
15. A. N. Andriotis, N. N. Lathiotakis, and M. Menon, *Europhys. Lett.*, 36 (1), 37–42, 1996.
16. A. D. Becke, *Phys. Rev.*, A38, 3098, 1988.
17. J. P. Perdew, *Phys. Rev.*, B33, 8822, 1986.
18. W. J. Hehre, R. Ditchfield, and J. A. Pople, *J. Chem. Phys.*, 56, 2257, 1972.
19. D. R. Lide (ed.), *CRC Handbook of Chemistry and Physics*, 75th edn., CRC Press, Boca Raton, FL, 1994.
20. P. Fuentealba and Y. Simón-Manso, *Chem. Phys. Lett. s*, 314 (1) 108–113, 1999.
21. D. Marx, in: Computational nanoscience: Do it yourself, *NIC Series*, 31, 195, 2006.

22. V. Bazterra, M. Caputo, M. Ferraro, and P. Fuentealba, *J. Chem. Phys.*, 117, 11158, 2002.
23. M. Iwamatsu, *J. Chem. Phys.*, 112 (10), 976, 2000.
24. C. Roberts, R. L. Johnston, and N. T. Wilson, *Theor. Chem. Acc.*, 104, 123, 2000.
25. V. E. Bazterra, M. B. Ferraro, and J. C. Facelli, *J. Chem. Phys.*, 116, 5984, 2002.
26. T. Bredow, G. Geudtner, and K. Jug, *J. Comput. Chem.*, 22, 861, 2001.
27. G. Igell-Mann, H. Stoll, and H. Preuss, *Mol. Phys.*, 65, 1321, 1988.
28. A. Sadlej, *Collect. Czech Chem. Commun.*, 53, 1995, 1998.
29. O. Oña, V. Bazterra, M. Caputo, J. Facelli, P. Fuentealba, and M. Ferraro, *Phys. Rev.*, A73, 053203, 2006.
30. Q. Sun and Y. Kawozoe, *Phys. Rev. Lett.*, 90, 135503, 2003.
31. N. Metropolis, N. Rosenbluth, M. Rosenbluth, and E. Teller, *J. Chem. Phys.*, 21, 1087, 1953.
32. E. Aarts and H. Landhooven, *Simulating Annealing, Theory and Application*, Springer, New York, 1987.
33. J. Perez, E. Florez, C. Hadad, P. Fuentealba, and A. Restrepo, *J. Phys. Chem. A*, 112, 5749, 2008.
34. B. Temelso and D. Sherrill, *J. Chem. Phys.*, 122, 064315, 2005.
35. A. Alexandrova and A. Boldyrev, *J. Chem. Theory Comput.*, 1, 566, 2005.
36. E. Florez and P. Fuentealba, *Int. J. Quantum Chem.*, 109, 1080, 2009.
37. R. H. Leary, *J. Global Optim.*, 11, 35–53, 1997.
38. K. A. Jackson, M. Horoi, I. Chaudhuri, T. Frauenheim, and A. A. Shvartsburg, *Phys. Rev. Lett.*, 93, 013401, 2004.
39. J. Centeno and P. Fuentealba, *Int. J. Quantum Chem.*, accepted for publication (2010).
40. R. O. Jones, A. I. Lichtenstein, and J. Hutter, *J. Chem. Phys.* 106(11), 15, 1997.
41. G. Gardet, F. Rogemond, and H. Chermette, *J. Chem. Phys.* 105, 9933, 1996.
42. A. Becke and K. Edgecombe, *J. Chem. Phys.*, 92, 5397, 1990.
43. A. Savin, O. Jepsen, J, Flad, O. Anderson, H. Preuss, and H. von Schnering, *Ang. Chem.*, 31, 187, 1992.
44. P. Fuentealba, E. Chamorro, and J. C. Santos, in: *Theoretical Aspects of Chemical Reactivity*, A. Toro (ed.), Elsevier, Amsterdam, 2007, p. 57.
45. B. Silvi and A. Savin, *Nature*, 371, 683, 1994.
46. B. Silvi, E. Fourre, and M. Alikhani, *Monatsh. Chem.*, 136, 855, 2005.
47. J. C. Santos, W. Tiznado, R. Contreras, and P. Fuentealba, *J. Chem. Phys.*, 120, 1670, 2004.
48. R. G. Parr and W. Yang, *Density Functional Theory of Atoms and Molecules*, Oxford Press, New York, 1990.
49. J. C. Santos, J. Andres, A. Aizman, and P. Fuentealba, *J. Chem. Theory Comput.*, 1, 83, 2005.
50. P. Fuentealba and A. Savin, *J. Phys. Chem. A*, 105, 11531, 2001.
51. M. Galvan, A. dal Pino, and J. Joannopoulus, *Phys. Rev. Lett.*, 70, 21, 1993.
52. S. Chrékien, M. Gordon, and H. Metiu, *J. Chem. Phys.*, 121, 3756, 2004.
53. S. Chrékien, M. Gordon, and H. Metiu, *J. Chem. Phys.*, 121, 9931, 2004.
54. R. Parr and W. Yang, *J. Am. Chem. Soc.*, 105, 4049, 1984.
55. W. Tiznado, O. Oña, V. Bazterra, M. Caputo, J. Facelli, M. Ferraro, and P. Fuentealba, *J. Chem. Phys.*, 123, 214302, 2005.
56. E. Florez, F. Mondragon, and P. Fuentealba, *J. Phys. Chem. B*, 110, 13793, 2006.
57. C. Bréchignac, Ph. Cahuzac, F. Carlier, and J. Leygnier, *Chem. Phys. Lett.*, 164, 433, 1989.
58. K. Selby, V. Kresin, J. Masui, M. Vollmer, W. A. de Heer, A. Scheidemann, and W. D. Knight, *Phys. Rev. B*, 43, 4565, 1991.
59. G. F. Bertsch, A. Bulgac, D. Tomanek, and Y. Wang, *Phys. Rev. Lett.*, 67, 2690, 1991.

60. G. Barton and C. Eberlein, *J. Chem. Phys.*, 95, 1512, 1991.
61. P. Zhang and T. K. Sham, *Phys. Rev. Lett.*, 90, 245502, 2003.
62. P. Crespo, R. Litrán, T. C. Rojas, M. Multigner, J. M. de la Fuente, J. C. Sánchez-López, M. A. García, A. Hernando, S. Penadés, and A. Fernández, *Phys. Rev. Lett.*, 93, 087204, 2004.
63. H. Hori, Y. Yamamoto, T. Iwamoto, T. Miura, T. Teranishi, and M. Miyake, *Phys. Rev. B*, 69, 174411, 2004.
64. I. Carmeli, G. Leitus, R. Naaman, S. Reich, and Z. Vager, *J. Chem. Phys.*, 118 (10), 372, 2003.
65. E. C. Stoner, *Proc. R. Soc.*, 154, 656–678, 1936; *Proc. R. Soc.*, 169, 339–371, 1939.
66. M. B. Knickelbein, *Phys. Rev. B*, 71, 184442, 2005.
67. J. Cox, J. G. Louderback, S. E. Apsel, and L. A. Bloomfield, *Phys. Rev. B*, 49, 12295, 1994.
68. L. A. Bloomfield, J. Deng, H. Zhang, and J. W. Emmert, in: *Proceedings of the International Symposium on Cluster and Nanostructure Interfaces*, P. Jena, S. N. Khanna, and B. K. Rao (eds.), World Publishers, Singapore, 2000.
69. M. B. Knickelbein, *Phys. Rev. Lett.*, 86, 5255, 2001.
70. M. B. Knickelbein, *Phys. Rev. B*, 70, 014424, 2004.
71. J. Cox, J. G. Louderback, and L. A. Bloomfield, *Phys. Rev. Lett.*, 71, 923, 1993.
72. D. Gatteschi and R. Sessoli, *Angew. Chem. Int. Ed.*, 42, 268–297, 2003.
73. R. J. Magyar, V. Mujica, M. Marquez, and C. Gonzalez, *Phys. Rev. B*, 75, 144421, 2007.
74. Perdew, J. P., Burke, K., and Ernzerhof, M., *Phys. Rev. Lett.*, 77, 3865, 1996.
75. A. D. Becke, *J. Chem. Phys.*, 98, 5648, 1993.
76. J. P. Perdew, K. Burke, and M. Ernzerhof, *Phys. Rev. Lett.*, 78, 1396, 1997.
77. A. Schaefer, C. Huber, and R. Ahlrichs, *J. Chem. Phys.*, 100, 5829, 1994.
78. W. R. Wadt and P. J. Hay, *J. Chem. Phys.*, 82, 284, 1985.
79. F. Michael, C. Gonzalez, V. Mujica, M. Marquez, and M. Ratner, *Phys. Rev. B*, 76, 224409, 2007.
80. C. Gonzalez, Y. Simón-Manso, M. Marquez, M. Ratner, and V. Mujica, *J. Phys. Chem. B*, 110, 687–691, 2006.
81. L. Puerta, H. J. Franco, J. Murgich, C. Gonzalez, Y. Simón-Manso, and V. Mujica, *J. Phys. Chem. A*, 112, 9771–9783, 2008.
82. J. Gräfenstein, E. Kraka, M. Filatov, and D. Cremer, *Mol. Phys.*, 99, 1899–1940, 2001; D. Cremer, *Int. J. Mol. Sci.*, 3, 360–394, 2002.
83. Y. Simón-Manso, M. Marquez, V. Mujica, and C. Gonzalez, *Altria-INEST Meeting*, Williamsburg, VA, May 17–19, 2005.
84. E. Artacho, D. Sanchez-Portal, P. Ordejon, A. Garcia, and J. M. Soler, *Phys. Stat. Sol. (b)*, 215, 809, 1999.
85. N. Troullier and J. L. Martins, *Phys. Rev. B*, 43, 1993, 1991.
86. O. F. Sankey and D. Niklewski, *Phys. Rev. B*, 40, 3979, 1989.

Index